Analog Circuits and Designs

Analog Circuits and Designs

Edited by **Gus Winters**

NY RESEARCH
P R E S S

New York

Published by NY Research Press,
23 West, 55th Street, Suite 816,
New York, NY 10019, USA
www.nyresearchpress.com

Analog Circuits and Designs
Edited by Gus Winters

International Standard Book Number: 978-1-63238-523-9 (Hardback)

The publisher's policy is to use permanent paper from mills that operate a sustainable forestry policy. Furthermore, the publisher ensures that the text paper and cover boards used have met acceptable environmental accreditation standards.

Trademark Notice: Registered trademark of products or corporate names are used only for explanation and identification without intent to infringe.

Printed in the United States of America.

Contents

Preface

This book will elucidate new techniques and their applications in a multidisciplinary approach. It will discuss the fundamental and modern approaches of analog circuits. As a branch of engineering, analog circuit refers to the signals which vary from zero to full power supply voltage. They are also referred to as linear signals because they provide continuous signal range which is not present in other circuits like digital circuits. This book includes detailed explanations of the various methods and theories related to analog circuits and their designing. It also provides interesting topics for research which readers can take up. The researches presented in this extensive text deal with the core subjects related to this branch. Those interested in this field will find this book full of crucial information. It is a vital tool for all researching and studying this field.

After months of intensive research and writing, this book is the end result of all who devoted their time and efforts in the initiation and progress of this book. It will surely be a source of reference in enhancing the required knowledge of the new developments in the area. During the course of developing this book, certain measures such as accuracy, authenticity and research focused analytical studies were given preference in order to produce a comprehensive book in the area of study.

This book would not have been possible without the efforts of the authors and the publisher. I extend my sincere thanks to them. Secondly, I express my gratitude to my family and well-wishers. And most importantly, I thank my students for constantly expressing their willingness and curiosity in enhancing their knowledge in the field, which encourages me to take up further research projects for the advancement of the area.

Editor

Functional Testbench Qualification by Mutation Analysis

Kai Huang,[1] Peng Zhu,[2] Rongjie Yan,[3] and Xiaolang Yan[2]

[1]*Department of Information Science and Electronic Engineering, Zhejiang University, Hangzhou 310027, China*
[2]*Institute of Very Large Scale Integrated Circuit Design, Zhejiang University, Hangzhou 310027, China*
[3]*Laboratory of Computer Science, Institute of Software, Chinese Academy of Sciences, Beijing 100080, China*

Correspondence should be addressed to Rongjie Yan; yrj@ios.ac.cn

Academic Editor: Avi Ziv

The growing complexity and higher time-to-market pressure make the functional verification of modern large scale hardware systems more challenging. These challenges bring the requirement of a high quality testbench that is capable of thoroughly verifying the design. To reveal a bug, the testbench needs to activate it by stimulus, propagate the erroneous behaviors to some checked points, and detect it at these checked points by checkers. However, current dominant verification approaches focus only on the activation aspect using a coverage model which is not qualified and ignore the propagation and detection aspects. Using a new metric, this paper qualifies the testbench by mutation analysis technique with the consideration of the quality of the stimulus, the coverage model, and the checkers. Then the testbench is iteratively refined according to the qualification feedback. We have conducted experiments on two designs of different scales to demonstrate the effectiveness of the proposed method in improving the quality of the testbench.

1. Introduction

Functional verification of modern hardware systems always requires the largest amount of resources and human efforts during the design cycle [1]. Simulation based verification is the predominant approach in hardware verification [1], which uses a testbench to verify the design under verification (DUV). There are three main components in a testbench as shown in Figure 1: (1) stimulus to activate the DUV; (2) a set of coverpoints to observe the behavior of the DUV and collect coverage information; and (3) a set of checkers to check internal states and outputs of the DUV. To reveal a bug, the corresponding circuits must be activated first, and then the erroneous results must be propagated to some ports which are properly checked by checkers. To thoroughly verify a hardware system, we need sufficient stimulus to activate all corners of the DUV, a sufficient coverage model to ensure that all important functions of the DUV are executed, and a sufficient set of checkers to guarantee that all results of the executed functions are adequately checked. All the three components should be considered to build a high quality testbench [2–4].

A coverage model, built according to coverage metrics such as code coverage and user defined functional coverage [5, 6], is the main way to evaluate the thoroughness of verification. Unfortunately, these coverage metrics focus only on the quantity of the activated coverpoints and ignore the propagation and the sufficiency of the checkers [4]. Meanwhile, the quality of the functional coverage model and the checkers heavily depends on the experience of verification engineers. It is easy to omit some corner scenarios in the coverage model, and it is practically impossible to encode all correct behaviors in the checkers. It is even harder to build a good checker, for some modern hardware systems that may exhibit a degree of indeterminism, and the golden output is not always known [7, 8]. Therefore, building a high quality functional coverage model and a set of high quality functional checkers is a challenging task.

Mutation analysis is originally used to design new test data and evaluate the quality of existing test data in software testing [9]. It modifies a program syntactically, for example, by using a wrong operator. Each mutated program is a mutant. Mutants are created from well-defined mutation operators that mimic typical programming errors. Tests kill

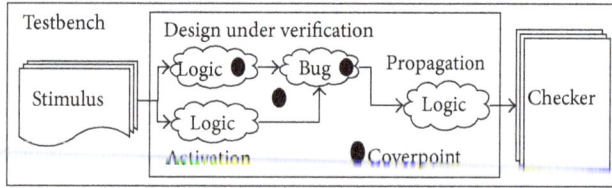

FIGURE 1: Simulation based verification testbench.

mutants by capturing the behavior of the original version that is different from that of the mutant. Error detecting ability of test suites is measured by the percentage of killed mutants. Note that mutation analysis assumes that the quality of the checkers is sufficient which usually is not real. The effectiveness of mutation analysis in hardware verification has been confirmed by recent works in diverse fields and levels of abstraction, such as RTL [10–12], SystemC [13–19], and embedded C software [20]. Commercial EDA tool for hardware description languages (HDL) mutation analysis has already been available [21].

To address aforementioned weaknesses of existing functional verification techniques, this paper presents a method to qualify the three components in the testbench by mutation analysis. The main contributions of our work include three aspects: (1) assessing both the integrity and quality of the coverage model and the set of checkers, (2) proposing a testbench quality metric which quantifies the quality of the testbench, and (3) proposing a simulation mechanism to accelerate the process of testbench qualification.

The rest of this paper is organized as follows. Section 2 reviews the existing related work. Section 3 explains the testbench qualification technique in detail. Section 4 presents experimental results to demonstrate the usefulness of the proposed technique. Section 5 concludes the paper.

2. Related Work

Mutation analysis has been actively studied for over three decades in software testing community and it has been applied to various software programming languages [9].

In recent years, mutation analysis has been applied to languages for system-level hardware modeling and verification, such as SystemC [13–19]. Several works have applied fault models for SystemC, such as perturbing SystemC TLM descriptions in [16, 17], introducing mutation operators for concurrent SystemC designs [18]. Other works have looked into error injection and fault localization in SystemC. For example, SCEMIT injects errors automatically into SystemC models [14]. Le et al. introduce an automatic fault localization approach for TLM designs based on bounded model checking [15]. Verification quality on SystemC is also investigated. The work in [13] develops mutation analysis based coverage metrics to attack the verification quality problem for concurrent SystemC programs. Functional qualification is introduced to measure the quality of functional verification of TLM models [19].

Additional to the application in system-level modeling and verification, mutation analysis has been applied also

to HDL, such as Verilog and VHDL [10–12, 22]. The work in [10] qualifies the error detecting ability of test cases by mutation analysis and automatically improves validation data. Mutation analysis performed at TLM is reused at RTL to help designers in optimizing the time spent for simulation at RTL and improving the RTL testbench quality [11]. Liu and Vasudevan measure branch coverage through mutated guard in the symbolic expression during symbolic simulation [12]. HIFSuite [22] provides a framework that supports many fundamental activities including mutation analysis.

Recently, mutation analysis is applied to guide the process of stimulus generation for hardware verification to efficiently kill more mutants [23, 24]. Xie et al. propose a search based approach and defined an objective cost function to solve the problem of automatic simulation data generation targeting HDL mutants [23]. They extend the work in [23] by representing a simulation flow with two phases towards an enhanced mutation analysis score [24].

Coverage discounting technique [2–4] maps survived mutants to deficiently covered coverpoints and removes these coverpoints from coverage results, which leads to a decreased but better quality coverage results. One of the shortcomings of this technique is that it concerns the quality of testbench but focuses only on the quality of the existing coverage results and ignores the integrity of the coverage model, which will be addressed in this paper.

Observability-based coverage is first introduced in [25] to address the activation-only nature of other coverage metrics. Since then, attempts have been tried to generalize or extend the technique [26, 27]. Observability-based coverage metrics are a form of implicit metric which introduces data-flow evaluation. We consider statements to be covered only if they are executed and the result of that execution has a dynamic data flow to the output. However, it is always an assumption that the output is properly checked, and the presence and quality of the checker are never actually evaluated. Observability-based coverage metrics require specialized simulation tools with extensive instrumentation to record dynamic data-flows that are not available in most cases. As the observability concept only works with implicit metrics, it cannot be combined with arbitrary functional coverpoints.

3. Testbench Qualification

This section is organized as follows. The first subsection presents the basic idea of testbench qualification. The second and the third subsections describe refinement of the coverage model and the checkers, respectively. Then we propose a testbench quality metric by considering all three components in the testbench. The last subsection presents a simulation mechanism to accelerate the process of testbench qualification.

3.1. The Basic Idea. Verification engineers (VE) build the testbench manually according to the test plan and design specification, which is error prone and may lead to low quality elements in the testbench. According to the types of faults, we can divide a coverage model (resp., a set of checkers)

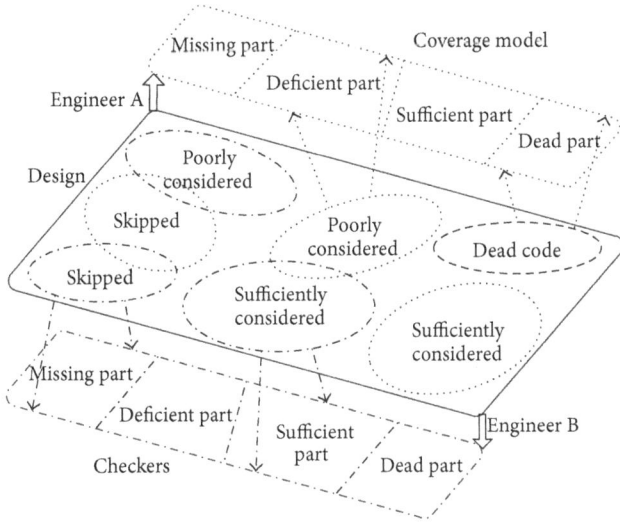

FIGURE 2: Fault classification of the coverage model and the checkers.

into four parts, as shown in Figure 2 (some dotted lines are not depicted for clarity): (1) the missing part (some relevant features of the design are skipped by the verification engineers because of careless), (2) the deficient part (some features of the design are poorly considered by the verification engineers), (3) the sufficient part (features of the design those are adequately handled by the verification engineers), and (4) the dead part (some features of the design can be dead codes which will never be executed). Although the partitions of the coverage model and the checkers are identical, a specific partition of the coverage model and the checkers can base on different parts of the design which usually overlap with each other, as shown in Figure 2. Because the coverage model and the checkers are possibly built by different engineers who can treat the design differently, or by the same engineer at different time, it is almost impossible to consider the design exactly the same at different time. The testbench qualification method aims at improving the testbench by eliminating the missing part and refining the deficient part.

Injecting error into the design may corrupt some combinational logics or/and sequential logics in the design. In the simulation results of the mutant, comparing to the simulation results of the original design, some unactivated functions may activate and some activated functions may be suppressed. With a well-built coverage model, all relevant functions are monitored by coverpoints. Therefore, with the error, some unfired coverpoints may fire and some fired coverpoints may not fire, which results in changes in the coverage results. In more detail, these changes can be one fired coverpoint being suppressed, or several unfired coverpoints being fired, or the former two events happen together. While with a deficient coverage model, it monitors only a subset of the relevant functions, which possibly not include the corrupted functions. In this situation, the coverage results may remain unchanged. For convenience, we use two terms, fluctuation and stable, in the rest of this paper. We define fluctuation as changes in the coverage results. If the coverage results remain identical, we define this is a stable coverage results.

TABLE 1: Reactions from the coverage model and checkers to the injected error.

Reactions from a coverage model	Reactions from checkers	Possible reasons
Fluctuated	Killed	Sufficient part
Fluctuated	Survived	Deficient checkers
Stable	Killed	Missing coverpoints
Stable	Survived	Several possible reasons

In a well-built testbench, there should be a corresponding relationship between the features of a design, the set of coverpoints in the coverage model, and the set of checkers. That is, for every feature in the design, there should be some coverpoints to ensure that it is actually implemented and executed and some checkers to ensure that the behavior of the feature is correct. Therefore, if some errors are injected into a design and the injection disturbs some functions of the design, these disturbed functions should result in fluctuation in the coverage results and be detected by the checkers. In other words, the design, the coverage model, and the set of checkers should behave consistently. If the expected result is not observed, either the coverage model or the checkers may be inadequate. Therefore, the quality of the coverage model and the checkers can be analyzed according to their reaction to the error injection during mutation analysis.

We summarize four possible combinations of reactions from the coverage model and the checkers when simulating a mutant. In Table 1, the first column lists possible reactions from a coverage model, which may lead to either fluctuated or stable coverage results. The second column presents possible reactions from checkers, which may either kill the mutant or let it survive. The last column provides possible reasons of each combination. In the first row, when a mutant is killed and the coverage results fluctuate, the related coverpoints and checkers are sufficient because they behave consistently. In the last row, when the mutant is alive and the coverage results remain stable, there are several possible reasons: (1) the mutant is not activated, (2) the coverage model and the set of checkers are both deficient, and (3) the error is injected into meaningless dead code. Complete analysis of this situation is quite heavy and left as a future work. We will mainly investigate the combinations of the two situations in the second and third rows, for example, missing coverpoints and deficient checkers, to refine the coverage model and the set of checkers.

3.2. Coverage Model Refinement. When a mutant is killed by the checkers but coverage results remain stable, some coverpoints are missing in the coverage model. As shown in Figure 3, only two coverpoints (CP1 and CP2) are used while there are more functions that need coverpoints to monitor. The coverage results before and after the injection of the error may always record CP1 and CP2 being covered, since the functions monitored by CP1 and CP2 are possibly not infected by the error. However, the checkers kill the mutant. A killed mutant means some functions of the design are indeed corrupted by the injected error, as the

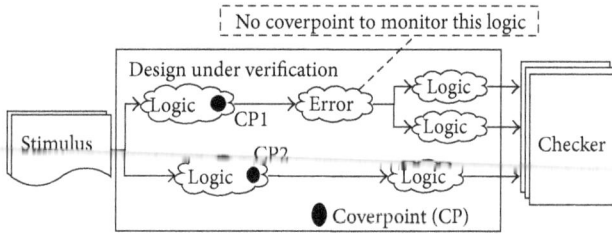

FIGURE 3: Insufficient coverage model.

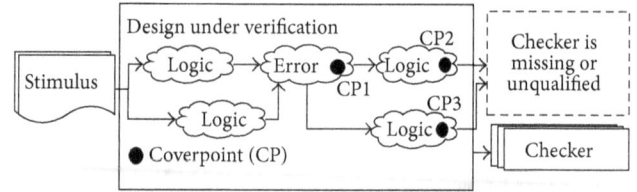

FIGURE 4: An insufficient checker.

```
(1)  input x; output y;
(2)  always @ (x) begin
(3)      if (x > 0) //the mutant replaces the > by <
(4)          y = 1 + x; //coverpoint2 is added here
(5)      else if (x == 0)
(6)          y = 2; //monitored by coverpoint1
(7)      else
(8)          y = 1 − x; //coverpoint3 is added here
(9)  end
(10) assertion: y > 1; //the checker
(11) //test vectors: x1 = 0; x2 = 1; x3 = 2;
(12) //original outputs: y1 = 2; y2 = 2; y3 = 3;
(13) //mutant outputs: y1 = 2; y2 = 0; y3 = −1;
```

LISTING 1: Coverage model qualification example.

checkers have detected it. The stable coverage results from the corrupted functions come from the missing coverpoints to monitor these corrupted functions. The coverage model can be qualified and improved by finding out and eliminating these missing coverpoints.

A well-built coverage model is constructed by a set of coverpoints which encode the entire features of the design at a certain level of abstraction that is determined by the verification requirement. The set of coverpoints should be able to monitor all behaviors of the design and be sensitive enough to the injected errors which cause wrong behaviors at that certain level of abstraction. For instance, the nontrivial functions that we plan to verify are a set $\{F1, F2, \ldots, Fn\}$. Accordingly, the coverage model should consist of a set of coverpoints $\{C1, C2, \ldots, Cn\}$ to monitor and ensure that all functions in $\{F1, F2, \ldots, Fn\}$ are implemented and exercised. The coverage model refinement is a repetitive process of qualification and improvement of the existing coverage model, involving the following steps: (1) simulating the original design and producing the original coverage results, (2) simulating mutants and producing coverage results from the mutants, (3) adding missing coverpoints if a mutant is killed but coverage results remain stable, and (4) repeating previous steps until the quality of the testbench satisfies the predefined threshold.

Consider Listing 1 as an example, which is a code block of the DUV with one checker (line 10), and the coverage model only contains coverpoint1 to monitor line (6). After simulation of the test vector ($x1 = 0; x2 = 1; x3 = 2$) against the original design, the output is $y1 = 2; y2 = 2; y3 = 3$. The

coverage model records that coverpoint1 is covered and the checker does not fire.

Then an error is injected into line (3) which replaces ">" by "<". After simulation of the test vectors ($x1 = 0; x2 = 1; x3 = 2$) against the mutant, the output is $y1 = 2; y2 = 0; y3 = −1$. The coverage model records that coverpoint1 is covered just the same as the original design. But the checker does fire when $x2 = 1$ and $y2 = 0$. Therefore, this mutant is killed. In this example, the coverage results remain stable, but the checker fires and kills the mutant. The discrepancy between reaction of the coverage model and reaction of the checkers indicates that some coverpoints are missing. Because the code is simple, it is straightforward to know that coverpoints should be added to line (4) and line (8). After improving the coverage model by adding coverpoint2 and coverpoint3 to line (4) and line (8), respectively, the firing of the checker is consistent with the fluctuation of coverage results. That is, the original coverage results record that coverpoint1 and coverpoint2 are covered, but the mutant coverage results record that coverpoint1 and coverpoint3 are covered.

3.3. Checker Refinement.

The features of the design, the coverpoints in the coverage model, and the set of checkers are tightly related. Anything happened in one component can either trace back to some reasons or result in some consequences in other components. Being at the later stage in the testbench, a perfect set of checkers should be capable of reflecting any activity happened in the design and the coverage model.

When a mutant survives but the coverage results fluctuate, the set of checkers is insufficient. More precisely, the fluctuated coverage results reflect that some functions of the design are corrupted by the injected errors, but the checkers are incapable of detecting the wrong behavior. The possible reasons can be either missing checkers or low quality checkers. As shown in Figure 4, three coverpoints (CP1, CP2, and CP3) are used to monitor the functions that are tightly related to the injected error. With the original design, the coverage results may recode that CP1 and CP2 are covered. However, the injection of the error can cause misbehavior of the design and lead to a fluctuated coverage results that recode CP1 and CP3 being covered. While the coverage results fluctuated, the mutant is alive, because the checker is either missing or unqualified.

As it is impractical to encode all functions of a design into a perfect set of checkers at the beginning, we also need an iterative improvement process to deliver adequate

```
(1)  input x; output y;
(2)  always @ (x) begin
(3)      if (x > 0) //mutant1 replaces the > by <
(4)          y = 1 + x; //monitored by coverpoint1
(5)      else if (x == 0)//mutant2 replaces the condition by true
(6)          y = 2; //monitored by coverpoint2
(7)      else
(8)          y = 1 − x; //monitored by coverpoint3
(9)  end
(10) Original assertion: none;
(11) The first added assertion: y > 1;
(12) The final assertion: (x == 0)? y == 2: y == 1 + abs(x);
(13) //test vectors: x1 = −2; x2 = 0;
(14) //original outputs: y1 = 3; y2 = 2;
(15) //mutant1 outputs: y1 = −1; y2 = 2;
(16) //mutant2 outputs: y1 = 2; y2 = 2;
```

LISTING 2: Checker qualification example.

checking ability. Sometimes, the link between checkers and features of the design can be missing or weak that they can only check partial properties of a function rather than the entire function. These missing checkers should be added, and weak checkers should be refined until they can reveal both functional and performance bugs.

A set of well-built checkers should be as sensitive as a well-built coverage model to the injected errors. The checkers refinement process is as follows: (1) simulating the original design and producing the original coverage results, (2) simulating mutants and producing coverage results from the mutants, (3) adding the missing checkers or improving the deficient checkers if a mutant is alive but coverage results fluctuate, and (4) repeating previous steps until the quality of the testbench satisfies the predefined threshold.

Listing 2 is another example to explain the checker refinement technique. The code block in Listing 2 is the same as that in Listing 1 except that there are more coverpoints and without any checkers at the beginning, after simulation of the test vector ($x1 = −2$; $x2 = 0$) against the original design, the coverage model records that coverpoint2 and coverpoint3 are covered.

In mutant1, an error is injected into line (3) which replaces ">" by "<". After simulation of the test vector ($x1 = −2$; $x2 = 0$) against mutant1, the output is $y1 = −1$; $y2 = 2$. The coverage results are changed to coverpoint1 and coverpoint2 which are covered. But the mutant is survived because there is no checker to kill it. Therefore, some checkers are missing, and we add a checker ($y > 1$) as line (11) in Listing 2 to kill mutant1.

In mutant2, instead of the error injected in line (3), another error is injected into line (5), which replaces the condition express ($in == 0$) with true. After simulation of the test vector ($x1 = −2$; $x2 = 0$) against mutant2, the output is $y1 = 2$; $y2 = 2$. The checker in line (11) ($y > 1$) cannot kill this mutant, which needs refinement. A more elaborate checker in line (12) can kill this mutant. After the refinement, the checker encodes the entire function of the design and deliveries good error detecting ability.

3.4. Testbench Quality Metric.
After refining the coverage model and the set of checkers, we propose a testbench quality metric to qualify the whole testbench. A metric that qualifies the whole testbench should consider (1) the stimulus, (2) the coverage model, and (3) the set of checkers. The metric is defined in (1), where $quality_{tb}$, $quality_{sti}$, $quality_{cov}$, and $quality_{che}$ are quality of testbench, stimulus, coverage model, and the set of checkers, respectively, and w_{sti}, w_{cov}, and w_{che} are weights that reflect different importance of various components with the constraint $w_{sti} + w_{cov} + w_{che} = 1$. The weights are selected feasibly to meet the current verification requirement. A lower weight is set for a component when you are confident about its quality. Otherwise, a higher weight is set

$$quality_{tb} = w_{sti} \times quality_{sti} + w_{cov} \times quality_{cov} + w_{che} \times quality_{che}, \tag{1}$$

$$quality_{sti} = compact\_factor_{sti} \times integrity\_factor_{sti}, \tag{2}$$

$$quality_{cov} = \frac{sufficient\_factor}{base}, \tag{3}$$

$$quality_{che} = compact\_factor_{che} \times integrity\_factor_{che}. \tag{4}$$

The $quality_{sti}$ is defined as the product of $compact\_factor_{sti}$ and $integrity\_factor_{sti}$ as in (2). Being the ratio of the number of tests having activated different coverpoints over the total number of tests, $compact\_factor_{sti}$ reflects the compactness of the stimulus. $integrity\_factor_{sti}$ is the ratio of the number of covered coverpoints over the total number of coverpoints, which represents the completeness of the stimulus. The range of $quality_{sti}$ lies between 0 and 1. When it approaches 0, all the tests are useless and none of them can active any coverpoints. When it approaches 1, the test suit is compact and complete.

The $quality_{cov}$ is defined as the ratio of the sufficient_factor over the base as in (3). sufficient_factor is the number of coverpoints whose coverage results fluctuate when simulating killed mutants. It reflects the size of the sufficient part of the coverage model that is properly checked. base is the total number of coverpoints. The range of $quality_{cov}$ is also between 0 and 1. When it approaches 0, all the existing coverpoints are useless and none of relevant functions are monitored. When it approaches 1, all the existing coverpoints are useful and monitoring some relevant functions.

The $quality_{che}$ is defined as the product of compact_factor$_{che}$ and integrity_factor$_{che}$ as in (4). The former, being the ratio of number of the fired checkers over the total number of checkers, reflects the compactness of the set of checkers. And its value lies between 0 and 1, where 0 means that all the checkers have not taken effect, and 1 means that all the checkers have indeed taken effect and there is no redundant checkers at all. The latter is the ratio of the number of killed mutants with fluctuated coverage results over the total number of mutants with fluctuated coverage results. integrity_factor$_{che}$ reflects the completeness of the set of checkers and the value lies between 0 and 1, where 0 means that all the existing checkers are useless, and 1 means that all the existing checkers are useful.

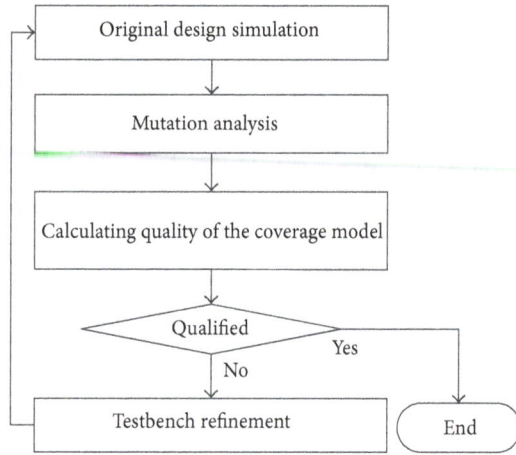

FIGURE 5: The work flow.

As the sum of weighted $quality_{sti}$, $quality_{cov}$, and $quality_{che}$, the range of $quality_{tb}$ is also between 0 and 1, where 0 means that the testbench is of low quality, and 1 means that the testbench is of high quality.

3.5. Simulation Mechanism. The overall work flow of the proposed method is depicted in Figure 5. First, normal simulation of the original design is conducted, and the $quality_{sti}$ can be determined by the collected coverage results. Then errors are injected and mutants are simulated during mutation analysis, and the $quality_{che}$ can be determined by mutation score and coverage results of mutant simulation. In this stage, we check the simulation results of the mutants to determine the possibility of coverage model refinement and checks refinement. If any one of the two refinements is possible, we record the type of the refinement, the test case, and the mutant, which will be used in the testbench refinement stage. Subsequently, we calculate the $quality_{cov}$ to determine the $quality_{tb}$. Finally, the process terminates if the quality of the testbench satisfies the predefined threshold. Otherwise, we refine the testbench and repeat the previous steps. In the testbench refinement stage, we investigate the possible refinements recorded in the mutation analysis stage. We refine the coverage model as discussed in Section 3.2 if the coverage results remain stable while the mutant is killed or refine the checkers as discussed in Section 3.3 if the coverage results fluctuated while the mutant is alive. There can be one or more possible refinements which may lead to addition of one or more coverpoints or checkers. But the adding is not always happening, since the refinement is manually done. In the iterative work flow, with the refined checkers produced in the previous iteration, we can further refine the coverage model in the current iteration. Because a better checker can possibly kill more mutants, which can result in more instances that the coverage results remain stable while the mutant is killed. Similarly, we can further refine the checkers with the better coverage model given by the previous iteration. So the coverage model and the checkers are incrementally refined iteration by iteration.

Usually, a quality threshold is predefined and the testbench qualification process can stop if it is achieved. Instead of calculating the exact $quality_{cov}$, approximating it as fast as possible can achieve the threshold earlier and accelerate the testbench qualification process. We provide two criteria to accelerate the calculation of an approximated $quality_{cov}$. (1) Criterion to select stimulus: select the one covering the most coverpoints. This criterion provides high probability to increase the numerator of $quality_{cov}$ as fast as possible. (2) Criterion to select mutant: select the one that is killed and activated by the least stimulus. Such a mutant usually is related to some corner features, and they can lead to higher $quality_{cov}$ when the coverage results reach the ceiling.

4. Experimental Results

4.1. Experimental Setup. We use two designs with different scales to show the effectiveness and scalability of the proposed method. The first is an interconnection module (ICM) that connects four cores and maintains cache coherence operations in a multicore processor. The second is a commercial RISC CPU. The fault injection and mutation analysis are performed with self-built scripts. We select a set of mutation operators to limit the population of mutants. According to the mutation operators, the fault injection script parses the design files to locate syntaxes to inject errors first and then inject errors at these located syntaxes to produce mutants. The mutation analysis script simulates every test case in the test suit on each mutant first and then checks the simulation results against the simulation results from the original design to determine whether the refinement of the coverage model or the checkers is possible. If possible, we record the type of refinement (the coverage model or the checkers), the test case, and the mutant in a text file which is used in the testbench refinement stage. Finally, according to the flow given in Figure 5, we build a top level script, which is responsible for calculating $quality_{sti}$, $quality_{cov}$, $quality_{che}$, and $quality_{tb}$, based on the fault injection script and the mutation analysis script. The weights w_{sti}, w_{cov}, and w_{che} are assigned with 0.2, 0.4, and 0.4, respectively. The values of w_{cov} and w_{che} are set higher than that of w_{sti}, because this paper focuses on the qualification of the coverage model and the checkers. The threshold of the testbench quality is 0.9. We carry out the experiments on a computer with an Intel i5 dual-core CPU at 1.8 GHZ and 8 G RAM.

4.2. Constrained Random Stimulus on the ICM. In this experiment, we have used 20 constrained random test cases. Each random test case consists of 100 memory accessing instructions in every thread. There are 400 instructions in a test case, for the setup with 4 threads for 4 cores. All instructions in these test cases access a limited range of address space to increase the probability of cache coherence interaction between different cores. The testbench aims at verifying all kinds of cache coherence operations supported by the ICM. A well-built coverage model that satisfies the verification requirements should consist of coverpoints (bins or sequences in system Verilog) to monitor all possible cache

TABLE 2: Results under random stimulus.

Iteration	0	1	2	3	4	5	6	7	8	9
Coverpoints	12	17	23	30	33	41	48	53	58	60
Checkers	4	5	7	8	9	11	12	13	15	16
Testbench quality	0.51	0.55	0.58	0.62	0.69	0.76	0.79	0.82	0.88	0.91

TABLE 3: Results under directed stimulus.

Iteration	0	1	2	3	4	5	6
Coverpoints	12	23	32	40	49	57	62
Checkers	4	6	7	9	11	14	16
Testbench quality	0.51	0.59	0.65	0.72	0.79	0.86	0.92

coherence operations between every individual core and the ICM. The total number of coverpoints is 64 in the 4-core processor. But only a subset of well-built coverage model is developed at the beginning of the experiments, which will be subsequently refined during testbench qualification. A well-built set of checkers should be able to thoroughly check results of all kinds of cache coherence operations. The total number of checkers is 16 in the 4-core processor. We develop a subset of well-built checker set at the beginning of the experiments for the same reason of the coverage model. There are 623 mutants in total.

During the testbench qualification process, coverpoints are added when a mutant is killed but the coverage results remain stable. And checkers are added or refined when a mutant survives but the coverage results fluctuate. With the improved testbench, more mutants are killed. Table 2 presents the experimental results. The first row is the index of each iteration, and iteration 0 is the initial state. The second and third rows show the numbers of the coverpoints and checkers, respectively. The number of coverpoints does not reach 64. The reason is that some coverpoints are interrelated, and injected errors can lead to the fluctuation in the coverage results with a subset or the entire set of coverpoints. Along with the improvement of the coverage model and the set of checkers, the quality of the testbench is improved, as shown in the last row. Finally, we take 9 iterations to satisfy the testbench quality threshold. The simulation time is a little more than 5 hours, which excludes the manual refinement time.

4.3. Directed Stimulus on the ICM. A good quality metric should be able to distinguish the quality of stimulus. A stimulus with high quality is elaborately designed to activate, propagate, and detect bugs in the DUV. Therefore, dedicated directed stimulus has higher probability to expose the weakness of the testbench than random stimulus. Therefore, we manually build 20 directed test cases each with 400 instructions, which are dedicated to test cache coherence operations. The testbench qualification process is reconducted on the ICM and we expect that the quality threshold should be satisfied with fewer iterations. The experimental results under directed stimulus is given in Table 3. The directed stimulus takes 6 iterations which cost less than 3 hours in simulation to satisfy the threshold and this confirms the expectation.

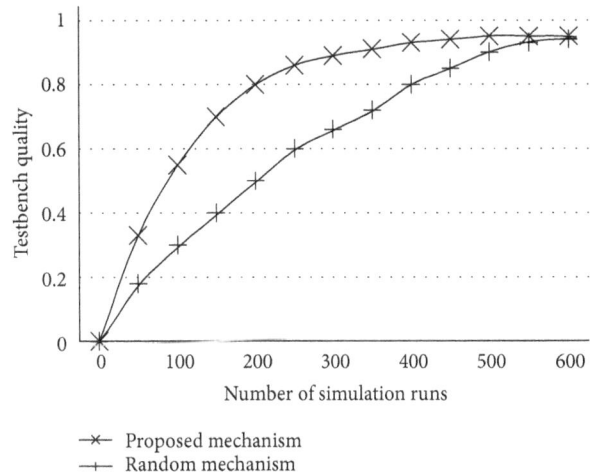

FIGURE 6: Comparisons of different simulation mechanisms.

4.4. Comparison of Different Simulation Mechanism on the ICM. We use a random selection mechanism as the comparison, which randomly selects test cases and mutants when calculating the quality of the coverage model. The difference between the speeds of the testbench quality improvement with different simulation mechanism will demonstrate the efficiency of the proposed simulation mechanism.

This experiment uses 50 random test cases to stimulate the ICM, and each test case consists of 100 instructions for each thread. We present the trends of the testbench quality under different simulation mechanism with 50 random test cases in Figure 6. The testbench quality under the proposed simulation mechanism increases much faster at the beginning stage than that under the random selection mechanism, meaning that the proposed simulation mechanism can achieve the quality threshold with fewer simulation runs.

4.5. Experiments on a Commercial CPU. To show that the proposed method is feasible for larger scale design, we adopt a commercial RISC CPU whose instruction set architecture consists of more than 150 instructions. We intend to use 20 constrained random test cases to thoroughly verify all kinds of instruction types, and each test case consists of 100 instructions. Operands of instruction are constrained as some rare values to incur interesting corner scenarios, for example, the address of memory accessing instruction is close to page boundary.

In this case, a sufficient coverage model should have coverpoints to monitor each individual instruction type, and a sufficient set of checkers should properly investigate outputs of all execution units. Only a subset of the sufficient one is

TABLE 4: Results from the commercial CPU.

Iteration	0	1	2	3	4	5	6	7	8	9	10
Coverpoints	30	39	52	61	74	81	91	105	123	131	143
Checkers	10	13	15	16	18	20	22	24	26	29	32
Testbench quality	0.48	0.55	0.60	0.63	0.68	0.71	0.75	0.79	0.83	0.86	0.91

constructed in the initial stage of the experiment to show the effect of the qualification. The improving progresses of the coverpoints, the checkers, and the quality of testbench are provided in Table 4. We can see that when the proposed method is applied to the CPU, it is as effective as the case to the ICM. Simulation in this experiment consumes almost 7 hours.

5. Conclusion

The growing complexity of modern hardware systems and time-to-market pressure require more efficient and high quality functional verification. These requirements necessitate a high quality testbench that can thoroughly verify the DUV in limited time. Currently, the thoroughness of function verification is dominantly measured by structural code coverage and user defined functional coverage. However, all of these coverage metrics concentrate on activation of the design but ignore the propagation and detection aspects that are indispensable to expose bugs. This paper qualifies the whole testbench by considering activation, propagation, and detection process. In particular, we have (1) improved the integrity and quality of the coverage model and the set of checkers through mutation analysis, (2) presented a metric to measure the quality of the testbench, and (3) proposed a simulation mechanism to satisfy the metric faster. Experimental results demonstrate the effectiveness of the proposed method. The future work can be extended in the following directions: (1) detail discussion about the situation that a mutant is survived and the coverage results remain stable and (2) formalizing the relationship between the design, the coverage model, and the set of checkers.

Conflict of Interests

The authors declare that there is no conflict of interests regarding the publication of this paper.

References

[1] Y. Shuo, R. Wille, and R. Drechsler, "Improving coverage of simulation-based verification by dedicated stimuli generation," in *Proceedings of the 17th Euromicro Conference on Digital System Design (DSD '14)*, pp. 599–606, Verona, Italy, 2014.

[2] P. Lisherness and K.-T. Cheng, "Improving validation coverage metrics to account for limited observability," in *Proceedings of the 17th Asia and South Pacific Design Automation Conference (ASP-DAC '12)*, pp. 292–297, Sydney, Australia, February 2012.

[3] P. Lisherness, N. Lesperance, and K.-T. Cheng, "Mutation analysis with coverage discounting," in *Proceedings of the 16th Design,*

Automation and Test in Europe Conference and Exhibition (DATE '13), pp. 31–34, IEEE, Grenoble, France, March 2013.

[4] P. Lisherness and C. Kwang-Ting, "Coverage discounting: a generalized approach for testbench qualification," in *Proceedings of the IEEE International High Level Design Validation and Test Workshop (HLDVT '11)*, pp. 49–56, Napa Valley, Calif, USA, November 2011.

[5] R. Grinwald, E. Harel, M. Orgad et al., "User defined coverage—a tool supported methodology for design verification," in *Proceedings of the Design Automation Conference (DAC '98)*, pp. 158–163, San Francisco, Calif, USA, 1998.

[6] S. Tasiran and K. Keutzer, "Coverage metrics for functional validation of hardware designs," *IEEE Design & Test of Computers*, vol. 18, no. 4, pp. 36–45, 2001.

[7] A. Meixner and D. J. Sorin, "Dynamic verification of memory consistency in cache-coherent multithreaded computer architectures," *IEEE Transactions on Dependable and Secure Computing*, vol. 6, no. 1, pp. 18–31, 2009.

[8] A. Adir, A. Nahir, and A. Ziv, "Concurrent generation of concurrent programs for post-silicon validation," *IEEE Transactions on Computer-Aided Design of Integrated Circuits and Systems*, vol. 31, no. 8, pp. 1297–1302, 2012.

[9] J. Yue and M. Harman, "An analysis and survey of the development of mutation testing," *IEEE Transactions on Software Engineering*, vol. 37, no. 5, pp. 649–678, 2011.

[10] Y. Serrestou, V. Beroulle, and C. Robach, "Functional verification of RTL designs driven by mutation testing metrics," in *Proceedings of the 10th Euromicro Conference on Digital System Design Architectures, Methods and Tools (DSD '07)*, pp. 222–227, Lübeck, Germany, August 2007.

[11] V. Guarnieri, G. Di Guglielmo, N. Bombieri et al., "On the reuse of TLM mutation analysis at RTL," *Journal of Electronic Testing*, vol. 28, no. 4, pp. 435–448, 2012.

[12] L. Liu and S. Vasudevan, "Efficient validation input generation in RTL by hybridized source code analysis," in *Proceedings of the 14th Design, Automation and Test in Europe Conference and Exhibition (DATE '11)*, pp. 1–6, Grenoble, France, March 2011.

[13] A. Sen and M. S. Abadir, "Coverage metrics for verification of concurrent SystemC designs using mutation testing," in *Proceedings of the 15th IEEE International High Level Design Validation and Test Workshop (HLDVT '10)*, pp. 75–81, IEEE, Anaheim, Calif, USA, June 2010.

[14] P. Lisherness and K.-T. Cheng, "SCEMIT: a systemc error and mutation injection tool," in *Proceedings of the 47th ACM/IEEE Design Automation Conference (DAC '10)*, pp. 228–233, ACM, Anaheim, Calif, USA, June 2010.

[15] H. M. Le, D. Grosse, and R. Drechsler, "Automatic TLM fault localization for SystemC," *IEEE Transactions on Computer-Aided Design of Integrated Circuits and Systems*, vol. 31, no. 8, pp. 1249–1262, 2012.

[16] N. Bombieri, F. Fummi, and G. Pravadelli, "A mutation model for the SystemC TLM 2.0 communication interfaces," in *Proceedings of the Design, Automation and Test in Europe (DATE '08)*, pp. 396–401, Munich, Germany, March 2008.

[17] N. Bombieri, F. Fummi, and G. Pravadelli, "On the mutation analysis of systemC TLM-2.0 standard," in *Proceedings of the 10th International Workshop on Microprocessor Test and Verification (MTV '09)*, pp. 32–37, Austin, Tex, USA, December 2009.

[18] A. Sen, "Mutation operators for concurrent systemC designs," in *Proceedings of the 10th International Workshop on Microprocessor Test and Verification: Common Challenges and Solutions (MTV '09)*, pp. 27–31, Austin, Tex, USA, December 2009.

[19] N. Bombieri, F. Fummi, G. Pravadelli et al., "Functional qualification of TLM verification," in *Proceedings of the Design, Automation and Test in Europe Conference and Exhibition (DATE '09)*, pp. 190–195, Nice, France, 2009.

[20] S. Bouvier, N. Sauzede, F. Letombe et al., "A practical approach to measuring and improving the functional verification of embedded software," in *Proceedings of the Design Verification Conference (DVC '12)*, June 2012.

[21] M. Hampton and S. Petithomme, "Leveraging a commercial mutation analysis tool for research," in *Proceedings of the Testing: Academic and Industrial Conference Practice and Research Techniques—MUTATION (TAICPART-MUTATION '07)*, pp. 203–209, Windsor, UK, September 2007.

[22] N. Bombieri, G. Di Guglielmo, M. Ferrari et al., "HIFSuite: tools for HDL code conversion and manipulation," *Eurasip Journal on Embedded Systems*, vol. 2010, Article ID 436328, 2010.

[23] T. Xie, W. Mueller, and F. Letombe, "HDL-mutation based simulation data generation by propagation guided search," in *Proceedings of the 14th Euromicro Conference on Digital System Design (DSD '11)*, pp. 608–615, Oulu, Finland, September 2011.

[24] X. Tao, W. Mueller, and F. Letombe, "Mutation-analysis driven functional verification of a soft microprocessor," in *Proceedings of the IEEE International SOC Conference (SOCC '12)*, pp. 283–288, IEEE, Niagara Falls, NY, USA, September 2012.

[25] F. Fallah, S. Devadas, and K. Keutzer, "OCCOM-efficient computation of observability-based code coverage metrics for functional verification," *IEEE Transactions on Computer-Aided Design of Integrated Circuits and Systems*, vol. 20, no. 8, pp. 1003–1015, 2001.

[26] T. Lv, J.-P. Fan, X.-W. Li, and L.-Y. Liu, "Observability statement coverage based on dynamic factored use-definition chains for functional verification," *Journal of Electronic Testing*, vol. 22, no. 3, pp. 273–285, 2006.

[27] P. Lisherness and K.-T. Cheng, "An instrumented observability coverage method for system validation," in *Proceedings of the IEEE International High Level Design Validation and Test Workshop (HLDVT '09)*, pp. 88–93, San Francisco, Calif, USA, November 2009.

The Design of Low Noise Amplifiers in Deep Submicron CMOS Processes: A Convex Optimization Approach

David H. K. Hoe[1] and Xiaoyu Jin[2]

[1]*Department of Engineering, Loyola University Maryland, Baltimore, MD 21210, USA*
[2]*Department of Electrical Engineering, University of Texas at Tyler, Tyler, TX 75799, USA*

Correspondence should be addressed to David H. K. Hoe; dhhoe@loyola.edu

Academic Editor: Roc Berenguer

With continued process scaling, CMOS has become a viable technology for the design of high-performance low noise amplifiers (LNAs) in the radio frequency (RF) regime. This paper describes the design of RF LNAs using a geometric programming (GP) optimization method. An important challenge for RF LNAs designed at nanometer scale geometries is the excess thermal noise observed in the MOSFETs. An extensive survey of analytical models and experimental results reported in the literature is carried out to quantify the issue of excessive thermal noise for short-channel MOSFETs. Short channel effects such as channel-length modulation and velocity saturation effects are also accounted for in our optimization process. The GP approach is able to efficiently calculate the globally optimum solution. The approximations required to setup the equations and constraints to allow convex optimization are detailed. The method is applied to the design of inductive source degenerated common source amplifiers at the 90 nm and 180 nm technology nodes. The optimization results are validated through comparison with numerical simulations using Agilent's Advanced Design Systems (ADS) software.

1. Introduction

The low-noise amplifier (LNA) is the critical component in the analog front-end of a radio frequency (RF) receiver. The LNA is responsible for providing sufficient amplification of weak input signals while minimizing the amount of added electronic noise and distortion. As a result, the characteristics of the LNA set the upper limit on the performance of the overall communication system. The optimization of the LNA is a complex task involving tradeoffs that must be made among several competing parameters including noise figure, linearity, and impedance matching [1]. While bipolar technologies have traditionally dominated RF designs due to their superior switching frequency and gain, they are not particularly suited for low power design and are not directly compatible with scaled digital CMOS processes. In the late 1990s, the transit frequency of CMOS devices reached the 40 GHz range, which enabled the design and implementation of RF CMOS circuits that could process signals on the order of 4 GHz [2]. A combination of improved

processing technology, suitable MOS circuit architectures, and amenable wireless standards have helped push CMOS technology to the forefront for RF circuit implementations [3–5]. Implementing high quality RF analog circuits on scaled digital processes is a desired goal for a couple of reasons. First, a prime motivation is the increased integration densities and resulting lower costs. Second, it allows the realization of a complete RF transceiver on a systems-on-a-chip (SoC) implementation, which would further improve design flexibility and system optimization [6]. An SoC design with digital, analog, and RF components will pave the way for radio systems that are largely software-controlled digital devices [3]. While the digital components dominate in a software-defined radio (SDR) architecture, the analog front-end including the LNA will remain the critical component that determines the overall system performance.

In the rapidly growing consumer demand for portable wireless devices with long battery life, obtaining sufficient receiver sensitivity while minimizing power dissipation is a major design objective. As process scaling continues to shrink the dimensions of the CMOS transistors, RF circuits will benefit from the improved switching frequencies. However, the main issue will be the reduction in performance due to increased thermal noise from MOSFETs implemented in scaled digital CMOS processes and lower gain and signal swing headroom as voltage supplies are inevitably decreased. With reduction in the analog voltage supplies, an increase in power dissipation is required in order to maintain constant performance [7]. Hence, innovations in circuit topologies and optimization methods will be required to maintain low power and high performance as device scaling continues deep into the nanometer-scale dimensions.

A common design methodology is to determine the minimum noise that can be obtained given constraints on impedance matching and power dissipation. Using classical two-port noise theory, the optimum impedance that must be presented at the source of the LNA in order to achieve the minimum noise figure (NF) can be calculated. An appropriate matching network is inserted between the source and the LNA to help achieve this goal as illustrated in Figure 1. It is well known that maximum power transfer between the source and the amplifier input occurs when the complex conjugate of Z'_{in} matches the source impedance Z_s [1]. Common-source (CS) amplifiers have been the most popular CMOS LNA circuit topologies due to their good low-noise performance and high gains. However, when using a CS amplifier, the input impedance Z_{in} presented by the MOSFET makes it difficult to optimally match with the external impedance Z_s, which is generally resistive in nature [8]. Hence, tradeoffs in terms of noise performance and gain must be made with this architecture.

The basic amplifier architecture illustrated in Figure 2 allows the noise figure to be minimized while achieving input matching under power constrained conditions. The addition of inductive source degeneration and an input impedance matching inductive component, denoted by L_s and L_g, respectively, allows improved impedance matching to be obtained over a narrow band of interest. The addition of the capacitance C_e gives the added design flexibility of meeting a given power dissipation and input matching specification while maintaining a very low noise performance [9]. The cascode device M_2 provides isolation with the load.

Efficient and accurate optimization techniques for implementing analog integrated circuits are a critical facet of a CAD-based design flow. This is essential when the goal is to minimize the time-to-market for a product, and thus have working designs on first silicon [10]. While the noise analysis of a linear two-port network provides some insight into how to optimize the noise figure (NF) of an amplifier [11], this classical approach does not provide any guidance on the sizing of the devices. Various approaches that incorporate suitable FET device characteristics and noise models into the design process have been developed [12–14]. In order to account for second order effects in devices as scaling occurs, two general optimization strategies can be used: simulation-based and equation-based methods. A simulation-based approach allows more general topologies and circuit parameter variations to be explored. However, there are a couple of drawbacks to this approach. First, as metaheuristic algorithms, such as

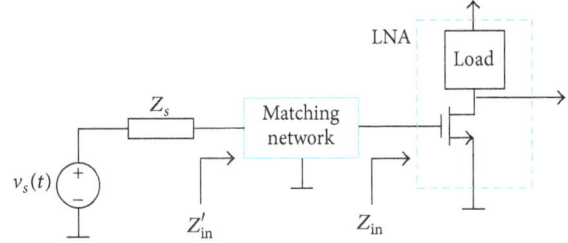

FIGURE 1: Generalized block diagram of the low noise amplifier as the key front-end component.

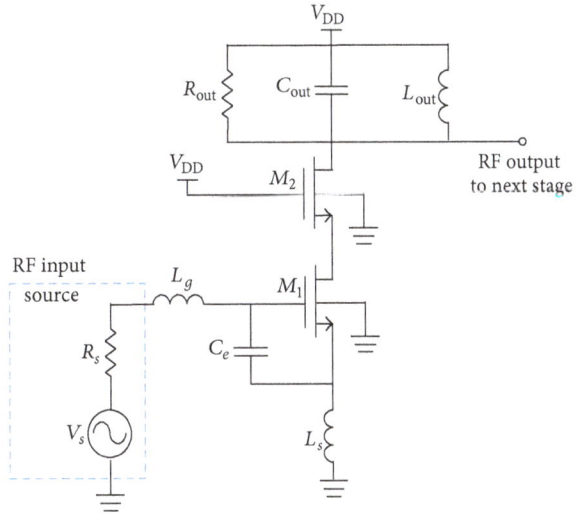

FIGURE 2: Schematic of a CMOS cascode low noise amplifier with inductive source degeneration.

genetic algorithms and other evolutionary techniques, are often used, this approach is very computationally intensive. This is due to the large number of iterations that require detailed circuit simulations to be executed [15–18]. Second, there is no guarantee that a globally optimum solution is found. On the other hand, equation-based methods attempt to formulate a solution for a more restricted set of parameters and usually for a predetermined circuit topology. Geometric programming methods are able to find a globally optimum solution very efficiently for a well-formulated problem [19]. The trade-off in this case is that certain approximations must be made to ensure that the device equations are in a form suitable for numerical optimization. However, once the system is set-up, a globally optimum design can quickly be found.

In this paper, the focus is on the optimization of a specific topology, the CS CMOS LNA, when short-channel effects such as excess thermal noise must be taken into consideration. As such, we will make use of convex optimization, a form of geometric programming. Geometric programming (GP) has been previously employed to optimize a variety of integrated circuit designs, including both analog and digital circuits [20–25]. For a comprehensive overview and list of GP applications, see [19]. The approach in this paper is to optimize the NF of the CS LNA subject to various constraints such as input circuit quality factor, power consumption, and input impedance matching, similar to [12]. The optimization procedure will allow the globally optimum selection of device parameters. Geometric programming has been used to optimize the design of RF CMOS low-noise amplifiers at the 0.35 μm technology node [26] using the design proposed in [9]. While the work by [26] results in globally optimum solutions with an extremely small computational cost, it does not take into account MOSFET submicron device characteristics and the issue with excess noise in the nanoscale regime is not addressed.

The main contribution of this study is the incorporation of important MOSFET short-channel effects including excess noise into a GP framework to enable the optimization of LNAs designed in deep submicron processes [27]. An extensive review and evaluation of the various approaches used to model the excess noise in nanoscale devices is given. In addition, the approximations required to convert the relevant device equations into a form required by the GP algorithms while minimizing the amount of accuracy lost in the noise and current-voltage model equations are detailed. This paper is outlined as follows. The second section provides the background theory on modeling MOSFET noise, including the various models used to explain the sources of excess thermal noise in MOSFETs with nanoscale dimensions. The third section describes the optimization method for designing the RF LNA using geometric programming. The fourth section presents the results from applying geometric programming to obtain the globally optimal solution for RF LNA designs in 90 nm and 180 nm processes. The generated optimal solutions are compared with results from Agilent's Advanced Design Systems (ADS) software. Finally, the implication of geometric programming for short-channel CMOS designs is discussed and future work in this area is described.

2. Noise in Deep Submicron CMOS Processes

While MOSFETs in aggressively scaled CMOS processes have sufficiently high transit frequencies (f_T) for RF circuit applications, issues with increased noise levels may prevent low noise operation. This section discusses the issues with excess noise that have been experimentally observed. A review of basic noise theory is first undertaken, followed by a discussion of the relevant observations in the literature regarding the issue of excess MOSFET noise. This section concludes with a summary of the key parameters used to model this excess thermal noise in deep submicron processes.

2.1. Basic MOSFET Noise Theory. An expression for the power spectral density due to the thermal noise in a MOSFET is derived in this subsection. The relationship between the channel current and the local channel conductivity of the MOSFET is considered first. The drain current of a MOSFET can be expressed by the following relationship:

$$I_d = g\left(V\left(x\right)\right) \cdot \frac{dV\left(x\right)}{dx}, \tag{1}$$

where $V(x)$ is the channel potential at position x in the device's channel as shown in Figure 3, $dV(x)$ is the dc voltage

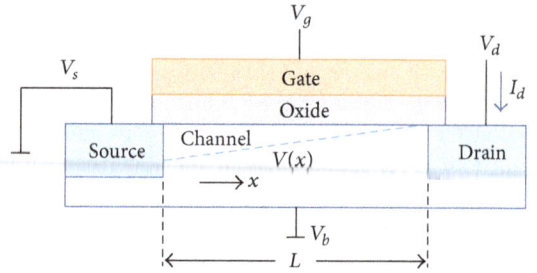

FIGURE 3: Cross-section of an n-channel MOSFET transistor.

difference in the electron quasi-Fermi level in the inversion layer and the hole quasi-Fermi level in the substrate at position x, and $g(V(x))$ is the local channel conductivity.

For a simple long-channel MOSFET using the gradual channel approximation, the following relationships can be written [8]:

$$g\left(V\left(x\right)\right) = \mu C_{\mathrm{ox}} W\left(V_{\mathrm{od}} - V\left(x\right)\right),$$
$$V_{\mathrm{od}} = V_{\mathrm{gs}} - V_{\mathrm{th}}, \tag{2}$$

where V_{gs} is the gate-to-source voltage, V_{th} is the threshold voltage, V_{od} is the gate overdrive voltage, W is the width of the MOSFET, μ is the carrier mobility, and C_{ox} is the oxide capacitance per unit area.

Assuming a differential segment Δx of the channel, a small noise voltage contribution $v(x)$ across the segment Δx is observed, which is added to the dc voltage $V(x)$. This voltage can cause noise in the drain current, which leads to a change in the dc current through the MOSFET. In the following analysis, a couple of assumptions are made. First, noise sources of the different channel segments are local and not correlated. Second, the charge carriers are in thermal equilibrium. The boundary conditions of the small voltage contribution $v(x)$ are $v(x)|_{x=0,L} = 0$ [8]. The power spectral density S_{i_d} for the thermal noise of a long-channel MOSFET is then expressed by the Klaassen-Prins equation [28]:

$$S_{i_d} = \frac{4kT}{L^2 I_d} \int_0^{V_{ds}} g^2\left(V\right) \cdot dV, \tag{3}$$

where I_d is the drain current of the device. The impact of hot electron effects can be modeled by replacing the lattice temperature with carrier temperature, $T_e(x)$ [8]:

$$S_{i_d} = \frac{4kT}{L^2 I_d} \int_0^{V_{ds}} \frac{T_e\left(x\right)}{T} g^2\left(V\right) \cdot dV. \tag{4}$$

For noise analysis, it is often convenient to treat the MOSFET as a resistive element:

$$S_{i_d} = 4kT\gamma g_{d0}. \tag{5}$$

The parameter γ is known as the white noise gamma factor, given the relationship between the thermal noise power spectral density and the output conductance at different bias conditions [27]:

$$\gamma = \frac{1}{g_0 I_d L} \int_0^{V_{ds}} \frac{T_e\left(x\right)}{T} g^2\left(V\right) \cdot dV. \tag{6}$$

In (6), g_0 is the channel conductance per unit length at the source and g_{d0} is the channel conductance at zero drain bias. The value of γ is unity for zero drain bias in long-channel devices and decreases toward 2/3 in saturation. The expression in (6) is commonly used to express the thermal noise in long-channel MOSFETs. In practice, the white noise gamma parameter continues to be used as a common metric to allow experimental or theoretical results to be compared from different research groups when describing the degree of excess channel thermal noise in short-channel transistors [29].

The variation of the channel charge due to thermal noise is capacitively coupled to the gate terminal, resulting in a noisy gate current. Just as the white noise gamma parameter provides a convenient way to express the power spectral density of the thermal noise, the introduction of a beta parameter allows the induced gate noise to be expressed in a similar manner [8]:

$$S_{i_g} = 4kT\beta g_g. \tag{7}$$

The parameter β is basically independent of the substrate conductivity, and its value is 4/3 in the saturation region for long-channel MOSFETs. The conductance g_g is given by

$$g_g = \frac{\omega^2 C_{gs}^2}{5g_{d0}}, \tag{8}$$

where C_{gs} is the intrinsic gate capacitance of the transistor. In the circuit model representation illustrated in Figure 4, the conductance g_g is connected between the gate and source shunted by the gate noise current $\bar{i}_{ng}$. From (8), it can be observed that the conductance g_g increases with frequency, indicating that the induced gate noise can dominate at radio frequencies. The conductance g_g is also proportional to the square of C_{gs}, so a small value of C_{gs} will favor a lower induced gate noise.

Since the induced gate noise is correlated with the drain thermal noise, the correlation coefficient is defined as [8]:

$$c = \frac{\overline{i_{ng} \cdot i_{nd}^*}}{\sqrt{\overline{i_{ng}^2} \cdot \overline{i_{nd}^2}}}, \tag{9}$$

where $\overline{i_{ng} \cdot i_{nd}^*}$ is the spectrum of the cross-correlation of the drain thermal noise and the induced gate noise. The complex correlation coefficient c is theoretically $0.395j$ for long-channel MOSFETs as noted in Appendix C [8].

The finite resistance of the gate material also contributes to this noisy gate current and can become the dominant source of gate thermal noise in short-channel MOSFETs [30]. Two factors tend to minimize this source of gate noise. First, modern CMOS processes use silicided gate material which helps reduce the resistance in the gate. Second, for wide MOSFETs, a multifinger layout can be used whereby several devices (i.e., "fingers") are connected in parallel, giving the gate resistance as [31]:

$$R_g = \frac{R_{sh}}{12 \cdot n_f^2} \frac{W}{L}, \tag{10}$$

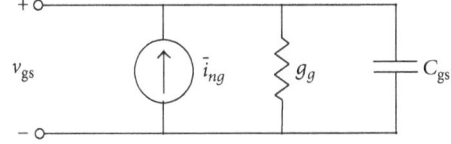

FIGURE 4: Circuit model for the gate noise [1, 8].

where R_{sh} is the sheet resistance of the gate material, n_f is the number of fingers, and the factor of 12 is due to the distributed nature of the gate resistance when it is contacted on both ends [12, 32]. As devices scale to submicron dimensions, the interface resistance between the silicide and polysilicon layers becomes an important component of R_g which is not significantly impacted by layout optimizations [33].

2.2. Modeling Thermal Noise in Short-Channel Devices. Excess thermal noise for scaled devices must be taken into account when designing LNAs operating at RF frequencies. In this subsection, we review recent methods used to model this noise. This provides the background on the current understanding of the excess thermal noise issue in deep submicron devices. While this issue is still a matter of open debate among researchers [34, 35], the development of models will provide the reader with the context to understand the comparisons in the following subsection as well as the rationale for the use of an empirical fit to the data for our optimization method. Here the emphasis is on extending the classical theory of thermal noise to submicron devices by including short-channel effects such as velocity saturation, channel-length modulation, and hot carriers. The Klaassen-Prins equation for the noise power spectral density in (3) can be modified to include channel-length modulation and velocity saturation effects as follows:

$$S_{i_d} = \frac{4kT}{L_{elec}^2 I_d} \int_0^{V_{ds}} g_c^2(V) \cdot dV, \tag{11}$$

where L_{elec} is the electrical channel length of the MOSFET, replacing the effective channel length L_{eff} in the long-channel expression [29, 30, 35]. The parameter L_{elec} is defined as $L_{elec} = L_{eff} - \Delta L$ where ΔL is the length of the velocity saturated region. The parameter g_c is the revised conductivity taking velocity saturation into consideration. The noise contribution of the pinch-off region is assumed to be negligible as experimental evidence indicates that the channel thermal noise is practically independent of the drain-to-source voltage beyond the saturation voltage [30].

The approach by Han et al. [36, 37] is to consider the effects of velocity saturation and carrier heating. While the carrier mobility is considered independent of the bias conditions and is usually modeled as a constant in long-channel MOSFETs, it is degraded in short-channel devices due to the high lateral electric field from drain to source [38] and is thus dependent on the bias conditions. The impedance field method [39] was used to recalculate the thermal noise for short-channel MOSFETs. The drain current of a MOSFET with the effect of mobility degradation is then given by [37]

$$I_d = g_0(V) \frac{(dV/dx)}{1 + (dV/dx)/E_C},$$ (12)

where the local channel conductance $g_0(V) = \mu_{\text{eff}} W C_{\text{ox}}(V_{\text{od}} - \alpha V)$. The parameter $E_C = 2v_{\text{sat}}/\mu_{\text{eff}}$ is the critical field at which velocity saturation occurs, v_{sat} is the saturation velocity of carriers, μ_{eff} is the effective mobility, and α is a coefficient describing the bulk-charge effect. The bulk-charge effect is the variation of threshold voltage caused by nonuniform channel depletion and the dependence of the threshold voltage on the channel potential. The impact of the carriers in the velocity saturation region on the drain thermal noise current is ignored in this analysis. Applying a similar procedure as [30], the channel noise of the MOSFET takes the form of [37]

$$S_{i_d} = \frac{4kT}{L_{\text{elec}}^2 I_d \left(1 + V_{\text{ds}}/(L_{\text{elec}} E_C)\right)^2} \cdot \int_0^{V_{\text{ds}}} g_0^2(V)\left(1 + \frac{E}{E_C}\right) \cdot dV,$$ (13)

where the electrical channel length of the MOSFET is $L_{\text{elec}} = L_{\text{eff}} - \Delta L$. In order to obtain a compact analytical equation, a closed-form expression is given by [37]

$$S_{i_d} \approx 4kT g_{d0} \frac{1 - u + u^2/3}{1 - u/2},$$ (14)

where g_{d0} is the drain conductance at $V_{\text{ds}} = 0$ V, $u = \alpha V_{\text{ds}}/V_{\text{od}}$. The coefficient of the bulk-charge effect α has a typical value of 1.2 [40].

Based on [36], the longitudinal electric field (E) along the channel was examined by Deen et al. [31]. The longitudinal electric field (E) is now expressed as a function of the position x along the channel instead of simply being constant:

$$E(x) = \frac{E_C V_d}{\left[(2V_{\text{od}} - V_d)^2 - 4\alpha E_C V_d x\right]^{1/2}},$$ (15)

where $V_d = I_d/(W C_{\text{ox}} v_{\text{sat}})$ and V_{od} is the gate overdrive voltage given in (2). The revised total channel charge can be obtained by integrating the drain current from 0 to L_{elec} with the expression of $E(x)$ in (15). The total drain current noise power spectral density is then obtained:

$$S_{i_d} = 4kT \frac{4V_{\text{od}}^2 + V_d^2 + V_{\text{od}} V_d}{3V_{\text{od}}^2 (V_{\text{od}} - V_d)} \alpha I_d.$$ (16)

An analytical thermal noise model following [41] was developed by Jeon et al. [42], which includes short-channel effects such as channel-length modulation, velocity saturation, and hot carrier effects. The ac conductance g_{ac} is a small signal conductance with the consideration of velocity saturation. It expresses the current noise source spectrum of a small segment of the channel length Δx:

$$\overline{\Delta i_n^2} = 4kT_c g_{\text{ac}} \Delta f,$$ (17)

where T_c is the carrier effective temperature. The carrier temperature has shown a dependency on the electric field when a high electric field is present in short-channel MOSFETs. The relation of T_c and the electric field is given as

$$\frac{T_c}{T_0} = \left(1 + \frac{E}{E_c}\right)^n,$$ (18)

where T_0 is the lattice temperature. When $n = 0$, the carrier is in thermal equilibrium without any carrier heating effect while the heating effect is considered for $n = 1$ or $n = 2$ [8]. Experimental measurements with devices having channel lengths of 130 nm indicate that the carrier heating effect with $n = 2$ gives the most accurate results [42].

2.3. Results and Comparisons of Modeling Noise in Short-Channel Devices. While the expressions for the power spectral density in (11) and (13) include short-channel effects, they are not compatible with the form required by geometric programming. A simpler noise formula which captures the essence of the noise issues at the deep submicron technology nodes is required. As previously noted, the channel thermal noise can be conveniently expressed using the white noise gamma expression given in (6). Since this expression is a simple closed-form equation, it has been widely used for noise analysis by circuit designers. For long-channel MOSFETs, the theoretical values of γ are well known. It is equal to unity at zero drain bias and 2/3 in the saturation region.

The analysis and experimental measurements by Scholten et al. [30] have shown that the channel thermal noise constant γ and the gate current noise parameter β are independent of the operating frequencies up to moderately high frequencies (around 10 GHz), and they are not very sensitive to bias conditions for high bias voltages. However, both parameters are expected to increase as channel lengths scale down in the submicron range. The values of γ are expected to be larger than their theoretical long-channel values due to excess channel thermal noise discussed previously for short-channel MOSFETs. Due to induced gate noise related to channel noise and the increased significance of the resistivity of the gate material at short-channel lengths, the parameter β will experience a similar increase in value.

Based on the measurements of Jeon et al. [43], the channel thermal noise power spectral density can still be expressed by use of the white noise parameter γ when short-channel effects account for

$$\gamma = \frac{g_{\text{ds}}}{g_{d0}} \left(1 + \frac{\overline{E}}{E_C}\right),$$ (19)

where g_{ds} is the conductance of the channel, and $\overline{E}$ is the average longitudinal electric field which is equal to $V_{\text{ds}}/L_{\text{elec}}$. The parameter E_C is the critical electric field, which is equal to $2v_{\text{sat}}/\mu_{\text{eff}}$. Based on the model of (19), γ is a function of the drain bias for different channel lengths.

For nanoscale devices with feature sizes below 100 nm, it is still debated whether short-channel effects, as discussed above, are adequate for describing the effects of short-channel noise [44]. Some researchers have suggested that

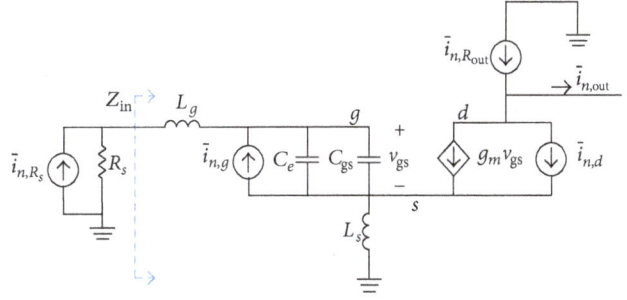

FIGURE 6: Small signal circuit for noise analysis.

TABLE 1: Thermal noise comparison of different analytical noise models.

Gate length	Power spectral density of channel thermal noise (A²/Hz)		
	Deen et al. [31]	Scholten et al. [30]	Jeon et al. [43]
90 nm	9.07×10^{-24}	1.04×10^{-23}	1.07×10^{-23}
180 nm	4.22×10^{-24}	4.63×10^{-24}	4.54×10^{-24}
350 nm	9.86×10^{-25}	1.17×10^{-24}	1.41×10^{-24}

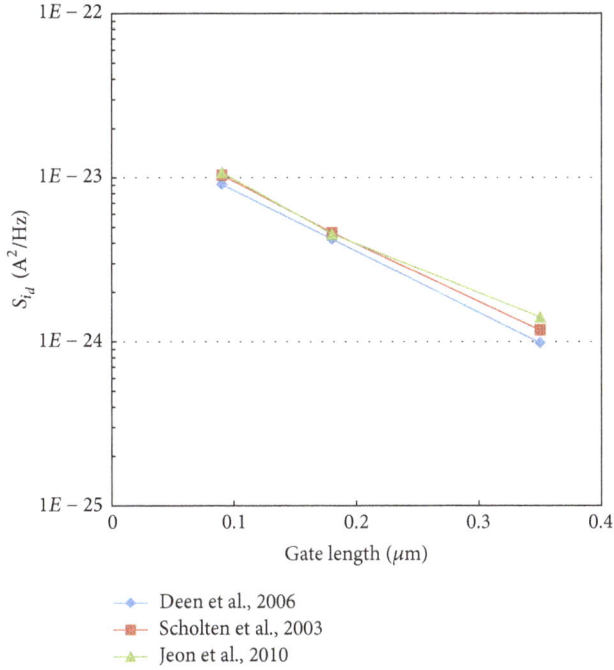

FIGURE 5: Thermal noise comparison of different analytical noise models.

shot noise is better able to describe the noisy behavior for FETs below 40 nm [34, 45, 46]. As this study focuses on LNA design optimization down to the 90 nm node, it will be assumed that excess thermal noise can be adequately handled through modification of the white noise gamma parameter γ. Experimental results from a number of researchers appear to support this approach [31, 34, 43]. A comparison between the expression for the channel thermal noise in (16) [31] with the thermal noise calculation using the two γ models from [30, 43] have been made. As shown in Figure 5 with the numerical data in Table 1, their results are comparable with a similar trend regarding different channel lengths. Since Scholten et al. [30] and Jeon et al. [43] have completed a relatively in-depth study of the noise parameters and there is relatively good agreement of their work with the analytical model of Deen et al. [31], the noise calculations in this work are carried out based upon the results of [30, 43].

3. Optimization Methods

The optimization of the CMOS LNA design in terms of minimizing its noise figure as the main cost function is considered in this section. The maximum allowed power dissipation is used as the main design constraint as this is a chief concern for modern systems, especially those intended for mobile electronic systems. The influence of other design constraints, such as the quality factor of the input circuit and the input impedance matching requirement, is taken into account during the optimization process. The noise analysis of the LNA and the parameters used to model the noise characteristics of submicron MOSFETs are considered first. Then the device equations needed to model the drain current I_{ds} as well as the transconductance g_m and the output conductance g_{d0} are described. Finally, the overall method used to optimize the LNA design within a geometric programming framework is detailed.

3.1. Noise Analysis of the LNA. This subsection describes how the noise figure of the LNA given in Figure 2 can be calculated by small signal analysis. Also, the design parameters used to describe the noise characteristics of short-channel MOSFETs are given. The thermal noise is the major concern at RF intermediate frequencies for MOSFETs. Four noise sources have been considered in this study: the thermal noise of the source resistance ($\bar{i}_{n,R_s}$), the channel thermal noise ($\bar{i}_{n,d}$), the gate noise ($\bar{i}_{n,g}$), and the thermal noise of the output resistance ($\bar{i}_{n,R_{out}}$). These are depicted in Figure 6. The noise contributions due to the gate resistance are factored into the elevated value for the parameter β as discussed below [30]. Neglecting the effect of the gate-to-drain capacitance C_{gd} on the noise calculations introduces a small error but allows closed-form equations to be derived. This error is minimized through the use of a cascode topology, where M_2 mitigates the Miller effect of C_{gd} [12]. The noise contributions of the cascode device M_2 in Figure 2 are considered to be negligible compared to the contributions of the main FET M_1. Following the observations by [30], the noise contributions of the MOSFET source and bulk resistance are taken to be minimal and are neglected in this analysis.

The contributions of these four noise sources referred to the output are denoted by $\bar{i}_{n,o,R_s}$, $\bar{i}_{n,o,d}$, $\bar{i}_{n,o,g}$, and $\bar{i}_{n,o,R_{out}}$, respectively. Table 2 summarizes the expressions for these noise sources [9, 47].

TABLE 2: Output-referred noise equations.

Noise source	Expression	Output-referred expression
R_s	$\overline{i_{n,R_s}^2} = 4kT\dfrac{1}{R_s}\Delta f$	$\bar{i}_{n,o,R_s} = \dfrac{g_m}{j2\omega_0 C_{\text{tot}}}\bar{i}_{n,R_s}$
$\bar{i}_{n,d}$	$\overline{i_{n,d}^2} = 4kT\gamma g_{d0}\Delta f$	$\bar{i}_{n,o,d} = -\dfrac{1}{2}\bar{i}_{n,d}$
$\bar{i}_{n,g}$	$\overline{i_{n,g}^2} = 4kT\beta g_g\Delta f$	$\bar{i}_{n,o,g} = \dfrac{g_m}{j\omega_0 C_{\text{tot}}}\dfrac{1 - jR_s\omega_0 C_{\text{tot}}}{j2R_s\omega_0 C_{\text{tot}}}\bar{i}_{n,g}$
R_{out}	$\overline{i_{n,R_{\text{out}}}^2} = 4kT\dfrac{1}{R_{\text{out}}}\Delta f$	$\bar{i}_{n,o,R_{\text{out}}} = \bar{i}_{n,R_{\text{out}}}$

The correlation between the induced gate noise and the channel thermal noise is represented by $\bar{i}_{n,\text{corr}}$. Therefore, the LNA noise factor can be expressed as

$$F = \frac{\overline{i^2}_{n,o,R_s} + \overline{i^2}_{n,o,d} + \overline{i^2}_{n,o,g} + \overline{i^2}_{n,o,\text{corr}} + \overline{i^2}_{n,o,R_{\text{out}}}}{\overline{i^2}_{n,o,R_s}}. \tag{20}$$

Then, the noise factor at resonance is obtained as

$$F = 1 + \frac{(1/4)\,\gamma g_{d0} + g_m^2\left(C_{\text{gs}}/C_{\text{tot}}\right)^2\left(Q^2 + 1/4\right)\beta/\left(5g_{d0}\right) + g_m c\left(C_{\text{gs}}/C_{\text{tot}}\right)\sqrt{(\gamma \cdot \beta)/20} + 1/R_{\text{out}}}{g_m^2 R_s Q^2}, \tag{21}$$

where γ is defined by (5), β is the gate noise parameter, c is the correlation coefficient, C_{gs} is the intrinsic gate capacitance, C_{tot} is the sum of C_{gs} and C_e, and Q is the quality factor of the input circuit.

Based on the studies by [30, 31, 43], the white noise factor γ is assumed to be independent of the operating frequencies up to 10 GHz and to be independent of bias conditions. A comparison of the values γ versus FET channel length is given in Figure 7.

The measured and analytical γ compare favorably when observed at various gate lengths (e.g., 90 nm, 180 nm, and 350 nm), as shown in Figure 7. As expected, the white noise factor γ increases when the channel length decreases. For long-channel devices (channel lengths greater than 1 μm), the traditional value for γ is 2/3.

Numerical values for the gate noise parameter β and correlation coefficient c are estimated from [30, 43] and are summarized, along with the parameter γ, in Table 3. There is a significant increase in the value of the parameter β as the channel length decreases due to the contribution from the gate resistance, which consists of the resistance of the vias, the effective resistance of the silicide, and the contact resistance between the silicide and polysilicon layers [30, 33]. The value of β is close to 4/3 for long-channel devices, but more than doubles in value for 180 nm devices. Therefore, a significant increase is predicted for devices at the 90 nm node. The magnitude of the correlation coefficient is 0.395 for long-channel devices [8], and it decreases due to larger γ and β when channel length reduces in size as can be inferred from (C.3) in Appendix C. A reasonable approximation is that the values for the parameters β and c are relatively independent of frequency and variations with bias conditions for strong inversion. Scholten et al. [30, 48] have shown that modeling the gate noise power spectral density S_{i_g} with a constant value for β using (7) gives a good fit to experimentally measured results for short-channel devices over a range of applied voltages up to 10 GHz. They also show that the correlation coefficient c is relatively independent of frequency and bias voltage.

In order to determine the sensitivity to γ and β in the calculation of the minimum noise figure, the effect of varying these parameters was analyzed (see Appendix D for further details). When a $\pm 10\%$ variation is applied to γ, a small percentage of variation (around 4%) occurs to the minimum noise figure. Similarly, less than 4% variation occurs on the minimum noise figure when a $\pm 10\%$ change is applied to β. This gives confidence to the assumption that the parameters γ and β can be modeled as constants for a given technology node without adversely affecting the optimization results.

3.2. Device Equations for Submicron FETs. This subsection outlines how the device models that take into account short-channel effects can be developed in a form suitable for geometric programming. As device geometries approach submicron dimensions and below, various high field effects such as velocity saturation and channel length modulation must be taken into consideration. A piece-wise model of the drain current I_{ds} which includes these effects has been used in this analysis [38]:

$$I_{ds} = \mu_{\text{eff}}C_{\text{ox}}\left(\frac{W}{L}\right) \cdot \frac{V_{\text{od}}V_{ds} - (m/2)V_{ds}^2}{1 + \left(\mu_{\text{eff}}V_{ds}\right)/\left(2v_{\text{sat}}L\right)},$$

$$I_{d\text{sat}} = \frac{\mu_{\text{eff}}C_{\text{ox}}\left(W/L\right)\left(V_{\text{od}}^2/(2m)\right)\left(1 + \lambda V_{ds}\right)}{1 + \mu_{\text{eff}}V_{\text{od}}/(2mv_{\text{sat}}L)},$$

$$\mu_{\text{eff}} = \frac{\mu_0}{1 + \theta V_{\text{od}}}, \tag{22}$$

where m is the body effect factor, v_{sat} is velocity saturation, μ_{eff} is effective mobility in m^2/V, μ_0 is normal field mobility, and θ is normal field mobility degradation factor in V^{-1}. In (21), the transconductance g_m and the output conductance g_{d0} are the two main technology-dependent parameters. Analytical solutions are obtained for g_m and g_{d0} by taking the derivative of the closed-form analytical drain current solutions for

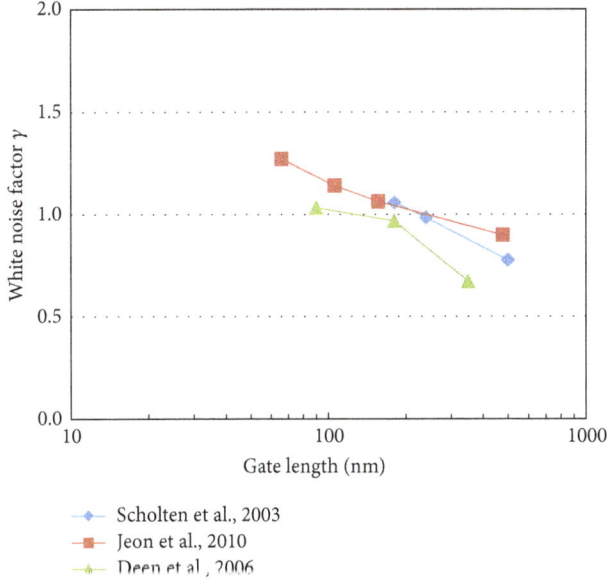

FIGURE 7: White noise factor γ versus gate length.

TABLE 3: Noise parameters for the noise analyses for 90 nm and 180 nm CMOS processes.

Parameters	90 nm design	180 nm design
White noise factor (γ)	1.2	1.05
Gate noise parameter (β)	7.5	3.8
Correlation coefficient (c)	0.2	0.2

short-channel CMOS transistors (see Appendix A), yielding the following equations:

$$g_m$$
$$= \frac{C_{ox}\mu_0 W v_{sat} \cdot (1 + \lambda V_{ds}) \left(4mv_{sat}LV_{od} + (2mv_{sat}L\theta + \mu_0) V_{od}^2\right)}{\left(2mv_{sat}L + (2mv_{sat}L\theta + \mu_0) V_{od}\right)^2}, \quad (23)$$

$$g_d = \frac{\mu_{eff}C_{ox} (W/L) \left[V_{od} - mV_{ds} - (m/2) \left(\mu_{eff}/ (2v_{sat}L)\right) V_{ds}^2\right]}{\left(1 + \left(\mu_{eff}V_{ds}\right)/ (2v_{sat}L)\right)^2}.$$

Then the channel conductance at zero bias condition is

$$g_{d0} = \mu_{eff} C_{ox} \left(\frac{W}{L}\right) V_{od}. \quad (24)$$

The final expressions for g_m and g_{d0} will need to be placed in a form suitable for geometric programming as described below.

3.3. Geometric Programming Optimization of the LNA.

A geometric program solves an optimization problem where the objective function is in the form of a posynomial function and the constraints are expressed as posynomial inequalities and monomial equalities. All design parameters are nonnegative variables. A monomial function has the following form [19]:

$$g(x) = cx_1^{a_1} x_2^{a_2} x_3^{a_3} \cdots x_n^{a_n}, \quad (25)$$

where c is a positive constant ($c > 0$), $x_1, x_2, \ldots,$ and x_n are real positive variables, and $a_1, a_2, \ldots,$ and a_n are constants known as the exponents of the monomial. Any positive constant is a monomial. Monomials are closed under multiplication and division. A posynomial function is a sum of one or more monomial functions as shown in the following equation:

$$f(x) = \sum_{k=1}^{K} c_k x_1^{a_{1k}} x_2^{a_{2k}} x_3^{a_{3k}} \cdots x_n^{a_{nk}}, \quad (26)$$

where $c_k > 0$. Note that posynomial functions are also closed under addition and multiplication. A standard form for a geometric programming can be defined as an optimization problem with the following form:

Minimize an objective function: $f_0(x)$

subject to constraints:

$$\begin{aligned} f_i(x) &\leq 1, \quad i = 1, \ldots, m, \\ g_i(x) &= 1, \quad i = 1, \ldots, p, \end{aligned} \quad (27)$$

where $x = (x_1, \ldots, x_n)$ is a vector with components x_i, $f_0(x)$ is an objective function with the form of a posynomial function, $f_1(x), f_2(x), \ldots, f_m(x)$ are posynomial functions, $g_1(x), g_2(x), \ldots, g_p(x)$ are monomial functions, and x_i are the optimization variables (x_i are always greater than zero) [19].

The objective function for this optimization problem is to minimize the noise figure NF, which is already in a posynomial form. Most of the design constraints are either in a posynomial form or monomial form. The main challenge is to translate the analytical expressions for the device transconductance (g_m) and output conductance (g_{d0}) in (23) into a form suitable for geometric programming. Following the work of [19], a curve-fitting approach is used to obtain monomial expressions for g_m and g_{d0}:

$$\begin{aligned} g_m &= A_0 L^{A_1} W^{A_2} I_{ds}^{A_3}, \\ g_{d0} &= B_0 L^{B_1} W^{B_2} I_{ds}^{B_3}. \end{aligned} \quad (28)$$

The details on the curve fitting and the resulting fitting parameters are given in Appendix B.

Process-dependent parameters for 90 nm and 180 nm technology nodes were derived from the SPICE model files provided by a predictive technology model (PTM) [49, 50]. Furthermore, the vertical field mobility degradation factor θ, the channel-length modulation parameter λ, and the body effect coefficient m were extracted from the device characterizations provided by running SPICE simulations using the PTM models. The relevant parameters are summarized in Table 4.

In addition to the noise figure, the major design constraints for LNAs include the quality factor, input impedance matching, and power consumption. Due to the resonant behavior of the circuit, the quality factor of the input circuit at the resonant frequency ω_0 is given by

$$Q = \frac{1}{R_{tot}\omega_0 C_{tot}} = \frac{1}{2R_s\omega_0 C_{tot}}. \quad (29)$$

TABLE 4: Technology parameters for 90 nm and 180 nm CMOS processes.

Parameters	90 nm	180 nm
Electron mobility μ_0	0.0179 m^2/V	0.0288 m^2/V
Electron velocity saturation ν_{sat}	1.10×10^5 m/s	9.10×10^5 m/s
Oxide capacitance per unit area C_{ox}	0.014 F/m^2	0.00857 F/m^2
Body effect coefficient m	1.21	1.18
Vertical field mobility degradation factor θ	0.3 V^{-1}	0.2 V^{-1}
Channel-length modulation parameter λ	0.4 V^{-1}	0.3 V^{-1}

To maximize the power transfer, the input impedance of the LNA is required to match the source input impedance, which is assumed to be 50 Ω. The impedance matching constraints can be expressed as

$$\omega_0 = \frac{1}{\sqrt{L_{tot} \cdot C_{tot}}},$$

$$R_s = \frac{g_m}{C_{tot}} L_s = 50 \text{ ohms,} \tag{30}$$

where L_{tot} is the sum of L_g and L_s.

The optimization problem using geometric programming can then be expressed as follows:

Minimize an objective function: Noise factor F in (20)

subject to design constraints:

$$L = L_{\text{feature size}},$$

$$1\,\mu m \leq W \leq 100\,\mu m,$$

$$\frac{C_{gs}}{C_{tot}} \leq 1,$$

$$\frac{3}{2}\frac{C_{gs}}{C_{ox}WL} = 1, \tag{31}$$

$$\frac{g_m L_s}{C_{tot}} = 50\,\Omega,$$

$$I_{ds} \cdot V_{DD} \leq P_{Dmax},$$

$$g_m = A_0 L^{A_1} W^{A_2} I_{ds}^{A_3},$$

$$g_{d0} = B_0 L^{B_1} W^{B_2} I_{ds}^{B_3}.$$

For the 90 nm process, $L_{\text{feature size}} = 90$ nm, $V_{DD} = 2$ V, and the maximum power dissipation P_{Dmax} is set at 1 mW. For the 180 nm process and $L_{\text{feature size}} = 180$ nm, $V_{DD} = 3$ V and $P_{Dmax} = 1.5$ mW. The current I_{ds} is the drain-to-source current through device M_1 in this design.

4. Results and Discussion

The optimal design of the CMOS LNA has been computed using CVX, a package for specifying and solving geometric

TABLE 5: Optimal design results for low-noise amplifier when input circuit quality factor $Q = 4$ and output circuit quality factor $Q_{out} = 5$.

Parameters	90 nm	180 nm
Output conductance (g_{d0})	0.0082 S	0.0063 S
Transconductance (g_m)	0.0069 S	0.0052 S
Gate width (W)	22.172 μm	27.006 μm
Gate length (L)	90 nm	180 nm
P factor ($P = C_{gs}/C_{tot}$)	0.1128	0.1681
Gate intrinsic capacitance (C_{gs})	18.696 fF	27.87 fF
Additional capacitance (C_e)	0.147 pF	0.13792 pF
Source inductor (L_s)	1.2063 nH	1.5828 nH
Gate inductor (L_g)	25.32 nH	24.943 nH
Drain current (I_{ds})	0.5 mA	0.5 mA
Minimum noise figure (F_{min})	0.6076 dB	0.8229 dB

programming problems [51]. The average execution time was about 1.45 seconds on a 3.23 GHz PC with 4 GB memory. The resulting optimal design parameters are shown in Table 5.

The results from the optimal design using geometric programming have been compared with results from Agilent's Advanced Design System (ADS) software, a numerical simulation tool used for RF design. The input FET M_1 was biased at 0.5 mA and the power supply was set to 2 V with the values of L_g, L_s, and C_e determined by the constraints used in the GP optimization. The output parallel RLC values are calculated by the output circuit quality factor, which is given as 5 in this study. For the 90 nm design, ADS simulations indicate that the minimum noise figure is 0.2799 dB for a gate width of 27 μm, while the optimal width from the optimization of geometric programming is 22.172 μm with a minimum noise figure of 0.6076 dB. For the 180 nm design, a minimum noise figure of 0.7708 dB was obtained for a gate width of 20 μm, while the optimal width from the optimization of geometric programming is 27.006 μm with a minimum noise figure of 0.8229 dB. As shown in Figure 8, the minimum noise figures from the ADS simulations are smaller than the minimum noise figures from the GP results. These discrepancies likely are caused by the lack of implementation of the excess thermal noise in the BSIM3 MOSFET models. The 90 nm design displays relatively larger differences than the 180 nm design, which is not unexpected as excess noise is more significant in shorter channel devices. The optimal widths for minimizing the NF from the GP optimization and ADS simulations are not an exact match, but the overall trends are fairly close. This indicates that geometric programming, which can rapidly find an optimal point, can be used to guide the design of short-channel CMOS LNAs. A good design methodology will then use detailed circuit simulations to fine tune the design and verify its performance. As current simulation models do not adequately account for excess thermal noise, some additional analysis based on experimentally determined FET noise characteristics will be required by the designer to ensure that the optimal design is found.

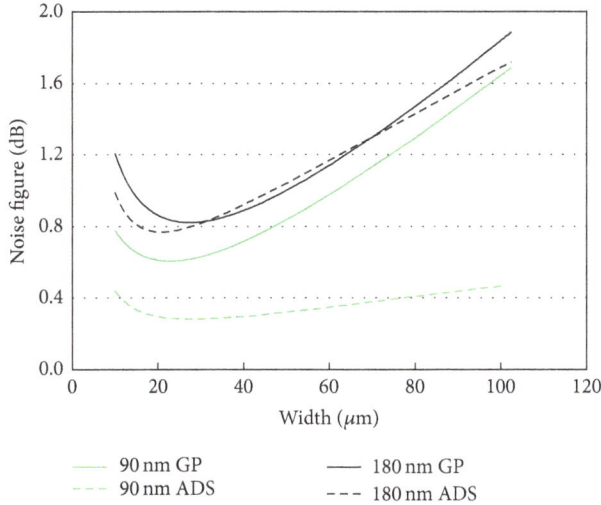

FIGURE 8: Variations of noise figure with different gate width and $Q = 4$.

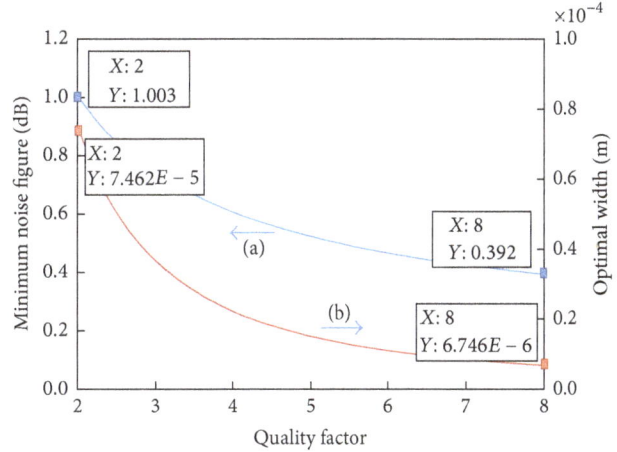

FIGURE 9: (a) Variation of minimal noise figure with different input quality factors Q and (b) variation of the optimal width with different quality factors.

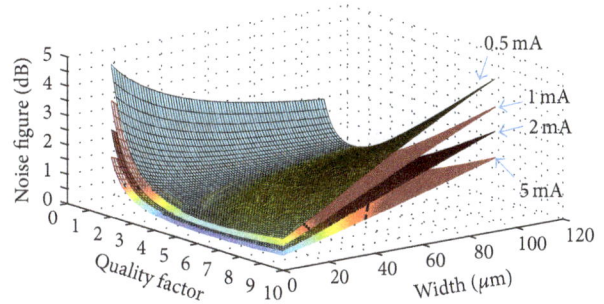

FIGURE 10: Effect of drain current and channel width on the noise figure (90 nm).

It should be noted that the inductor value of 25 nH for L_g would not be economical in terms of area when implemented as an on-chip planar spiral inductor. A prudent design choice would be to implement part of the inductance on the chip and the rest through the bond wire; alternatively, one could use the bond wire plus an external inductor on the printed circuit board [1]. Also advances in materials and fabrication technologies have made it possible to embed high quality inductors on the order of 20 nH to 30 nH in a package substrate that are suitable for RF applications [52, 53].

Tradeoff analyses were performed to examine the influence of the quality factor and drain current on the design of short-channel CMOS LNAs. As the optimization results for LNAs designed in 90 nm and 180 nm processes are similar, the trade-off analysis for the 90 nm case is presented in this paper. An inverse relationship is observed between the quality factor and the minimum noise figure, as seen in Figure 9(a). When the input quality factor increases from 2 to 8, the minimum noise figure decreases from 1 dB to 0.39 dB. The quality factor not only affects the minimal noise figure but also influences the optimal width of the LNAs. When the quality factor varies from 2 to 8, the optimal width changes almost 10 times from 75 μm to 6.7 μm, as seen in Figure 9(b). This considerable change in the optimal width indicates the importance of the quality factor in determining the optimal width of M_1.

The drain current appears to have great influence on the noise figure when the drain current is at a relatively small scale (i.e., less than 0.5 mA). However, there is not much variation in the noise figure when the drain current increases from 1 mA to 5 mA, as shown in Figure 10. Such an observation is true at different levels of channel width. This suggests that, for this 90 nm process, the best balance between power dissipation, area, and noise figure exists when the LNA is biased with 0.5 to 1.0 mA of current. When the channel width is set to 20 μm, the optimal range for the input circuit quality factor is from 4 to 6. This observation is consistent with the results reported in [9].

Variations in the frequency of operation also have a significant influence on the noise figure (Figures 11 and 12). In many applications, an RF LNA will be optimized for a particular narrowband of operation, for example, at 2.4 GHz. Therefore, the influence of operational frequency on the noise figure will be limited, and there is a clear choice for the optimum device width for minimizing the noise figure.

In sum, our results show that the use of geometric programming allows the global optimal design optimization of an LNA to be obtained with great efficiency. This study has focused on the common LNA configuration that uses source inductive degeneration. Short-channel effects have been taken into account when modeling the electronic noise in the MOSFETs as well as in the device characteristics. While some approximations must be made to put the equations in the proper form required by a GP framework, the results are guaranteed to return a globally optimum solution. Various trade-off analyses can be efficiently run as well under given constraints, such as power dissipation and input quality factor. For example, the input circuit quality factor has a great influence on not only the minimum noise figure but also the optimal width. Our results, in general, align well with other results in the literature. In the particular case of the 90 nm technology node used in this study, one can

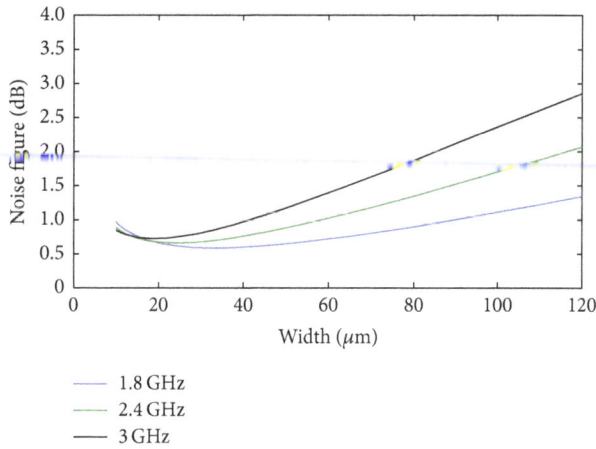

FIGURE 11: Effect of channel width on the noise figure at different frequencies.

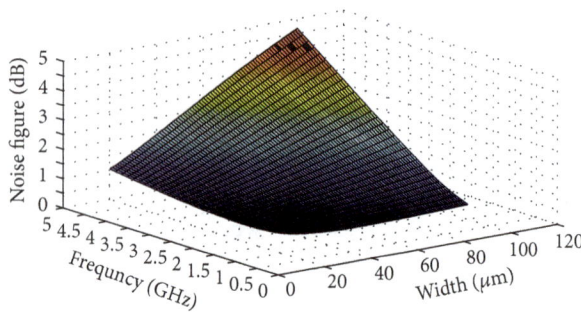

FIGURE 12: Effect of operational frequency and channel width on the noise figure.

quickly determine the "sweet spot" in the design. The trade-off analyses in this case indicate that the best designs in terms of power and noise figure for the LNA design occur when the drain current is in the range of 0.5 mA to 1 mA with an input circuit quality factor around 5.

5. Summary and Future Directions

This paper has examined the use of geometric programming for obtaining the globally optimum design of RF CMOS LNAs implemented with short-channel devices. The main contribution of this work has been the development of a framework for noise modeling of short-channel devices by including short-channel effects including velocity saturation and channel-length modulation. This noise model forms the basis of the objective function for geometric programming to minimize the noise figure of CMOS LNAs. In addition, the noise figure is minimized subject to the design constraints of input circuit quality factor, power consumption, and input impedance matching. Specific results from the optimization procedure are applied at the 90 nm and 180 nm technology nodes to determine the optimal channel width and noise figure for RF CMOS LNAs. Trade-off analysis indicates some important relationships among the design parameters such as the inverse relationship between noise figure and input circuit

quality factor. The relationship between the noise figure and channel width at a given power dissipation and the input circuit quality factor are consistent with simulations from Agilent's ADS software. The overall design trends are also consistent with other studies reported in the literature. Hence, this study has validated the use of geometric programming as an efficient method to guide the optimal design of CMOS LNAs targeted for implementation at nanoscale technology nodes.

Future work will focus on the enhancement of noise modeling for short-channel CMOS LNAs. For example, the noise contributions from the gate inductor (L_g) and the source inductor (L_s) due to their finite quality factor caused by parasitic effects should be included in the analysis. As devices continue to scale to deep submicron nodes, the doping concentration in the substrate will increase. This affects how the device characteristics are modelled such as the relationship between carrier mobility and diffusivity. In addition, quantum effects should be included when modeling the noise in the channel current [31]. It is expected that more sophisticated equivalent circuit models will be required to model the physical effects of nanoscale devices. The effect of the substrate as a source of noise and the back-gate transconductance in the small signal model should be considered. The thinning of the gate oxide at aggressively scaled technologies may make gate leakage effects an important consideration. Other sources of noise, such as shot noise, should also be taken into consideration below the 40 nm node. The existing noise optimization framework using GP can be modified to include these effects. In addition, the application of GP optimization for other topologies, such as the shunt-series feedback amplifier, will be considered in future work. Finally, with the trend towards biasing analog circuits in the weak to moderate inversion regions to reduce power dissipation, it would be interesting to explore GP methods as outlined in this paper to optimize these circuits.

Appendices

A. Expressions for MOSFET Output Conductance and Transconductance

In this appendix, analytical expressions for the output conductance and transconductance are discussed for both long-channel devices and short-channel devices.

A.1. Derivations of g_{d0} and g_m for Long-Channel Devices. For long-channel devices, the well-known expressions of the drain current in both the triode region and saturation region are given as

$$I_{d\text{triode}} = \mu_0 C_{\text{ox}} \frac{W}{L} \left(V_{\text{od}} \cdot V_{\text{ds}} - \frac{1}{2} V_{\text{ds}}^2 \right),$$

$$I_{d\text{sat}} = \frac{1}{2} \mu_0 C_{\text{ox}} \frac{W}{L} V_{\text{od}}^2, \tag{A.1}$$

where $V_{od} = V_{gs} - V_{th}$. By definition, the output conductance g_d is

$$g_d = \left.\frac{\partial I_{dtriode}}{\partial V_{ds}}\right|_{V_{gs}} = \mu_0 C_{ox} \frac{W}{L}\left(V_{od} - V_{ds}\right). \qquad (A.2)$$

Therefore, the output conductance at zero bias (i.e., $V_{ds} = 0$), can be expressed by

$$g_{d0} = \left. g_d\right|_{V_{ds}=0} = \mu_0 C_{ox} \frac{W}{L} V_{od} = \sqrt{2\frac{W}{L}\mu_0 C_{ox} I_{dsat}}. \qquad (A.3)$$

The transconductance of a long-channel device in saturation is given as

$$g_m = \left.\frac{\partial I_{dsat}}{\partial V_{gs}}\right|_{V_{ds}} = \mu_0 C_{ox} \frac{W}{L} \cdot V_{od} = \sqrt{2\frac{W}{L}\mu_0 C_{ox} I_{dsat}}. \qquad (A.4)$$

For long-channel devices, it is obvious that the output conductance at zero bias g_{d0} has the same form as the transconductance in saturation in terms of V_{od} or I_{dsat}.

A.2. Derivations of g_{d0} and g_m for Short-Channel Devices.

The drain current for short-channel devices is expressed differently than for the long-channel devices. By taking some important short-channel effects into account, such as velocity saturation and channel-length modulation, the expressions of

the analytical drain current model in both the triode region and saturation region are given by [38]

$$I_{dtriode}$$
$$= \mu_{eff} C_{ox}\left(\frac{W}{L}\right)\cdot\frac{\left(V_{gs} - V_{th}\right)V_{ds} - (m/2)V_{ds}^2}{1 + \left(\mu_{eff} V_{ds}\right)/\left(2v_{sat}L\right)},$$

$$I_{dsat} \qquad (A.5)$$
$$= \mu_{eff} C_{ox}\left(\frac{W}{L}\right)\frac{\left(V_{gs} - V_{th}\right)^2/(2m)}{1 + \mu_{eff}\left(V_{gs} - V_{th}\right)/\left(2mv_{sat}L\right)}$$
$$\cdot\left(1 + \lambda V_{ds}\right),$$

where [38, 41]

$$\mu_{eff} = \frac{\mu_0}{1 + \theta\left(V_{g0} - V_{th}\right)}, \qquad (A.6)$$

$$\theta = \frac{\beta_\theta}{t_{ox}}, \qquad (A.7)$$

$$m = 1 + \frac{\sqrt{\varepsilon_{si}qN_{ch}/(4\Psi_B)}}{C_{ox}}, \qquad (A.8)$$

$$\Psi_B = \left(\frac{kT}{q}\right)\ln\left(\frac{N_{ch}}{n_i}\right). \qquad (A.9)$$

After applying the quotient rule, the output conductance g_d can be expressed as

$$g_d = \left.\frac{\partial I_{dtriode}}{\partial V_{ds}}\right|_{V_{gs}} = \mu_{eff} C_{ox}\left(\frac{W}{L}\right)\frac{\left(V_{od} - mV_{ds}\right)\cdot\left(1 + \mu_{eff}V_{ds}/\left(2v_{sat}L\right)\right) - \left(V_{od}V_{ds} - (m/2)V_{ds}^2\right)\cdot\left(\mu_{eff}/\left(2v_{sat}L\right)\right)}{\left(1 + \mu_{eff}V_{ds}/\left(2v_{sat}L\right)\right)^2}$$

$$= \frac{\mu_{eff}C_{ox}\left(W/L\right)\left[V_{od} - mV_{ds} - (m/2)\left(\mu_{eff}/\left(2v_{sat}L\right)\right)V_{ds}^2\right]}{\left(1 + \mu_{eff}V_{ds}/\left(2v_{sat}L\right)\right)^2}. \qquad (A.10)$$

Therefore, the output conductance at zero bias ($V_{ds} = 0$) can be expressed by

$$g_{d0} = \left. g_d\right|_{V_{ds}=0} = \frac{\mu_0}{1 + \theta V_{od}}C_{ox}\frac{W}{L}V_{od}. \qquad (A.11)$$

By substituting the effective mobility equation into the saturation drain current formula, the equation of I_{dsat} for short-channel devices can be rewritten as

$$I_{dsat}$$
$$= \mu_{eff} C_{ox}\left(\frac{W}{L}\right)\frac{\left(V_{gs} - V_{th}\right)^2/(2m)}{1 + \mu_{eff}\left(V_{gs} - V_{th}\right)/\left(2mv_{sat}L\right)} \qquad (A.12)$$
$$\cdot\left(1 + \lambda V_{ds}\right).$$

Using (A.6) and (A.12), the transconductance of a short-channel device in saturation is given as

$$g_m = \left.\frac{\partial I_{dsat}}{\partial V_{gs}}\right|_{V_{ds}}$$

$$= C_{ox}\mu_0\left(\frac{W}{L}\right)\frac{1}{2m}\left(1 + \lambda V_{ds}\right)$$

$$\cdot\left[\frac{V_{od}^2}{\left(1 + \theta V_{od} + \left(\mu_0/2mv_{sat}L\right)V_{od}\right)}\right]' \qquad (A.13)$$

$$= C_{ox}\mu_0 W v_{sat}\cdot\left(1 + \lambda V_{ds}\right)$$

$$\cdot\frac{4mv_{sat}LV_{od} + \left(2mv_{sat}L\theta + \mu_0\right)V_{od}^2}{\left(2mv_{sat}L + \left(2mv_{sat}L\theta + \mu_0\right)V_{od}\right)^2}.$$

B. Monomial Expressions for g_m and g_{d0}

This appendix describes how a curve-fitting approach is used to determine monomial expressions for the transconductance (g_m) and output conductance (g_{d0}) from the analytical expressions derived in Appendix A. Monomial expressions of transconductance (g_m) and output conductance (g_{d0}) are given by

$$g_m = A_0 L^{A_1} W^{A_2} I_{ds}^{A_3};$$
$$g_{d0} = B_0 L^{B_1} W^{B_2} I_{ds}^{B_3}. \tag{B.1}$$

The geometry ranges specified for the devices for the monomial curve-fitting are given in Table 6(a). Additionally, the bias conditions are chosen to ensure the transistors operate in the saturation regions; for example, $V_{ds} \geq V_{od}$ as shown in Table 6(a). The fitting parameters that were determined from the above process are listed in Table 6(b) for both the 90 nm and 180 nm CMOS processes used in this study.

The accuracy of the curve fitting has been examined by comparing the estimated transconductance (g_m) and output conductance (g_{d0}) from the monomial expressions with calculated values from the analytical solutions.

The curve fitting results for the 90 nm process are shown in Figures 13 and 14. The coefficient of determination (R^2 value) for the transconductance curve fitting is 0.9999, indicating that the regression fits extremely well with the data compared with the analytical solutions in (23). The maximum relative error from curving fitting is about 2.56% (Figure 13(a)). Furthermore, 98.2% of the curve fitting data has a relative error less than 1.0% (Figure 13(b)).

The coefficient of determination for the output conductance is 1.0, suggesting that the curve fitting is close to perfect. The accuracy of curve fitting is shown in Figure 14(a) with a maximum relative error of 0.97%. Moreover, among this curve fitting data, 99.99% of the points have a relative error of less than 0.96% (Figure 14(b)).

The curve fitting results are shown in Figures 15 and 16 for the 180 nm process. The coefficients of determination (R^2 value) for these two curve fittings are very close to 1 and more than 97% of curve fitting data have a relative error less than 1.0% for both cases.

C. Expression for the Correlation Coefficient

This appendix describes the calculation of correlation coefficient c following [8]. Since the induced gate noise is correlated with the drain thermal noise, the correlation coefficient is defined as

$$c = \frac{\overline{i_{ng} \cdot i_{nd}^*}}{\sqrt{\overline{i_{ng}^2} \cdot \overline{i_{nd}^2}}}, \tag{C.1}$$

where $\overline{i_{ng} \cdot i_{nd}^*}$ is the spectrum of the cross-correlation of the drain thermal noise and the induced gate noise, $\overline{i_{n,d}^2}$, is the spectrum of the drain thermal noise and $\overline{i_{n,g}^2}$ is the spectrum

TABLE 6: (a) Ranges of devices geometry and bias conditions for calculation of g_m and g_{d0} for 90 nm and 180 nm CMOS processes. (b) Fitting parameters of monomial expressions of g_m and g_{d0} for 90 nm and 180 nm CMOS processes.

(a)

Parameters	90 nm	180 nm
Gate length L	$0.09\,\mu m \leq L \leq 0.45\,\mu m$	$0.18\,\mu m \leq L \leq 0.9\,\mu m$
Gate width W	$1\,\mu m \leq W \leq 100\,\mu m$	$1\,\mu m \leq W \leq 100\,\mu m$
Overdrive voltage V_{od}	$0.1\,V \leq V_{od} \leq 0.4\,V$	$0.1\,V \leq V_{od} \leq 0.5\,V$
Drain to source voltage V_{ds}	$0.5\,V \leq V_{ds} \leq 1.0\,V$	$0.6\,V \leq V_{ds} \leq 1.2\,V$

(b)

Parameters	90 nm	180 nm
A_0	0.0423	0.0463
A_1	−0.4578	−0.4489
A_2	0.5275	0.5311
A_3	0.4725	0.4689
B_0	0.0091	0.0096
B_1	−0.5637	−0.5595
B_2	0.5305	0.5194
B_3	0.4695	0.4806

of the induced gate noise. In a long-channel device, they are given as [8]

$$\overline{i_{ng} \cdot i_{nd}^*} = 4kT \cdot \frac{1}{9} j\omega \left(C_{ox} WL \right) \cdot \Delta f,$$
$$\overline{i_{n,d}^2} = 4kT \gamma_{long} g_{d0} \Delta f, \tag{C.2}$$
$$\overline{i_{n,g}^2} = 4kT \beta_{long} g_g \Delta f,$$

where g_g is given by (8) and $C_{gs} = (2/3) C_{ox} WL$. By substitution of (C.2) into (C.1), the correlation coefficient c for long-channel can be calculated as

$$c = \frac{1}{6\sqrt{(1/5)\,\beta_{long} \cdot \gamma_{long}}} j. \tag{C.3}$$

Substituting β_{long} and γ_{long} with their corresponding long-channel values of 4/3 and 2/3 yields $c = \sqrt{5/32}\,j = 0.395\,j$.

D. Sensitivity of the γ and β Parameters

This appendix shows the sensitivity of the γ and β parameters on the calculation of the minimum noise figure. The effect of varying the γ parameter is shown in Figure 17. When a ±10% variation is applied to γ, a small percentage of variation (around 4%) occurs to the minimum noise figure. Similarly, less than 4% variation occurs on the minimum noise figure when a ±10% change is applied to β, as illustrated in Figure 18.

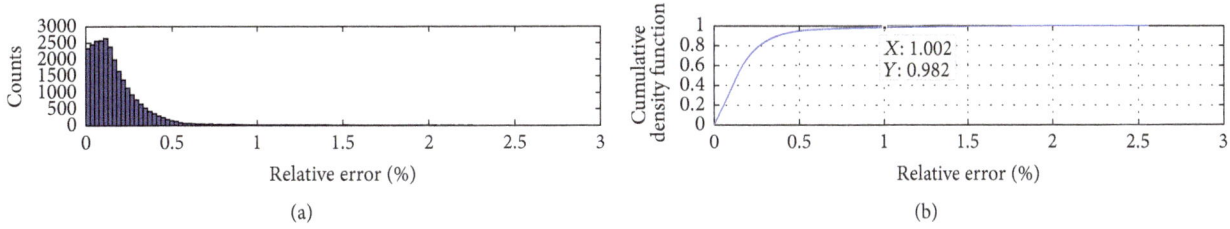

(a)

(b)

FIGURE 13: (a) Histogram of relative error for curve fitting of g_m for 90 nm. (b) Cumulative density function of relative error for curve fitting of g_m for 90 nm.

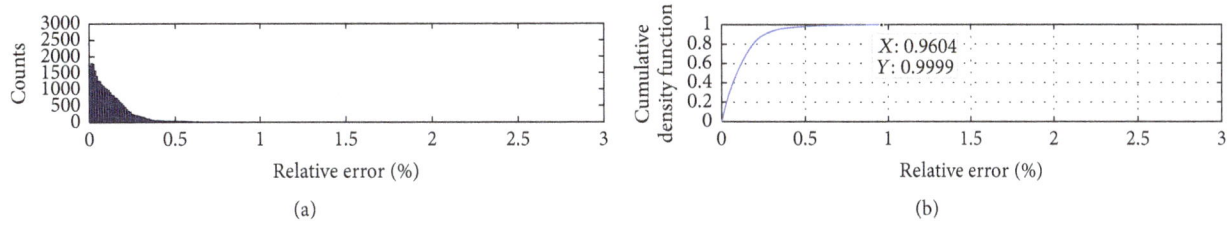

(a)

(b)

FIGURE 14: (a) Histogram of relative error for curve fitting of g_{d0} for 90 nm. (b) Cumulative density function of relative error for curve fitting of g_{d0} for 90 nm.

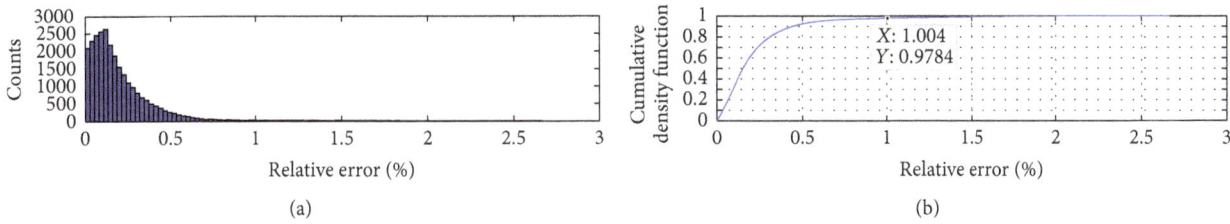

(a)

(b)

FIGURE 15: (a) Histogram of relative error for curve fitting of g_m for 180 nm. (b) Cumulative density function of relative error for curve fitting of g_m for 180 nm.

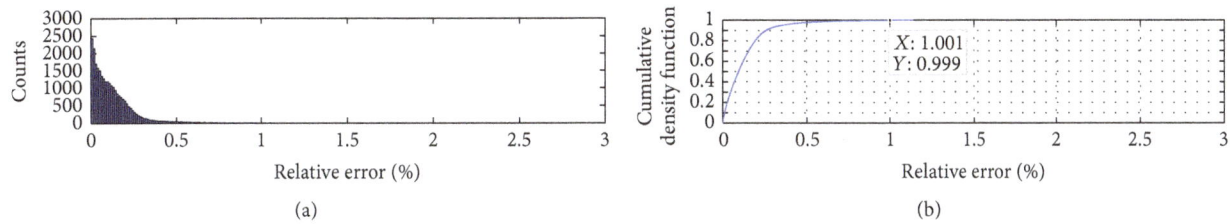

(a)

(b)

FIGURE 16: (a) Histogram of relative error for curve fitting of g_{d0} for 180 nm. (b) Cumulative density function of relative error for curve fitting of g_{d0} for 180 nm.

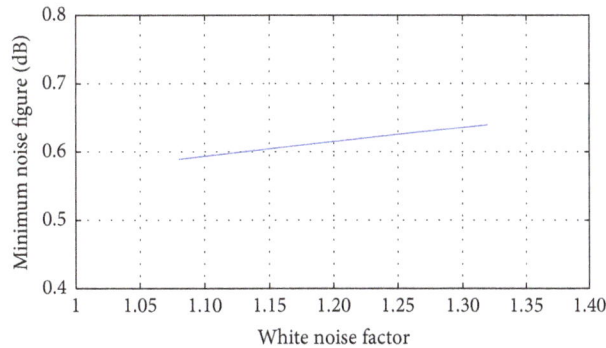

FIGURE 17: Variation of γ factor on the minimum noise figure for a nominal value of $\gamma = 1.2$.

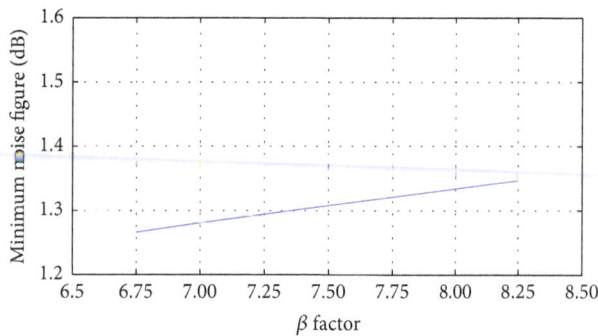

FIGURE 18: Variation of β factor on the minimum noise figure for a nominal value of $\beta = 7.5$.

This gives confidence to the assumption that the parameters γ and β can be modeled as constants for the purposes of optimization.

Conflict of Interests

The authors declare that there is no conflict of interests regarding the publication of this paper.

References

[1] T. H. Lee, *The Design of CMOS Radio-Frequency Integrated Circuits*, Cambridge University Press, Cambridge, UK, 2nd edition, 2004.

[2] M. Hammes, C. Kranz, and D. Seippel, "Deep submicron CMOS technology enables system-on-chip for wireless communications ICs," *IEEE Communications Magazine*, vol. 46, no. 9, pp. 154–161, 2008.

[3] A. A. Abidi, "RF CMOS comes of age," *IEEE Microwave Magazine*, vol. 4, no. 4, pp. 47–60, 2003.

[4] T. H. Lee, "From oxymoron to mainstream: the evolution and future of RF CMOS," in *Proceedings of the IEEE International Workshop on Radio-Frequency Integration Technology (RFIT '07)*, pp. 1–6, IEEE, Singapore, December 2007.

[5] H. S. Bennett, R. Brederlow, J. C. Costa et al., "Device and technology evolution for Si-based RF integrated circuits," *IEEE Transactions on Electron Devices*, vol. 52, no. 7, pp. 1235–1258, 2005.

[6] P.-H. Bonnaud, M. Hammes, A. Hanke et al., "A Fully Integrated SoC for GSM/GPRS in 0.13 μm CMOS," in *Proceedings of the IEEE International Solid-State Circuits Conference (ISSCC '06)*, pp. 1942–1951, IEEE, San Francisco, Calif, USA, February 2006.

[7] A.-J. Annema, B. Nauta, R. van Langevelde, and H. Tuinhout, "Analog circuits in ultra-deep-submicron CMOS," *IEEE Journal of Solid-State Circuits*, vol. 40, no. 1, pp. 132–143, 2005.

[8] A. van der Ziel, *Noise in Solid State Devices and Circuits*, John Wiley & Sons, New York, NY, USA, 1986.

[9] P. Andreani and H. Sjöland, "Noise optimization of an inductively degenerated CMOS low noise amplifier," *IEEE Transactions on Circuits and Systems II: Analog and Digital Signal Processing*, vol. 48, no. 9, pp. 835–841, 2001.

[10] R. A. Rutenbar, G. G. E. Gielen, and J. Roychowdhury, "Hierarchical modeling, optimization, and synthesis for system-level analog and RF designs," *Proceedings of the IEEE*, vol. 95, no. 3, pp. 640–669, 2007.

[11] H. A. Haus, W. R. Atkinson, W. H. Fonger et al., "Representation of noise in linear twoports," *Proceedings of the IRE*, vol. 48, no. 1, pp. 66–74, 1960.

[12] D. K. Shaeffer and T. H. Lee, "A 1.5-V, 1.5-GHz CMOS low noise amplifier," *IEEE Journal of Solid-State Circuits*, vol. 32, no. 5, pp. 745–759, 1997.

[13] J.-S. Goo, H.-T. Ahn, D. J. Ladwig, Z. Yu, T. H. Lee, and R. W. Dutton, "A noise optimization technique for integrated low-noise amplifiers," *IEEE Journal of Solid-State Circuits*, vol. 37, no. 8, pp. 994–1002, 2002.

[14] T.-K. Nguyen, C.-H. Kim, G.-J. Ihm, M.-S. Yang, and S.-G. Lee, "CMOS low-noise amplifier design optimization techniques," *IEEE Transactions on Microwave Theory and Techniques*, vol. 52, no. 5, pp. 1433–1442, 2004.

[15] P. Vancorenland, C. De Ranter, M. Steyaert, and G. Gielen, "Optimal RF design using smart evolutionary algorithms," in *Proceedings of the 37th Design Automation Conference (DAC '00)*, pp. 7–10, June 2000.

[16] M. Chu and D. J. Allstot, "Elitist nondominated sorting genetic algorithm based RF IC optimizer," *IEEE Transactions on Circuits and Systems I: Regular Papers*, vol. 52, no. 3, pp. 535–545, 2005.

[17] X. Xia, Y. Li, W. Ying, and L. Chen, "Automated design approach for analog circuit using genetic algorithm," in *Proceedings of the 7th International Conference on Computational Science (ICCS '07), Beijing, China, May 2007, Part IV*, vol. 4490 of *Lecture Notes in Computer Science*, pp. 1124–1130, Springer, 2007.

[18] A. Somani, P. P. Chakrabarti, and A. Patra, "An evolutionary algorithm-based approach to automated design of analog and RF circuits using adaptive normalized cost functions," *IEEE Transactions on Evolutionary Computation*, vol. 11, no. 3, pp. 336–353, 2007.

[19] S. Boyd, S.-J. Kim, L. Vandenberghe, and A. Hassibi, "A tutorial on geometric programming," *Optimization and Engineering*, vol. 8, no. 1, pp. 67–127, 2007.

[20] M. D. Hershenson, S. P. Boyd, and T. H. Lee, "Optimal design of a CMOS op-amp via geometric programming," *IEEE Transactions on Computer-Aided Design of Integrated Circuits and Systems*, vol. 20, no. 1, pp. 1–21, 2001.

[21] P. K. Meduri and S. K. Dhali, "A methodology for automatic transistor-level sizing of CMOS opamps," in *Proceedings of the 24th International Conference on VLSI Design (VLSI Design '11)*, pp. 100–105, IEEE, Chennai, India, January 2011.

[22] M. D. M. Hershenson, A. Hajimiri, S. S. Mohan, S. P. Boyd, and T. H. Lee, "Design and optimization of LC oscillators," in *Proceedings of the IEEE/ACM International Conference on Computer-Aided Design. Digest of Technical Papers*, pp. 65–69, IEEE, ACM, San Jose, Calif, USA, November 1999.

[23] S. S. Mohan, M. D. M. Hershenson, S. P. Boyd, and T. H. Lee, "Simple accurate expressions for planar spiral inductances," *IEEE Journal of Solid-State Circuits*, vol. 34, no. 10, pp. 1419–1420, 1999.

[24] B. Swahn and S. Hassoun, "Gate sizing: FinFETs vs 32nm bulk MOSFETs," in *Proceedings of the 43rd IEEE/ACM Design Automation Conference (DAC '06)*, pp. 528–531, 2006.

[25] K. Kasamsetty, M. Ketkar, and S. S. Sapatnekar, "A new class of convex functions for delay modeling and its application to the transistor sizing problem," *IEEE Transactions on Computer-Aided Design of Integrated Circuits and Systems*, vol. 19, no. 7, pp. 779–788, 2000.

[26] W.-T. Cheung and N. Wong, "Optimized RF CMOS low noise amplifier design via geometric programming," in *Proceedings of the International Symposium on Intelligent Signal Processing and Communications (ISPACS '06)*, pp. 423–426, Yonago, Japan, December 2006.

[27] X. Jin and D. H. K. Hoe, "Optimization of short channel CMOS LNAs by geometric programming," in *Proceedings of the IEEE 55th International Midwest Symposium on Circuits and Systems (MWSCAS '12)*, pp. 9–12, IEEE, Boise, Idaho, USA, August 2012.

[28] F. M. Klaassen and J. Prins, "Thermal noise of MOS transistors," *Philips Research Reports*, vol. 22, pp. 505–514, 1967.

[29] A. J. Scholten, R. van Langevelde, L. F. Tiemeijer, and D. B. M. Klaassen, "Compact modeling of noise in CMOS," in *Proceedings of the IEEE Custom Integrated Circuits Conference (CICC '06)*, pp. 711–716, San Jose, Calif , USA, September 2006.

[30] A. J. Scholten, L. F. Tiemeijer, R. van Langevelde, R. J. Havens, A. T. Zegers-van Duijnhoven, and V. C. Venezia, "Noise modeling for RF CMOS circuit simulation," *IEEE Transactions on Electron Devices*, vol. 50, no. 3, pp. 618–632, 2003.

[31] M. J. Deen, C.-H. Chen, S. Asgaran, G. A. Rezvani, J. Tao, and Y. Kiyota, "High-frequency noise of modern MOSFETs: compact modeling and measurement issues," *IEEE Transactions on Electron Devices*, vol. 53, no. 9, pp. 2062–2081, 2006.

[32] B. Razavi, R.-H. Yan, and K. F. Lee, "Impact of distributed gate resistance on the performance of MOS devices," *IEEE Transactions on Circuits and Systems I: Fundamental Theory and Applications*, vol. 41, no. 11, pp. 750–754, 1994.

[33] A. Litwin, "Overlooked interfacial silicide-polysilicon gate resistance in MOS transistors," *IEEE Transactions on Electron Devices*, vol. 48, no. 9, pp. 2179–2181, 2001.

[34] V. M. Mahajan, P. R. Patalay, R. P. Jindal et al., "A physical understanding of RF noise in bulk nMOSFETs with channel lengths in the nanometer regime," *IEEE Transactions on Electron Devices*, vol. 59, no. 1, pp. 197–205, 2012.

[35] J. C. J. Paasschens, A. J. Scholten, and R. van Langevelde, "Generalizations of the Klaassen-Prins equation for calculating the noise of semiconductor devices," *IEEE Transactions on Electron Devices*, vol. 52, no. 11, pp. 2463–2472, 2005.

[36] K. Han, J. Gil, S.-S. Song et al., "Complete high-frequency thermal noise modeling of short-channel MOSFETs and design of 5.2-GHz low noise amplifier," *IEEE Journal of Solid-State Circuits*, vol. 40, no. 3, pp. 726–734, 2005.

[37] K. Han, H. Shin, and K. Lee, "Analytical drain thermal noise current model valid for deep submicron MOSFETs," *IEEE Transactions on Electron Devices*, vol. 51, no. 2, pp. 261–269, 2004.

[38] Y. Taur and T. H. Ning, *Fundametals of Modern VLSI Devices*, Cambridge University Press, Cambridge, UK, 2nd edition, 2009.

[39] W. Shockley, J. A. Copeland, and R. P. James, "The impedance field method of noise calculation in active semiconductor devices," in *Quantum Theory of Atoms, Molecules and Solid State*, pp. 537–563, Academic Press, New York, NY, USA, 1966.

[40] B. J. Sheu, D. L. Scharfetter, P.-K. Ko, and M.-C. Jeng, "BSIM: berkeley short-channel IGFET model for MOS transistors," *IEEE Journal of Solid-State Circuits*, vol. 22, no. 4, pp. 558–566, 1987.

[41] Y. Tsividis, *Operation and Modeling of the MOS Transistor*, Oxford University Press, New York, NY, USA, 2nd edition, 2003.

[42] J. Jeon, J. D. Lee, B.-G. Park, and H. Shin, "An analytical channel thermal noise model for deep-submicron MOSFETs with short channel effects," *Solid-State Electronics*, vol. 51, no. 7, pp. 1034–1038, 2007.

[43] J. Jeon, B.-G. Park, and H. Shin, "Investigation of thermal noise factor in nanoscale MOSFETs," *Journal of Semiconductor Technology and Science*, vol. 10, no. 3, pp. 225–231, 2010.

[44] V. M. Mahajan, R. P. Jindal, H. Shichijo, S. Martin, F.-C. Hou, and D. Trombley, "Numerical investigation of excess RF channel noise in sub-100 nm MOSFETs," in *Proceedings of the 2nd International Workshop on Electron Devices and Semiconductor Technology (IEDST '09)*, pp. 1–4, Mumbai, India, June 2009.

[45] R. Navid and R. Dutton, "The physical phenomena responsible for excess noise in short-channel MOS devices," in *Proceedings of the International Conference on Simulation of Semiconductor Processes and Devices (SISPAD '02)*, pp. 75–78, Kobe, Japan, 2002.

[46] J. Jeon, J. Lee, J. Kim et al., "The first observation of shot noise characteristics in 10-nm scale MOSFETs," in *Proceedings of the Symposium on VLSI Technology, Technical Digest*, pp. 48–49, Honolulu, Hawaii, USA, June 2009.

[47] X. Jin, *Optimization of short channel RF CMOS low noise amplifiers by geometric programming [M.S. thesis]*, University of Texas, Tyler, Tex, USA, 2012.

[48] A. J. Scholten, L. F. Tiemeijer, R. van Langevelde et al., "Compact modelling of noise for RF CMOS circuit design," *IEE Proceedings—Circuits, Devices and Systems*, vol. 151, no. 2, pp. 167–174, 2004.

[49] W. Zhao and Y. Cao, "New generation of predictive technology model for sub-45 nm early design exploration," *IEEE Transactions on Electron Devices*, vol. 53, no. 11, pp. 2816–2823, 2006.

[50] Predictive Technology Model website, http://ptm.asu.edu/.

[51] M. Grant and S. Boyd, "CVX: Matlab software for disciplined convex programming, version 1.21," http://cvxr.com/cvx/.

[52] S. Dalmia, F. Ayazi, M. Swaminathan et al., "Design of inductors in organic substrates for 1–3 GHz wireless applications," in *Proceedings of the IEEE MTT-S International Microwave Symposium Digest*, vol. 3, pp. 1405–1408, June 2002.

[53] K. K. Samanta and I. D. Robertson, "Advanced multilayer thick-film system-on-package technology for miniaturized and high performance CPW microwave passive components," *IEEE Transactions on Components, Packaging and Manufacturing Technology*, vol. 1, no. 11, pp. 1695–1705, 2011.

## 3

On the Use of an Algebraic Signature Analyzer for Mixed-Signal Systems Testing

Vadim Geurkov and Lev Kirischian

Department of Electrical and Computer Engineering, Ryerson University, 350 Victoria Street, Toronto, ON, Canada M5B 2K3

Correspondence should be addressed to Vadim Geurkov; vgeurkov@ee.ryerson.ca

Academic Editor: M. Renovell

We propose an approach to design of an algebraic signature analyzer that can be used for mixed-signal systems testing. The analyzer does not contain carry propagating circuitry, which improves its performance as well as fault tolerance. The common design technique of a signature analyzer for mixed-signal systems is based on the rules of an arithmetic finite field. The application of this technique to the systems with an arbitrary radix is a challenging task and the devices designed possess high hardware complexity. The proposed technique is simple and applicable to systems of any size and radix. The hardware complexity is low. The technique can also be used in arithmetic/algebraic coding and cryptography.

1. Introduction

Signature analysis has been widely used for digital and mixed-signal systems testing [1–12]. Mixed-signal systems consist of both digital and analog circuits; however the signature analysis method is only applicable to the subset of these systems that have digital outputs (such as analog-to-digital converters, measurement instruments, etc.). Signature analysis can be employed as an external test solution or can be embedded into the system under test. In the built-in implementation, a circuit under test (CUT) of digital or mixed-signal nature is fed by test stimuli, while the output responses are compacted by a signature analyzer (SA), as illustrated in Figure 1. The actual signature is compared against the fault-free circuit's signature and a pass/fail decision is made. A signature of a fault-free circuit is referred to as a reference signature. If the CUT is of a digital nature, the SA essentially constitutes a circuit that computes an *algebraic remainder*. The reference signature has only one, *punctual* value, and the decision making circuit consists of a simple digital comparator. If the CUT is of a mixed-signal nature, the SA computes an *arithmetic residue*. In this case, the reference signature becomes an *interval* value and the decision making circuit uses a window comparator.

Design methods for an algebraic signature analyzer have been well developed in error-control coding [13]. A remainder calculating circuit for an arbitrary base (binary or nonbinary) can be readily designed for a digital CUT of any size. In contrast, it is much harder to design a residue calculating circuit, specifically for a nonbinary base [14]. Furthermore, due to the presence of carry propagating circuitry, the implementation complexity and error vulnerability of the residue calculating circuit are higher compared to the remainder calculating circuit.

We propose an approach to design of an algebraic signature analyzer that can be used for mixed-signal systems testing. Due to an algebraic nature, the analyzer does not contain carry propagating circuitry. This helps to improve its error immunity, as well as performance.

2. A Conventional Signature Analyzer

An algebraic signature analyzer is designed on the basis of a polynomial division circuit, as shown in Figure 2 [3, 13, 15]. This circuit divides the incoming sequence of nonbinary symbols (digits), $a_{m-1}, \ldots, a_1, a_0$, treated as a polynomial:

$$a(y) = a_{m-1} y^{m-1} + \cdots + a_1 y + a_0 \qquad (1)$$

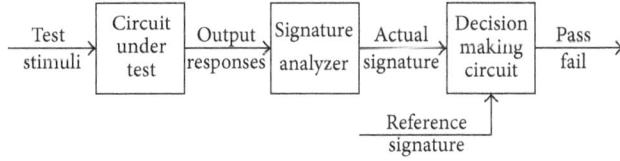

FIGURE 1: Built-in signature analysis of a circuit under test.

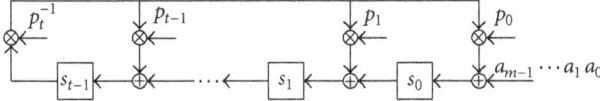

FIGURE 2: A t-stage polynomial division circuit.

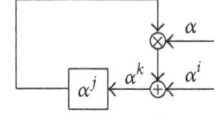

FIGURE 3: A symbolic presentation of a one-stage algebraic signature analyzer.

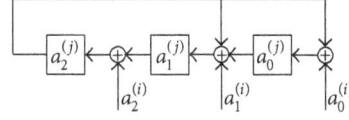

FIGURE 4: A logic level presentation of the algebraic 3-input signature analyzer.

by the polynomial

$$p(y) = p_t y^t + \cdots + p_1 y + p_0, \quad t \ll m. \tag{2}$$

The remainder

$$s(y) = s_{t-1} y^{t-1} + \cdots + s_1 y + s_0 \tag{3}$$

constitutes a CUT *signature*.

Each digit, a_i, $0 \leq i \leq m - 1$, consists n bits and is considered to be an element of the field $GF(2^n)$. The degree of the polynomial (2), or the number of stages, t, in Figure 2, depends on the desired probability of undetected error in the sequence of incoming digits. For long sequences with independent errors, this probability is estimated as $P_{nd} \approx 2^{-tn}$. In practice, $n \geq 8$ and even for the one-stage circuit, $P_{nd} \leq 2^{-(1 \times 8)} = 0.0039$, which is quite low. Therefore, a multiple-input signature analyzer normally contains only one stage. Such an analyzer is presented in Figure 3 [14], where α is a primitive element of the field $GF(2^n)$, that is, a root of a primitive polynomial $g(x) = g_{n-1} x^{n-1} + \cdots + g_1 x + g_0$. Each element of the field can be represented by a power of α. Let α^i be the incoming digit and α^j the content of the analyzer. Then, each operational cycle of the analyzer is described by the expression

$$\alpha^j \alpha \oplus \alpha^i = \alpha^k. \tag{4}$$

Without a loss of generality, we will consider a 3-bit signature register ($n = 3$), with α being a primitive element of $GF(2^3)$, in particular a root of a primitive polynomial $g(x) = x^3 + x + 1$. Then, a symbolic scheme of Figure 3 will transfer to the logic level circuit of Figure 4, where

$$\alpha^l = a_2^{(l)} x^2 + a_1^{(l)} x + a_0^{(l)}, \quad a_i^{(l)} \in \{0, 1\}, \tag{5}$$
$$0 \leq i \leq 2, \quad 0 \leq l \leq 6.$$

This expression indicates the relationship between the power and vector representations of a field element, as reflected in Table 1 (where $x = \alpha$).

If the preliminary "cleared" analyzer receives, for example, the following sequence of 3-bit output responses from

TABLE 1: Three representations for the elements of $GF(2^3)$ generated by $g(x) = x^3 + x + 1$. Here $g(\alpha) = 0$.

Power representation α^l	Polynomial representation $a_2^{(l)}\alpha^2$		$+$	$a_1^{(l)}\alpha^1$	$+$	$a_0^{(l)}\alpha^0$	Vector representation $a_2^{(l)} a_1^{(l)} a_0^{(l)}$
0				0			0 0 0
α^0						α^0	0 0 1
α^1				α^1			0 1 0
α^2	α^2						1 0 0
α^3				α^1	$+$	α^0	0 1 1
α^4	α^2	$+$		α^1			1 1 0
α^5	α^2	$+$		α^1	$+$	α^0	1 1 1
α^6	α^2				$+$	α^0	1 0 1

a digital CUT, α^5, α^6, α^4, α^2, α^1, α^0, then after the 6th shift its content will become

$$\left(\left(\left(\left(\left(0 \cdot \alpha + \alpha^5\right)\alpha + \alpha^6\right)\alpha + \alpha^4\right)\alpha + \alpha^2\right)\alpha + \alpha^1\right)\alpha + \alpha^0 = \alpha. \tag{6}$$

The power representation of the field element, α, corresponds to the vector representation, 010, which is the actual signature of the CUT.

In contrast to a digital CUT, the output responses of a mixed-signal CUT are distorted even in a fault-free case. Small permissible variations in the responses cause a significant deviation of the final signature. For example, if in the above sequence of output responses the least significant bit in the first response changes from 1 to 0 (i.e., the vector 111 changes to 110, or power α^5 changes to α^4), then the actual signature will change from 010 to 101 (or from α to α^6 in power form).

Apparently, the conventional SA represented in Figures 3 and 4 cannot be employed for mixed-signal circuits testing.

In the known methods, output responses of mixed-signal circuits are compacted by a circuit referred to as a modulo adder (or accumulator, or digital integrator) [4–8]. It should be noted that a modulo adder is a special case of a *residue computing circuit* [14]. A residue computing circuit is represented in Figure 5. Here a_j is the current content of

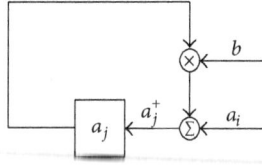

FIGURE 5: A symbolic presentation of a one-stage arithmetic signature analyzer.

the register, a_i is the incoming (arithmetic) symbol, and b is the base of the system. This circuit divides the incoming sequence of symbols, $a_{m-1}, \ldots, a_1, a_0$, treated as a number:

$$a = a_{m-1}b^{m-1} + \cdots + a_1 b + a_0 \qquad (7)$$

by the modulus

$$p = p_{t-1}b^{t-1} + \cdots + p_1 b + p_0, \quad t \ll m. \qquad (8)$$

As in the case with the algebraic SA, we consider a single-stage device; that is, $t = 1$, $p = p_0 < b = 2^n$, where n is the number of bits occupied by the symbol. The residue, s_0, constitutes a signature.

An operational cycle of the circuit in Figure 5 can be described by the expression

$$a_j b + a_i = a_j^+ \,(\mathrm{mod}\, p). \qquad (9)$$

Although the circuits of Figures 3 and 5 look similar, their implementation is quite different. In general case, the designing procedure for the arithmetic circuits is more complicated and their hardware complexity is greater.

As an example, Figure 6 represents the circuit that computes a modulo 5 residue of the incoming sequence of 3-bit symbols treated as an octal number [14]. Here a_i is the incoming octal digit and C is a combinational circuit which generates the following next state signals:

$$c_2 = a_0^j \overline{a_1^i} a_0^i a_1^i a_2^i + a_1^j \left(a_0^j a_1^i \oplus a_2^i + a_0^j a_0^i \overline{a_2^i} \right),$$

$$c_1 = a_2^j \overline{a_2^i} \left(\overline{a_0^i} + \overline{a_1^i} \right) + \overline{a_2^j} a_2^i \left(\overline{a_0^j} a_0^i + \overline{a_1^j} a_1^i \right)$$

$$\quad + a_1^j \left(\overline{a_0^j} + \overline{a_2^i} \right) + a_0^j \left(a_1^i \oplus a_2^i \right), \qquad (10)$$

$$c_0 = a_0^j a_1^i \left(\overline{a_1^j \oplus a_2^i} \right) + a_1^i \overline{a_2^i} \left(\overline{a_0^j} + \overline{a_0^j a_1^i} \right)$$

$$\quad + a_2^j + \overline{a_1^j} a_2^i \left(\overline{a_0^j a_1^i} + a_0^i \overline{a_1^i} + a_0^j \overline{a_0^i} \right).$$

Each shift of this circuit implements the operation $a_j \times 8 + a_i$ (mod 5).

In addition to high hardware complexity, the arithmetic compactor contains carry propagating circuitry (shown in red color in Figure 6) that delays the operation and aggravates the effect of a single fault.

Below, we design an algebraic circuit that can be employed for mixed-signal data compaction. It does not contain carry propagating circuitry.

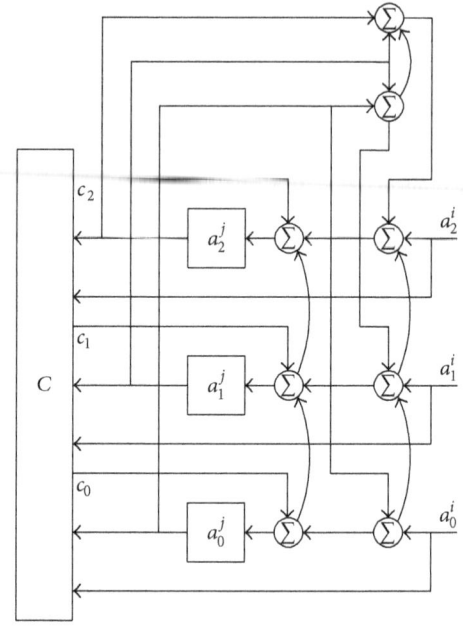

FIGURE 6: A 3-input arithmetic compactor.

3. A Novel Approach

Polynomial (1) in conjunction with the reference signature can be considered as a code word of the code whose minimal distance is defined by the $g(x)$. The distance here is the Hamming distance. This distance characterizes algebraic error-detecting properties of the code and is not convenient for arithmetic errors that occur in mixed-signal systems. Indeed, a small permissible deviation of the data to be compacted causes the reference signature to span the entire space. Under these conditions, the decision making circuit in Figure 1 must be able to compare the actual signature with the entire set of possible reference signatures. This increases the analyzer complexity.

To decrease the complexity, an arithmetic SA treats the sequence of output responses from a mixed-signal circuit as a number (7). In conjunction with the reference residue, this is considered as a code word of an arithmetic error-control code. The properties of this code depend on the arithmetic minimal distance which in turn depends on the modulus p. The arithmetic residue calculating analyzer does not search the entire space, since the space of arithmetic reference signatures is now contiguous. To make a decision, it employs a window comparator. This simplifies the circuitry. However, the hardware complexity of the arithmetic SA can still be quite high, as it was illustrated above.

In the rest of this paper, we will show how to design an algebraic SA, which generates a contiguous space of algebraic reference signatures.

In order to be contiguous, the space of signatures must be ordered. A signature can be represented in the vector or power forms. We will use the power exponent as the criterion for ordering the signature set. The distance between two vectors (signatures) will be evaluated as the arithmetic difference between the corresponding exponents. For example,

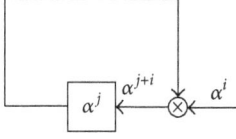

FIGURE 7: A symbolic form of an algebraic SA for a mixed-signal CUT.

the distance between the signatures 010 and 101 will be 5, because the exponents of powers α^6 and α differ by 5. We can interpret these exponents as output responses of a mixed-signal CUT, since they possess arithmetic properties. At the same time, the corresponding vectors (signatures) possess algebraic properties. Therefore, arithmetic data is mapped into algebraic data. Figure 7 represents the circuit which performs the mapping and computes an algebraic signature.

The circuit of Figure 7 can be obtained from the circuit of Figure 3 by the following transform:

$$\alpha^j \alpha^i = \left(\alpha^j \alpha\right) \alpha^{i-1} = \left(\alpha^j \alpha\right) \overbrace{\frac{\alpha^{i-1}}{(1 + \alpha^k)}}$$

$$= \alpha^j \alpha + \alpha^{j+1+k} = \alpha^j \alpha + \alpha^l. \tag{11}$$

Since the finite field $GF(2^n)$ is closed and errors are independent, this mapping will not change the probability of undetected error.

The logic level implementation of the circuit of Figure 7 is more complex compared to the circuit of Figure 3, but it is less complex than that of the circuit of Figure 5.

Prior to designing the circuit, we have to make a few observations.

The *first* observation is that

$$\alpha^j \alpha^i = \left(\cdots \left(\alpha^j \underbrace{\alpha\right) \alpha \cdots \right) \alpha}_{i}. \tag{12}$$

Let us denote an output response from a mixed-signal CUT by i. The *second* observation is that the response i can be considered as an exponent of the power, that is, α^i. Essentially, this means that the arithmetic values i are mapped into algebraic values α^i.

Based on these observations, we can design a signature analyzer in the way shown in Figure 8. Here α is a primitive element of a finite field $GF(2^n)$; n coincides with the bit-length of the output responses. The lower and upper inputs of the multiplexer in Figure 8 are connected together, since $\alpha^{2^n-1} = \alpha^0$ in $GF(2^n)$.

Considering the case when the analyzer is fed by 3-bit data, its more detailed implementation will have the form of Figure 9.

Here the buses consist of 3 lines, as indicated by the appropriate number. The initial content of the SA before the shift is α^j, or $a_2 x^2 + a_1 x + a_0$ in the polynomial form (we have omitted the superscripts for the sake of simplicity). The notations a_k and a_k^+, where index k can be one of 0, 1, and 2, indicate the present and next states, respectively.

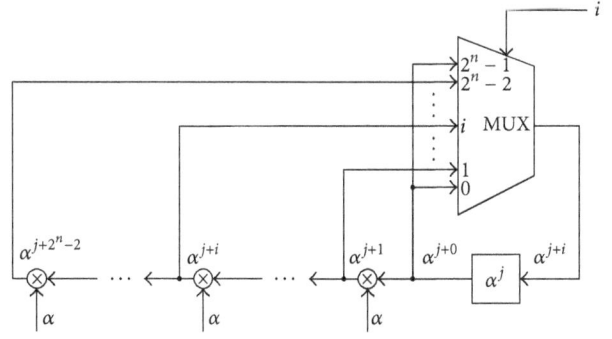

FIGURE 8: A more detailed symbolic form of the SA.

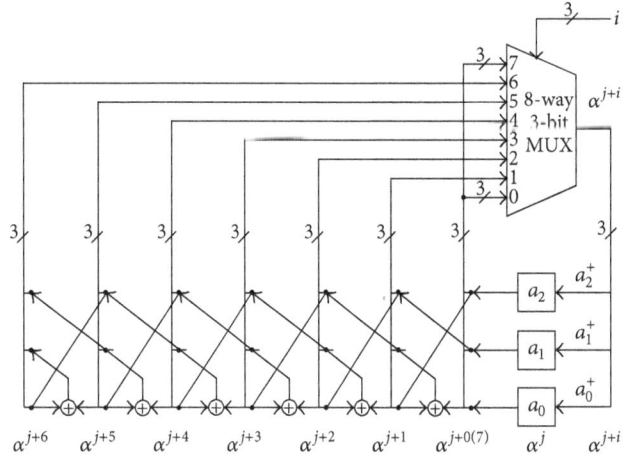

FIGURE 9: A register transfer level implementation of the SA.

A multiplier by α in $GF(2^3)$ is realized bearing in mind that $g(x) = x^3 + x + 1$, α corresponds to x, and

$$\left(a_2 x^2 + a_1 x + a_0\right) x \bmod g(x)$$

$$= \left(a_2 x^3 + a_1 x^2 + a_0 x\right) \bmod g(x)$$

$$= a_2 (x + 1) + a_1 x^2 + a_0 x \tag{13}$$

$$= a_1 x^2 + \left(a_2 + a_0\right) x + a_2.$$

This operation is shown by crosslines in Figure 9. The multiplexer inputs "0" and "7" are tied together, because $\alpha^7 = \alpha^0$ in the field $GF(2^3)$.

In order to demonstrate how to use this analyzer, we will assume that it receives only two values from a CUT, in particular j and i. Since the CUT is of a mixed-signal nature, there is an unavoidable (and thereby permitted) deviation of these values by ±1 (the greater tolerances can also be considered). The analyzer will map the received data into $\alpha^{j\pm1}$ and $\alpha^{i\pm1}$, respectively. If we assume that the initial content of the SA is 001 (versus 000 for a conventional SA), then after the first shift the content becomes $\alpha^0 \alpha^{j\pm1} = \alpha^{j\pm1}$. After the second shift, it changes to $\alpha^{j\pm1} \alpha^{i\pm1} = \alpha^{j+i\pm2}$. This expression is derived using the interval arithmetic rules. It states that for the fault-free CUT the actual result must match one of the values from the interval $[\alpha^{j+i-2}, \alpha^{j+i+2}]$, that is, one of the following:

$$\alpha^{j+i-2}, \alpha^{j+i-1}, \alpha^{j+i}, \alpha^{j+i+1}, \alpha^{j+i+2}. \tag{14}$$

To further simplify the SA operation, we will assume that instead of α^0 (i.e., 001) the initial SA content is $\alpha^{-(j+i)}$. We will refer to this value as the *seed* value. Then, by the same reasoning, the SA content after two shifts will match one of the following powers:

$$\alpha^{-2}, \alpha^{-1}, \alpha^0, \alpha^1, \alpha^2. \tag{15}$$

Due to the closure property of the field $GF(2^3)$, this power set is equivalent to

$$\alpha^5, \alpha^6, \alpha^0, \alpha^1, \alpha^2. \tag{16}$$

Consequently, the decision making circuit in Figure 3 will work as follows. If the actual signature does not match any value from set (16), the CUT is considered to be faulty. Since these values are ordered (and surround the power α^0), the decision making circuit can employ a comparator, thereby reducing the hardware complexity of the SA.

As in any signature analyzer, some errors in the CUT output responses may escape detection. The aliasing rate can be estimated as described in [16] and will coincide with the aliasing rate of the conventional analyzer.

Example. Let us consider a 3-bit CUT, which is fed by two input stimuli. Under the fault-free operation, the CUT produces the output responses $j = 101 \pm 1$ and $i = 110 \pm 1$. Therefore, the seed value will be $\alpha^{-(j+i)} = \alpha^{-(5+6)} = \alpha^{-11} = \alpha^3$, or 011 in the vector form. If the CUT is fault-free, then after 2 shifts the SA content must match one of the elements in set (16). For example, if the actual responses are $101 + 1 = 110$ (or α^6) and $110 + 1 = 111$ (or α^7) (i.e., the variations are within the tolerance bounds), the signature will be $\alpha^3 \alpha^6 \alpha^7 = \alpha^2$ which belongs to set (16). And the decision making circuit will generate a *pass* signal. The validity of such a decision is determined by the aliasing rate.

Let us assume that a fault in the CUT has made the following changes in the output responses: $110 \rightarrow 011$ ($\alpha^6 \rightarrow \alpha^3$) and $111 \rightarrow 100$ ($\alpha^7 \rightarrow \alpha^4$). Then the actual signature will become $\alpha^3 \alpha^3 \alpha^4 = \alpha^3$. This element does not belong to set (16), so the fault is detected.

There are two distinct ways of designing the decision making circuit depending on the optimization criteria (time or hardware overhead).

Hardware Overhead. If performance is paramount and time overhead is not desirable, the following approach can be employed. Let m be the number of output responses. All of the $2m + 1$ α-multiplier outputs (see Figure 8) that belong to set (16) are connected to the first inputs of the $2m+1$ comparators of a similar type. The second inputs of these comparators are shared and fed by the vector $0 \cdots 01$. If the CUT is fault-free, one of the comparators will produce a logic "1" signal. The logic OR of the comparator outputs will constitute a *pass/fail* signal.

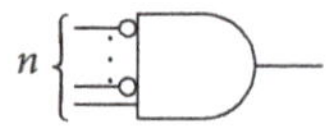

FIGURE 10: An n-bit comparator.

The above procedure is based on the fact that the fault-free CUT produces one of the signatures from set (16). If the actual signature is α^0, the comparator connected directly to the signature register produces a logic "1," thus indicating that the CUT is fault-free. If the actual signature is α^6, then the product $\alpha^6 \alpha$, generated at the output of the first α-multiplier, equals 1, which is detected by the next comparator. The same reasoning applies to the rest of the signatures from set (16). The logic diagram of the n-bit comparator is shown in Figure 10.

Time Overhead. If time overhead is allowed, the hardware complexity can be further reduced. In terms of implementation, it is more convenient to use the following seed value: $\alpha^{-(j+i+m+1)}$, where m is the number of output responses. For the above example, $\alpha^{-(11+3)} = \alpha^0$, and set (16) will transform to

$$\alpha^2, \alpha^3, \alpha^4, \alpha^5, \alpha^6. \tag{17}$$

After the last output response has been shifted in, the SA continues to shift its content $2m + 1$ more times, while the input i is forced to 1. This ensures that the SA content is multiplied by α with each shift. For the above example, $2m + 1 = 5$. If, within this time, the match with an element of set (17) has been determined, the CUT is considered to be fault-free. Otherwise, it is faulty.

If the CUT is fault-free and its output responses have not exceeded their tolerances, then while cycling through the states during the extra $2m + 1$ shifts, the output of the multiplexer in Figure 8 will go through the power α^0 or vector $0 \cdots 01$. The match with the vector $0 \cdots 01$ is detected by the comparator of Figure 10 connected to the multiplexor's output. The comparator output is actually producing a *pass/fail* signal.

The implementation complexity of the circuit of Figure 8 increases significantly with the growth of the data width, n. Therefore, this circuit can only be implemented for the output responses with relatively low values of n. For greater values of n, we will modify the circuit of Figure 8 to the one shown in Figure 11. The modified circuit contains binary-weighted stages and is more economical in terms of hardware. The complexity of the multiplier $\times \alpha^i$ is comparable with that of the multiplier $\times \alpha$, whereas the number of multipliers drops from 2^n to n. The economy increases with the growth of n.

For the case of 3-bit data, the circuit of Figure 11 transfers to the one shown in Figure 12. This circuit operates much in the same way. The α^i-multipliers structure is determined from the following expressions:

$$x \left(a_2 x^2 + a_1 x + a_0 \right) \bmod g(x)$$
$$= a_1 x^2 + \left(a_2 + a_0 \right) x + a_2,$$

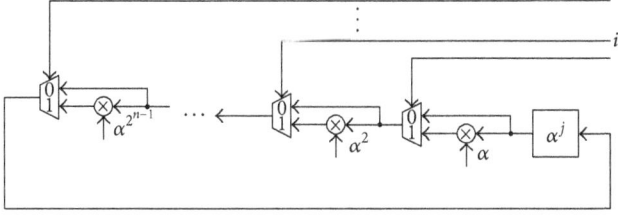

FIGURE 11: A binary-weighted version of the SA.

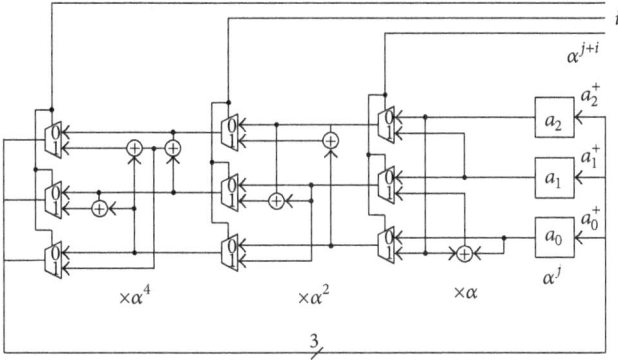

FIGURE 12: A register transfer level implementation of the 3-bit SA.

$$x^2 \left(a_2 x^2 + a_1 x + a_0 \right) \bmod g(x)$$

$$= \left(a_2 + a_0 \right) x^2 + \left(a_2 + a_1 \right) x + a_1,$$

$$x^4 \left(a_2 x^2 + a_1 x + a_0 \right) \bmod g(x)$$

$$= \left(a_2 + a_1 + a_0 \right) x^2 + \left(a_1 + a_0 \right) x + \left(a_2 + a_1 \right). \tag{18}$$

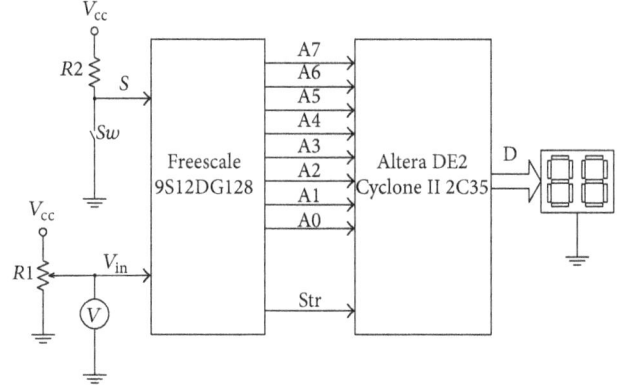

FIGURE 13: The experimental setup.

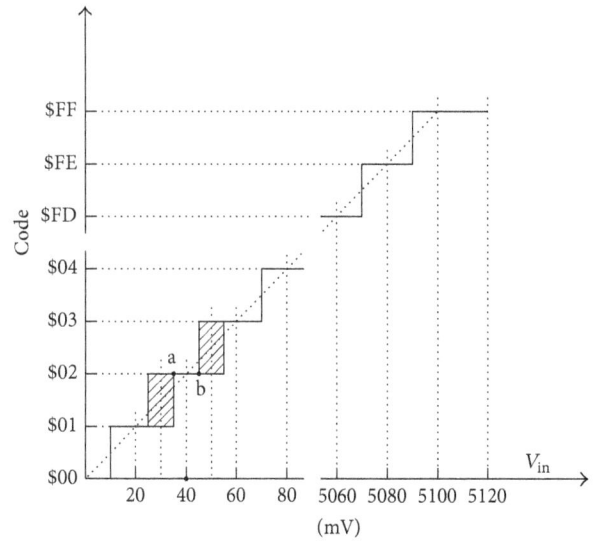

FIGURE 14: 9S12 ADC transfer function.

4. Experimental Results

The experimental setup to test the proposed method of signature analysis is shown in Figure 13. The setup includes the microcontroller system board Adapt9S12D (Technological Arts Inc.) based on Freescale's 9S12DG128 microcontroller and the Altera DE2 Development board based on the Cyclone II EP2C35F672C6 field-programmable gate-array (FPGA) device. We have selected 16 input test stimuli (voltages V_{in}) equally distributed over the range $(0 \sim 5.12)V$ and applied them to the analog-to-digital converter (ADC) of the 9S12 microcontroller (which served as a mixed-signal system). Each input voltage, V_{in}, was measured by a high-precision voltmeter and regarded as a nominal test input value.

The circuit in Figure 13 operates as follows. Every time the switch Sw is closed, the system performs 8 measurements of the same test signal and averages the result by accumulating the sum of the eight 8-bit measurements and shifting it right three times, which eliminates noise. The ADC transfer characteristic is presented in Figure 14 [17]. According to this characteristic, each conversion result for a properly operating device can deviate from the nominal value by ± 1, which is an implication of the fact that the permissible

differential nonlinearity can range from -0.5 to $+0.5$ LSB (see shadowed boxes in Figure 14). For example, if $V_{in} = 40$ mV, the conversion result can be $\$01$, $\$02$, or $\$03$ (in the worst case, the points a and b coincide). Therefore, each of the thirty-two 8-bit average results contains an error of at most ± 1 count. The test stimuli have been selected equal to the midpoints of the quantization bins, thereby increasing the uncertainty and worsening the probability of undetected error. If the test stimuli would have been selected at the transition points of the characteristic, the probability of undetected error (aliasing rate) would improve. This follows from the observation that each conversion would result in 2 possible values as opposed to 3 possible values in the previous case.

As soon as average values of the conversion results are computed by the microcontroller, they are transferred to the DE2 board. The transfer of each datum is accompanied by a high-to-low transition of the strobe signal Str. The Str signal serves as a *clock* for the state machine that implements the signature analyzer (in its 8-bit configuration). The signature, D,

TABLE 2: Relationship between input test stimuli and output responses.

Input voltage	Output code				
mV	Min	Nom.	Max	No fault	Fault
80	3	4	5	3	3
400	19	20	21	21	21
720	35	36	37	37	37
1040	51	52	53	53	53
1360	67	68	69	68	70
1680	83	84	85	85	85
2000	99	100	101	99	99
2320	115	116	117	117	117
2640	131	132	133	133	133
2960	147	148	149	148	150
3280	163	164	165	165	165
3600	179	180	181	179	179
3920	195	196	197	197	197
4240	211	212	213	212	240
4560	227	228	229	229	230
4880	243	244	245	244	244

is displayed on a two-digit 7-segment display in hexadecimal form.

The first experiment was performed on the properly operating device. In the second experiment, the average results were corrupted digitally in the microcontroller (thereby simulating random faults in the ADC) and sent to the analyzer. The analyzer has correctly identified the faulty device.

The relationship between input voltages and output codes is presented in Table 2. Based on this table and taking into consideration that $g(x) = x^8 + x^4 + x^3 + x^2 + 1$, the seed value is calculated as follows:

$$4 + 20 + \cdots + 244 = 1984 = 199 \bmod \left(2^8 - 1\right) = 199,$$

$$\alpha^{-199} = \alpha^{56} = 01011101, \tag{19}$$

$$\text{Seed Value} = \alpha^{56}\alpha^{-16} = \alpha^{40} = 01101010 = 106.$$

In addition to test experiments, the operation of the analyzer (the DE2 part of the test setup) was simulated using Altera Quartus II software. Based on the two experiments represented in Table 2, the signatures that correspond to fault-free and faulty ADCs are, respectively, 233 and 201 (in decimal form). The process of calculation of these signatures is demonstrated in Figures 15 and 16. Figures 17 and 18 represent the fault detection process. The actual final signatures are shifted additionally 32 times. If the value 1 appears in the analyzer during these shifts, the system is fault-free. Otherwise it is faulty.

The simulation results matched the experimental results.

5. Conclusion

We examined an algebraic signature analysis method that can be employed for mixed-signal circuits testing. We demonstrated how to design the appropriate device. This device does

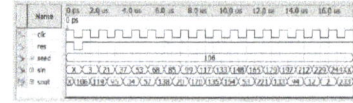

FIGURE 15: All output code deviations are within the tolerance bounds.

FIGURE 16: Some of the output code deviations exceed the tolerance bounds.

FIGURE 17: The combination "1" is detected: ADC is operating properly.

FIGURE 18: The combination "1" is not detected: ADC is faulty.

not produce arithmetic carries and is therefore less prone to errors. The absence of carry propagating circuitry also contributes to the higher performance of the device.

The proposed scheme can also be used in arithmetic and algebraic error-control coding, as well as cryptography.

Conflict of Interests

The authors declare that there is no conflict of interests regarding the publication of this paper.

References

[1] R. Frohwerk, "Signature analysis: a new digital field service method," *Hewlett-Packard Journal*, vol. 28, no. 9, pp. 2–8, 1977.

[2] G. J. Starr, J. Qin, B. F. Dutton, C. E. Stroud, F. F. Dai, and V. P. Nelson, "Automated generation of built-in self-test and measurement circuitry for mixed-signal circuits and systems," in *Proceedings of the 15th IEEE International Symposium on Defect and Fault Tolerance in VLSI Systems (DFT '09)*, pp. 11–19, October 2009.

[3] D. K. Pradhan and S. K. Gupta, "A new framework for designing and analyzing BIST techniques and zero aliasing compression," *IEEE Transactions on Computers*, vol. 40, no. 6, pp. 743–763, 1991.

[4] C. Stroud, J. Morton, T. Islam, and H. Alassaly, "A mixed-signal builtin self-test approach for analog circuits," in *Proceedings of the Southwest Symposium on Mixed-Signal Design*, pp. 196–201, 2003.

[5] N. Nagi, A. Chatterjee, H. Yoon, and J. A. Abraham, "Signature analysis for analog and mixed-signal circuit test response compaction," *IEEE Transactions on Computer-Aided Design of Integrated Circuits and Systems*, vol. 17, no. 6, pp. 540–546, 1998.

[6] N. Nagi, A. Chatterjee, and J. A. Abraham, "Signature analyzer for analog and mixed-signal circuits," in *Proceedings of the IEEE International Conference on Computer Design: VLSI in Computers and Processors*, pp. 284–287, October 1994.

[7] S. Mir, M. Lubaszewski, V. Liberali, and B. Courtois, "Built-in self-test approaches for analogue and mixed-signal integrated circuits," in *Proceedings of the IEEE 38th Midwest Symposium on Circuits and Systems*, vol. 2, pp. 1145–1150, Rio de Janeiro, Brazil, August 1995.

[8] J. Rajski and J. Tyszer, "The analysis of digital integrators for test response compaction," *IEEE Transactions on Circuits and Systems II: Analog and Digital Signal Processing*, vol. 39, no. 5, pp. 293–301, 1992.

[9] L. Wei and L. Jia, "An apprach to analong and mixed-signal BIST based-on pseudorandom testing," in *Proceedings of the IEEE International Conference on Communications, Circuits and Systems, (ICCCAS '08)*, pp. 1192–1195, Fujian, China, May 2008.

[10] F. Corsi, C. Marzocca, and G. Matarrese, "Defining a BIST-oriented signature for mixed-signal devices," in *Proceedings of the IEEE Southwest Symposium on Mixed-Signal Design*, pp. 202–207, 2003.

[11] S. Demidenko, V. Piuri, V. Yarmolik, and A. Shmidman, "Bist module for mixed-signal circuits," in *Proceedings of the IEEE International Symposium on Defect and Fault Tolerance in VLSI Systems*, pp. 349–352, 1998.

[12] T. Damarla, "Implementation of signature analysis for analog and mixed signal circuits," U.S. Patent 6 367 043, 2002.

[13] W. W. Peterson and J. Weldon, *Error Correcting Codes*, MIT Press, Cambridge, Mass, USA, 2nd edition, 1972.

[14] V. Geurkov, "Optimal choice of arithmetic compactors for mixed-signal systems," in *Proceedings of the IEEE International Symposium on Defect and Fault Tolerance in VLSI and Nanotechnology Systems (DFT '12)*, pp. 182–186, October 2012.

[15] S. Lin and D. Costello, *Error Control Coding*, Pearson Education, Upper Saddle River, NJ, USA, 2004.

[16] V. Geurkov, V. Kirischian, L. Kirischian, and R. Sedaghat, "Concurrent testing of analog-to-digital converters," *i-manager's Journal on Electronics Engineering*, vol. 1, no. 1, pp. 8–14, 2010.

[17] MC9S12DT128 Device User Guide, V02.11, Motorola, May 2004, http://www.freescale.com/.

## 4

Design of Synthesizable, Retimed Digital Filters Using FPGA Based Path Solvers with MCM Approach: Comparison and CAD Tool

Deepa Yagain and A. Vijaya Krishna

Department of ECE, PESIT, Bangalore 560085, India

Correspondence should be addressed to Deepa Yagain; deepa.yagain@gmail.com

Academic Editor: Jose Silva-Martinez

Retiming is a transformation which can be applied to digital filter blocks that can increase the clock frequency. This transformation requires computation of critical path and shortest path at various stages. In literature, this problem is addressed at multiple points. However, very little attention is given to path solver blocks in retiming transformation algorithm which takes up most of the computation time. In this paper, we address the problem of optimizing the speed of path solvers in retiming transformation by introducing high level synthesis of path solver algorithm architectures on FPGA and a computer aided design tool. Filters have their combination blocks as adders, multipliers, and delay elements. Avoiding costly multipliers is very much needed for filter hardware implementation. This can be achieved efficiently by using multiplierless MCM technique. In the present work, retiming which is a high level synthesis optimization method is combined with multiplierless filter implementations using MCM algorithm. It is seen that retiming multiplierless designs gives better performance in terms of operating frequency. This paper also compares various retiming techniques for multiplierless digital filter design with respect to VLSI performance metrics such as area, speed, and power.

1. Introduction

High level synthesis is the process of converting behavioral description or an algorithm to structural level specification. In behavior description or an algorithm, the input and output behavior is described in terms of data transfers and operations without any implementation details. Structural description maps this data transfers and operations into combinational functional units and registers on to hardware. High level synthesis of DSP algorithms is very much useful as it reduces time to market window. Various optimization methods are available in literature for sequential synthesis [1]. Though synthesis of combinational logic has attained a significant level of maturity, sequential circuit synthesis has been lagging behind [2] in terms of frequency performance. DSP algorithms are repetitive and periodically iterations must be repeated to execute the computations [3]. Here, iteration period is the minimum time needed for computation and this

is limited by critical path. Critical path can be altered by redistributing the delays such that functionality is preserved. Retiming algorithm [4] is used to redistribute the delays without altering [5] the functionality.

A great amount of research has been done on retiming [5, 6]. The retiming technique is the valuable optimization technique in problems of digital filters which can be represented as data flow graphs (DFGs). Efficient filter systems are needed to decrease the overall computation time since scientific applications can be recursive, nonrecursive, and iterative. Retiming transformation along with other high level transforms like multiple constant multiplication approach for filters in high level synthesis aids in reducing area of the filter circuit and most importantly decreases the clock period. Critical path and shortest path computations consume most of the time in retiming computation. The retiming minimizes the overall clock period, thereby increasing the clock frequency by reducing the filter critical path. In the general purpose processor where actual retiming vectors are computed for digital filters, the speed with which the retiming transformation is performed suffers since the entire transformation code will be written as software. Hence, FPGA based path solver architecture is designed in this paper which addresses the frequency issue in retiming and reduces the burden on general purpose processors.

A computer aided design (CAD) tool framework called DiFiDOT is developed which generates the synthesizable hardware descriptions of chosen digital filter with specified user constraints such as area, speed, and power. Since the digital filters are composed of adders/subtractors, multipliers, and delay elements, DiFiDOT picks the best choice of adders and multipliers as per users design constraints. Also, multiplication operation is expensive in terms of area, power, and delay. Exchanging multipliers with adders is advantageous because adders weigh less than multipliers in terms of silicon area [7]. Since the coefficients to be multiplied are known beforehand, the full flexibility of multiplier is not necessary in the design. So a multiplierless design in digital filter is proposed under multiple constant multiplications architecture and an option is included in DiFiDOT for generating multiplierless hardware descriptions. This significantly reduces the area of filters when compared to those designed using multiplier blocks. Here, sharing of partial terms in multiple constant multiplications (MCMs) concept [8] is used which reduces area and covers all possible partial terms that may be used to generate the set of coefficients in the MCM instance. For simulations, the authors have used some of ACM/SIGDA benchmark circuits.

2. Background

High level transformation techniques are applied to get optimal speed in sequential filter systems. For the designed optimization environment, input is considered as data flow graphs. This section introduces the data flow graphs (DFGs) and problem definitions and gives an overview on previously proposed retiming transformation algorithms with their drawbacks.

2.1. Data Flow Graphs (DFGs). Digital filters are important part of digital signal processor. Their extraordinary performance is one of the parameters that made DSP become so popular [9]. Filters are used in audio processing, speech processing (detection, compression, and reconstruction), modems, motor control algorithms, video and image processing, and so forth. Retiming is important step [10] in high level synthesis (HLS) of digital filters. HLS is nothing but to map behavioural descriptions of algorithms to physical realizations. All digital filters which are iterative, recursive, and nonrecursive can be represented using data flow graphs (DFGs) [11, 12]. Any digital filter can be realized by functional blocks such as adder/subtractor and multiplier with delay elements. A filter DFG consists of such functional blocks and connectivity information for the data flow. This example is a 4th-order low pass elliptic filter block. High level transformations operate on the filter functional blocks for better

performance. This can be done by changing the execution order or by altering the number of functional blocks in the critical path of retiming [3] process. Performance can also be improved by altering the architecture of functional block's implementation characteristics without altering the functionality of the filter in the optimization environment. In this paper, MCM technique is used to achieve this. For applying all these transformations, input is given in the form of DFGs. The input DFGs can also be represented in the form of matrices for further computations. In this paper, 4th-order low pass elliptic filter is used as an application example to explain the present work. The elliptic filter response is given by

$$|H(jw)|^2 = \frac{1}{1 + \epsilon^2 R_n^2(\omega, \gamma)}, \quad (1)$$

where $R_n^2(\omega, \gamma)$ is the nth order Chebyshev rational function with ripple parameter γ. Let Ap be the maximum pass-band loss and let As be the minimum stop-band loss in decibels. Depending on the need, we can define ω_p and ω_s which are pass-band cutoff frequency and stop-band cutoff frequency. The selectivity factor k is computed using ω_p and ω_s. With these parameters, modular constant $q = u + 2u^5 + 15u^9 + 150u^{13}$ is computed where u is

$$q = \frac{1 - \sqrt[4]{1 - k^2}}{2\left(1 + \sqrt[4]{1 - k^2}\right)} \quad (2)$$

The discrimination factor D and the filter order n are used to obtain A_s which is given by

$$A_s = 10\log\left(1 + \frac{10^{A^p/10} - 1}{16q^n}\right). \quad (3)$$

The second order elliptic filter block is as shown in Figure 1(a). The typical DFG for the filter is shown in Figure 1(b). For the designed optimization environment, the filter information is given in the form of matrices. There are two matrices which represent the filter information. The node-weight matrix represents node weights that are nothing but computation time unit delays in the filter graph. The computational complexity of the adder is $O(n)$, whereas for multiplication it is $O(n^2)$ for two n digit numbers. Multiplication computation complexity is higher when compared to addition. Hence, in the present work, time delay considered for multiplication in retiming is twice that of addition. Incidence matrix defines the edge weights between all the nodes which represent connectivity information. The number of delay elements present in between the computation nodes (adder and multiplier) is considered as edge weight. If there are no delay elements and adder or multiplier node is directly connected to another node, then the edge weight will have zero value. The node weight matrix and incidence matrix are used as the inputs for optimization environment where high level transformations are applied to obtain performance improvement. The critical path and shortest path of the filter are computed for retiming which is one of the efficient

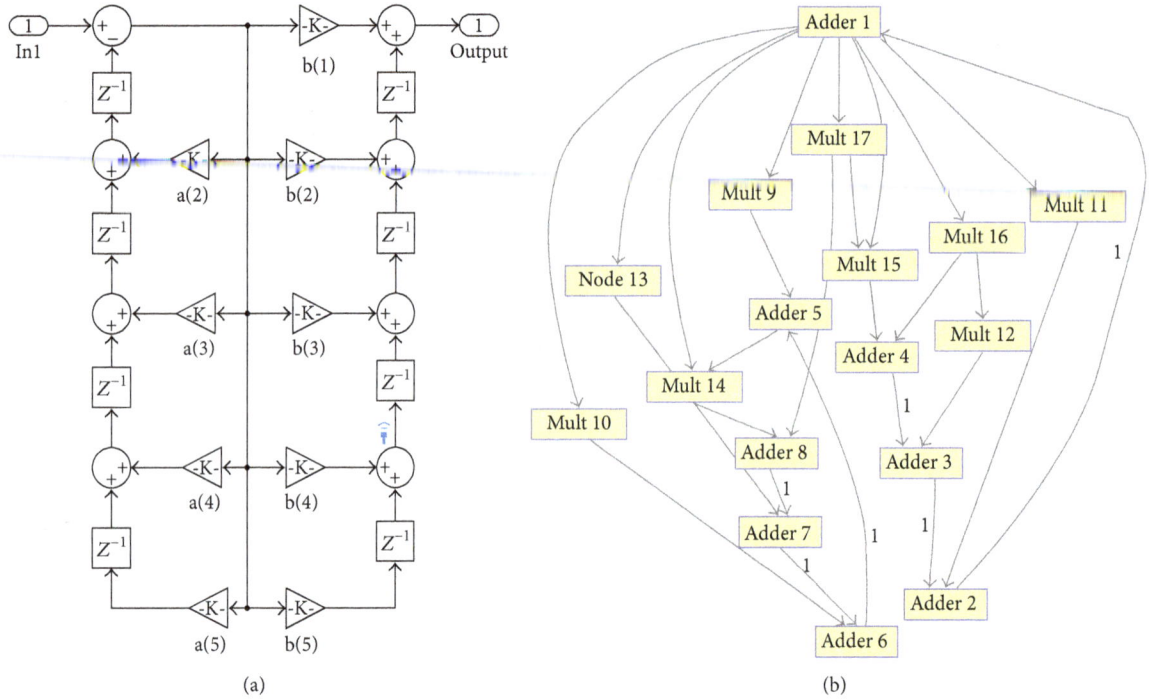

FIGURE 1: (a) Block diagram of 4th-order low pass elliptic filter; (b) DFG of elliptic filter block DFG.

optimization techniques to obtain a filter solution with reduced clock period which in turn increases the filter speed. The critical path can be obtained by observing Figure 1(b) DFG. The critical path is node $1 \rightarrow 9 \rightarrow 5 \rightarrow 14 \rightarrow 8$.

Retiming Transformation. Retiming is a high level transformation technique in which the location of the registers is altered in such a way that the overall clock period reduces, thereby increasing the clock frequency [13]. This happens due to reduction in the critical path which bounds the speed of the design. Due to intelligent placement of registers, the clock period gets minimised without altering the filter functionality. Critical path is the longest computation path in between computational elements [14] or delay elements. The critical path can also be minimized by inserting the delay elements on the primary inputs of the filter circuit and retiming the circuit. This is called automatic pipelining technique. Both the methods are used to find the best optimal solution in the present work. Retiming for filter optimization is found to be NP complete problem, and time to find the solution increases as the problem size increases. There are two ways of applying retiming transformation:

(i) retiming using clock period minimization method,

(ii) retiming using register minimization method.

The retiming algorithm for clock period minimization is efficient in terms of clock frequency improvement. Its computational complexity is $O(n^3 \log n)$, where n is the number of nodes which are nothing but computation elements such as adders and multipliers. The algorithm starts by building a new graph from the original DFG. The new graph can give us a set of inequalities called the critical path constraints.

The original DFG also presents a set of equalities called the feasibility constraints. A constraint graph can be built from the critical path constraints and the feasibility constraints. The retiming values for each node can be derived by applying a Floyd-Warshall shortest path algorithm to the constraint graph. The weight for each edge in the retimed DFG can be calculated using the original weight and the retiming values of the two nodes are connected by this edge.

(i) Calculate $M = t_{\max_n}$, where n represents the number of nodes in the original DFG G and $t_{\max}$ is the maximum computation time of all the nodes in the DFG. Also compute the critical path which defines the required clock period of original graph.

(ii) A new DFG G^* can be created from G. G^* has the same nodes and edges as G. For each edge in G^*, the edge weight is $W^*(e) = MW(e) - t(u)$, where $W(e)$ is the edge weight of the same edge in G; $t(u)$ is the computation time of the node initiating this edge.

(iii) We then apply the Floyd-Warshall shortest path algorithm to compute S^*_{UV}, which represents the shortest path from node U to node V.

(iv) From S^*_{UV}, W_{UV} and D_{UV} are calculated. If $U \neq V$, then $W_{UV} = S^*_{UV}/M$ and $D_{UV} = MW_{UV} - S^*_{UV} + t(V)$. If $U = V$, then W_{UV} and $D_{UV} = t(u)$. Here, $t(U)$ and $t(V)$ represent the computation times of node U and node V, respectively.

(v) We then find the maximum value of D_{UV} and the minimum values of D_{UV}. We check all the possible clock periods starting from maximum value of D_{UV} to minimum value of D_{UV} one by one. If we find a clock period that can give us a feasible solution, we stop and find the minimal clock period by solving for solutions critical path. The solution contains the retiming values for all nodes. Here, 4th-order low pass IIR elliptic filter is designed and retimed using the above mentioned clock period minimization algorithm. The retimed data flow graph with reduced clock period is shown in Figure 2.

After applying retiming transformation to the filter, the critical path changes to $2 \rightarrow 1 \rightarrow 9$.

Since each delay element occupies about one-third of the binary adder, it is important to reduce the number of delay elements [11]. In retiming using register minimization, we can obtain the digital filter that uses minimum number of registers and satisfies the clock period constraints [15]. Here, forward splitting or register sharing [12] is used. If the node has several output edges carrying the same signal, the number of registers required to implement these edges is the maximum number of registers on any one of the edges. Consider Figure 3. The maximum number of registers required in Figure 3(a) is 6 whereas after register sharing, this gets reduced to 3 as shown in Figure 3(b).

The number of registers needed to construct this output edges (e) in retimed graph W_r and the total cost are

$$R_v = \text{Max}\left(W_r(e)\right), \qquad \text{Cost} = \sum R_v. \qquad (4)$$

The cost is WRT:

(i) fan-out constraints: $R_V \geq W_r$ for all V and all edges $V \xrightarrow{e} any \ other \ vertex$

(ii) feasibility constraints: $r(U) - r(V) \geq W(e)$ for every edge $U \xrightarrow{e} V$

(iii) clock period constraints: $r(U) - r(V) \geq W(U,V) - 1$ for all vertices such that $D(U,V) \geq c$, where c is the clock period.

This method makes use of gadgets to represent the nodes with multiple edges. The register minimization retiming can be modeled as linear programming problem. A dummy node with zero computation time will be introduced in this. The weight of the edge e_i is defined to be $W(e_i) = W_{\text{max}} \neg W(e_i)$, where $W_{\text{max}} = \max(W(e_i))$, where $1 \leq i \leq K$ where k is the number of edges available. Also β parameter is used which is the breadth associated to model the memory required by edge e_i. The breadth of each edge is inverse of k. A binary search is performed for clock period and below is the procedure used while performing retiming using register minimisation. The register minimization retiming values can be obtained as below.

(i) Use the gadget model of the graph to compute the cost function.

(ii) Calculate S' by using shortest path Floyd-Warshall algorithm.

(iii) Compute $D(U,V)$ and $W(U,V)$ matrices from the original graph and S' matrix.

(iv) Perform LP formulation such that the cost function gets minimized which is subjected to feasibility and clock period constraints.

This LP problem is solved to obtain the retiming solution which minimizes the number of registers by satisfying the clock period. Figure 4 shows the DFG of 4th-order low pass IIR elliptic filter. It is observed that the register minimum retimed solution provides the filter solution with reduced register count for reduced clock period. However, in some cases, it is found that clock period minimization efficiency reduces in comparison to clock period minimization retiming technique as the priority is given to the register count. For the considered elliptic filter for a clock period of 4 units, it is found that the register count gets minimized to 9. After applying register minimization retiming transformation to the filter, the critical path changes to $1 \rightarrow 9 \rightarrow 5$.

Problem Formulation. Critical path and shortest path solving contribute to most of the computation time in retiming.

Definition 1 (the path solver problem). Let $S = \{s_0, s_1, s_2, s_3, \ldots, s_k\}$, where k is the maximum number of feasible solutions available for retiming of a considered filter DFG. During retiming of digital filters in high level synthesis, the shortest path between the nodes must be computed for $(k + 1)$ times where k is the number of feasible solutions available for the DFG which is nothing but unique entries in path delay D matrix. Similarly, the critical path must be computed for $(k + 1)$. General purpose processors (GPPs) where retiming algorithm is implemented are fully programmable but are less efficient in terms of power and performance. Hence, the problem is to improve the performance and power of retiming using FPGA based path solvers. Further, along with retiming, high level transformation technique called automatic pipeline is applied to improve the filter speed.

Definition 2 (multiple constant multiplication in digital filters). For the considered filter coefficient constant T in the retimed filters, find the set of multiplierless operations $\{O_1, O_2, O_3, \ldots, O_n\}$ with minimum number of addition, subtraction, and shift operations using multiple constant multiplier architecture to optimize the filter architecture further.

Definition 3 (optimization and automation of filter HDL). An environment needs to be developed to obtain HDLs of retimed filters in which user can choose different data path element architectures depending on the specifications. This reduces time to market and helps to evaluate a lot of hardware implementation trade-offs. Filter equivalence checking after applying high level transformation needs to be done which needs to be developed as a part of the optimization environment.

Principle of Shortest Path and MCM Algorithm. Several FPGA synthesis algorithms have been proposed specifically for sequential circuits. In [16], authors have proposed how to

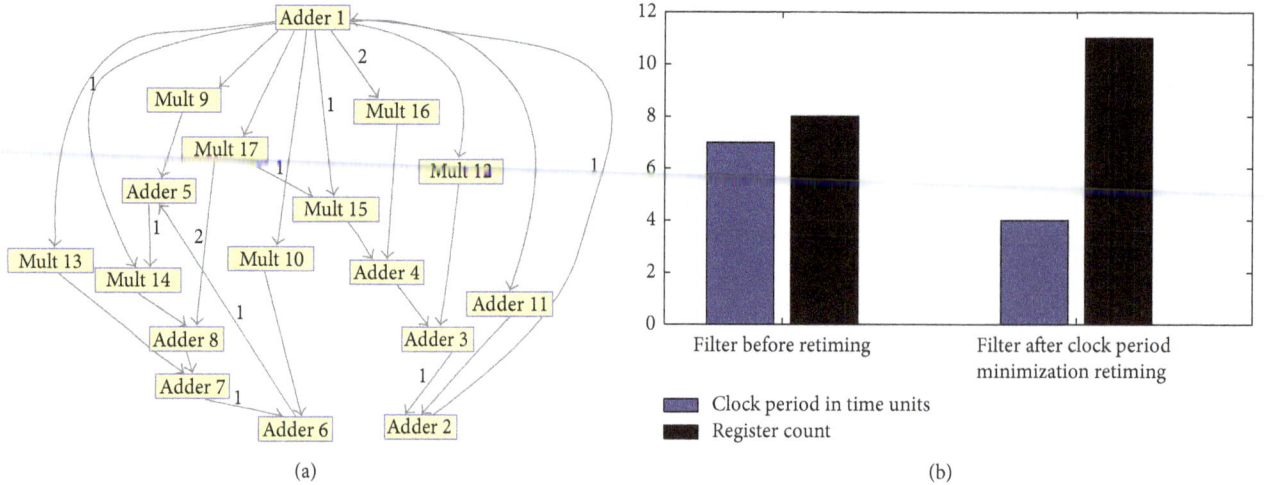

FIGURE 2: 4th-order elliptic filter after clock period minimization retiming; (a) DFG after retiming; (b) clock period and register count before and after retiming.

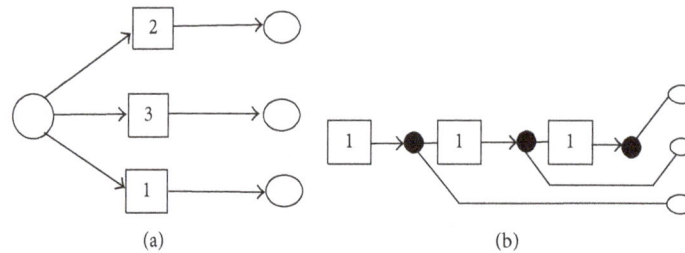

FIGURE 3: (a) Graph before register sharing. (b) Graph after register sharing.

map retimed circuits on to FPGAs efficiently. However, in this paper, authors suggest a method for efficient retiming process using FPGA based path solvers. This can be applied to any retiming techniques available in literature. Shortest path is solved in filter DFG using Floyd-Warshall algorithm. The Floyd-Warshall algorithm uses an approach of dynamic programming to solve the shortest-paths problem on a DFG. The Floyd-Warshall Algorithm can solve the shortest path problem in $O(n^3)$ time where n is the number of nodes in the DFG. Let $d_{ij(k)}$ denote the weight of the shortest path from i to j such that all intermediate vertices are contained in the set $\{1, 2, \ldots, k\}$. That is, the path p is decomposed into $i \rightarrow k \rightarrow j$. Let the vertices in the graph be numbered from $1, 2, \ldots, n$. Consider the subset $\{1, 2, \ldots, k\}$ of these n vertices. Find the shortest path from vertex i to vertex j that uses vertices in the set $\{1, 2, \ldots, k\}$ only. Then, there are two situations possible:

(i) k is an intermediate vertex on the shortest path,

(ii) k is not an intermediate vertex on the shortest path.

If the vertex k is not an intermediate vertex on p, then

$$d_{ij}(k) = d_{ij}(k-1) \text{ else } d_{ij}(k) = d_{ik}(k-1) + d_{kj}(k-1).$$

$$(5)$$

In either case, the subpaths contain nodes from $\{1, 2, \ldots, (k-1)\}$. Therefore,

$$d_{ij}(k) = d_{ij}(k-1) + d_{kj}(k-1). \qquad (6)$$

When $k = 0$, then

$$d_{ij}(0) = \{W_{ij}\},$$

and if $k0$ then $d_{ij}(k) = \min\{d_{ij}(k-1) + d_{ij}(k-1)\}.$

$$(7)$$

Let D be the incidence matrix with the graph edge weight information W initially. D is then updated with the calculated shortest paths; see Algorithm 1.

The final D matrix will store all the shortest paths. This algorithm is extended for retiming of digital filters.

The multiple constant multiplication (MCM) problem is addressed in the literature [14] using either graph based methods or using common subexpression elimination method. In common subexpression elimination algorithm, all possible subexpressions are extracted for a variable. But this is possible only if it is defined as minimum signed digit and as canonical signed digit. Then the subexpression is found such that it can be shared by multiple constant multiplication values. In this paper, the above two concepts are extended for automatic pipelining and retiming of digital filters in high

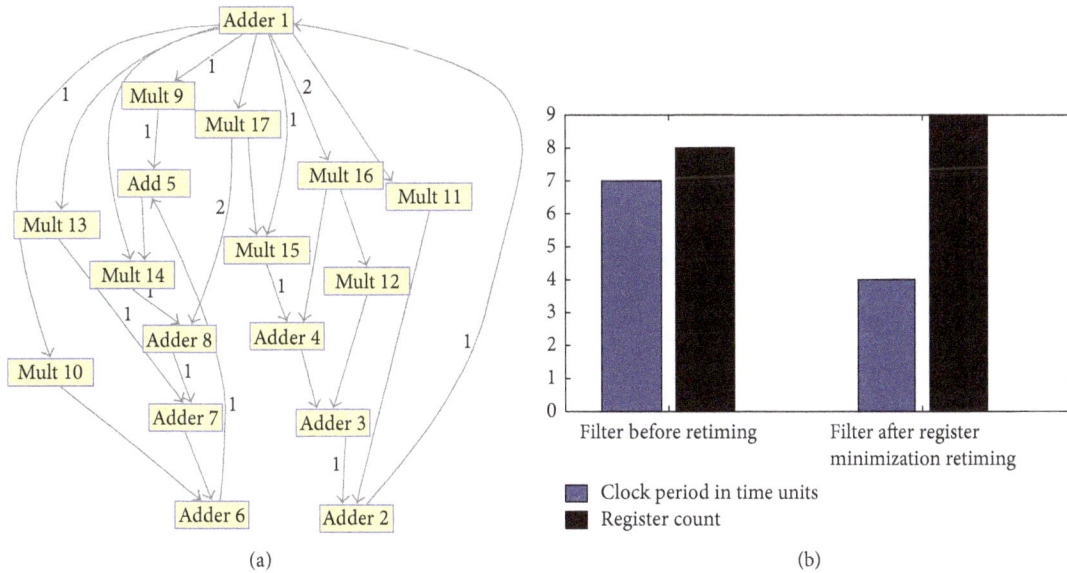

FIGURE 4: 4th-order elliptic filter after register minimization retiming; (a) DFG after retiming; (b) clock period and register count before and after retiming.

```
(1) n = # of rows in W,  D⁰ = W
(2) for(k=1 to n)
(3)         for(i=1 to n)
(4)                 for(j=1 to n)
(5)                         dᵢⱼᵏ = min{dᵢⱼ^(k-1), dᵢₖ^(k-1)  + dᵢₖ^(k-1)}
(6)                 end for
(7)         end for
(8) end for
(9) return Dⁿ
```

ALGORITHM 1

level synthesis. In all the digital filters, the filter coefficients are known beforehand. Hence, full flexibility of the multiplier is not necessary and we can make use of MCM designs. This method is more efficient when compared to shift and add multiplications as intermediate results can be shared which reduces the area of multiplierless implementation of digital filters. The sharing of intermediate result will provide potential area saving with increased filter order (Figure 5).

Consider the filter coefficient set which is to be used for the filter design given by $T = \{c_1, c_2, c_3, \ldots, c_n\}$, we need to find the smallest set S given by $\{a_1, a_2, a_3, \ldots, s_1, s_2, s_3, \ldots, \}$ where a (adders/Subtractors) & s (shifts) $< S$ such that the set is made of adderssubtracters, shifters, and A operations. Here, shift operations also can be shared across multiple points so that the output set is optimum. Here H_{cub} algorithm [8] is used to generate corresponding DFG for the multiplier block implementing the parallel multiplications $c_1 * x, c_2 * x, \ldots, c_n * x$. The only operations used in the generated DAG and input design matrices are additions, subtractions, shifts, and negations. In this paper, performance of MCM based filter designs is further improved by combining this approach with retiming. The multiplierless filter circuit is further retimed

to reduce the overall clock period which increases the clock frequency.

Consider l_1 and l_2 as two integers which specifies left shifts and $r \geq 0$ specifies right shift and let s be the sign bit which can be $\{0, 1\}$. An A operation is an operation with two integer inputs u and v and one fundamental output which is defined as

$$A_p(u, v) = \left| (u \ll l_1) + (1) s (v \ll l_2) \right|$$

$$\gg r = 2^{l_1} u + (-1)^s 2^{l_2} v \mid 2^{-r},$$

(8)

where $\ll$ is a left binary shift, $\gg$ is a right binary shift, and $p = \{l_1, l_2, r, s\}$ is the parameter set or the A configuration of A_p. To preserve all significant bits of the output, 2^r must divide $2^{l_1} u + (-1)^s 2^{l_2} v$. The left shifts are limited to the bit width of the target. All A operations are used to build $A - graph$. For a given set of target filter coefficients C, we can find set S such that multiplierless digital filter is designed.

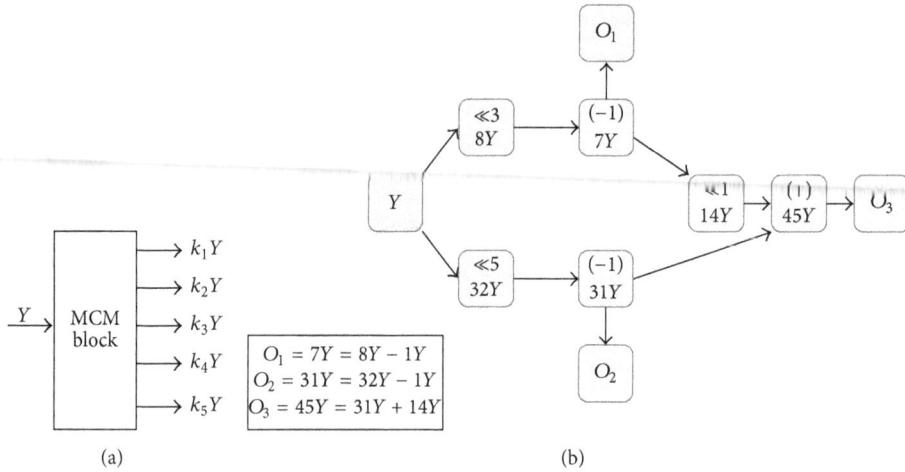

FIGURE 5: Example for addressing MCM problem in digital filters.

3. Design and Analysis

Each DSP filter block is associated with the critical path which limits maximum iteration period in the filter design [12]. This can be reduced by retiming where the clock period gets reduced and increases the clock speed. To reduce the critical path, we need to find the original critical path of the circuit using critical path solving algorithm and then apply retiming transformation to digital filter. While retiming, shortest path algorithm is required for solving the system inequalities. FPGAs are nothing but set of configurable logic blocks with configurable interconnects. Designer can program it to work like a specific hardware. These give great speedup over general purpose processors for many long running algorithms. Hence, for high performance systems, FPGAs become a better choice. In the present work, path solvers are implemented on FPGA to increase the performance.

3.1. Critical Path Solver Algorithm: Design and Analysis. The critical path is defined as maximum delay path between the output node and node causing the state change of the output node with zero delay. The significance of the critical path is that it determines the operating frequency of the design. In retiming, which is one among the steps in high level synthesis, it is imperative that we find the critical path [17] in real time. To speed up this process, the use of a dedicated FPGA hardware can speed up the process with low power. Consider α = Number of adder elements and β = Number of multiplier elements in the considered digital filter. Let = $\{n_1, n_2, \ldots, n_i\}$, where $i = \alpha + \beta$, which is maximum combinational adder and multiplier elements. Consider $O = o_1, o_2, \ldots, o_j$ where O is the set of output nodes in the filter circuit and $I = \{i_1, i_2, \ldots, i_k\}$, where I is the set of input nodes in the filter circuit such that IN and ON. The critical path of the circuit is defined in terms of γ_{n_i} which is the delay of individual combinational block. In this procedure of computing critical path on FPGA, it sorts the vertices such that vertices occurring early in the list are connected to vertices later in the list by edges having zero delays. While

sorting, if the vertex is connected to previous one, then path length is sum of its time with the sum of all the vertices found in the path; otherwise, path length of the node is equal to its own computation time. We need this for constructing the retimed graph as well as verifying the retimed graph result. The equation of the critical path is

$$\gamma_m = \sum_{i=1}^{i=N} t_{m_1}, \tag{9}$$

where N is the sum of adder and multiplier elements in the topologically sorted vertices connected with zero delay edges. The delay of the circuit is given by $t_d = \max\{\gamma_m\}$ where t_d is the delay of the critical path. Algorithm 2 shows the critical path formulation. In the considered optimization environment, the below steps are used for critical path computation.

(i) The filter network graph is considered as input to critical path solver algorithm.

(ii) All the zero-weight edges in the network graph are found and a matrix of their source and destination nodes is formed.

(iii) For each row in the above matrix, if the destination node of any zero-weight edge path is the same as the source node of the zero-weight edge path, the two paths are joined. This step is repeated to obtain a matrix whose rows will have nodes of all the possible zero-weight edge paths in the graph.

(iv) The computational time of each zero-weight edge path from this matrix is calculated.

(v) The zero-weight edge path with the greatest computational time is found. This is the critical path and its computational time is the critical path delay.

A critical path solver algorithm is designed in the present work on FPGA. The state diagram for the implemented critical path solver is given in Figure 6. In $S0$, the filter graph or matrix is given as input to the critical path solver module.

Design of Synthesizable, Retimed Digital Filters Using FPGA Based Path Solvers with MCM Approach: Comparison...

41

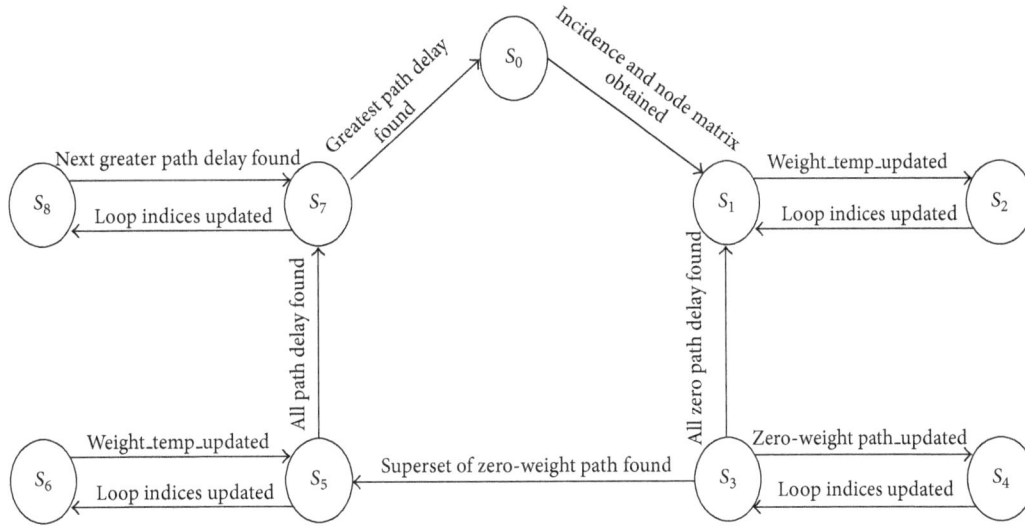

FIGURE 6: Critical path solver state diagram.

```
(1) //Algorithm for computing the critical path
(2) Input: a DFG of G = (V,E,t,d) Where c is the
(3) computation time of the node and d
(4) is the initial delay on edge E
(5) Output: Critical path C
(6) Sort all the vertices topologically in the DFG G
(7) with v fallowing u
(8) if there is a zero delay edge from u → v
(9) For all vertices from the sorted list
(10) If non zero delay on the edge E in G then
(11) γ_i = t_v
(12) else
(13) γ_i = t_v + max(γ_i) ∈ edge e : u → v in G with d_e = 0
(14) end if;
(15) γ = γ_1, γ_2,...,γ_m
(16) where m = number of entries in the topologically sorted list
(17) end for;
(18) compute γ = max{γ}
```

ALGORITHM 2

Since HDL does not provide a method to represent infinity, some number, say 255, can be chosen which is always greater than any other weight in the incidence matrix. Also since edge weight 0 is a valid input, any negative number, say −1, can be used to denote the uninitialized matrix element. In state S1, all the zero weight edges in the DFG are found along with their source and destination nodes and are stored in a matrix called zero_weight_path. The zero_weight_path matrix contains two columns. The first column contains the source node of a directed zero-weight edge while the second column has the destination node of the directed zero-weight edge. Simultaneously, we will keep a count on the number of zero-weight edges.

The state S2 is provided to enable looping action and for updating of all the signals. In state S3, in each row of zero_weight_path matrix, the module will find the next node with a zero-weight edge connecting it to the node in the previous column (if it exists). Thus, if the destination node in any zero-weight path is same as the source node in another zero-weight path, the two paths are concatenated, that is, if the destination node in path a is the source node in path b, then we make the destination node in b as the destination node of a. The state S4 is provided to enable looping action and for updating all the signals. At the end of this state, the zero_weight_path matrix will contain only those superset paths that are a superset of the remaining zero weight paths. In state S5, the module calculates the sum of all the node weights through each of these paths. State S6 is provided for looping action and for updating all signals.

In state S7 the path with the highest node weights sum is found, which is the critical path of the DFG. All the nodes in this path are then stored in order in a matrix called the critical

```
(1) Algorithm for computing the shortest path
(2) Input: a DFG of G = (V,E,t,d) Where c is the computation time of the node and d
(3) is the initial delay on edge E
(4) Output: All pair shortest path matrix M
(5) for i = 1 to N
(6)    for j = 1 to N
(7)       if i = j, then
(8)         M[i,j] = (0,0)
(9)       else M[i,j] = inf
(10)   end for
(11) end for
(12) for all the edges e : u → v, M[u,v] = d for edge e
(13) for k → 1 to N
(14)   for i → 1 to N
(15)     for j → 1 to N
(16)        if M[i,j] > M[i,k] + M[k,j]
(17)          M[i,j] = M[i,k] + M[k,j]
(18)    end for
(19)    end for
(20) end for
(21) Output shortest path matrix M
```

ALGORITHM 3

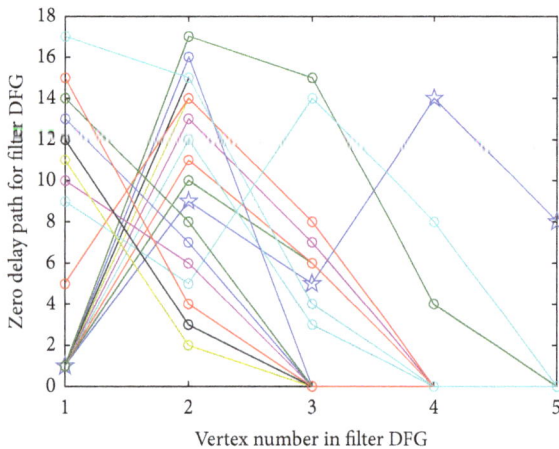

FIGURE 7: Zero path delays and critical path for 4th-order low pass elliptic filter.

path matrix. These signals, in this matrix, are output as the critical path. The state $S8$ is provided to enable looping action and for updating all the signals. The state machine then goes back to state $S0$ and awaits new inputs. Next, algorithm to find the shortest path between two nodes in a graph is described. For retiming technique in high level synthesis, we need the shortest path to solve system of inequalities. It is seen that time needed to compute critical path on FPGA is reasonably less when compared to computation on general purpose processor. This also reduces the retiming computation time. The zero delay paths are computed for 4th-order elliptic filter shown in Figure 7. The highlighted path delay is from $1 \rightarrow 9 \rightarrow 5 \rightarrow 14 \rightarrow 8$ where nodes $1, 5, 8$ are adders and $9, 14$

are multipliers. Maximum path delay which is highlighted is considered to be the critical path.

3.2. Shortest Path Solver Algorithm and State Diagram. Let $D(u, v)$ be the maximum delay between nodes u and v and let $T(u, v)$ be total computation time of zero delay path from u to v. We can check the condition $T(u, v) - \min\{t(u), t(v)\} > \{derived\ clock\ period\}$ then select those paths to retime so that computation time in this path can be reduced. We have to retime the edges by constructing system of linear inequalities. This can be done using Floyd-Warshall shortest path algorithm Algorithm 3. This can be used for retiming the graph further (Figure 6).

Floyd-Warshall all pair shortest path algorithm is designed and implemented as a part of path solvers on FPGA [17] which reduces the computational burden of general purpose processor where actual retiming has been carried out. The speed of computation is also increased by a larger extent. The HDL program for the shortest path solver on FPGA was designed based on the state diagram shown in Figure 8. Updating of the looping variables is done in $S1$ and then transition from $S1$ to $S0$ occurs. The transition from $S0$ to $S2$ occurs after the incidence matrix is completely copied to the signal weight_temp. In state $S2$, the signal weight_temp is operated upon to obtain the pair wise shortest path matrix with state $S3$ enabling looping action. Transition from $S2$ to $S3$ takes place after each pair wise path distance is found. Updating of the looping variables is done in $S3$ and then transition from $S3$ to $S2$ occurs. The transition from $S2$ to $S4$ occurs after all the pair wise shortest paths are stored in the signal weight_temp. In the state $S4$, the elements of the signal matrix weight_temp are copied to the output matrix.

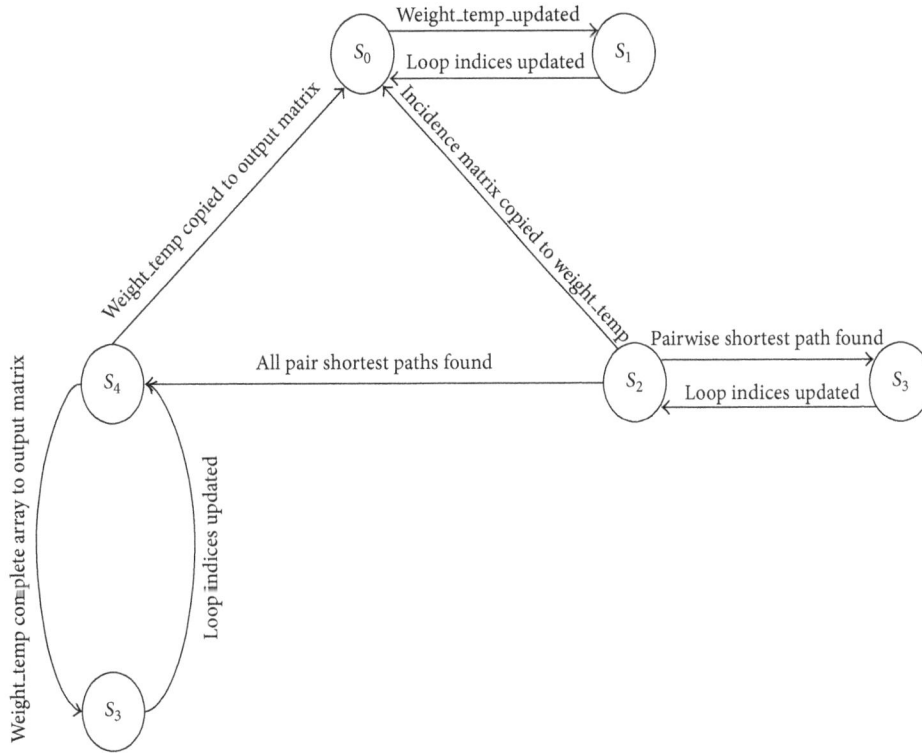

FIGURE 8: Shortest path solver state diagram.

The state $S5$ enables looping action for $S4$. Transition from $S4$ to $S0$ occurs after the output matrix is available with all the pair wise shortest paths. The state machine is then initialized and awaits new inputs.

$$
SPM = \begin{bmatrix}
inf & 0 & 0 & 0 & 0 & 0 & 0 & 0 & 0 & 0 & 0 & 0 & 0 & 0 & 0 & 0 & 0 \\
1 & 1 & 1 & 1 & 1 & 1 & 1 & 1 & 1 & 1 & 1 & 1 & 1 & 1 & 1 & 1 & 1 \\
2 & 1 & 2 & 2 & 2 & 2 & 2 & 2 & 2 & 2 & 2 & 2 & 2 & 2 & 2 & 2 & 2 \\
3 & 2 & 1 & 3 & 3 & 3 & 3 & 3 & 3 & 3 & 3 & 3 & 3 & 3 & 3 & 3 & 3 \\
inf & inf & inf & inf & 3 & 2 & 1 & 0 & inf & inf & inf & inf & inf & 0 & inf & inf & inf \\
inf & inf & inf & inf & 1 & 3 & 2 & 1 & inf & inf & inf & inf & inf & 1 & inf & inf & inf \\
inf & inf & inf & inf & 2 & 1 & 3 & 2 & inf & inf & inf & inf & inf & 2 & inf & inf & inf \\
inf & inf & inf & inf & 3 & 2 & 1 & 3 & inf & inf & inf & inf & inf & 3 & inf & inf & inf \\
inf & inf & inf & inf & 0 & 2 & 1 & 0 & inf & inf & inf & inf & inf & 0 & inf & inf & inf \\
inf & inf & inf & inf & 1 & 0 & 2 & 1 & inf & inf & inf & inf & inf & 1 & inf & inf & inf \\
1 & 0 & 1 & 1 & 1 & 1 & 1 & 1 & 1 & 1 & 1 & 1 & 1 & 1 & 1 & 1 & 1 \\
2 & 1 & 0 & 2 & 2 & 2 & 2 & 2 & 2 & 2 & 2 & 2 & 2 & 2 & 2 & 2 & 2 \\
inf & inf & inf & inf & 2 & 1 & 0 & 1 & inf & inf & inf & inf & inf & 2 & inf & inf & inf \\
inf & inf & inf & inf & 3 & 2 & 1 & 0 & inf & inf & inf & inf & inf & 3 & inf & inf & inf \\
3 & 2 & 1 & 0 & 3 & 3 & 3 & 3 & 3 & 3 & 3 & 3 & 3 & 3 & 3 & 3 & 3 \\
4 & 3 & 2 & 1 & 4 & 4 & 4 & 4 & 4 & 4 & 4 & 4 & 4 & 4 & 4 & 4 & 4 \\
3 & 2 & 1 & 0 & 3 & 3 & 2 & 1 & 3 & 3 & 3 & 3 & 3 & 3 & 0 & 3 & 3
\end{bmatrix} \quad (10)
$$

3.3. Multiplierless Digital Filters.

The digital FIR filters and the transposed IIR filters will have block of multipliers in the filter structure. This is shown in Figure 9.

For a target set $T = \{t_1, t_1, \ldots, t_n\}$ in digital filter, we have to find the ready set $R = \{r_0, r_1, \ldots, r_m\}$ that is small and *Aoperation* composed of minimum number of addition, subtraction, and shift operations. After this target set is obtained, multiplierless multiple constant multiplication filters can be designed with this target set. Multiple constant multiplication (MCM) is an efficient way of implementing

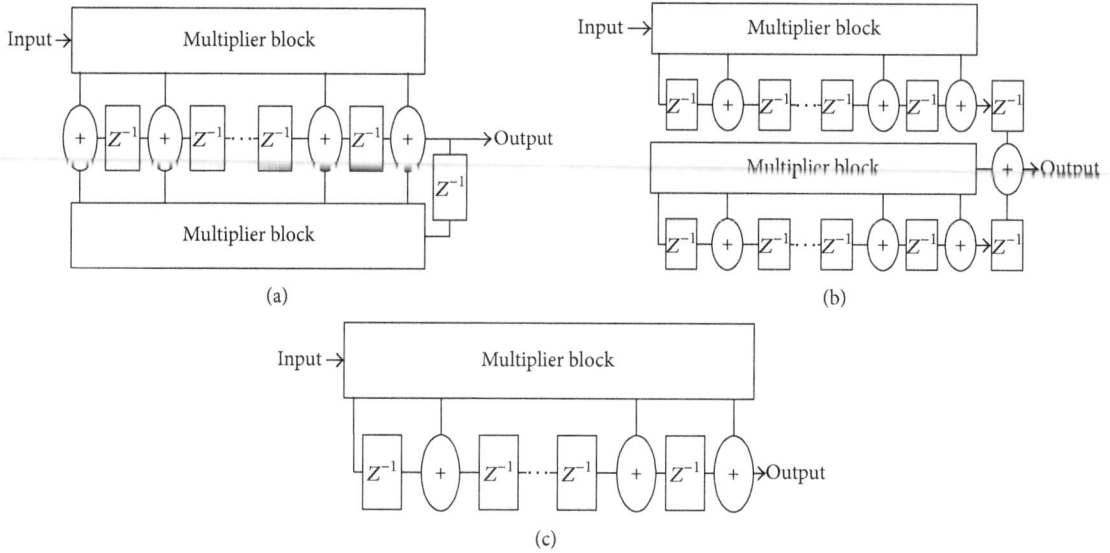

FIGURE 9: General structure of MCM block for (a) FIR filter, (b) transposed direct form-I IIR filter, and (c) transposed direct form-II IIR filter.

several constant multiplications with the input data [18, 19]. The coefficients are implemented using shifts, adders, and subtracters. By removing the redundancy between the coefficients, the number of adders and subtracters is reduced which results in a low complexity implementation. Retiming for multiplierless MCM filters is still unexplored in the literature and authors have combined retiming for multiplierless MCM filters which shows decrease in the combinational path delay. For filter graph G, multiplierless MCM filter can be designed using target set and *Aoperations*, and multiplierless MCM filter graph G_i is obtained. This is again retimed to increase the speed performance of G_i by modifying the critical path of the filter. The graph after retiming of multiplierless MCM filter is considered as G_r. In the present work, H_{cub} algorithm is used for G_i computation. The input to the H_{cub} algorithm is target set T and algorithm computes a ready set R which is the output solution. The R set computation requires multiple iterations and, in each iteration, successor set S of R is chosen as the next fundamental based on the heuristic. Here, S which is set of constants of distance 1 from R is given as

$$S = \{s \mid \text{dist}(R, s) = 1\} = A_s(R, R). \quad (11)$$

For the target set of constants T for the considered filter graph G, using H_{cub} algorithm compute set $R = \{r_1, r_2, \ldots, r_m\}$ with $T \in R$. If the targets are found in the S, then it is optimal synthesis. Here, heuristic function $H(R, S, T)$ of an algorithm can be chosen when no more targets are found in S. This can happen when all the targets are more than one *Aoperation* away. The optimal part is when $(T \cap S \neq \phi)$, then there is a target in the successor set and it can be synthesized. Optimal set is the one in which the entire target is synthesized in this way and the solution is optimal. In heuristic part, the computation can be done by two ways:

(i) maximum benefit,

(ii) cumulative benefit.

To build the heuristic, we can define the benefit function as $B(R, s, t)$:

$$B(R, s, t) = \text{dist}(R, t) - \text{dist}(R + s, t). \quad (12)$$

A successor $s \in S$ needs to be picked which is closest to the target set to minimize the cost. This is possible if we can compute or estimate the A-Distance. It is useful to also take into account the current estimate of the distance between R and T. Thus, to build the heuristic, we must first define the benefit function $B(R, s, t)$ to quantify to what extent adding a successor s to the ready set R improves the distance to a fixed, but arbitrary, target t. However, for remote targets, the estimate becomes less accurate; hence, we can have weighted benefit function given as

$$B_b(R, s, t) = 10^{\text{dist}(R+s,t)} (\text{dist}(R, t) - \text{dist}(R + s, t)), \quad (13)$$

where $10^{\text{dist}(R+s,t)}$ is a weight factor and decreases exponentially as t grows. The benefit function for different targets t can be added and joint optimization can be achieved by using cumulative benefit which is used in the present work. Hence, heuristic function for cumulative benefit is given by

$$H_{cub}(R, S, T) = \arg\left[\max\left[\sum_{t \in T} B_b(R, s, t)\right]\right]. \quad (14)$$

Here, cumulative benefit heuristic adds up the weighted benefit considering all the targets. With this particular method, target set is calculated. With this target set, filter graph which is multiplierless MCM based can be designed. It is found that multiplierless designs reduce the combinational path delays and due to sharing of intermediate results in the MCM approach. The performance can be further improved by retiming G_i to give G_r. These two different optimization techniques reduce the combination delay and critical path

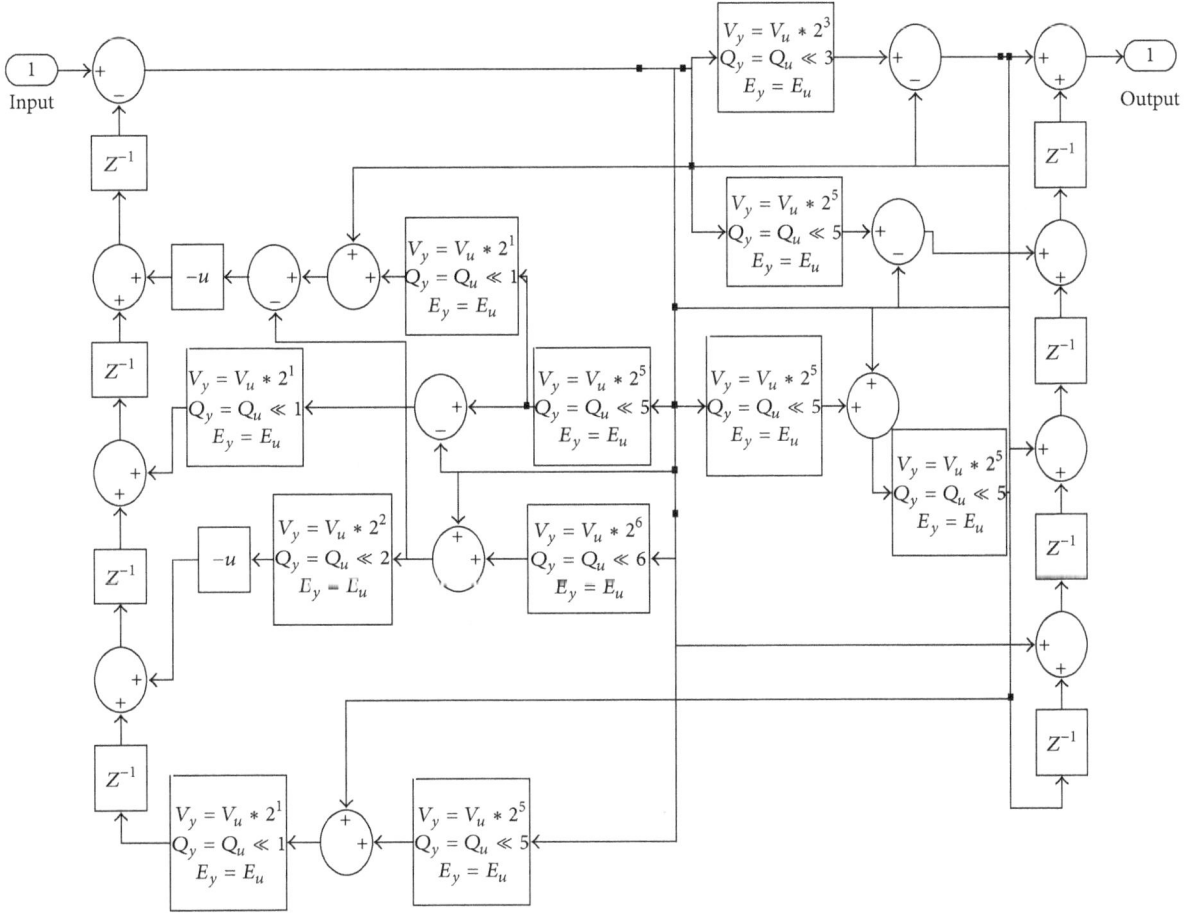

FIGURE 10: Multiplierless MCM based 4th-order elliptic filter.

without changing the functionality which further increases the clock speed. The 4th-order lattice filter with multiplierless MCM concept using H_{cub} algorithm is shown in Figure 10. It is seen from the synthesis that combination delay is reduced. It is further retimed either for clock period minimization or register minimization. This requires solving a set of linear inequalities with a computation complexity of $O(n^3)$ where n is the number of nodes using the Floyd-Warshall algorithm, where n is the number of nodes [8]. The clock period minimization and register minimization retiming algorithms are designed and implemented with FPGA based path solvers which reduces computation time when compared to previous methods [8, 16] to design multiplierless digital filters.

The algorithm starts by building a new graph from the original DFG. The new graph can give us a set of inequalities called the critical path constraints. The original DFG also presents a set of equalities called the feasibility constraints. A constraint graph can be built from the critical path constraints and the feasibility constraints. The retiming values for each node can be derived by applying a Floyd-Warshall shortest path algorithm to the constraint graph. The weight for each edge in the retimed DFG can be calculated using the original weight and the retiming values of the two nodes connected by this edge. The improvement in the clock frequency is shown in Figure 11. Here, 4th-order lattice filter is considered. *Design1* is the filter with multipliers and without retiming, *Design2* is multiplierless MCM based filter without retiming, *Design3* is the filter with multipliers with retiming, and *Design4* is multiplierless MCM based lattice filter with retiming. The maximum operating frequency of the filter has increased by 19.6% in multiplierless MCM approach as multipliers will get eliminated and get replaced by adders which have much less computation delay. Further, it is observed that by combining this approach with retiming, operating frequency increases by 35.4% which is a significant increase. However, with this technique, the number of registers increases from 9 to 11.

Hence, when the filter is designed without multipliers (that is using only adders/subtractors and shifters) along with the retiming technique, operating clock speed is found to increase which gives a greater speed advantage for the design under consideration.

3.4. Computer Aided Design Tool. This section presents the DiFiDOT tool which is designed as the part of research work. Initially, the design of filters is performed using retimed architecture where user can choose either clock period

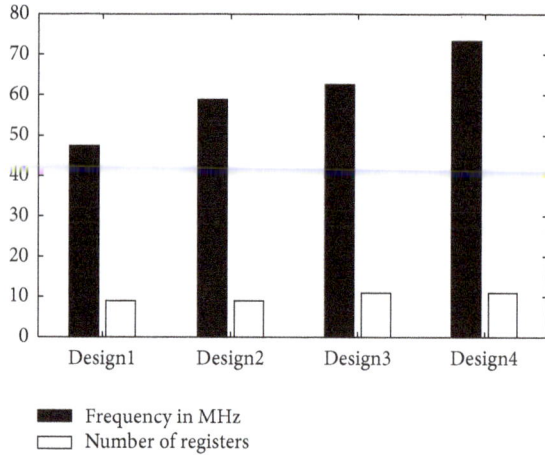

FIGURE 11: Comparison of operating frequency and number of registers for different filter designs of 4th-order elliptic filter.

minimization or register minimization retiming as per his need. The tool will retime the digital filter by optimizing the critical path and generate verilog/VHDL based filter RTL for the same. The performance of a filter can also be increased by varying the choice combinational adder and multiplier elements in the RTL filter description. A graphical user interface (GUI) is created in DiFiDOT using Nokia QT 4.8.0 for component selection and optimization of digital filters. Here, user has to input the HDL file which was automatically generated after retiming for further component optimization. The user can choose adders and multipliers of his choice according to the design requirements for the retimed digital filters using drop down menu. The original HDL is automatically modified with respect to the components chosen which is again synthesizable and is given as the output to the user. This easy to use GUI helps designer to optimize and generate digital filter RTL with the adder and multipliers of his choice. With this, designer can conveniently explore the solution space of possible architectures and also analyze the trade-offs in the energy-area-performance space [20]. The different adder and multipliers considered in the tool are as below.

Multiplier Architecture. The most critical function carried out by any filter is multiplication. Digital multiplication [19] is the most extensively used operation in signal processing. Innumerable schemes have been proposed for realization of the operation. In this paper, we consider three types of multipliers.

Array Multiplier. It is the basic type of multiplier. Consider two binary numbers A and B, of n bits, respectively. The multiplication is given as

$$A = \sum_{i=0}^{n-1} A^i 2^i, \quad B = \sum_{j=0}^{n-1} A^i B^i 2^{i+j}$$

$$P = \sum_{i=0}^{n} \sum_{j=0}^{n-1} A^i B^i 2^{(i+j)} A^i B^i 2^{i+j}. \tag{15}$$

In each stage, the partial products P_i are generated that are added to obtain final product P. In general, for $m * n$ array multiplier, we need $m * n$ AND gates, n half adders, and $(m - 2) * n$ full adders.

Radix 4 Booth Multiplier. It has the advantage of lesser area and faster multiplication compared with array multiplication. Radix 4 Booths Algorithm can scan strings of three bits and is converted depending on modified Booth encoder table. The design of Booths multiplier in this project consists of four Modified Booth Encoders (MBE), four sign extension correctors, four partial product generators (comprises of 5 : 1 multiplexer), and finally a Ripple carry Adder. This Booth multiplier technique is to increase speed by reducing the number of partial products by half. Since a 32-bit booth multiplier is used in this project, there are only sixteen partial products that need to be added instead of 32 partial products generated using conventional multiplier.

Vedic Multiplier. It is used for faster multiplication operations in higher order bits. It has less combinational path delay [21] compared with others when the bit size is higher. However, it consumes more area than Booth multiplier and array multiplier. The multiplier is based on an algorithm Urdhva Tiryakbhyam (vertical & crosswise) Sutra which is a general multiplication formula applicable to all cases of multiplication. It means vertically and crosswise. It is based on a novel concept through which the generation of all partial products can be done with the concurrent addition of these partial products. The speed advantage is compromised with increased power dissipation and area. Due to its regular structure, layout of this can be easily generated.

The different multipliers are designed for different bit sizes and results are compared. This is as shown in Table 1.

3.5. Adders. In this paper, qualitative evaluations of the classified binary adder architectures are performed since adder is another basic component of FIR filter. Here, Ripple-carry adder, BruntKung adder, and Ling adder are considered to emphasize the performance properties. Adders affect the critical path delay and area.

Ripple Adder. It is the basic adder type. This is composed of cascaded full adders for n-bit adder. It is constructed by cascading full adder blocks in series. The carry-out of one stage is fed directly to the carry-in of the next stage. For an n-bit parallel adder, it requires n full adders.

Parallel-Prefix Adders. Parallel prefix adders [22] offer a highly efficient solution to the binary addition problem. Among all the parallel prefix adders, Brunt Kung adder has a good balance between area, power, and performance. It is found that Ling adder using Kogge-Stone parallel prefix adder is also having the advantage of faster addition operation [22], but it consumes more power than Brunt Kung Adder.

TABLE 1: Comparison of multipliers for delay, power, and area.

Type of multiplier	Delay in ns			Power in mW			Number of LUTs		
	32 bit	16 bit	8 bit	32 bit	16 bit	8 bit	32 bit	16 bit	8 bit
Array	76.1	39.9	21	21	11	7	1519	375	91
Booth	86.1	27.99	14.9	25	15	12	1277	317	77
Vedic	70.7	39.02	24.4	28	18	12	2378	565	126

The basic equations used in parallel prefix adders are given below. The equations of bit generate and propagate are

$$G_{0:0} = G_0 = c_{in}$$

$$P_{0:0} = P_0 = 0 \qquad (16)$$

$$G_{i:j} = G_{i:k} + p_{i:k} * g_{(k-1):j} P_{i:j} = P_{i:k} * p_{(k-1):j}.$$

The sum generation is given by

$$S_i = P_i \, \mathrm{XOR} \, G_{(i-1):0}. \qquad (17)$$

Different Adders are designed for different bit sizes and their VLSI design metrics are compared as shown in Table 2. The delay generated is based on the combinational path delay after synthesis. It is measured in *ns*.

In the GUI, an option is crested for particular adder and multiplier combination also depending on whether the performance parameter is speed, power, or area and also based on the bit size. For example, if the design constraint that user chooses is power then Brent-Kung adder and array multiplier pair are considered as the best combination to implement the filter in the design optimization GUI. User can also choose any one of his choice among area, power, or speed constraint for digital filter HDL generation. Along with this, an option is created for multiplierless filter design description as well based on MCM approach. It is seen that the retimed MCM circuits outperform the existing MCM methods [23] in terms of speed. Using this tool, user can design retimed digital filter which has combination elements of his choice which are specific to particular design constraint and generate the RTL for the same. The obtained RTL can be synthesized with any of the commercially available synthesis tools. The GUI designed is shown in Figure 12. A H_{cub} based algorithm is considered for implementing MCM blocks in multiplierless digital filters for specific user defined option in DiFiDOT. Since all the multipliers can be realised as a block in transposed IIR and FIR filters, they are well suited for MCM implementation. After retiming, the multiplier blocks in digital filter can be replaced by a block constructed by adderssubtractors, negation operations, and shifters in multiplierless design approach. The generated MCM block will have tree depth in terms of different components and this depth in our work is assumed to be infinity. The tool DiFiDOT automatically generates the HDL of retimed digital filter which is under consideration which can be directly synthesizable. With this tool and automation, even if reiteration of the design cycle happens due to specification change, time taken to reiterate is very little.

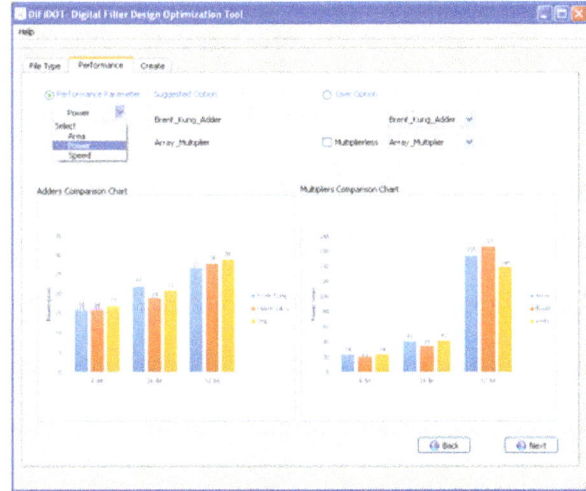

FIGURE 12: GUI for dDesign optimization environment created to generate synthesizable retimed digital filter HDL optimized for VLSI design metrics.

4. Experimental Results

This section is divided in to three parts: the first part presents the results of retiming with FPGA based path solvers, second part presents comparison of various retiming techniques, and third part presents the timing results of retimed filter structures with MCM blocks.

4.1. Results on Path Solvers for Retiming. The main idea of implementing path solver algorithms on FPGA is to speed up the results for retiming purposes. The inputs are passed to the FPGA based path solver block by a processor where retiming algorithm is implemented. The computations are performed in FPGA based block and shortest path along with critical path is computed and communicated back to the processor where retiming will be performed. For comparison, a set of designs is used to test the path solver algorithms. The designs are a diverse set of DSP functions of varying complexity which includes recursive and nonrecursive filter structures. The considered target device for path solver implementation is Spartan6 family based XC6SSLX16. The simulation and synthesis of path solvers are performed using Xilinx ISE tool suit and the synthesis and the timing results after synthesis are shown in Table 1. The FPGA based path solver computes critical path and shortest path and communicates the results to the processor where retiming is performed. This reduces the burden on main processor (Table 3).

TABLE 2: Comparison of adders for delay, power, and area.

Type of adder	Delay in ns			Power in mW			Number of LUTs		
	32 bit	16 bit	8 bit	32 bit	16 bit	8 bit	32 bit	16 bit	8 bit
Ling	8.854	15.24	20.21	6	9	18	23	53	107
BrentKung	10.4	18.39	25.83	4	6	9	15	30	63
Ripple	12.12	20.63	37.6	2	7	14	9	18	36

TABLE 3: Device utilization and timing summary of path solvers.

Path solver name	Device utilization summery		Timing summery			Max. frequency (Hz)
	Logic utilization	Used	Min period in ns	Setup time in ns	Hold time in ns	
Critical path solver	Number of slices	5804	9.068 ns	15.72 ns	6.141 ns	110.277
	Number of LUTs	10462				
	Number of slice Flipops	3664				
Shortest path solver	Number of slices	4147	14.089 ns	10.477 ns	4.114 ns	70.978
	Number of LUTs	7511				
	Number of slice Flipops	1496				

Here, various IIR and FIR filters have been considered to analyze the FPGA based path solvers and execution time of FPGA design is compared with the general purpose processor (GPP) based design. Also, GPP denotes the required CPU time in milliseconds of the path solver to find the minimum solution on a PC with Intel Pentium 5 machine at 2 GHz and 4 GB of memory. FPGA based design solves for critical path and shortest path in very less time when compared to the general purpose processor based path solvers. The time taken by the FPGA path solvers is compared in Table 4 to the time taken by the algorithms run using general purpose processor with Matlab environment. The time overhead needed for general purpose processor where retiming algorithm is implemented in MATLAB to communicate with the FPGA based path solvers is around 210 ns for each computation. Including this, the time gain achieved is quite substantial when compared to designs without FPGA based path solvers. These time gains are good and can really help speed up the results, which is crucial for retiming.

4.2. Comparison of Clock Period Minimization and Register Minimization Retiming Technique. Different filter structures are designed and they are compared with respect to the clock period and register count before and after retiming. It is observed that after retiming, the clock period gets reduced. The register count gets altered depending on the filters iteration bound. Here, three models are considered. *Model*1 is the filter without retiming and with adder, subtractor, multiplier, and delay elements. *Model*2 is retimed filter based on clock period minimization algorithm. *Model*3 is retimed filter based on register minimization algorithm. After retiming, the results are compared with the original circuit [24]. The comparison results are shown in Figure 13. After retiming, the finite state machine is extracted from the retimed circuit and it is compared with original circuit for its functionality. It is observed that clock period minimization retiming algorithm is efficient in terms of reduction critical path, thereby increase in the clock frequency. However, this

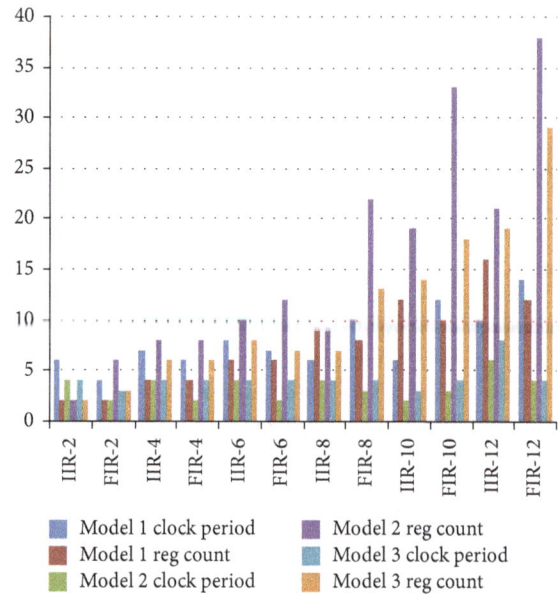

FIGURE 13: Clock period and register count before and after retiming for various digital filter blocks.

might increase the register count. In register minimization retiming [18], the number of registers after retiming will be reduced while compromising the clock period.

4.3. Area, Power, and Timing Results for Digital Filter before and after Retiming for Different Adder and Multiplier Combinations. The FIR and IIR filters are designed with respect to different adders and multipliers combinations. As an application example, IIR and FIR filters [25] of order 10 are considered. Table 5 shows the results of FIR/IIR filters before and after retiming for particular adder and multiplier combinations. User can choose any adder and multiplier for the filter circuit depending on the design requirement. In

TABLE 4: Computation time comparison.

Filter order	Critical path solver algorithm				Shortest path solver algorithm			
	IIR filter		FIR filter		IIR filter		FIR filter	
	FPGA based (ns)	GPP based (ms)	FPGA based (ns)	GPP based (ms)	FPGA based (ns)	GPP based (ms)	FPGA based (ns)	GPP based (ms)
2	4.60	13.8	9.06	12.83	2.78	3.05	3.05	12.80
4	15.71	15.78	16.31	14.46	3.68	13.91	13.91	13.19
6	29.98	19.18	19.23	15.47	3.98	15.42	15.42	17.31
8	31.62	21.90	29.71	16.42	4.52	25.23	25.23	32.94
10	39.81	26.27	36.53	18.61	5.36	42.93	42.93	45.34
12	46.72	31.42	43.28	23.52	6.71	55.34	55.34	51.61

TABLE 5: Comparison results of different adder/multiplier combinations for digital filters.

Filter block	Adder/multiplier combinations	Before retiming			After retiming		
		Number of LUTs	Max. operating freq in MHz	Power in mw	Number of LUTs	Max. operating freq in MHz	Power in mw
IIR-10	Brentkung Adder/Array Multiplier	2222	62.526	99	2411	76.977	89
	Ling Adder/Vedic Multiplier	2214	69.702	112	2193	95.381	94
	Ripple carry Adder/Booth Multiplier	2146	50.861	114	1809	65.248	95
FIR-10	Brentkung Adder/Array Multiplier	1736	62.526	94	1811	99.43	85
	Ling Adder/Vedic Multiplier	2162	72.493	111	2271	100.72	95
	Ripple carry Adder/Booth Multiplier	1637	52.302	105	1615	71.345	87

the GUI, particular adder and multiplier combination is considered depending on whether the performance parameter is delay, power, or area and also based on the bit size. If user does not want to use these in built combinations, user can choose any one of his choice among the available for FIR/IIR digital filter HDL generation with specific combinational components.

4.4. Results for Optimization of Latency, Multiplier Components, and Power in Multiplierless Multiple Constant Multiplication Based Filter Designs Using Retiming Algorithm. Table 6 presents the results of the filters designed using multiplierless MCM approach and optimization using retiming algorithm. Here, 3 models are used.

(i) *Model* 1: Filter with adder, multiplier, and delay elements.

(ii) *Model* 2: Filter based on multiplierless multiple constant multiplication approach.

(iii) *Model* 3: Retimed multiplierless multiple constant multiplication based filter.

All the three models are compared for the performance parameters such as area, power, and delay. Here, it is ensured that functionality of the circuits after and before retiming is retained. The frequency improvement seen for different filters by considering the above models is given in Figure 14. It is seen that frequency parameter is improved when retiming technique is applied for multiplierless MCM based digital filters.

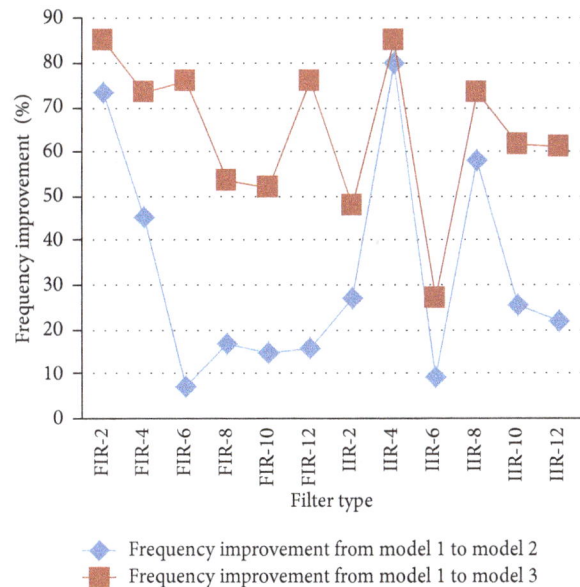

FIGURE 14: Frequency improvement in % factor.

5. Application Example

The electrocardiogram (ECG) is the most commonly used diagnostic method for heart diseases. Good quality ECG is utilized by physicians for interpretation and identification of physiological and pathological phenomena. ECG recordings

TABLE 6: Comparison of area, delay, and power for different models of various digital filters.

Filter block	Adder multipliers Flipflops			DelayMax Freq in MHz			Power in Watts		
	Model 1	Model 2	Model 3	Model 1	Model 2	Model 3	Model 1	Model 2	Model 3
FIR-2	5/2/3	5/0/3	5/0/4	51.54	192.14	340.62	0.056	0.063	0.065
FIR-4	10/3/5	11/0/5	11/0/8	59.41	108.41	222.04	0.047	0.057	0.060
FIR-6	7/2/7	17/0/7	17/0/14	62.91	67.64	259.47	0.051	0.062	0.064
FIR-8	15/5/9	22/0/9	22/0/16	54.82	65.92	117.91	0.054	0.058	0.065
FIR-10	18/6/11	25/0/11	25/0/11	48.22	56.37	100.72	0.058	0.061	0.063
FIR-12	20/7/13	29/0/13	29/0/13	46.34	54.86	193.40	0.060	0.063	0.067
IIR-2	9/4/3	11/0/3	11/0/3	55.03	75.53	89.10	0.047	0.050	0.050
IIR-4	16/7/5	20/0/5	19/0/6	22.78	113.88	151.65	0.059	0.062	0.063
IIR-6	24/11/7	35/0/7	35/0/8	38.71	42.54	53.142	0.051	0.059	0.058
IIR-8	30/11/10	33/0/10	33/0/7	29.46	70.14	110.21	0.044	0.064	0.081
IIR-10	37/16/13	54/0/13	54/0/14	36.43	48.85	95.381	0.051	0.067	0.085
IIR-12	42/20/17	63/0/17	63/0/19	39.73	50.74	101.52	0.063	0.071	0.088

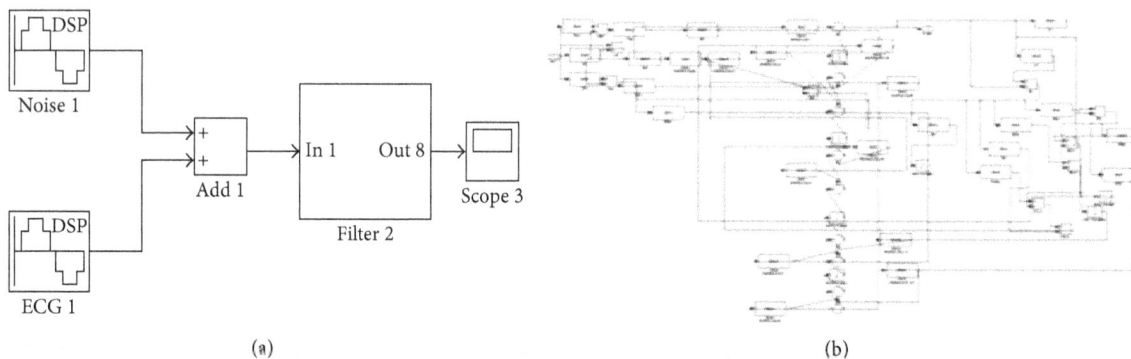

FIGURE 15: Structure of ECG block for power noise removal; (a) block diagram; (b) filter block expanded.

are often corrupted by high-frequency noises, such as power-line interference, electromyography (EMG) noise, and instrumentation noise. An ECG is usually affected by the 50/60 Hz noise in the power supply lines. This noise can be eliminated by using a digital filter. The model is constructed in matlab and tested for ECG signals for removing the noise. The constructed model uses retimed multiplierless MCM filter which is implemented on FPGA and tested for ECG signal which is corrupted by power-line noise. The filter efficiently filters out the noise and outputs the clean ECG signal. The ECG noise removal block using the optimized filter structure is shown in Figure 15.

6. Conclusions

In this paper, we introduced the retiming approach for designing multiplierless MCM based digital filters with speed and area as the constraint. The implementation cost at the gate level is reduced by using addition, subtraction, and shift operations instead of multiplication and by using register sharing and register minimization retiming algorithm approach. Since there are still instances with which multiplierless designs can not cope, we also proposed

the combination of adder and multiplier blocks which can be used in retimed filter design which is applicable for specific VLSI design constraint such as power, area, and timing. This yields the optimal clock speed and gate-level area in design and implementation of digital filters. This paper also introduced the design architectures for the digital filter and a CAD tool for the realization of retimed digital filters which can be either multiplierless MCM based or with adder/subtractor, multiplier, and delay elements. This tool directly gives the synthesizable filter RTL which reduces lot of designers' time and effort in the design cycle. The experimental results indicate that the retiming algorithm efficiency can be further increased by using FPGA based path solver algorithms proposed in this paper. It was shown that the realization of path solver architectures for solving critical path and shortest path in retiming computation and communicating the results to the processor where retiming algorithm is implemented yields significant increase in computation time gain when compared to the filter designs for which path solver algorithms are implemented as a part of retiming algorithm in the processor. It is observed that a designer can find the synthesizable digital filter RTL that fits best in an application.

Conflict of Interests

The authors declare that there is no conflict of interests regarding the publication of this paper.

References

[1] C. Soviani, O. Tardieu, and S. A. Edwards, "Optimizing sequential cycles through shannon decomposition and retiming," *IEEE Transactions on Computer-Aided Design of Integrated Circuits and Systems*, vol. 26, no. 3, pp. 456–467, 2007.

[2] S. Bommu, N. O'Neill, and M. Ciesielski, "Retiming-based factorization for sequential logic optimization," *ACM Transactions on Design Automation of Electronic Systems*, vol. 5, no. 3, pp. 373–398, 2000.

[3] K. K. Parhi, "A systematic approach for design of digit-serial signal processing architectures," *IEEE Transactions on Circuits and Systems*, vol. 38, no. 4, pp. 358–375, 1991.

[4] D. Yagain, A. V. Krishna, and S. Chennapnoor, "Design optimization platform for synthesizable high speed digital filters using retiming technique," in *Proceedings of the 10th IEEE International Conference on Semiconductor Electronics (ICSE '12)*, pp. 551–555, Kuala Lumpur, Malaysia, September 2012.

[5] N. Shenoy, "Retiming: theory and practice," *Integration, the VLSI Journal*, vol. 22, no. 1-2, pp. 1–21, 1997.

[6] C. E. Leiserson and J. B. Saxe, "Retiming synchronous circuitry," *Algorithmica*, vol. 6, no. 1–6, pp. 5–35, 1991.

[7] Y. Tsao and K. Choi, "Area-efficient VLSI implementation for parallel linear-phase FIR digital filters of odd length based on fast FIR algorithm," *IEEE Transactions on Circuits and Systems II: Express Briefs*, vol. 59, no. 6, pp. 371–375, 2012.

[8] K. K. Parhi, *VLSI Digital Signal Processing Systems: Design and Implementation*, John Wiley & Sons, 2007.

[9] K. K. Parhi, "Hierarchical folding and synthesis of iterative data flow graphs," *IEEE Transactions on Circuits and Systems II: Express Briefs*, vol. 60, no. 9, pp. 597–601, 2013.

[10] X. Zhu, T. Basten, M. Geilen, and S. Stuijk, "Efficient retiming of multirate DSP algorithms," *IEEE Transactions on Computer-Aided Design of Integrated Circuits and Systems*, vol. 31, no. 6, pp. 831–844, 2012.

[11] N. Liveris, C. Lin, J. Wang, H. Zhou, and P. Banerjee, "Retiming for synchronous data flow graphs," in *Proceedings of the Asia and South Pacific Design Automation Conference (ASP-DAC '07)*, vol. 7, pp. 480–485, Yokohama, Japan, January 2007.

[12] N. L. Passos, E. H. Sha, and S. C. Bass, "Optimizing DSP flow graphs via schedule-based multidimensional retiming," *IEEE Transactions on Signal Processing*, vol. 44, no. 1, pp. 150–155, 1996.

[13] J. R. Jiang and R. K. Brayton, "Retiming and resynthesis: a complexity perspective," *IEEE Transactions on Computer-Aided Design of Integrated Circuits and Systems*, vol. 25, no. 12, pp. 2674–2686, 2006.

[14] N. Maheshwari and S. Sapatnekar, "Efficient retiming of large circuits," *IEEE Transactions on Very Large Scale Integration (VLSI) Systems*, vol. 6, no. 1, pp. 74–83, 1998.

[15] D. Yagain and A. Vijaya Krishna, "High speed digital filter design using register minimization retiming & parallel prefix adders," in *Proceedings of the 3rd International Conference on Emerging Applications of Information Technology (EAIT '12)*, pp. 449–453, Kolkata, India, December 2012.

[16] J. Cong and C. Wu, "An efficient algorithm for performance-optimal FPGA technology mapping with retiming," *IEEE Transactions on Computer-Aided Design of Integrated Circuits and Systems*, vol. 17, no. 9, pp. 738–748, 1998.

[17] D. Yagain, A. Vijayakrishna, P. Nikhil, A. Adarsh, and S. Karthikeyan, "FPGA based path solvers for DFGs in high level synthesis," in *Proceedings of the 2nd International Conference on Advances in Computational Tools for Engineering Applications (ACTEA '12)*, pp. 273–278, IEEE, Beirut, Lebanon, December 2012.

[18] Y. Voronenko and M. Püschel, "Multiplierless multiple constant multiplication," *ACM Transactions on Algorithms*, vol. 3, no. 2, article 11, Article ID 1240234, 2007.

[19] K. Johansson, O. Gustafsson, and L. Wanhammar, "Multiple constant multiplication for digit-serial implementation of low power FIR filters," *WSEAS Transactions on Circuits and Systems*, vol. 5, no. 7, pp. 1001–1008, 2006.

[20] A. Baliga, "Design of high-speed adders for efficient digital design blocks," *ISRN Electronics*, vol. 2012, Article ID 253742, 9 pages, 2012.

[21] H. D. Tiwari, G. Gankhuyag, C. M. Kim, and Y. B. Cho, "Multiplier design based on ancient indian vedic mathematics," in *Proceedings of the International SoC Design Conference (ISOCC '08)*, vol. 2, pp. II65–II68, Busan, Republic of Korea, November 2008.

[22] G. Dimitrakopoulos and D. Nikolos, "High-speed parallel-prefix VLSI ling adders," *IEEE Transactions on Computers*, vol. 54, no. 2, pp. 225–231, 2005.

[23] L. Aksoy, E. da Costa, P. Flores, and J. Monteiro, "Exact and approximate algorithms for the optimization of area and delay in multiple constant multiplications," *IEEE Transactions on Computer-Aided Design of Integrated Circuits and Systems*, vol. 27, no. 6, pp. 1013–1026, 2008.

[24] M. N. Mneimneh, K. A. Sakallah, and J. Moondanos, "Preserving synchronizing sequences of sequential circuits after retiming," in *Proceedings of the Asia and South Pacific Design Automation Conference*, pp. 579–584, IEEE Press, 2004.

[25] D. Yagain and K. A. Vijaya, "Fir filter design based on retiming and automation using vlsi design metrics," in *Proceedings of the International Conference on Technology, Informatics, Management, Engineering, and Environment (TIME-E '13)*, pp. 17–22, IEEE, 2013.

## 5

A Modularized Noise Analysis Method with Its Application in Readout Circuit Design

Xiao Wang, [1,2,3] **Zelin Shi,** [1,3] **and Baoshu Xu** [1,3]

[1] *Shenyang Institute of Automation, Chinese Academy of Sciences, Shenyang 110016, China*
[2] *University of the Chinese Academy of Sciences, Beijing 100049, China*
[3] *Key Laboratory of Opto-Electronic Information Processing, Chinese Academy of Sciences, Shenyang 110016, China*

Correspondence should be addressed to Xiao Wang; wangxiao@sia.cn

Academic Editor: Chang-Ho Lee

A readout integrated circuit (ROIC) is a crucial part that determines the quality of imaging. In order to analyze the noise of a ROIC with distinct illustration of each noise source transferring, a modularized noise analysis method is proposed whose application is applied for a ROIC cell, where all the MOSFETs are optimized in subthreshold region, leading to the power dissipation 2.8 μW. The modularized noise analysis begins with the noise model built using transfer functions and afterwards presents the transfer process of noise in the form of matrix, through which we can describe the contribution of each noise source to the whole output noise clearly, besides optimizing the values of key components. The optimal noise performance is obtained under the limitation of layout area less than 30 μm × 30 μm, resulting in that the integration capacitor should be selected as 0.74 pF to achieve an optimal noise performance, the whole output noise reaching the minimum value at 74.1 μV. In the end transient simulations utilizing Verilog-A are carried out for comparisons. The results showing good agreement verify the feasibility of the method presented through matrix.

1. Introduction

An infrared detector has a wide range of applications in areas of military, research, and manufacture, whose core part is an infrared focal plane assembly [1, 2]. The assembly mainly consists of two parts, a focal plane array (FPA) that functions to convert radiation to current signal and a readout integrated circuit (ROIC) that is responsible for realization of serial processing of signals sampled from the FPA [3].

As an essential part of the signal chain, the research on integrators and correlated double sampler (CDS) has been highlighted since they were invented, alongside with the character of noise transferring from end to end, which is affected by various factors like amplifying, sampling, filtering, and so on. An integrator conducts current signals through an integration capacitor to be transformed into voltages, which inevitably induces strong reset noise. So a CDS has to be used to suppress it, which also subtracts the reference voltage from the output to present a net integration voltage.

Applications of FPAs with large format and high resolution put forward more harsh demands on the small layout area. In order to give an optimization for noise performance of a ROIC, a modularized noise analysis method is proposed for a readout cell, consisting of an integrator and a new type of correlated double sampler (CDS), both of which are switched capacitor circuits, where all the MOSFETs are optimized in subthreshold region to be acceptable to the power limitation of large format FPAs [4, 5]. The modularized noise analysis begins with the architecture of the readout cell that is for pixel level ROICs, operating in snapshot mode. Afterwards the noise model is built using transfer functions and the method of presenting the transfer process of noise in the form of matrix is given, through which we can describe the contribution of each noise source to the whole output noise distinctly. In the end, calculations of noise for the proposed readout cell are carried out by the transfer function method and transient simulations utilizing Verilog-A, respectively. The results due to the two methods show good agreement.

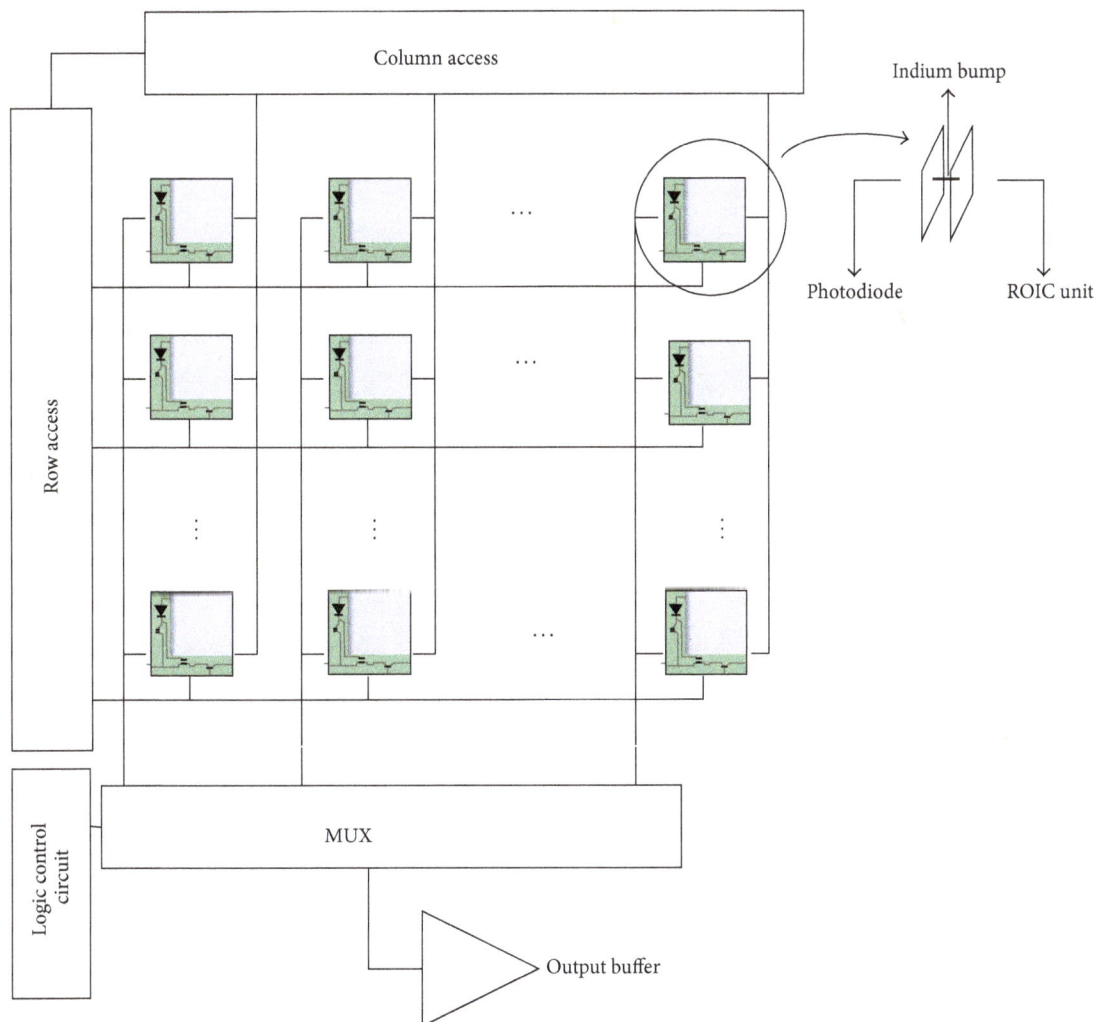

FIGURE 1: Architecture of snapshot ROIC.

2. Architecture of the ROIC

Figure 1 depicts the architecture of the ROIC, whose readout timing operates on pixel level. Snapshot mode is employed. Each pixel corresponds to each readout unit cell, which are connected electrically through indium bump [6]. All the pixels are exposed to radiation simultaneously and transfer the generated photocurrent to the corresponding readout unit cells. Utilizing the architecture provides the capability of expanding integration time, which can adapt to the situation of low intensity radiation [7]. After an integration period ends, with the control of digital logical signals, the integration voltage maintained in each readout unit cell is transmitted serially through a multiplexer and then delivered outside the ROIC through a buffer.

3. Design of the Readout Unit Cell

The structure of the proposed readout unit cell is shown in Figure 2, constituted by two parts, an integrator and a CDS. The integrator adopts the CTIA structure, which provides

nearly 100% injection efficiency and linearity. Meanwhile the CDS is composed of three OPs, A_2, A_3, and A_4, connected as buffers, two sampling capacitors, and four complementary switches, S_1, S_2, S_3, and S_4. From those A_2 and A_3 act to isolate switching operations between the former and latter circuit part.

Herein the structure we proposed is compared with four different ROIC unit cells in [8–11]. The added CDS part provides the capability of suppressing low frequency noise from reset operation and photocurrent for the proposed ROIC one which in aspect of noise performance surpasses those referred to in [8–10] without CDS circuits. The linearity of the BDI structure in [11] can reach 99% which is high enough for the type of structure while for the CTIA type like the one we proposed and these in [8–10] it is easy to achieve more than 99.5% linearity; besides, the power dissipation of the two types makes no difference if the static power of the OP involved is the same. Thus how to choose the most suitable OP and optimize the power dissipation of the OPs involved is another crucial problem.

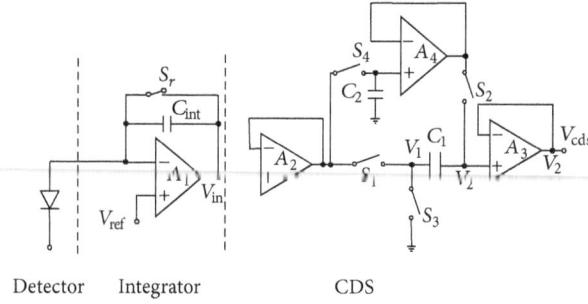

FIGURE 2: Structure of the proposed readout unit cell.

TABLE 1: Performance comparison.

	Gain	Output range	Speed	Power	Noise
Cascade OP	Medium	Medium	High	Low	Low
Folded cascade OP	Medium	Medium	High	Medium	Medium
Two-stage OP	High	High	Low	Medium	Low
Gain boost OP	High	Medium	Medium	High	Medium

Several types of structure of OP are able to be used for the proposed unit cell, such as cascade OP, folded cascade OP, two-stage OP, and gain boost OP. Table 1 gives the comparison among them [12]. From Table 1 we see that, except for the speed disadvantage, the two-stage OP is the best choice, as shown in Figure 3, which just maintains medium power obtaining high gain which can guarantee low static error. On account of the application of the proposed ROIC unit for infrared image acquisition, the clock operations are not in a high frequency, so the speed of the OPs is easy to achieve in either of the OP types.

Introduction of operational amplifiers (OPs) which consumes more power will sacrifice the performance of power dissipation of circuits inevitably. Especially for the structure of the ROIC unit cell we proposed, four OPs are needed for an individual unit cell. To overcome the shortcoming, subthreshold technology is applied [4, 5, 13]. Subthreshold technology is operating transistors in subthreshold region by providing gate-to-source voltage lower than threshold voltage ($V_{gs} < V_{th}$). Ideally, when V_{gs} is lower than V_{th}, the channel between source and drain is shut down. Nevertheless, some electrons still flow across the two ports, known as subthreshold current. Research demonstrates that the subthreshold current is increasing exponentially with the V_{gs} increase, like the current in BJT. The relationship can be expressed as

$$I_{sub} = I_0 e^{(V_{gs}-V_{th})/nV_T} \left(1 - e^{-V_{ds}/V_T}\right), \tag{1}$$

where I_0 is the drain current when $V_{gs} = V_{th}$, V_T is the thermal voltage, and V_{ds} represents the drain-to-source voltage. MOSFETs operating in subthreshold region have

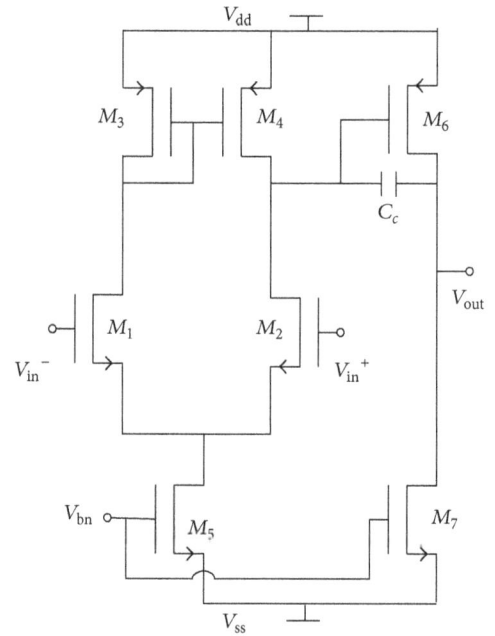

FIGURE 3: Structure of the OPs used in the proposed ROIC.

larger gm-to-channel current ratio than those in saturation region, which implies that subthreshold technology can be applied to optimization power dissipation of analog ICs with the guarantee for sufficient gain.

It is worth noting that the behavioural model of MOSFETs in subthreshold region is not accurate enough when the process of ICs goes into deep submicron, like 0.18 μm. To do the calculation precisely, all the parameters adopted should be those obtained through simulations. Figure 4 shows the qualitative illustration of the real gm-to-channel current ratio compared with the theoretical one, (a) for NMOS and (b) for PMOS, from which it can be seen that although the ratio does not increase as fast as the theoretical curve expects when the $V_{gs} - V_{th}$ goes down, it is much larger in the subthreshold region than that in the saturation region. That is what we desire to improve efficiency of utilizing current.

Figure 5 shows the timing diagram of the readout unit cell. The switches turn on and off when the controlling signals are high and low, respectively. At "t_1" S_r and S_4 turn on, the integrator resets, and the voltage of "V_{in}" node is set to the reference voltage V_{ref}, which is sampled at the positive input

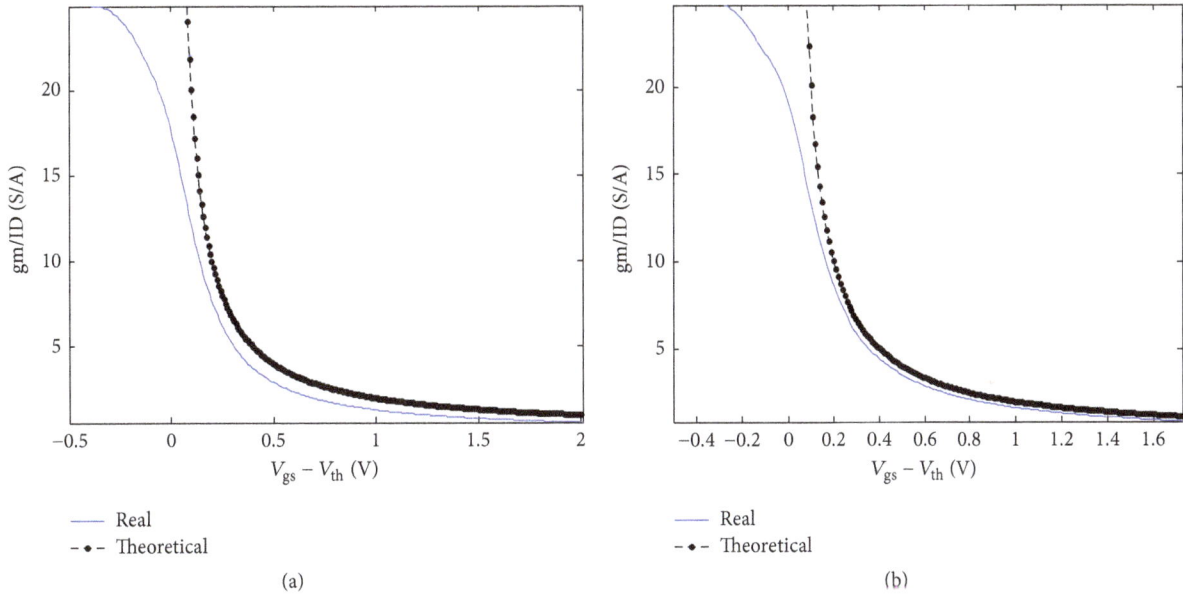

Figure 4: gm-to-channel current ratio.

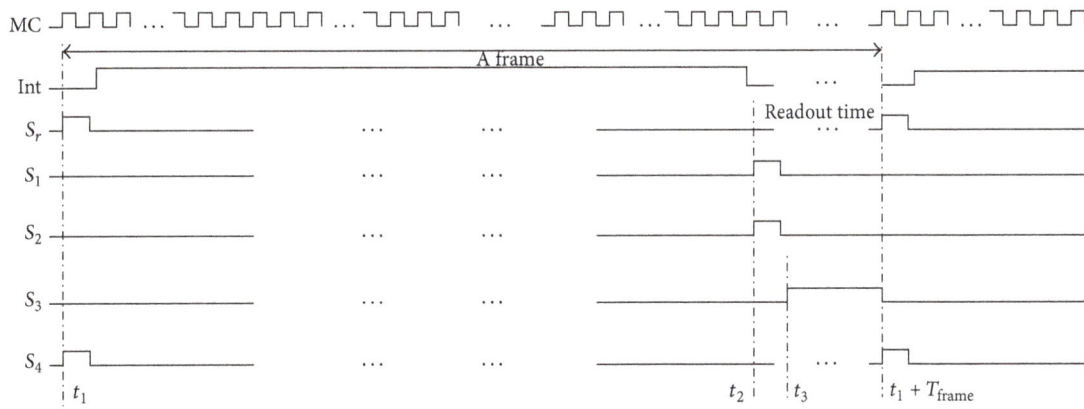

Figure 5: Timing diagram of the proposed readout unit cell.

port of A_4 by C_2 immediately. After the integration ends, the voltage of "V_{in}" node is given by

$$V_{in} = V_{ref} - \frac{I_L T_{int}}{C_{int}}, \tag{2}$$

where I_L is the photocurrent and C_{int} and T_{int} are integration capacitor and integration time, respectively. At "t_2" S_1 and S_2 turn on and the voltages of the two nodes "V_1" and "V_2" across C_1 become the V_{in} and V_{ref}, respectively. At "t_3" S_3 turns on and "V_1" node connects to ground. Based on the law of conservation of charge, the voltage of "V_2" node turns into

$$V_2 = V_{ref} - V_{in} = \frac{I_L T_{int}}{C_{int}}. \tag{3}$$

The output node "V_{cds}" should be equal to "V_2" under ideal condition, thus giving the net integration voltage and subtracting the charge injection error brought by the reset process of integrator. However the transmission of signal will

be influenced by noise, which is analyzed in the next section in detail.

The transient response of the readout unit cell is depicted in Figure 6. We can see that the reset voltage equal to the reference voltage is 2.5 V and the "V_{in}" node becomes 1.5 V from 2.5 V through integration. At "t_3" S_3 turns on, the "V_1" node connects to ground, and "V_2" sets to 1 V, which drives the output voltage to 1 V, the net integration voltage. Besides, the static power dissipation of the proposed readout unit cell reaches 2.8 μW with 700 nW per OP.

4. Noise Model Based on Transfer Function

4.1. The Integrator Part. The noise model of the integrator part is shown in Figure 7. There are three noise sources: photocurrent noise caused by photoelectric effect which cannot be avoided [14], the noise caused by the reset process of switch S_r, and the noise of the OP presenting as the input reference noise source.

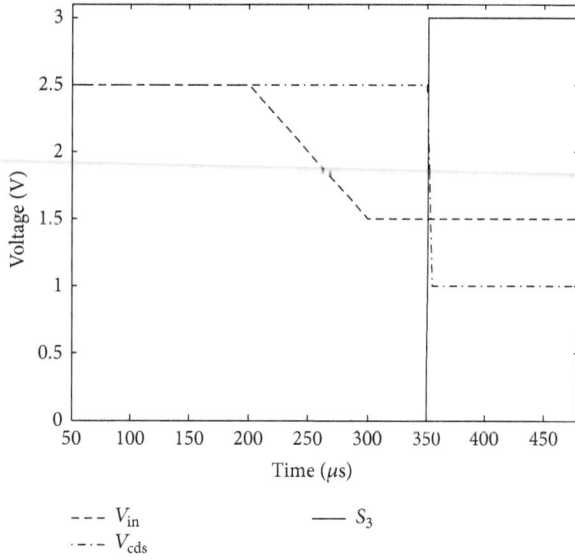

FIGURE 6: Signal transfer process.

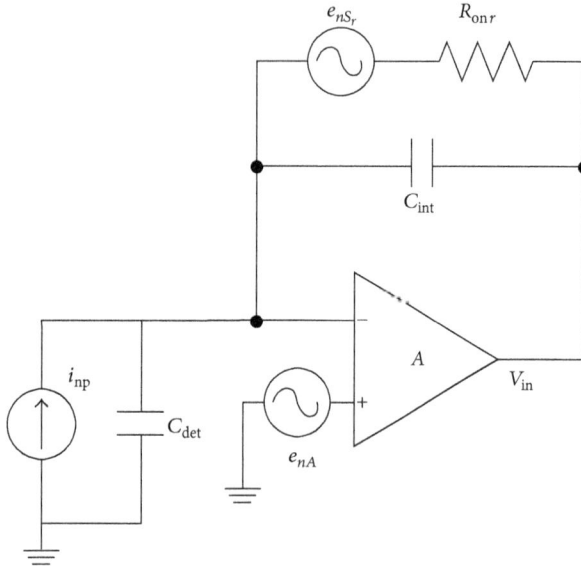

FIGURE 7: Noise model of integrator part.

For CTIA ROIC, photocurrent and its noise integrate on C_{int} directly, which can be expressed as

$$V_{out}\left(n\left(T + T_{int}\right)\right) = \frac{1}{C_{int}} \int_{nT}^{n(T+T_{int})} i\left(t\right) dt, \qquad (4)$$

where T represents the time length of a frame and (3) can be comprehended as $i(t)$ convolving a window function with T_{int} window length [15]. So in frequency domain the process can be expressed as

$$V_{out}\left(f\right) = \frac{1}{C_{int}} i\left(f\right) H_{square}\left(f\right)$$
$$= \frac{1}{C_{int}} i\left(f\right)\left(T_{int}\mathrm{sinc}\left(\pi f T_{int}\right)\right) = i\left(f\right) H_i\left(f\right), \qquad (5)$$

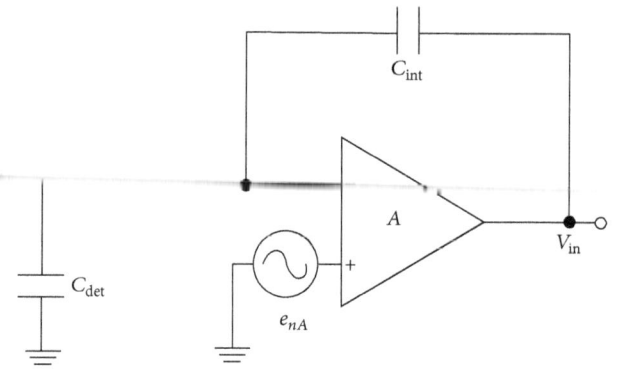

FIGURE 8: Noise model of OP.

where $H_i(f)$ is the transfer function of photocurrent noise source to the "V_{in}" node and the contribution to the output noise of the integrator part is given in the form of power spectrum density (PSD) by

$$e_{no1,inp}^2 = H_i^2\left(f\right) i_{np}^2, \qquad (6)$$

where $e_{no1,inp}^2$ represents the PSD of noise voltage on "V_{in}" node induced by i_{inp}^2, that is, the PSD of photocurrent noise from the corresponding pixel. The transfer function of the OP noise source enA, denoted by $H_A(f)$, can be derived by the model depicted in Figure 8:

$$H_A\left(f\right) = \frac{C_{int} + C_{det}}{C_{int} + \left(C_{int} + C_{det}\right)/A\left(f\right)}, \qquad (7)$$

where C_{det} is the shunt capacitor of the pixel and $A(f)$ is the open loop transfer function of the OP, which can be expressed as [11]

$$A\left(f\right) = \frac{A_0}{1 + jfA_0/\mathrm{GBW}}, \qquad (8)$$

where A_0 and GBW are the low frequency gain and gain bandwidth product of the OP.

We can get the transfer function of enSr by the same approach, which is given by

$$H_R\left(f\right) = \frac{1}{1 + jfR_{off}C_{int}}, \qquad (9)$$

where R_{off} is the off-resistance of the switch S_r. The output noise of integrator part can be expressed in matrix, which is given by

$$O_1 = \begin{bmatrix} e_{no1,inp}^2 \\ e_{no1,A}^2 \\ e_{no1,S_r}^2 \end{bmatrix}$$

$$= \begin{bmatrix} H_i^2\left(f\right), & 0, & 0 \\ 0, & H_A^2\left(f\right), & 0 \\ 0, & 0, & H_R^2\left(f\right) \end{bmatrix} \begin{bmatrix} i_{np}^2 \\ e_{nA}^2 \\ e_{nS_r}^2 \end{bmatrix}. \qquad (10)$$

4.2. CDS Part. The noise model of the CDS is shown in Figure 9, where all the OPs are connected as unity-gain buffers. Because of the noninverting structure, their output resistance is reduced by A_0 times which can be ignored based on the Thevenin equivalent.

When the integrator resets, the voltage on "V_{in}" node is sampled on C_2; after the integration, the voltage on "V_{in}" node is transferred to the "V_1" node at the moment S_1 and S_2 turn on. The voltage across C_1 is

$$V_{C_1}(nT) = V_2(nT) - V_1(nT). \tag{11}$$

For Δ time after S_1 and S_2 turn off and S_3 turns on, the voltage on "V_2" is

$$V_2(n(T+\Delta)) = V_{C_1}(n(T+\Delta)) - V_1(n(T+\Delta)). \tag{12}$$

The expression in frequency domain is given by

$$v_{n2}^2 = \int e_{n2}^2 df = \int \left(e_{nC_1}^2 + e_{n1,\Delta}^2 \right) df, \tag{13}$$

where e_{ncds}^2, $e_{nC_1}^2$, and $e_{n1,\Delta}^2$ stand for the PSD of noise voltage on "V_2," across C_1 at "nT" and on "V_1" at "$n(T+\Delta)$." So the latter two can be expressed as follows:

$$e_{n,C_1} = \left[H_{int}(f), H_{S_r}(f), H_{A_2}(f), H_{S_1}(f), H_{A_4}(f), \right.$$

$$\left. H_{S_2}(f) \right]
\begin{bmatrix}
e_{nint} \\
e_{on1,S_r} H_D(f) \\
e_{nA_2} \\
e_{nS_1} \\
e_{nA_4} \\
e_{nS_2}
\end{bmatrix} \tag{14}$$

$$e_{n1,\Delta}^2 = e_{nS_3}^2, \tag{15}$$

where $H_X(f)$ is the transfer function of the noise source X to the nodes across C_1, respectively. e_{nint} is the noise of "V_{in}" node at the moment an integration ends which is equal to $e_{on1,A} + e_{on1,inp} + e_{on1,S_r}$. $H_D(f) = \exp(-j\omega T_{int})$ represents the reset process that is before the integration starts [15]. Consider

$$e_{n,C_1}^2 = \left[H_{int}^2(f), H_{int}^2(f), \right.$$

$$\left| H_{int}(f) + H_{S_r}(f) H_D(f) \right|^2, H_{A_2}^2(f), H_{S_1}^2(f),$$

$$\left. H_{A_4}^2(f), H_{S_2}^2(f) \right]
\begin{bmatrix}
e_{on1,inp}^2 \\
e_{on1,A}^2 \\
e_{on1,S_r}^2 \\
e_{nA_2}^2 \\
e_{nS_1}^2 \\
e_{nA_4}^2 \\
e_{nS_2}^2
\end{bmatrix}. \tag{16}$$

$n(T+\Delta)$ marks the end of one period of correlated double sampling; at that time the noise on "V_2" and the noise of A_3 are transferred through A_3 with a transfer function denoted by $A_{UG}(f)$ to the "V_{cds}" node:

$$v_{on}^2 = \int e_{on}^2 df = \int \left(e_{n2}^2 + e_{nA_3}^2 \right) \left| A_{UG}^2(f) \right| df, \tag{17}$$

$$A_{UG}(f) = \frac{1}{1 + jf/\text{GBW}}. \tag{18}$$

The PSD of the noise of the output node "V_{cds}" of CDS can be written as the following equation:

$$O_2 =
\begin{bmatrix}
e_{on2,inp}^2 \\
e_{on2,A}^2 \\
e_{on2,S_r}^2 \\
e_{on2,A_2}^2 \\
e_{on2,S_1}^2 \\
e_{on2,A_4}^2 \\
e_{on2,S_2}^2 \\
e_{on2,S_3}^2 \\
e_{on2,A_3}^2
\end{bmatrix} = H_2 E_{cds} = H_2
\begin{bmatrix}
e_{on1,inp}^2 \\
e_{on1,A}^2 \\
e_{on1,S_r}^2 \\
e_{nA_2}^2 \\
e_{nS_1}^2 \\
e_{nA_4}^2 \\
e_{nS_2}^2 \\
e_{nS_3}^2 \\
e_{nA_3}^2
\end{bmatrix}, \tag{19}$$

where H_2 is a diagonal matrix whose diagonal line corresponds to the transfer functions of each noise element in E_{cds}. It can be seen that the first three elements in E_{cds} are noise sources from the integrator. To show the noise sources in CDS, we define

$$E_2 = \left[0, 0, 0, e_{nA_2}^2, e_{nS_1}^2, e_{nA_4}^2, e_{nS_2}^2, e_{nS_3}^2, e_{nA_3}^2 \right]^T. \tag{20}$$

So the following expression can be acquired:

$$E_{cds} = E_2 + T_1\{O_1\}, \tag{21}$$

where T converts O_1 to an enhanced length with 0, which means O_1 changes from $[e_{on1,inp}^2 \quad e_{on1,A}^2 \quad e_{on1,S_r}^2]^T$ to $[e_{on1,inp}^2 \quad e_{on1,A}^2 \quad e_{on1,S_r}^2 \quad 0 \quad 0 \quad 0 \quad 0 \quad 0 \quad 0]^T$. Written briefly, the transmission of noise through the two parts of the readout unit cell is

$$O_1 = H_1 E_1,$$
$$O_2 = H_2\left(E_2 + T_1\{O_1\} \right). \tag{22}$$

Every noise source in each part of the readout unit cell transmitting to the output node is expressed precisely and compactly by using (22). The design contains only two parts, so the numbers marking the parts are 1 and 2. Under the situation of much more constituent parts, the transmission of noise source can be written as

$$O_{n+1} = H_{n+1}\left(E_{n+1} + T_n\{O_n\} \right). \tag{23}$$

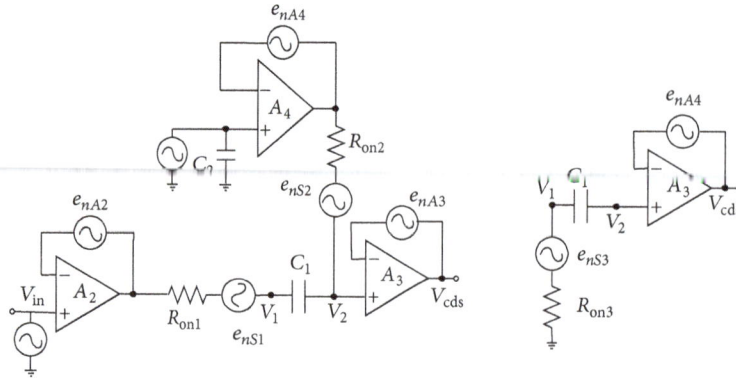

FIGURE 9: Noise model of CDS part.

5. Noise Computing

5.1. Noise Computing Based on Transfer Function. The photocurrent noise is related to the value of the generating photocurrent, which varies with the radiation intensity. Strictly speaking, it is pixel noise, which does not belong to the noise of readout circuits [15]. Thus the photocurrent noise will not be calculated in the paper. All the noise sources in the readout unit cell are induced by the MOSFETs. For the switches, MOSFETs operating in linear region produce thermal noise modeled by [12, 16]

$$e_{nS_i}^2 = 4kTR_{oni},\tag{24}$$

where k is the Boltzman constant, T is the absolute temperature, and R_{oni} represents the on-resistance of the switches. MOSFETs in the OP operate in the subthreshold region, producing thermal noise and $1/f$ noise; the whole noise of an OP can be represented as the reference input noise:

$$e_{nA_i}^2 = K_{1e} + \frac{K_{2e}}{f},\tag{25}$$

where K_{1e} and K_{2e} represent thermal noise factor and $1/f$ noise factor for the OPs, respectively. The detail of computing expression can be referred to in the chapter introducing OP of [12, 16], through which the specific values of the two parameter are obtained. The process on which the project is based is Globalfoundries 0.18 μm Industry Compatible Dual Voltage 1.8 V/3.3 V Process. The frequency character of the OP is simulated, which is shown in Figure 10. From it we can see that the OP achieve a 61-degree phase margin that leads to stability when countering an input of a step signal.

The switches in the proposed ROIC are implemented as CMOS switches, individual of which consists of a P type MOSFET, Mp, and an N type MOSFET, Mn, as shown in Figure 11. The W/L of both of the two MOSFETs are set to be 0.3 μm/2.35 μm. The simulation curve of resistance of the CMOS switch is shown in Figure 12, where the input signal spans from 0.5 V to 2.5 V. Figures 12(a) and 12(b) present the value of maximum on-resistance 2.2 kΩ and the value of minimum off-resistance 1 GΩ.

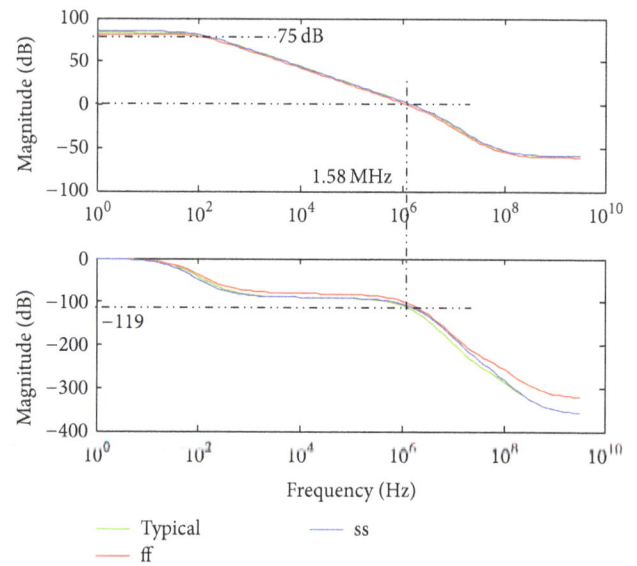

FIGURE 10: Frequency character of OP used in the proposed ROIC.

The parameters used for the calculation of noise are summarized in Table 2, where we assume that all the OPs have the same parameters and so do the switches.

For the application of high resolution, the area of a single pixel is restricted, leading to the limited layout area for the readout unit cell. With respect to the fact that capacitors consume the largest area resource, therefore, with the guarantee of sufficient layout area for the MOSFETs, more layout areas should be distributed to the three capacitors in the unit cell. Through calculations, effect on the output noise due to C_2 can be neglected, which should be designed small to save up layout area to allow larger C_{int} and C_1. With the limitation of 30 μm $\times$ 30 μm layout area, the maximum of the sum of the two capacitors is 2 pF.

The contribution of each noise source to the output noise is calculated using the transfer functions in Section 3, which is shown in Figure 13, where T_{int} is set as 100 μs. Figure 13(a) depicts the dependence of the output noise, integrator noise, and CDS noise on the C_{int}. The diagram reveals an opposite variation trend. The integrator noise is an increasing function

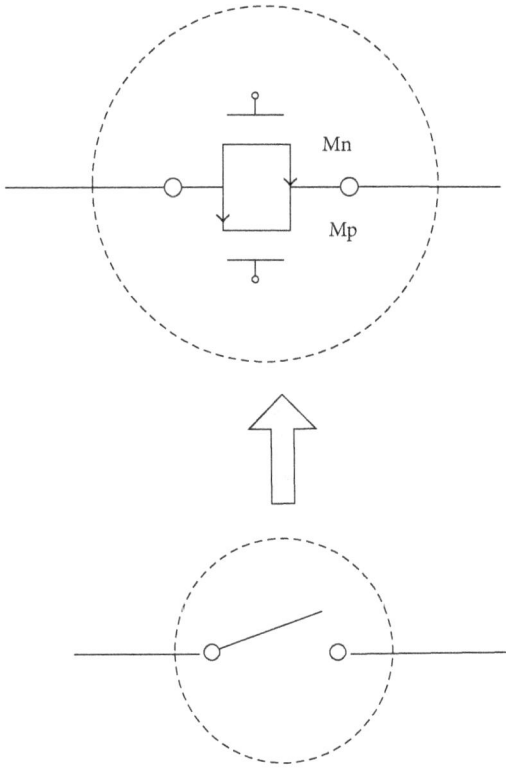

FIGURE 11: CMOS switch used in the proposed ROIC.

and the CDS noise is a decreasing function, which results from the limitation of the sum of capacitors. C_{int} increase leads to C_1 decrease inevitably, causing enlargement of the bandwidth of transfer functions of noise sources in CDS, which weakens the ability of low pass filtering. Figure 13(b) gives the curves of thermal noise, $1/f$ noise, and reset noise with the change of C_{int}. We can see that $1/f$ noise and reset noise increase with C_{int} increase while an optimal point appears for the thermal noise which is the majority, at 0.66 pF. When the C_{int} is smaller than 0.26 pF, the reset noise is smaller than $1/f$ noise. The output noise reaches the minimum at 74.1 μV when the C_{int} is 0.74 pF and C_1 is 1.26 pF.

Figure 14 shows the output noise as a function of T_{int} with three different C_{int}. It can be seen that the output noise increases when T_{int} increases, despite different C_{int}. This is mainly because the increasing of T_{int} results in increasing of interval of correlated double sampling, which weakens the capability of suppressing $1/f$ noise and reset noise. From the figure we also get that, with smaller C_{int}, the increase speed of output noise is faster.

5.2. Noise Calculation Based on Verilog-A. The transient analysis of the proposed readout unit cell is carried out in the section. HSPICE provides the possibility to simulate circuit noise in AC response by computing the PSD but cannot give the waveform of noise in transient response directly, whereas we carried out an approach to model time domain noise source using Verilog-A, which is described in detail in [13, 17] with both thermal and $1/f$ noise source modelling.

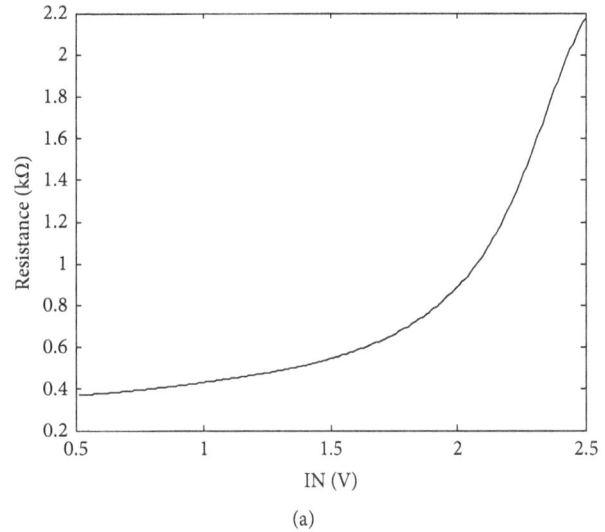

(a)

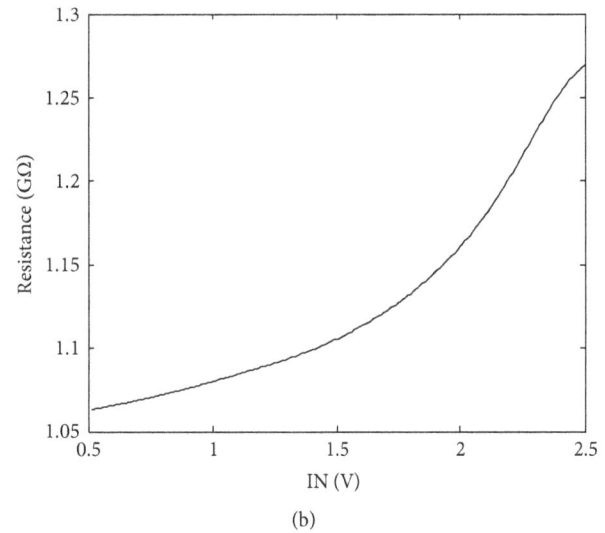

(b)

FIGURE 12: Simulation curve of resistance of the CMOS switch.

TABLE 2: Parameters of noise calculation.

R_{on} (on-resistance of switch)	2.2 kΩ
R_{off} (off-resistance of switch)	1 GΩ
K_{1e} (thermal noise parameter of OP)	4×10^{-17} V^2/Hz
K_{2e} ($1/f$ noise parameter of OP)	4×10^{-12} V^2/Hz
GBW (gain bandwidth product of OP)	1.58 MHz
A_0 (low frequency gain of OP)	75 dB

The averaged RMS of the output noise obtained by Verilog-A are presented in Tables 3 and 4 to compare with those that were calculated through transfer functions specified in Section 3, where the unit of noise is μV. In Table 3, T_{int} is assumed to be 100 μs and C_{int} is 0.74 pF in Table 4. With different C_{int} and T_{int}, the two sets of results are in good agreement, the difference between which is lower than 6%, therefore proving the feasibility of the method of transfer function noise analysis.

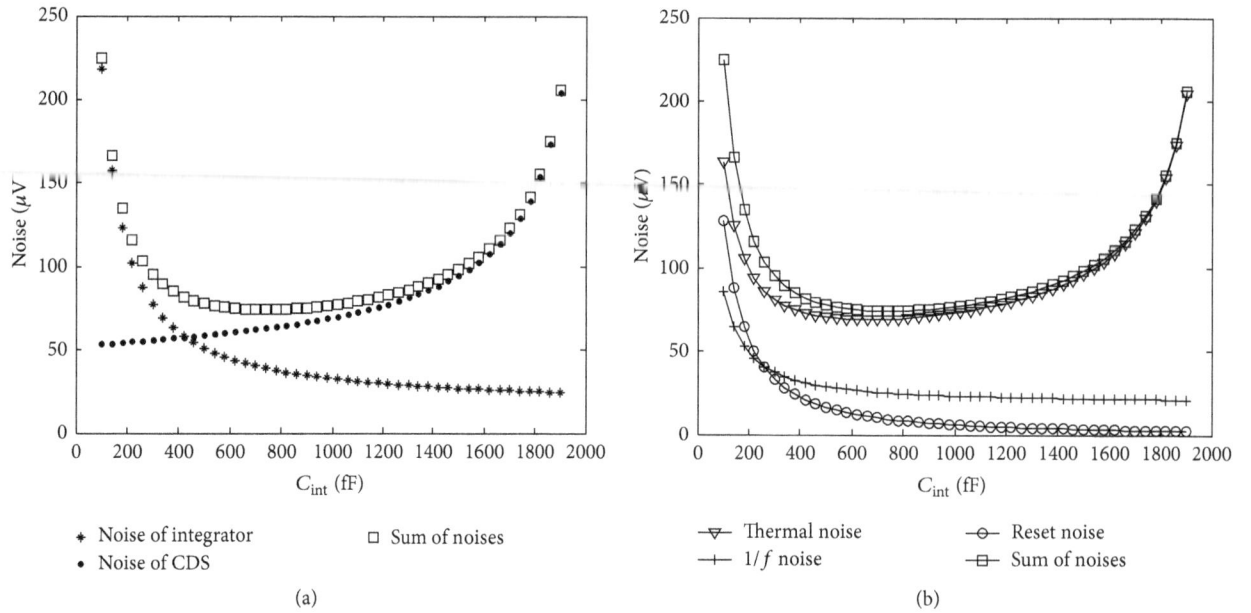

FIGURE 13: Noise voltage versus integration capacitor.

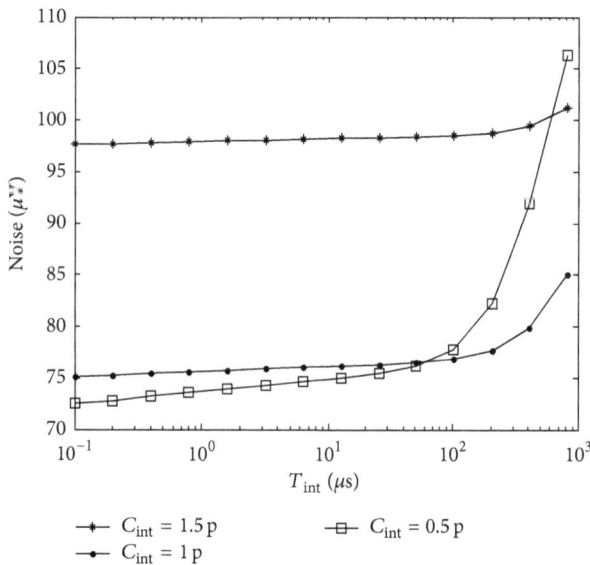

FIGURE 14: Noise voltage versus integration time.

TABLE 3: Comparison between transfer function based and Verilog-A based.

C_{int}	100 fF	460 fF	820 fF	1.18 pF	1.54 pF	1.9 fF
Theoretical results	224.7	79.5	74.4	81.5	101.9	129.2
Verilog-A results	212.1	81.9	77.0	79.8	108.2	123.8

6. Conclusion

To adapt the application of imaging for large format and high resolution, a new readout cell with low power and small layout area is proposed, consisting of an integrator and

TABLE 4: Comparison with different integration time.

T_{int}	200 ns	800 ns	3.2 us	12.8 us	51.2 us	204.8 us
Theoretical results	71.5	72.0	72.5	72.9	73.5	75.9
Verilog-A results	74.6	74.4	69.9	75.6	72.8	74.4

a new type of CDS, both of which are switched capacitor circuits where all the MOSFETs are optimized in subthreshold region. The integrator adopts CTIA structure and the CDS is comprised of four switches, three Ops, and two capacitors. The noise model is built using transfer function in Section 3 and a method of presenting the transfer process of noise in the form of matrix is given, through which we can describe the contribution of each noise source to the whole output noise distinctly. In Section 4, noise analysis of the proposed readout cell is carried out by the transfer function method and transient analysis utilizing Verilog-A, respectively. The results showing good agreement verify the feasibility of the method presented through matrix. Under the limitation of layout area less than $30 \, \mu m \times 30 \, \mu m$, the integrator noise is an increasing function of C_{int} and the CDS noise is a decreasing function. With respect to the types of noise, $1/f$ noise and reset noise increase with C_{int} increase while an optimal point appears for the thermal noise which is the majority. The integration capacitor should be selected as 0.74 pF to achieve an optimal noise performance, the whole output noise reaching the minimum value at $74.1 \, \mu V$.

Conflict of Interests

The authors declare that there is no conflict of interests regarding the publication of this paper.

Acknowledgment

The authors thank Key Laboratory of Opto-Electronic Information Processing for the continuous supporting of the research on noise of readout integrated circuits.

References

[1] D. A. Scribner, M. R. Kruer, and J. M. Killiany, "Infrared focal plane array technology," *Proceedings of the IEEE*, vol. 79, no. 1, pp. 66–85, 1991.

[2] J. Schuster, *Advanced numerical modeling and hybridization techniques for third-generation infrared detector pixel arrays* [Ph.D. thesis], Boston University, 2014.

[3] E. R. Fossum and B. Pain, "Infrared readout electronics for space-science sensors: state of the art and future directions," in *Infrared Technology XIX*, vol. 2020 of *Proceedings of SPIE*, pp. 262–285, International Society for Optics and Photonics, San Diego, Calif, USA, July 1993.

[4] B. H. Calhoun, A. Wang, and A. Chandrakasan, "Modeling and sizing for minimum energy operation in subthreshold circuits," *IEEE Journal of Solid-State Circuits*, vol. 40, no. 9, pp. 1778–1785, 2005.

[5] J. Chen, G. Lee, and S. Ren, "Ultra-low power read-out integrated circuit design," in *Proceedings of the IEEE National Aerospace and Electronics Conference (NAECON '12)*, pp. 144–148, July 2012.

[6] H. S. Gupta, A. S. K. Kumar, S. Chakrabarti et al., "Design of large dynamic range, low-power, high-precision ROIC for quantum dot infrared photo-detector," *Electronics Letters*, vol. 49, no. 16, pp. 1018–1020, 2013.

[7] B. Hu, P. Li, and A. Ruan, "A multifunction snapshot ROIC design for low and high background application," in *Proceedings of the International Conference on Optoelectronics and Image Processing (ICOIP '10)*, vol. 1, pp. 44–46, IEEE, November 2010.

[8] Y. Wang, J.-N. Chen, D.-M. Ke, and J. Hu, "Design of an interface circuit for integrated capacitive sensors," in *Proceedings of the International Conference on Electronics, Communications and Control (ICECC '11)*, pp. 1408–1411, IEEE, September 2011.

[9] P. Wang, G.-Q. Chen, L. Gao, J. Zhou, and R.-J. Ding, "Optimal design of high frequency readout IC for short-wave IRFPA," in *Infrared, Millimeter-Wave, and Terahertz Technologies II*, vol. 8562 of *Proceedings of SPIE*, International Society for Optics and Photonics, Beijing, China, November 2012.

[10] G. Chen, J. Zhang, P. Wang et al., "ROIC for HgCdTe e-APD FPA," in *Proceedings of the 5th International Symposium on Photoelectronic Detection and Imaging (ISPDI '13)*, 89120F, p. 10, International Society for Optics and Photonics, 2013.

[11] L.-C. Hao, R.-J. Ding, A.-B. Huang, H.-L. Chen, C. Zhou, and P. Wang, "A high performance readout circuit (ROIC) with BDI structure for SWIR FPAs," in *International Symposium on Photoelectronic Detection and Imaging: Advances in Infrared Imaging and Applications*, vol. 8193 of *Proceedings of SPIE*, International Society for Optics and Photonics, September 2011.

[12] B. Razavid, *Design of Analog CMOS Integrated Circuits*, McGraw-Hill, New York, NY, USA, 2000.

[13] X. Wang and Z. Shi, "A new CDS structure for high density FPA with low power," *VLSI Design*, vol. 2015, Article ID 767161, 7 pages, 2015.

[14] P. Norton, "HgCdTe infrared detectors," *Opto-Electronics Review*, vol. 23, no. 3, pp. 159–174, 2002.

[15] J. Lv, H. Zhong, Y. Zhou, B. Liao, J. Wang, and Y. Jiang, "Model-based low-noise readout integrated circuit design for uncooled microbolometers," *IEEE Sensors Journal*, vol. 13, no. 4, pp. 1207–1215, 2013.

[16] P. R. Gray, P. J. Hurst, R. Meyer G et al., *Analysis and Design of Analog Integrated Circuits*, Wiley, 2008.

[17] N. Kawai and S. Kawahito, "Noise analysis of high-gain, low-noise column readout circuits for CMOS image sensors," *IEEE Transactions on Electron Devices*, vol. 51, no. 2, pp. 185–194, 2004.

## 6

Ultra-Low-Voltage Self-Body Biasing Scheme and Its Application to Basic Arithmetic Circuits

Ramiro Taco, Marco Lanuzza, and Domenico Albano

Department of Computer Science, Modeling, Electronics and System Engineering, University of Calabria, Via P. Bucci 42C, 87036 Rende, Italy

Correspondence should be addressed to Ramiro Taco; taco@dimes.unical.it

Academic Editor: Jose Carlos Monteiro

The gate level body biasing (GLBB) is assessed in the context of ultra-low-voltage logic designs. To this purpose, a GLBB mirror full adder is implemented by using a commercial 45 nm bulk CMOS triple-well technology and compared to equivalent conventional zero body-biased CMOS and dynamic threshold voltage MOSFET (DTMOS) circuits under different running conditions. Postlayout simulations demonstrate that, at the parity of leakage power consumption, the GLBB technique exhibits a significant concurrent reduction of the energy per operation and the delay in comparison to the conventional CMOS and DTMOS approaches. The silicon area required by the GLBB full adder is halved with respect to the equivalent DTMOS implementation, but it is higher in comparison to conventional CMOS design. Performed analysis also proves that the GLBB solution exhibits a high level of robustness against temperature fluctuations and process variations.

1. Introduction

Ultra-low-voltage (ULV) operation is a popular design approach to achieve high energy efficiency. When the power supply voltage (V_{DD}) is scaled down, dynamic power consumption is considerably decreased; however, as V_{DD} approaches the transistor threshold voltage (V_{TH}), the delay starts to exponentially increase [1–6] and circuit performances become extremely sensitive to process variations and temperature fluctuations [7, 8]. In order to guarantee a widespread adoption of ULV designs, these issues have to be addressed [7].

To boost performances of ULV designs, while also improving robustness against process and temperature variations, the forward body biasing (FBB) technique can be effectively used [7–13]. The FBB can be applied (also dynamically) at different levels of granularity ranging from macroblock level to the transistor level. The key rationale for applying such a technique at the macroblock level is to amortize the silicon area and the body control signal routing complexity of a finer grained implementation. As a drawback, when V_{TH} is reduced at the block level to compensate for variations and/or to provide a temporary speed boost, leakage power is increased for all the gates in the block, while speed-up would be needed

only on timing critical gates. Better energy-delay trade-offs can be obtained by reducing the body-bias control granularity at the expense of larger silicon area occupancy [13].

Body biasing can be dynamically managed at the transistor level by exploiting the dynamic threshold voltage MOSFET (DTMOS) approach [8]. DTMOS logic uses transistors whose gates are tied to their bodies. As the substrate voltage varies with the gate voltage, the threshold voltage of the device is dynamically changed. When the device is turned ON, its threshold voltage is forced to drop, thus allowing a much higher ON current compared to a standard MOSFET [8]. On the contrary, in the OFF state, the characteristics of a DTMOS transistor become similar to those of a regular MOSFET. A major limitation for the use of bulk DTMOS devices is that a large distance between transistors controlled by different gate signals has to be maintained to ensure correct body isolation between differently body-biased devices [14, 15]. This causes not only a higher occupied silicon area but also longer interconnections which in turn degrade speed and energy performances. As an additional drawback, the large body capacitance and resistance [16] of devices provide an additional RC delay in charging the substrate and the input nodes of the DTMOS logic gates [17]. Moreover, the substrate bias voltage of DTMOS logic gates would change also when input transitions do not imply output switching. This would charge and discharge the large body capacitances, thus wasting precious dynamic energy [11]. All the above effects can erode the expected advantages of DTMOS circuits.

Recently, a gate level body biasing technique was proposed [11, 18] to overcome the speed and energy limits of DTMOS logic gates. Exploiting this solution, the RC delay in charging the body of the devices does not affect the speed of logic gates. Additionally, when input signals switch without changing the logic gate status, the body capacitances are no more uselessly charged/discharged.

In this work, an extended postlayout analysis of the potentiality of gate level body-biased (GLBB) nanoscaled designs in low voltage regime is presented. As a main result, we demonstrate that GLBB designs are fully functional, robust, fast, and energy efficient both in the subthreshold and near threshold regions. The benefits of the proposed scheme are initially evaluated by comparing the suggested approach with respect to zero body-biased (ZBB) CMOS and DTMOS solutions in the case of simple logic gates as NAND2 and NOR2. Afterwards, a mirror full adder (FA) [18] implemented according to the GLBB technique is compared to equivalent ZBB CMOS and DTMOS counterparts. All the FA designs, evaluated through a preliminary prelayout analysis in [18], were laid out exploiting the ST 45 nm CMOS triple-well technology. It is worth noting that postlayout analysis is strictly required when adaptive body biasing techniques are used in nanometer technologies. This is because the physical distances needed to provide correct body isolation between differently body-biased devices have a very large impact on delay and energy characteristics of the circuits. All the compared circuits were evaluated at ultra-low-voltage regime under different running conditions. Depending on power supply voltage level, the GLBB FA allows delay to be reduced in the ranges of 6%–34% and 24%–40% in comparison to the ZBB CMOS and DTMOS circuits, respectively. This is achieved also saving energy per operation. As an example, for an 80 FO4 clock cycle period and activity factor of 10%, the GLBB circuit reduces energy per operation in the ranges of 15%–27% and 47%–77% with respect to the ZBB CMOS and DTMOS FAs. Such energy and speed advantages are obtained at the expense of increased silicon area occupancy in comparison to a conventional ZBB CMOS design but reducing area occupancy of about two times with respect to the DTMOS implementation. Additionally, the GLBB FA maintains a high level of robustness against temperature and process variations.

The rest of the paper is organized as follows. Section 2 discusses the operating characteristics of the GLBB approach. The compared mirror full adder designs are discussed and postlayout is comparatively characterized in Section 3. Finally, Section 4 concludes the paper.

2. Operational Features of Gate Level Body-Biased Logic Gates

As shown in Figure 1(a), the generic GLBB logic gate consists of two circuit sections: the logic subcircuit which is responsible for the logical functionality and the body biasing

generator (BBG) which manages the body voltage (V_B) of all the devices belonging to the logic subcircuit. The BBG is a simple push-pull amplifier, which acts as a voltage follower for the output voltage V_{OUT} while decoupling the large body capacitances from the output node.

In Figures 1(b)-1(c), the transient behavior for the input voltage (V_{IN}), the output voltage (V_{OUT}), and the body voltage (V_B) is shown for the falling and rising output transitions, respectively. When V_{OUT} is equal to V_{DD} (0 V), the BBG transfers high (low) voltage on V_B net, thus preparing the pull-down (pull-up) network for a faster logic gate switching. Since the MOSFETs of the switching network (either pull-up or pull-down) are already forward body-biased before gate inputs' arrival, the output transition is largely favored by a switching current significantly higher in comparison to the case of ZBB CMOS scheme. Speed improvement also exists with respect to the DTMOS configuration. In fact, the transition of the input signals is not slowed down from the body-induced RC delay, as occurs in sub-DTMOS gates, whereas the high capacitive load seen by the BBG does not constitute a speed bottleneck, since V_B voltage is always established well before inputs' transition. Indeed, by inspecting the transient behavior reported in Figures 1(b)-1(c), it is easy to understand that the body-induced RC delay at the output of the BBG represents a benefit since it allows a slower transition for the body voltage and consequently a faster transition on the gate output.

In spite of the performance advantages previously discussed, GLBB logic gates show somewhat increased leakage current with respect to their ZBB CMOS and DTMOS counterparts. This is mainly due to the fact that the output voltage transition of the BBG is not rail-to-rail (a PMOS device is used to transfer a low voltage on V_B net, whereas NMOS transistor is used for transferring the high voltage). This causes the threshold voltage reduction of leaky devices belonging to the OFF network (either pull-down or pull-up) of the logic subcircuit during the idle status. An additional contribution to the static power consumption of the GLBB logic gates would be due to the static current flowing in the BBG circuit. However, such a current is inherently limited by the reverse body biasing of the BBG transistors ($V_{BS,N}$ and $V_{SB,P}$ are always <0 in the BBG circuit) and becomes negligible if reduced size devices are used for the BBG implementation.

In order to estimate the trade-off potentially offered by the proposed approach, leakage current (I_{leak}) versus delay curves are shown in Figures 2(a)-2(b) in the case of NAND2 and NOR2 logic gates and for the ZBB CMOS, DTMOS, and GLBB implementations, respectively. For a fair comparison, all the logic gates were sized for a W_P/W_N ratio of $k \cdot W/W$ between pull-up and pull-down networks, where k was chosen to assure symmetric switching delay under the typical NMOS, typical PMOS (TT) process corner, $V_{DD} = 300$ mV, $T = 27°C$, rise and fall times of 500 ps, and a capacitive load of 1.2 fF. The curves have been obtained varying the sizing factor W from 0.12 μm to 1.2 μm, with a step of 0.12 μm. Reported results are normalized to data obtained for minimum sized ZBB CMOS implementations.

As expected, at a parity of W, the GLBB technique shows leakage current higher than that which the other competitors

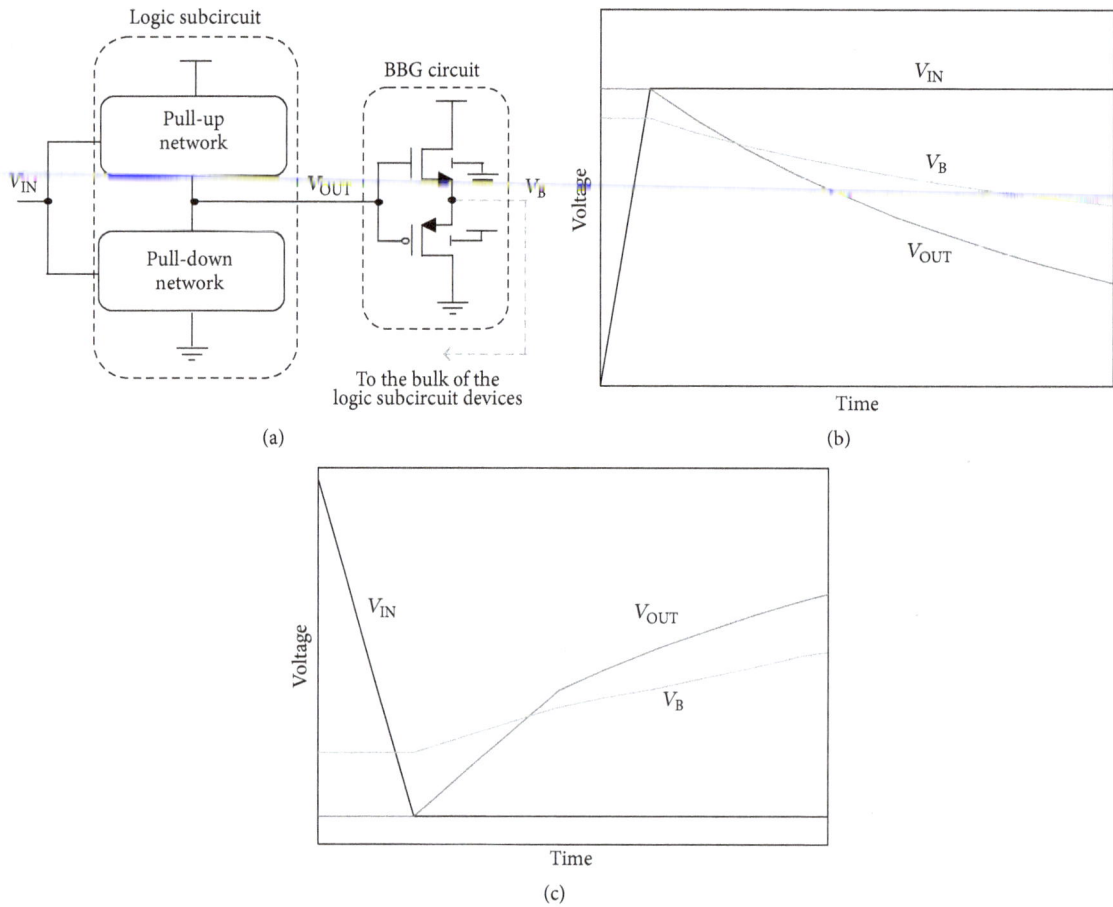

(a)

(b)

(c)

FIGURE 1: Logic gate with gate level dynamic body biasing (a) and transient behavior for output falling (b) and rising (c) voltage.

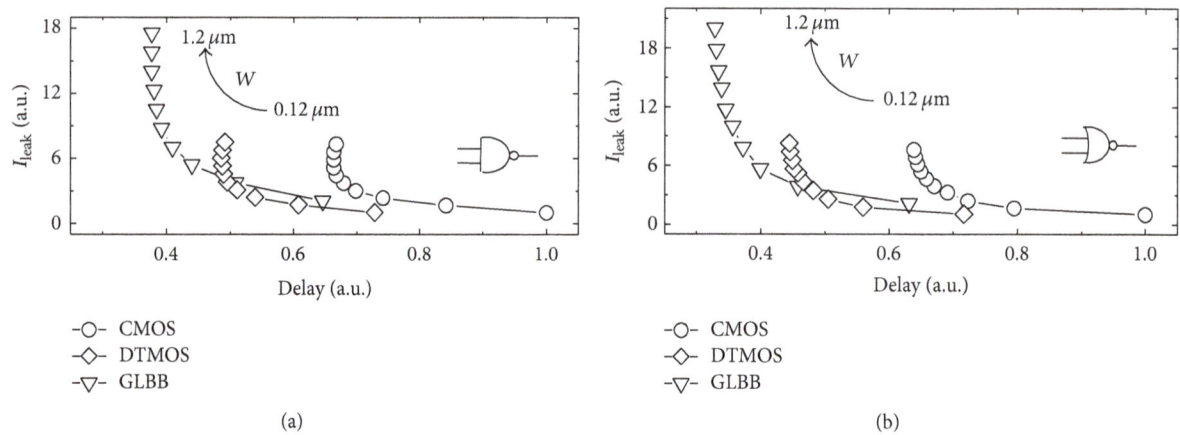

(a)   (b)

FIGURE 2: Leakage current delay plots for NAND2 (a) and NOR2 (b) logic gates.

show. This means that, among the different evaluated choices, the GLBB style is the less suitable if the minimization of static power is the main design target. On the contrary, if the speed requirement represents the main design aim, the GLBB style becomes the most reasonable choice allowing higher performance to be reached at the parity of leakage power consumption since the boosting action of the BBG allows the delay target to be reached using smaller transistors. Moreover, the GLBB technique allows performance ranges which are unaffordable for both ZBB CMOS and DTMOS configurations.

3. Benchmark Circuit and Postlayout Comparative Analysis

To further validate the GLBB design technique, the low-voltage mirror FA, shown in Figure 3, was designed and

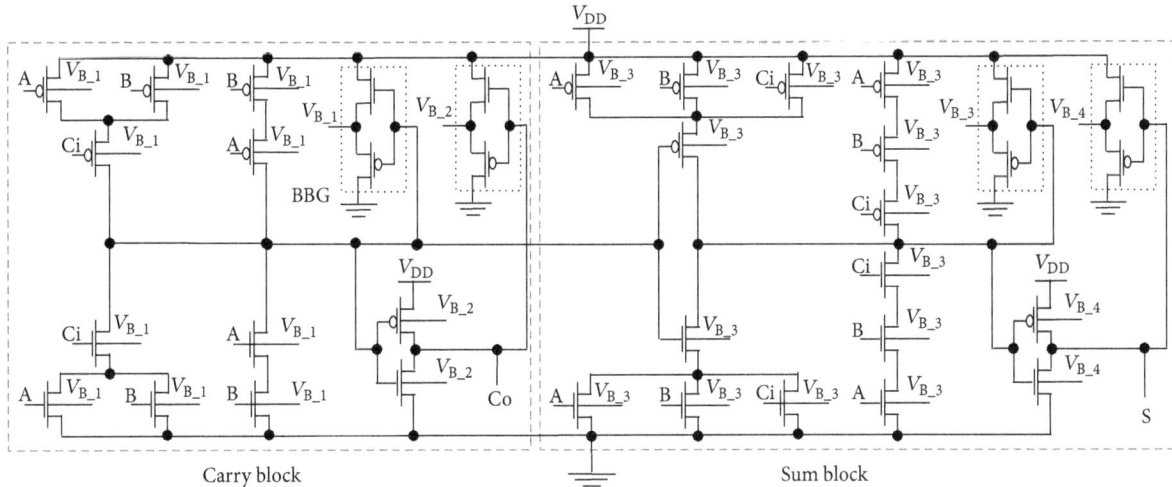

FIGURE 3: Low-voltage mirror FA designed according to the GLBB technique.

TABLE 1: Pull-up/pull-down width ratio.

Stack config.	ZBB	DTMOS	GLBB
1	$1.5W/W$	W/W	$1.1W/W$
2	$4.5W/2.5W$	$2.5W/2.5W$	$3.2W/2.5W$
3	$8.25W/4.5W$	$4.5W/4W$	$5.5W/4.5W$

TABLE 2: Comparison between ZBB, DTMOS, and GLBB schemes at nominal conditions (TT process corner, $V_{DD} = 0.3$ V, and $T = 27°C$).

	ZBB	DTMOS	GLBB
Silicon area [μm^2]	20.7	123.2	60.5
Delay [μs]	0.70	0.78	0.59
Leakage current [nA]	0.20	0.24	0.21
Energy per operation ($T = 80$ FO4, $\alpha = 0.2$) [fJ]	0.75	2.27	0.57

postlayout was characterized in comparison to correspondent ZBB CMOS and DTMOS designs. Four BBGs are used to speed up the switching of logic subcircuits. This translates in eight additional devices in comparison to CMOS and DTMOS circuits.

Devices belonging to the logic subsections of compared circuits were sized with minimum channel length (i.e., $L_{min} = 40$ nm), whereas the pull-up/pull-down channel width ratio was chosen to obtain comparable strength for $V_{DD} = 0.3$ V and $T = 27°C$, imposing equal width for series-connected transistors.

In Table 1, the width ratio between pull-up and pull-down networks is explicitly reported for the compared designs and for the different stacking configurations. The sizing factor W was chosen by iterative simulations, imposing similar leakage current at nominal conditions (i.e., TT process corner, $V_{DD} = 0.3$ V, and $T = 27°C$) for all the compared designs.

In order to correctly take into account the impact of layout parasitics on performance, the physical design of the compared circuits was carried out (see Figure 4) considering the design rules imposed by the ST 45 nm bulk CMOS triple-well technology. For DTMOS and GLBB designs, the deep N-well layer was used to shield N-channel devices from the P-type general substrate, thus obtaining P-well regions isolated from the underlying substrate. Each of these regions is vertically surrounded by an N-well region to provide also lateral isolation [14, 15]. Due to distances needed to provide correct body isolation between differently body-biased devices, implementations exploiting unconventional body biasing (i.e., DTMOS and GLBB) exhibit significantly

increased silicon area occupancy in comparison to the ZBB CMOS circuit. In an area optimized layout, the DTMOS implementation requires one isolated P-well region for each different transistor gate signal, thus requiring 5 different isolated P-well islands. On the contrary, in the proposed approach, the number of isolated P-type islands is reduced to 4 (i.e., one for each BBG). This, along with the reduced size of its transistors, leads the proposed implementation to reduce silicon area occupancy of more than 50% with respect to the DTMOS design. Table 2 reports postlayout comparison results under nominal simulation conditions.

Comparative postlayout delay results, evaluated for V_{DD} ranging from 0.2 V to 0.5 V with a voltage step of 0.05 V, are shown in Figure 5. Given results are normalized with respect to the delay of ZBB CMOS design. For $V_{DD} = 0.5$ V, the suggested approach allows delay to be reduced to 34% and 24% with respect to the standard CMOS and DTMOS implementations, respectively. Observing the insert of Figure 5, it is easy to note that as V_{DD} decreases below 0.45 V, the impact of FBB in boosting the performance is reduced but with a different rate on GLBB and DTMOS techniques. As final effect of this, the speed benefit brought by the suggested approach over the conventional CMOS circuit reduces down to 6% for the minimum considered power supply voltage (i.e., $V_{DD} = 0.2$ V). On the contrary, the speed advantages with respect to the DTMOS implementation become more pronounced coming up to 60% for $V_{DD} = 0.2$ V (the speed

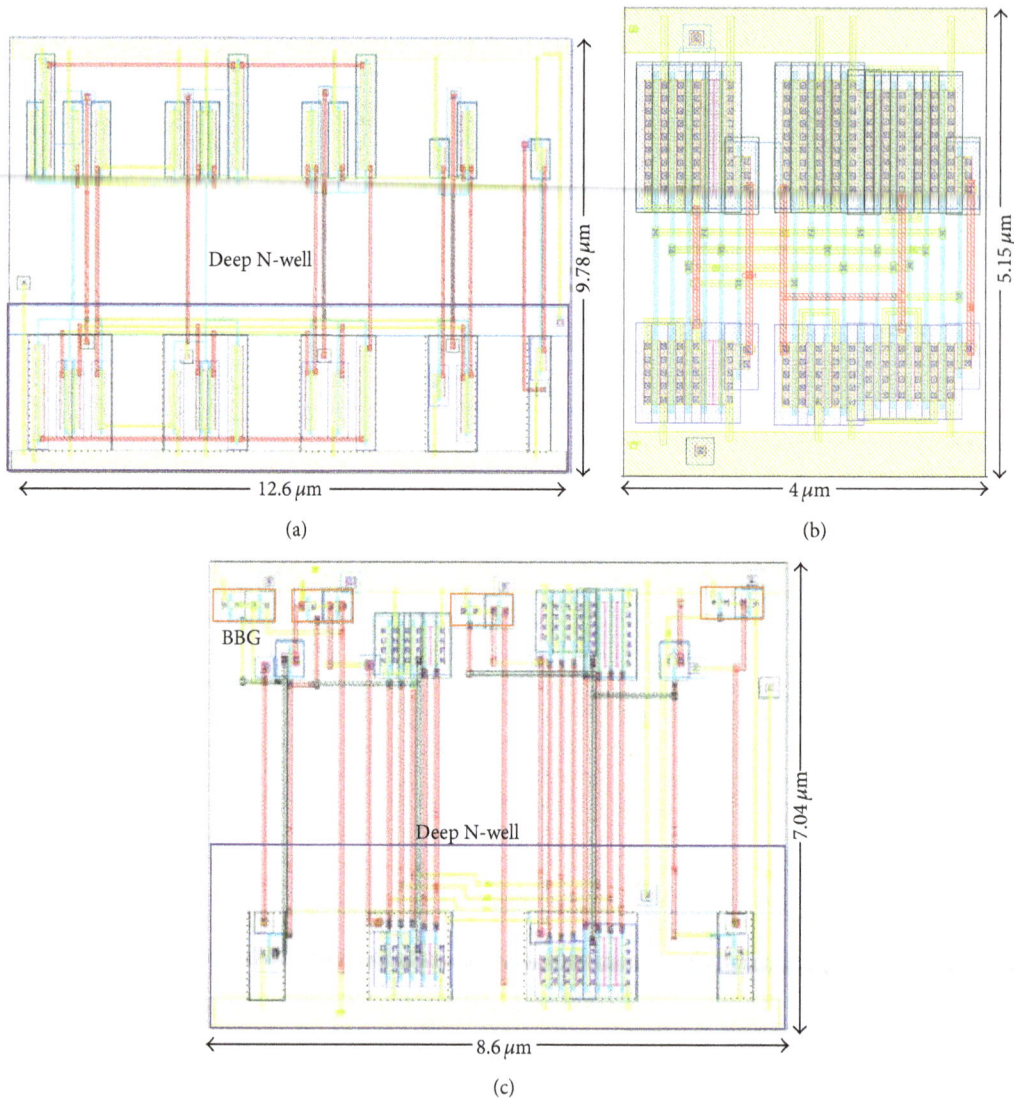

FIGURE 4: Layouts of FA for DTMOS (a), CMOS (b), and GLBB (c), respectively.

boosting on DTMOS due to the FBB is overcome by the negative impact of the body-induced RC delay).

Figure 6 reports I_{leak} versus V_{DD} for the three compared circuit topologies. Here, I_{leak} is normalized to the value of CMOS design for $V_{DD} = 0.3$ V. Due to the adopted sizing criterion, all the circuits have similar I_{leak} for $V_{DD} = 0.3$ V (see Table 2). However, this property is not maintained for different power supply voltage levels. As V_{DD} drops lower than 0.3 V, the proposed approach, which benefits from reduced transistors' sizing, leads to the lowest I_{leak}. On the contrary, the standard CMOS FA exhibits the lowest I_{leak} for $V_{DD} > 0.3$ V. Note that, for V_{DD} higher than 0.45 V, the parasitic pn junctions of DTMOS devices start to conduct a nonnegligible current which dramatically increases leakage power consumption of DTMOS-based FA.

Figures 7 and 8 compare the energy per operation (E_{OP}) behavior versus V_{DD} for the three compared circuit implementations, evaluated under different running conditions.

Results are normalized to energy data obtained for conventional CMOS circuit evaluated under the operating condition of $V_{DD} = 0.3$ V, activity factor (α) of 0.2, and clock cycle time (T_{clk}) of 80 FO4 (FO4 represents the delay of a CMOS inverter driving four identical inverters), which is typical of low power VLSI circuits [19]. More precisely, Figure 7 plots E_{OP} considering $T_{clk} = 80$ FO4 for $\alpha = 0.1, 0.2$, and 0.3.

Considering the lowest activity factor ($\alpha = 0.1$), the GLBB solution allows E_{OP} to be reduced in the ranges of 15%–27% and 47%–77% with respect to the ZBB CMOS and DTMOS designs, respectively. This is mainly due to the reduced transistors' sizes (see Table 1) of the GLBB circuit, which allow decreased total physical capacitances on the internal nodes of the circuit, even taking all the parasitics of the layout into account. Additionally, the proposed body biasing technique allows faster transitions of the gates which in turn diminish the short circuit component in dynamic energy. The above advantages are even emphasized for larger

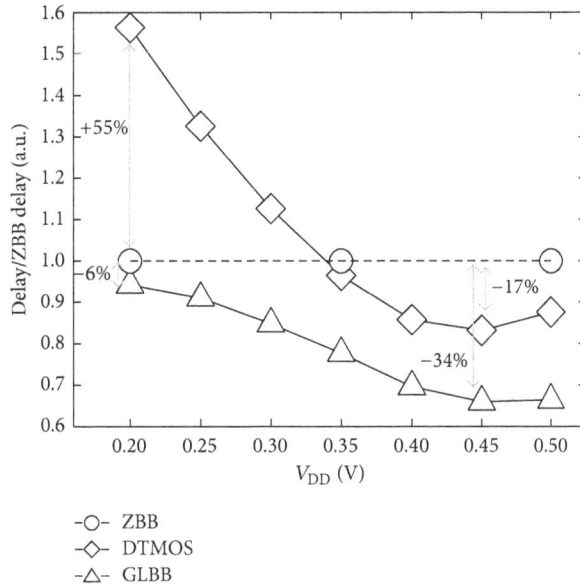

FIGURE 5: Delay versus V_{DD}.

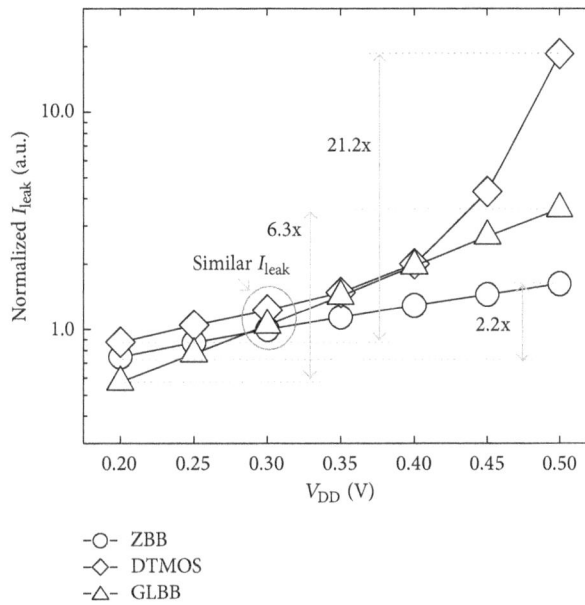

FIGURE 6: Leakage current (log scale) versus V_{DD}.

activity factors (i.e., when dynamic energy contribution in total E_{OP} increases). Due to the previous discussed input capacitive drawbacks, the larger devices and the longer interconnections, the DTMOS implementation results to be very energy hungry. Additionally, the bulk bias voltage of DTMOS devices can change also when input transitions do not imply switching of circuit internal nodes. This further increases the dynamic energy consumption due to unnecessarily charging/discharging the large body capacitances.

Figure 8 shows E_{OP} versus V_{DD} when $\alpha = 0.2$ and for $T_{clk} = 50$ FO4, 80 FO4, and 100 FO4. It should be noted that as the leakage energy contribution increases (i.e., when

T_{clk} increases), the suggested solution continues to maintain significant advantages in terms of total energy, also for V_{DD} higher than 0.3 V.

Figure 9 better emphasizes PDP and delay advantages of the proposed FA, when employed in a 16-bit ripple carry adder (RCA). The power of the FA under test is consequently evaluated for maximum frequency of the whole adder (to correctly take into account leakage contribution), whereas delay is related to the device under test in the FA chain. In the above scenario, the GLBB FA lowers minimum PDP point of 22% and 68% in comparison to the CMOS and DTMOS circuits, respectively. This is achieved with a speed

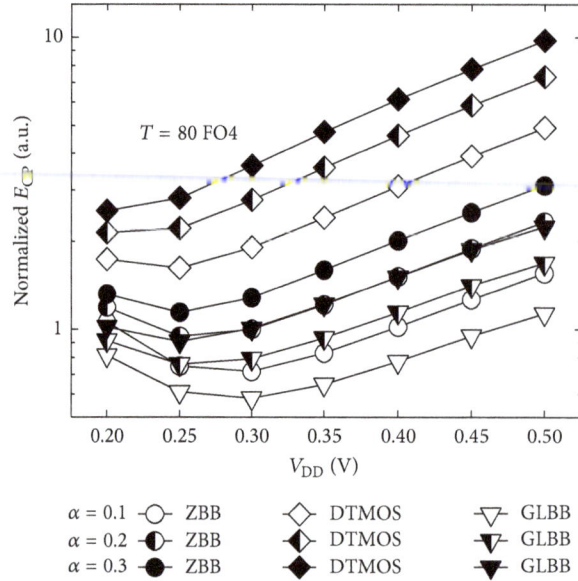

FIGURE 7: Energy per operation (log scale) for T_{clk} = 80 FO4 and for different activity factors.

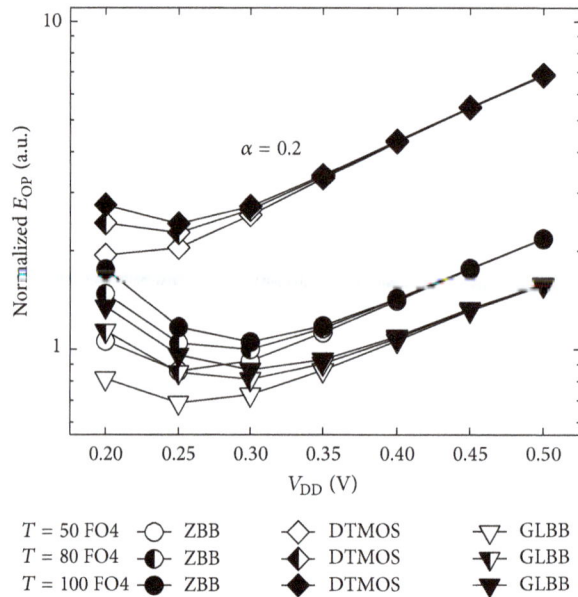

FIGURE 8: Energy per operation (log scale) for α = 0.2 and different clock cycle times.

boost of 17%/66% when compared to the CMOS/DTMOS implementations. Speed and PDP advantages are recorded in the whole power supply range.

Figure 10 describes the behavior of the compared circuits as the temperature varies from −25°C to 100°C for V_{DD} = 0.3 V. As shown in Figure 10(a), all the circuits demonstrate similar leakage currents at low operating temperatures (<25°C). However, as the temperature increases, the leakage current of the DTMOS circuit increases faster than its counterparts, becoming approximately 1.6 times higher for T = 100°C. Figure 10(b) demonstrates that the GLBB FA maintains its speed advantages in the whole considered operating temperature range.

The impact of process variability was investigated by performing Monte Carlo (MC) simulations on 1000 samples for V_{DD} = 0.3 V and T = 27°C. In this analysis, both interdie and intradie fluctuations were considered. MC leakage and delay results are given in Figures 11 and 12, respectively. When compared to its counterparts, the ZBB CMOS circuit exhibits the lowest mean leakage current (−19% and −9% in comparison to the DTMOS and GLBB designs, resp.) with a slight higher leakage current variability (σ/μ = 11% for the CMOS design against σ/μ = 8% and 10.4% for the DTMOS and GLBB solutions). On the other hand, the suggested approach results to be more robust in terms of delay. In fact, MC delay results reported in Figure 12 demonstrate

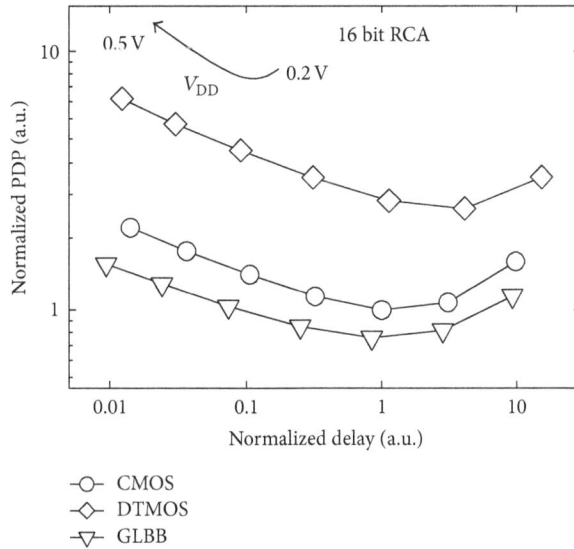

FIGURE 9: PDP (log scale) versus delay (log scale) for different $V_{DD}s$

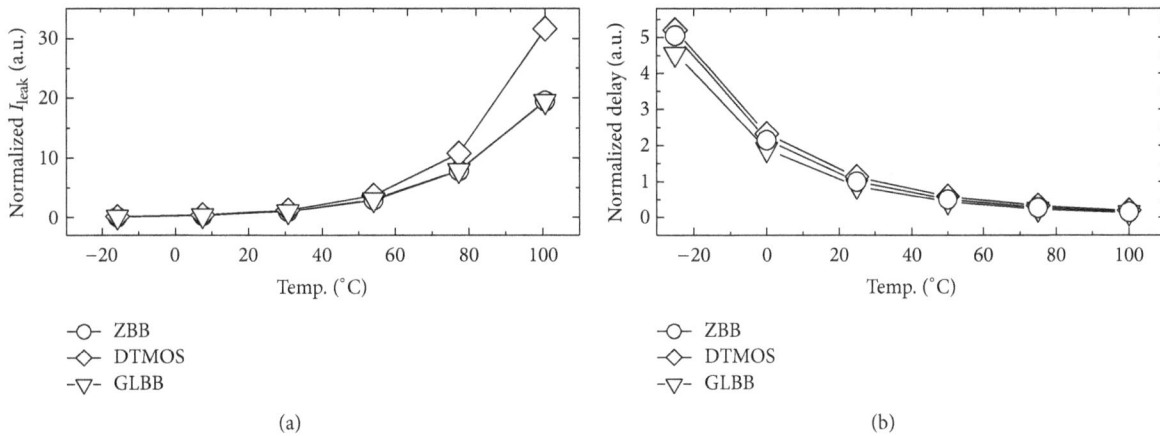

FIGURE 10: Temperature variation results (@V_{DD} = 300 mV): (a) leakage current versus temperature; (b) delay versus temperature.

that the mirror FA designed according to the proposed design style reaches a mean delay of only 0.5 us, which is about 20% and 28% lower than that of the standard CMOS (0.63 us) and DTMOS (0.7 us) implementations, respectively, while maintaining a delay standard deviation of about 0.21 μs.

4. Conclusion

In this work, the advantages of the recently introduced ULV gate level body biasing scheme were investigated. A preliminary analysis performed on simple logic gates demonstrates that the speed boosting provided by the suggested approach allows ULV GLBB circuits to reach performances which are unaffordable for both conventional CMOS and DTMOS configurations.

To take into account all the parasitic effects of the gate level body polarization in the case of more complex circuits, a

GLBB mirror full adder was laid out and compared against its conventional CMOS and DTMOS counterparts. Postlayout simulation results have shown that the GLBB design style is, at the parity of leakage power consumption, able to obtain significantly higher performance with reduced total energy per operation consumption in comparison to conventional CMOS and DTMOS implementations. The silicon area required by the GLBB full adder is halved with respect to the equivalent DTMOS implementation, but it is higher in comparison to conventional CMOS design. Finally, Monte Carlo simulations prove that the GLBB solution exhibits a high level of robustness against temperature fluctuations and process variations.

Conflict of Interests

The authors declare that there is no conflict of interests regarding the publication of this paper.

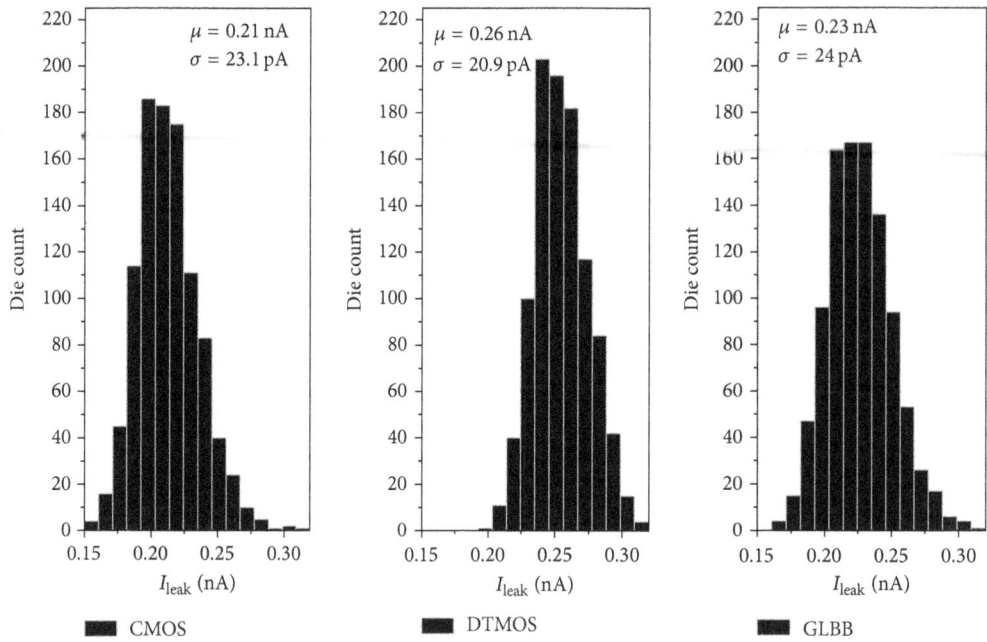

FIGURE 11: Monte Carlo leakage results (V_{DD} = 0.3 V, TT process corner, and T = 27°C).

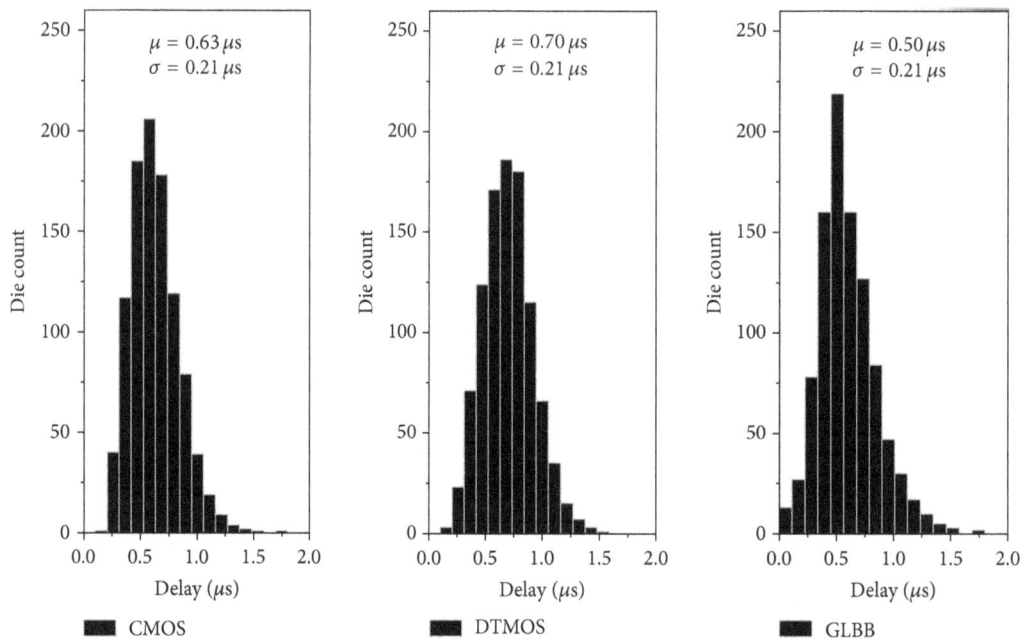

FIGURE 12: Monte Carlo delay results (V_{DD} = 0.3 V, TT process corner, and T = 27°C).

References

[1] A. Wang and A. Chandrakasan, "A 180-mV subthreshold FFT processor using a minimum energy design methodology," *IEEE Journal of Solid-State Circuits*, vol. 40, no. 1, pp. 310–319, 2005.

[2] B. Zhai, L. Nazhandali, J. Olson et al., "A 2.60 pJ/inst subthreshold sensor processor for optimal energy efficiency," in *Proceedings of the Symposium on VLSI Circuits (VLSIC '06)*, pp. 154–155, June 2006.

[3] J. Rabaey, J. Ammer, B. Otis et al., "Ultra-low-power design—the roadmap to disappearing electronics and ambient intelligence," *IEEE Circuits and Devices Magazine*, vol. 22, no. 4, pp. 23–29, 2006.

[4] S. Hanson, M. Seok, Y.-S. Lin et al., "A low-voltage processor for sensing applications with picowatt standby mode," *IEEE Journal of Solid-State Circuits*, vol. 44, no. 4, pp. 1145–1155, 2009.

[5] K. S. Chong, B. H. Gwee, and J. S. Chang, "A low energy FFT/IFFT processor for hearing aids," in *Proceedings of the IEEE International Symposium on Circuits and Systems (ISCAS '07)*, pp. 1169–1172, New Orleans, La, USA, May 2007.

[6] A. Kaizerman, S. Fisher, and A. Fish, "Subthreshold dual mode logic," *IEEE Transactions on Very Large Scale Integration (VLSI) Systems*, vol. 21, no. 5, pp. 979–983, 2013.

[7] S. Hanson, B. Zhai, M. Seok et al., "Exploring variability and performance in a sub-200-mV processor," *IEEE Journal of Solid-State Circuits*, vol. 43, no. 4, pp. 881–891, 2008.

[8] H. Soeleman, K. Roy, and B. C. Paul, "Robust subthreshold logic for ultra-low power operation," *IEEE Transactions on Very Large Scale Integration (VLSI) Systems*, vol. 9, no. 1, pp. 90–99, 2001.

[9] M. Radfar, K. Shah, and J. Singh, "A highly sensitive and ultra low-power forward body biasing circuit to overcome severe process, voltage and temperature variations and extreme voltage scaling," *International Journal of Circuit Theory and Applications*, vol. 43, no. 2, pp. 233–252, 2015.

[10] M. Meijer, J. Pineda de Gyvez, and A. Kapoor, "Ultra-low-power digital design with body biasing for low area and performance efficient operation," *Journal of Low Power Electronics*, vol. 6, no. 4, pp. 521–532, 2010.

[11] P. Corsonello, M. Lanuzza, and S. Perri, "Gate-level body biasing technique for high-speed sub-threshold CMOS logic gates," *International Journal of Circuit Theory and Applications*, vol. 42, no. 1, pp. 65–70, 2014.

[12] M. Meijer and J. Pineda de Gyvez, "Body-bias-driven design strategy for area- and performance-efficient cmos circuits," *IEEE Transactions on Very Large Scale Integration (VLSI) Systems*, vol. 20, no. 1, pp. 42–51, 2012.

[13] M. R. Kakoee and L. Benini, "Fine-grained power and body-bias control for near-threshold deep sub-micron CMOS circuits," *IEEE Journal on Emerging and Selected Topics in Circuits and Systems*, vol. 1, no. 2, pp. 131–140, 2011.

[14] M.-E. Hwang and K. Roy, "A 135 mV 0.13 μW process tolerant 6T subthreshold DTMOS SRAM in 90 nm technology," in *Proceedings of the Custom Integrated Circuits Conference (CICC '08)*, pp. 419–422, San Jose, Calif, USA, September 2008.

[15] H. Mostafa, M. Anis, and M. Elmasry, "A novel low area overhead direct adaptive body bias (D-ABB) circuit for die-to-die and within-die variations compensation," *IEEE Transactions on Very Large Scale Integration (VLSI) Systems*, vol. 19, no. 10, pp. 1848–1860, 2011.

[16] C. Wann, F. Assaderaghi, R. Dennard, C. Hu, G. Shahidi, and Y. Tan, "Channel profile optimization and device design for low-power high-performance dynamic-threshold MOSFET," in *Proceedings of the International Electron Devices Meeting (IEDM '96)*, pp. 113–116, San Francisco, Calif, USA, 1996.

[17] G. O. Workman and J. G. Possum, "A comparative analysis of the dynamic behavior of BTG/SOI MOSFETs and circuits with distributed body resistance," *IEEE Transactions on Electron Devices*, vol. 45, no. 10, pp. 2138–2145, 1998.

[18] M. Lanuzza, R. Taco, and D. Albano, "Dynamic gate-level body biasing for subthreshold digital design," in *Proceedings of the IEEE 5th Latin American Symposium on Circuits and Systems (LASCAS '14)*, pp. 1–4, Santiago, Chile, February 2014.

[19] M. Alioto, "Ultra-low power VLSI circuit design demystified and explained: a tutorial," *IEEE Transactions on Circuits and Systems I*, vol. 59, no. 1, pp. 3–29, 2012.

## 7

Investigation of a Superscalar Operand Stack Using FO4 and ASIC Wire-Delay Metrics

Christopher Bailey[1] and Brendan Mullane[2]

[1]*Department of Computer Science, University of York, Heslington, York YO10 5DD, UK*
[2]*Circuits and Systems Research Centre, University of Limerick, Limerick, Ireland*

Correspondence should be addressed to Christopher Bailey; chrisb@cs.york.ac.uk

Academic Editor: Kiyoung Choi

Complexity in processor microarchitecture and the related issues of power density, hot spots and wire delay, are seen to be a major concern for design migration into low nanometer technologies of the future. This paper evaluates the hardware cost of an alternative to register-file organization, the superscalar stack issue array (SSIA). We believe this is the first such reported study using discrete stack elements. Several possible implementations are evaluated, using a 90 nm standard cell library as a reference model, yielding delay data and FO4 metrics. The evaluation, including reference to ASIC layout, RC extraction, and timing simulation, suggests a 4-wide issue rate of at least four Giga-ops/sec at 90 nm and opportunities for twofold future improvement by using more advanced design approaches.

1. Introduction

Current trends in semiconductor technology, and in particular the International Technology Roadmap for Semiconductors [1], suggest that future concerns in microarchitecture at the VLSI level will pose significant challenges. These include increasing power density [2], progressively severe thermal hot spots in increasingly complex designs [3], the impact of growing static power [4], and the problem of wire versus gate-delay and power scaling [5, 6]. Such problems are often most acutely exposed in key mainstream processor components such as cache, register related logic such as reorder buffers, rename logic, and the register file itself. Any alternative scheme to the traditional register-based computing paradigm can therefore open up the possibility of new approaches to these problems. However, register files are so highly optimized that measuring alternatives now requires complete layout of an optimal design for comparison, followed by timing and power analysis and nothing as simple as functional comparison of abstract logic. This paper focuses upon one possible unexplored option for operand storage which is alternative in its structure to that of a register file. The questions we examine are (a) can a LIFO (last-in-first-out)

stack support superscalar operand access and (b) what is its performance relative to established mainstream approaches.

This work is undertaken with a 90 nm UMC CMOS process library; however, we ultimately utilize FO4 as a delay metric [7] in order to provide a general measure of performance that can be scaled to other process nodes. The work is undertaken using standard cell digital libraries and not at the transistor level. Although this is not therefore an optimal solution, it permits rapid assessment of multiple implementation schemes and semicustom design of the most promising candidate. This also means that performance results act as a conservative (lower) limit on potential performance.

Inevitably this work has some relevance to the age-old argument of stack processors versus register machines, particularly as significant improvements have been made in stack-processor code efficiency and code optimization [8–10] and in design of architectures capable of multiple instruction issue with out-of-order completion and/or previous attempts to parallelize stack structures [11–14].

However, the fundamental focus and aim of this paper is not to compare two competing processor paradigms but simply to establish in its own right the feasibility of a stack structure capable of permitting efficient access to operands in a superscalar access mode. It is also possible, for example, to have stacks employed in situations where they are not the basis for complete programmable processors. Therefore, the application space for such a solution is recognizably, but not exclusively, in the CPU design space.

In a traditional operand stack, multiple operands are held in a stack area organized notionally as a pushdown stack or LIFO buffer. Often the physical implementation employs a pointer into a small memory area. However, this leads to serious bottlenecks and there is little difference between this approach and a register file, especially where multiple operands need to be accessed in the same clock cycle (a requirement for superscalar operand access). In CPU oriented applications, it is not unusual to simply map a virtual stack onto an architectural register file, but undoubtedly this offers a poor presentation of a stack-oriented system since it is only an emulation of a true stack. It does however potentially allow superscalar execution since register files can be superscalar, though with significant hardware cost in terms of hazard avoidance and related logic. A true stack structure therefore appears at first sight to be restricted in its ability to act outside of its serial LIFO mode of operation. However, in this paper we demonstrate a discrete stack with multiport capability and superscalar functionality.

2. Superscalar Stack Issue Array (SSIA)

We propose a novel stack structure, the superscalar stack issue array (SSIA), whereby a small number of independent hardware stack cells are capable of being accessed collectively by multiple ports concurrently. However, such operations are inseparable from their stack effects (push, pop, or no effect). Consequently, collective stack reordering is performed as a single cycle internal operation, such that operands can be dispatched and stack-state kept coherent even with multiple actions in a group. The use of a novel tag scheme permits out-of-order write-back. All of these attributes can be achieved with basic logic structures. An unexpected benefit of destructive readout, usual in a stack structure, is that it eliminates the RAW/WAR dependency problem which hinders register-based processors to the extent as to demand renaming and reorder buffers and virtual registers with significant power, area, and thermal penalties. Therefore, an SSIA module can support superscalar operand issue and out-of-order completion without encountering RAW/WAR hazards or requiring a renaming scheme for its contents. These could conceivably be important benefits as described.

The tagging scheme allows delayed write-back to the stack structure, whereby any uncommitted stack cell content is supplanted by a unique tag which simply reserves and occupies the stack cell in question until the write-back can be completed. The unique aspect of this tagging scheme is that tags *move* with stack content, and therefore operands and results are actually *nomadic* in their behavior, and we refer to these incomplete contents as *nomads*. There is no possibility of multiple writes being able to target the same reserved space and hence no RAW/WAR.

If we consider how this looks at the logic level and how it differs from a standard stack, it may become clearer.

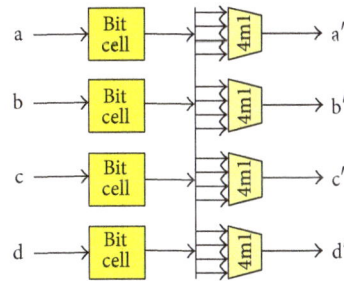

FIGURE 1: Fundamental stack logic scheme.

Figure 1 illustrates a fundamental stack implementation scheme. Each stack element bit is translated from its current state $\{a, b, c, d\}$ into its respective next state $\{a', b', c', d'\}$ by use of a multiplexer arrangement, which performs local reordering, to reflect the stack effect of the applied action. In a standard scalar stack, the next state is fed back to the state retention flip-flops. However, if multiple reorder-mux-stages are cascaded, then the state feedback represents the stack effects of a *group* of actions, and the intermediate stages provide multiple operand pairs simultaneously, as illustrated in Figure 2. It is also notable that only one state transition is observed at the state retaining flip-flop, no matter how many issue slots are cascaded. A superscalar stack will have significantly fewer such transitions than a serial/scalar stack given the same sequence of actions, in effect one that actually saves dynamic flip-flop power by going from a scalar to a superscalar structure. By adding logic to permit insertion of a tag-matched value from a common data bus (CDB), it is possible for stack cells to contain tags reserving locations for delayed write-back values and for these to "retire" when a CDB tag is matched.

The extra inputs on the 5-way multiplexers facilitate placement of an immediate operand d_n via the topmost multiplexer or selection of an optional "stack fill" value m_{fn} from memory in the bottom-most multiplexer. How such memory-to-stack data movement is handled has many options, and we do not explore these here; however, even trivial buffer schemes of small capacity or a memory queue approach is a feasible solution [13, 15, 16]. Thus, this scheme achieves the ability for *in-order issue* with *out-of-order completion*. This is investigated further in the literature [11, 13, 14, 17].

Analyzing this SSIA structure one can make several observations; first of all it appears that each issue slot contributes a logic delay to the overall cycle time of the system. The logic delay for cascaded SSIA structures is typically *linear with respect to issue width*, a fact that is not true of register files; indeed area growth in register files can be cubic or quadratic as a function of port count [18] and have considerable impact in design scaling [19]. A further observation is that SSIA operand access delays are unequal. Again this is not true of a register file where all operands are emitted from parallel read ports with approximately equal latency. For a superscalar stack scheme, each operand pair is accessed with a different latency, ranging from near zero latency to a larger latency at the final issue slot. This may or

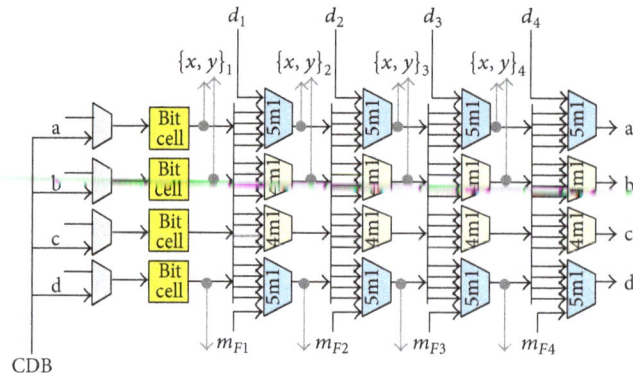

FIGURE 2: Superscalar stack issue array (issue width = 4).

may not be exploitable to some advantage; we simply observe the characteristic at this stage rather than speculating too far. However, what we would expect to observe is a similar story for power behavior where the *peak* power will not be a linear multiple of issue width, because each slot will peak at a *different* time and not at the *same* time. Total power over the whole cycle should however scale with issue width.

3. Implementation Scope and Variations

It should be clear, by analyzing these several cases of advanced stack operand stores, that the basic component building block of such a system is a multiplexer of some given characteristics, primarily the number of inputs or *input-ordinal*. It also follows that the best multiplexer design (in terms of speed, area, or power) will yield the best structure for the proposed stack-operand management scheme for the same optimization targets. An obvious choice is to use standard cell multiplexer components; however, this proves to be suboptimal where standard cell multiplexer choices are limited to a few cases and not necessarily implemented in the best way to facilitate ganging into suitable combinations. In order to evaluate this, it is necessary to consider the building block design at the logic level, its fan-out behavior, the logical effort, localized wire delay, and ultimately the full structure layout when a number of issue slots are cascaded and the logic is duplicated to represent the "n" bits of stack width (typically 16, 32, or 64 data bits per operand, for example).

We also note at this point that the depth of the discretely implemented stack is usually much greater than four elements. The top four elements only represent the "hot-zone" of the stack, where activity is more complex. Below this is a region we refer to as the "tidal zone" whereby all elements pop or push in unison, like ebb and flow of a tide, but are not reordered. Figure 3 highlights this difference in stack element zone behavior. It is noted that these deeper stack elements may be clock-gated or subject to dynamic power management when they are "empty." The nature of the stack makes this easily identifiable. We do not consider this in the analysis presented here however.

The tidal stack-element multiplexing arrangements are less complex. A 3-to-1 multiplexer selects either the existing

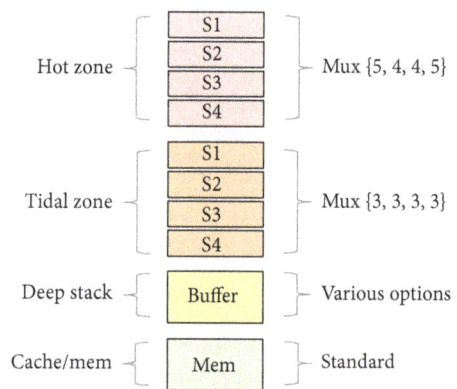

FIGURE 3: Functionally different regions of SSIA stack.

cell-state or one of two neighboring cells (above or below). How deep this tidal zone should be is a matter for further research. It should be as small as possible to reduce area/power cost, but a tidal zone of say 4 or 8 elements might not be sufficient to allow tags to achieve delayed write-back within the deeper stack before content migrates into memory or cache. This is an area for future research and we therefore assume, for the present, a discrete stack portion consisting of multiplexer ordinals as a set {3, 3, 3, 3, 5, 4, 4, 5} reading bottom to top. The {5, 4, 4, 5} portion being the "hot-zone" is symbolized by Figure 2 and the {3, 3, 3, 3} portion being a "tidal zone" is as in Figure 3. This configuration is assumed in the rest of the paper. There are numerous possibilities for implementing multiplexers, each has its own advantage and disadvantage. Some options are summarized below:

(1) combinational logic using AND/OR/MUX,

(2) tristate selection of multiple inputs,

(3) pass-gate transistor based solutions.

In this paper, we focus on cases (1) and (2), and we present timings as an FO4 delay metric and as an absolute timing. In both cases, we derive FO4 results from a 90 nm 1 volt process but also quote absolute timing for comparison between design choices. This allows FO4 delays to be comparable across process nodes up to a point. In later sections, we compare the SSIA with alternatives using similar process nodes.

In this paper, we identify several standard cells of interest for our design objectives and comparative implementations:

(i) AO22: dual two-input-AND feeding into an OR-gate,

(ii) AOI22: equivalent to AO22 but omits final inverter,

(iii) INV: an inverter,

(iv) INVT: inverter with tristate output,

(v) MUX2/MLX2: two input multiplexers, implemented using pass-gate method,

(vi) MUX3/MLX3: 3 input versions of MUX2 and MLX2, implemented as cascaded MUX2 and MLX2 circuit implementation styles,

(vii) QDBAH: active low data latch.

For our experiments we always select the X1 variation of the available standard cell. This is not the fastest option but achieves low area cost. The trade-off between larger faster cells and smaller slower cells is not straightforward: leakage-power differs, and larger cells will lead to longer wires in the full structure and alter RC loading effects. Using these cells, we evaluate the ganged tristate approach of Figure 4, and also four more typical implementations, based upon the schemes illustrated in Figures 5(a), 5(b), 6(a), and 6(b). We had predicted that the tristate selection method would perform better than combinational logic, based upon initial logical effort analysis. The tristate model assumes the preceding decode stage generates one-hot selection signals. Other cases assume encoded selection signals, generated in the same way. We can thus define five initial models for evaluation, which we will refer to using the notation given as follows:

(i) CEN: combinational logic, encoded select lines, and noninverting logic,

(ii) CEI: combinational logic, encoded select lines, and inverting logic where possible,

(iii) MEN: MUX2 based design, encoded select inputs, and noninverting cell outputs,

(iv) MEI: MLX2 inverting mux used where possible, encoded select inputs,

(v) TNN: tristate mux model with nonencoded select inputs, and noninverting output per stage.

In addition to the basic switching structure, each model has a constant delay associated with the state retention flip-flop/latch and tag-match data-insertion multiplexer. Referring back to Figure 2, it can be shown that each multiplexing stage cascades into its successor stage with fan-out of 5 (FO5C), with the exception of the final stage (or only stage in a nonsuperscalar case). The final stage feeds in to the storage bit cell, possibly via the tag match insert logic, which can be placed either immediately before the bit-cell input or immediately after the bit-cell output. Therefore, it has a fan-out of 1 rather than the fan-out of 5 encountered when cascading to further stages. We therefore have case FO1T (feeding a tag-logic stage) and case FO1L feeding a latch stage.

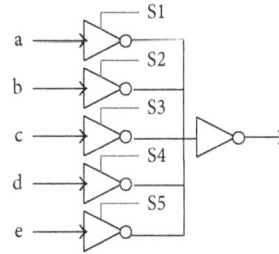

FIGURE 4: TNN-ganged tristate multiplexer scheme.

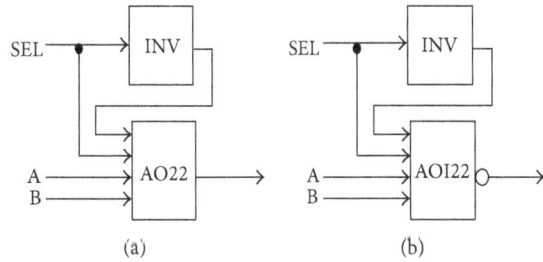

FIGURE 5: Combinational logic gate schemes. (a) CEN-two-input combinational Mux, (b) CEI-inverting equivalent.

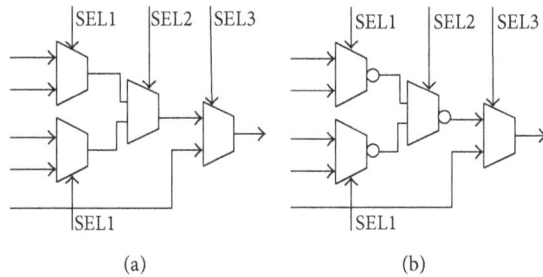

FIGURE 6: Cascaded multiplexer schemes. (a) MEN-two-input cascaded Mux, (b) MEI-inverting equivalent.

The loading effect of these choices then result in three fan-out loading cases that need to be calculated for each multiplexer style, relating to the position in the cascade chain:

FO5C: fan-out 5 cascaded,

FO1T: fan-out 1 to tag insert mux,

FO1L: fan-out 1 to Latch.

We evaluate this worst case here and results of our evaluations are presented in Table 1. From this initial analysis, it appears that the TNN model has the best speed advantage, and that either FO1T or FO1L variants are preferable for speed.

4. Tristate Multiplexer Models in Detail

In order to assess the tristate multiplexer strategy with as much accuracy as possible, we utilized several evaluations. First of all, evaluations based upon the 90 nm process data sheets were used to generate an initial timing estimate for

TABLE 1: Timing of 5ml component models.

	FO5C	FO1T	FO1L
CEN	291 ps	229 ps	270 ps
CEI	211 ps	149 ps	164 ps
MEN	225 ps	188 ps	208 ps
MEI	168 ps	130 ps	131 ps
TNN	125 ps	97 ps	97 ps

TABLE 2: Data-sheet predictions.

Gang	Delay	Tristate fF	Inv fF
1	89 ps	2.67 fF	13.05 fF
2	98 ps	4.41 fF	13.05 fF
3	107 ps	6.15 fF	13.05 fF
4	116 ps	7.90 fF	13.05 fF
5	125 ps	9.64 fF	13.05 fF
6	134 ps	11.38 fF	13.05 fF
7	143 ps	13.13 fF	13.05 fF
8	152 ps	14.87 fF	13.05 fF

ganged tristate multiplexers, as shown in Table 2, using the model introduced earlier in Figure 4.

The total load capacitance (CLOAD) driven by the active tristate in the group was calculated as a function of fan-out (F) and ganging (G) plus the next stage inverter, such that load capacitance equated to

$$\text{CLOAD} = 2.668 \times F + 1.743 \times (G - 1). \tag{1}$$

The value 2.668 refers to input capacitance of the assumed next stage (an Inverter INVX1 cell), whilst the value of 1.743 refers to the capacitance of the outputs of the other inverters (referred to in the data sheet as input capacitance). Both values are extracted from the cell-library data sheets. These estimates appear to be sufficient for a first-order comparison of implementation methods. However, this may not be wholly accurate: signal behavior, layout, and wire effects have some importance, as will be highlighted later in this paper.

Having evaluated the timing models in each case and repeated the analysis for each of the FO5C, FO1T, and FO1L output loading cases, we have enough data to perform an evaluation of delay variation as a function of issue width. However, a complete analysis still needs to incorporate the constant delay of latch and tag insertion timing parameters themselves and not just the effect they have on output loading of preceding stages. This is outlined in the following Sections 4.1 and 4.2.

4.1. Tag Insertion Multiplexer.
The MUX2 standard cell (X1 type) drives a single fan-out load for the data input of the latch QDBAH (X1 type). QDBAHX1 has an input capacitance of 1.976 fF for the D input node. This allows us to calculate a delay of 70 ps using the straight-line fit to MUX2 propagation delay data versus capacitive loading of the driven QDABHX1 latch input.

4.2. State Storage Latch.
The latch itself has several important timing requirements. Setup and hold time amounts to just

TABLE 3: Bit cell and tag timing data.

FO5	MUX2	MLX2	AO22	AOI22	INVT
D-Q	74 ps	76 ps	76 ps	76 ps	76 ps
Total	179 ps	181 ps	181 ps	181 ps	181 ps

TABLE 4: Example aggregate timing formulae.

IW	Delay summation
1	Latch/tag + CEN.FO1T
2	Latch/tag + CEN.FO5C + CEN.FO1T
3	Latch/tag + (2 × CEN.FO5C) + CEN.FO1T
4	Latch/tag + (3 × CEN.FO5C) + CEN.FO1T

TABLE 5: Cycle times.

IW	CEN	CEI	MEN	MEI	TNN
1	410	330	367	311	306
	(9.1)	(7.3)	(8.1)	(6.9)	(6.8)
2	701	541	592	479	431
	(15.5)	(12.0)	(13.1)	(10.6)	(9.5)
3	992	752	815	647	556
	(22.0)	(16.7)	(18.1)	(14.4)	(12.3)
4	1283	963	1042	815	681
	(28.5)	(21.4)	(23.1)	(18.1)	(15.1)

FO4 delays in brackets.

under 35 picoseconds. Data propagation from D to Q varies according to output load. This will be fan-out 5 in all cases but the target cell could be any one of MUX2, MLX2, AO22, AOI22, or INVT depending upon the implementation. We therefore have a further table (Table 3) for values derived for this latch propagation delay and also the final total with tag logic included.

Examination of Table 3 presents the data latch and total timing (tag stage plus latch timings). This shows that the variation in timing of the bit cell latch for different input variants is marginal.

5. First-Order Timing Projections

With timings projected for each component under all of the encountered input and output conditions, it is possible to assemble a timing analysis for a stack issue-logic slice for each of the component models introduced. At this stage, the subcomponent delays can simply be summed according to the configurations used. So, for example, the CEN model using combinational noninverting logic has configurations with respect to issue width as given in Table 4.

Replacing "CEN" with any of the other models allows the same cascading to be used to calculate the respective delays of each model. These are given in Tables 5 and 6 and plotted as graphs in Figures 7 and 8. Tables 5 and 6 show the absolute timing for the chosen 90 nm process, and the FO4 delay figure extracted on the basis that FO4 delay for 90 nm technology is 45 ps. By way of confirmation, our experimental timing measures derived an average FO4 delay of around

TABLE 6: Access times.

IW	CEN	CEI	MEN	MEI	TNN
1	0 ps	0 ps	0 ps	0 ps	0 ps
	0	0	0	0	0
2	291 ps	211 ps	225 ps	168 ps	125 ps
	6	5	5	4	3
3	582 ps	422 ps	450 ps	336 ps	250 ps
	6/13	5/9	5/10	4/7	3/5
4	873 ps	633 ps	675 ps	504 ps	375 ps
	6/13/19	5/9/14	5/10/15	4/7/11	3/5/8

FO4 delays shown beneath for issues slots 2/3/4.

43 ps for timing combined tpHL and tpLH. The results show that TNN is the best choice with this level of analytical detail. The combinational logic model CEN is by far the worst option even for a scalar issue model, and TNN is almost twice as fast as CEN for the highest issue width examined.

6. ASIC Layout and Core Evaluations

After performing timing estimates based primarily upon reported standard cell characteristics, it appeared that the tristate model offered the best delay characteristics without resorting to full custom cell design. This implementation model was then evaluated further during a collaborative visit to the Circuits and Systems Research Centre (CSRC), University of Limerick, by one of the authors.

VHDL coded descriptions of the 5m1 multiplexer, using tristate internal selection, were synthesized to create a standard-cell based mux core. After manually tidying up, the 5m1 cell appeared as illustrated in Figure 9.

Examining the cell critical wires as in Figure 10, wire lengths approximate to 13.7 μm (ganged wire), 2.7 μm (input), and 2.0 μm (output). After DRC and LVS checks, the 5m1 cell was cropped and rechecked by removal or addition of one or more tristates to create a range of cells from 2m1 through to 8m1. At this stage, a number of effects were then considered in order to get an accurate delay estimate for the building block. These are outlined as follows.

6.1. Slew Rates.
Our tests revealed that input slew rates of the input test signals have an appreciable effect on timings, adding more than 10 ps to measurements for a conservative slew rate of 100 ps. We used a buffer chain to condition the input test signals and give realistic signal properties for our tests.

6.2. Standard Cell RC Delay.
With RC behavior included, the delays for our multiplexer combinations are increased noticeably. The data sheets only provide for transistor characteristics and not layout related to wire and metal effects. Adding wire-related delays for the ganging interconnect in the multiplexer further increases the delay. Simulation data for these measurements is given in Table 7.

6.3. Fan-Out Conditions.
Delays were measured under a fan-out of five-tristate inverter load, to match the anticipated

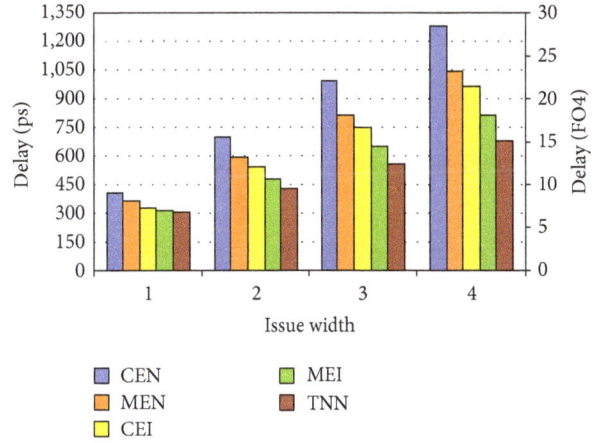

FIGURE 7: Issue width versus cycle time for various implementations.

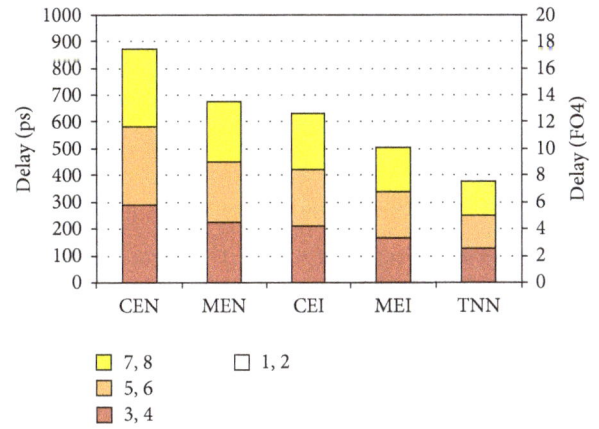

FIGURE 8: Issue width versus access time for stack using various implementation schemes. Showing operand access time for operand pairs 1-2, 3-4, 5-6, and 7-8, where operand pairs 1-2 have zero access time.

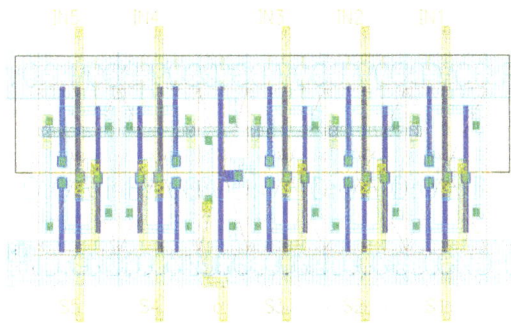

FIGURE 9: 5m1 tristate multiplexer layout.

output gate loading conditions. At this stage it becomes obvious that the initial timing projections given in Section 5 are conservative when compared against real layout and RC extracted timings (as shown in Table 7). Examination of the internal behavior of a 5m1 ganged tristate multiplexer core shows the effect of ganging and the unequal switching times for tPHL and tPLH transients. This is illustrated in Figure 11,

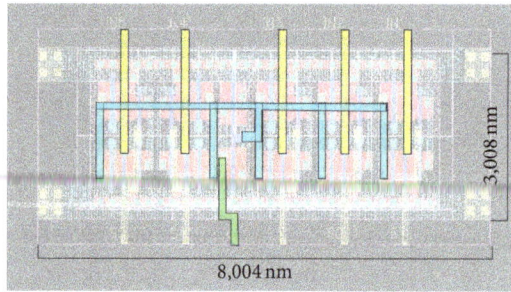

FIGURE 10: Significant wires in the 5ml tristate cell, with dimensions in nm.

TABLE 7: Delays with and without local wires, for TNN MUX.

	Data sheets (FO5)	Cell delays (FO5)	Cell and wire (FO5)
2ml	89 ps	105 ps	117 ps
3ml	98 ps	121 ps	136 ps
4ml	107 ps	134 ps	155 ps
5ml	116 ps	150 ps	173 ps
6ml	125 ps	164 ps	190 ps
7ml	143 ps	179 ps	208 ps
8ml	152 ps	192 ps	225 ps

which shows an input signal of matched slew rate, causing internal node switching on the ganged tristate, which in turn drives the final output inverter. Both tPHL and tPLH are shown; it is clear that the transient on the internal node is the critical issue for this design. Using a dynamic precharging method on the ganged bus (when all tristates are disabled) might have a significant impact on this problem, allowing faster operation. This would certainly be an area for further investigation.

6.4. Fully Cascaded Structure and Bus Interconnect. To make comparative analysis easier, we derived a model for tristate behavior under given output loading conditions and multiplexer input counts. We began with the plot shown in Figure 12, which shows schematic timings data (blue), RC extracted data (red), and our final model (green: Figure 13) incorporating a distribution bus metal structure suitable for driving a 5-4-4-5 multiplexer row, which is the required worst case for the cascaded n-wide issue structure described in this paper. We also performed simulations of multiplexer cores at schematic and layout levels, and with cascaded cells, to derive timings approaching those for a full ASIC implementation. The data in Figure 13 is used for the final performance estimates, leading to the revised component timings for the TNN model, as given in Tables 8(a), 8(b), and 8(c), with wire effects of Table 9 and predicted access and cycle times in Table 10. This incorporates schematic derived timing data for the latch and tag stage (see Tables 8(b) and 8(c)), and wire effects (Table 9).

The full layout structure is shown in several figures given as follows: a single row of hot-zone elements is shown in Figure 14. When several of these rows are stacked one below another, interlocking the input and output bus lines,

TABLE 8: (a) Timing models, various simulation modes. (b) Simulation timings for TNN multiplexer driving various next stages. (c) Schematic timings for LATCH and TAG-MUX components.

(a)

Mux	Data sheet	Schematic	Cell + wire	With bus
3ml	107 ps	121 ps	136 ps	156 ps
4ml	116 ps	134 ps	155 ps	176 ps
5ml	125 ps	150 ps	172 ps	195 ps

(b)

	3ml	4ml	5ml
FO5C	156 ps	176 ps	195 ps
	(3.4)	(3.9)	(4.3)
FO1T	127 ps	144 ps	159 ps
	(2.8)	(3.2)	(3.5)
FO1L	127 ps	144 ps	159 ps
	(2.8)	(23.2)	(3.5)

Includes std cell RC effects, intercell wires, and interrow/slot distribution bus. Bracketed data gives equivalent FO4 delays.

(c)

LATCH TO MUX2	106 ps	Total
MUX to {5, 4, 4, 5} row FO4	132 ps	**237 ps**
MUX2 to latch	102 ps	Total
LATCH to {5, 4, 4, 5} row	135 ps	**238 ps**

TABLE 9: Delay data for simple wire of length 0 um–60 um.

Wire	0 um	20 um	40 um	60 um
Delay	~	3.0 ps	6.2 ps	8.7 ps

TABLE 10: Cycle/access times versus issue width.

	IW 1	IW 2	IW 3	IW 4
Access Time	0 ps	195 ps	390 ps	585 ps
	(0)	(4.3)	(8.7)	(13.0)
Cycle Time	397 ps	592 ps	787 ps	982 ps
	(8.8)	(13.1)	(17.5)	(21.8)

the structure appears as in Figure 15, which represents a 4-wide "hot-zone" structure for a single bit of stack word width.

Hot-zone control wire supply lines are required for each issue slot, 20 per issue slot. These are capable of being routed over the cells via a higher metal layer (seen running vertically top to bottom in Figure 15). These control lines do not impact upon the structure's overall area and standard cell packing density and have no influence on the dimensions of buses used to connect critical data paths between stacked issue slots. Finally, Figure 16 shows the additional cells, added to the left hand side of the layout, representing the {3, 3, 3, 3} "tidal" stack zone attached to each hot-zone issue slot module, with shared control lines coming from the left-hand side in this case (though over-cell routing is possible). The practical consideration to be made here is what impact the interconnect buses (between hot-zone modules) have upon

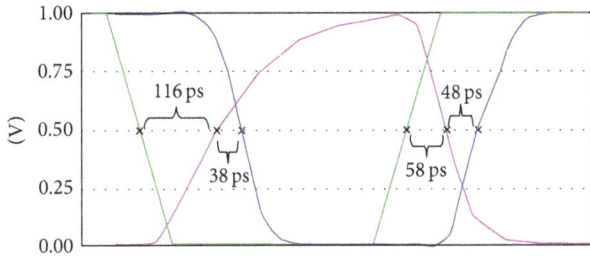

FIGURE 11: External and internal signals relationships for 5m1 RC extracted simulation. CADENCE timing data plot (redrawn for clarity).

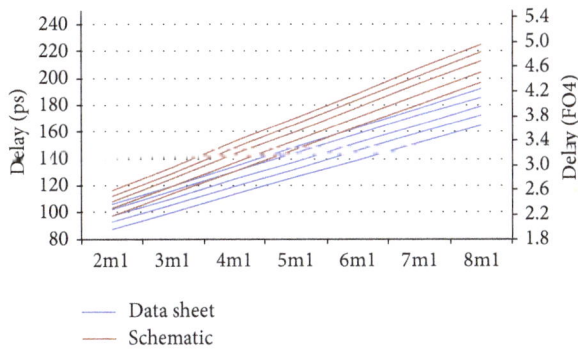

Data sheet
Schematic

FIGURE 12: Data sheet versus schematic driven timings. Data sheet (Blue) measured schematic (red).

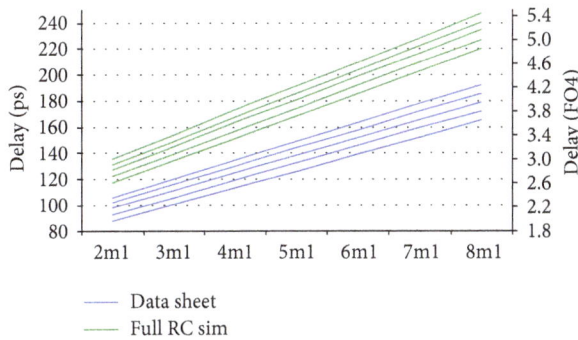

Data sheet
Full RC sim

FIGURE 13: Data sheet versus RC delay with bus metal. Data sheet (blue) distribution bus (green).

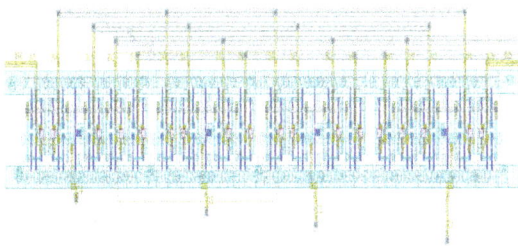

FIGURE 14: Hot-zone row {5, 4, 4, 5} configuration, and interconnect bus detail. Four outputs at the bottom are used to drive the four bus lanes of the next cell (equivalent to those at the top of this cell).

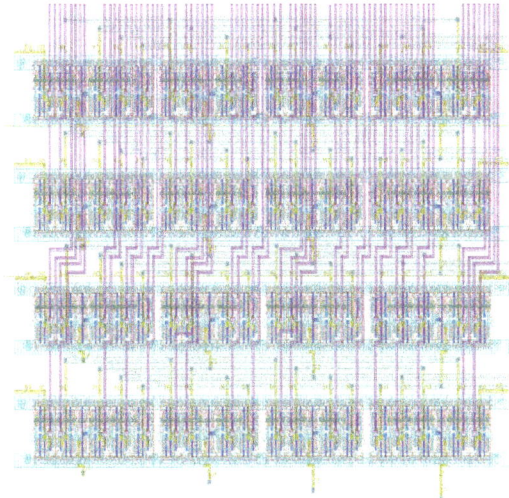

FIGURE 15: 4-wide {5, 4, 4, 5} issue structure.

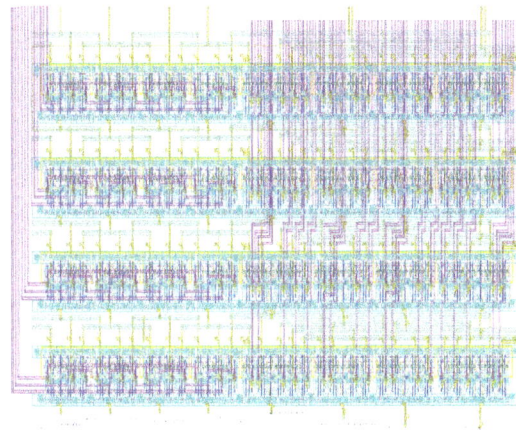

FIGURE 16: Complete {3, 3, 3, 3, 5, 4, 4, 5} × 4-slot structure, showing metal layers only.

signal delay due to added capacitance. These bus lines are singularly driven but fan out to 5 destinations. We performed simulation of cascaded rows using the bus structure shown in Figure 16 in order to account for this and found that bus related delays were typically of the order of 20 ps.

7. Comparative Performance

In previous sections, we developed a structural model for a superscalar operand stack and presented timing estimates for five different (relatively straightforward) implementation strategies. The next question to pose is to ask how fast a superscalar stack is with respect to register file alternatives. However, given the radically different nature of the superscalar operand store, as compared to the register file paradigm, blind like-for-like comparisons are difficult to assert. Consider a likely configuration of our superscalar stack unit, with perhaps four top-of-stack "hot-zone" elements (representing the {5, 4, 4, 5} multiplexer arrangement) and between four and eight deeper "tidal" stack elements implemented with 3-to-1 multiplexers. If we take the latter case then we have a total of 12 stack elements. The question is raised: *is this SSIA equivalent to a register file of 12 registers?* Our rationale for comparison is as follows.

7.1. Word Count. Such a stack as described above can push operands deeper than the 12 discrete elements implemented directly in logic, so it may not be accurate to describe it as having equivalent capacity to a register file of 12 words. Conversely, the number of "useful" registers in a register file under certain workloads is a small fraction of full capacity. Many registers are either awaiting write-back or contain dead data that will never be read again. Useful lifespans are typically measured in low 10's of clock cycles [39], so much so that some researchers have even proposed discharging unused registers to save power [40]. Stack content is almost exclusively live and useful, and it is much rarer to have dead "nomads"; thus, direct comparison is misleading. Empty stack elements are always identifiable easily and this correlates with our early comment about empty stack elements being able to be power or clock gated in a more sophisticated SSIA solution. However, we note that the fact that redundant register content is seen as a significant concern for static power wastage suggests that the stack approach has potential here as it rarely contains redundant content, which could be exploited for static/dynamic power reductions.

7.2. Port Count. Typical superscalar register files are arranged into "*n*" write ports and "*2n*" read ports. A 12-port register file is often organized as 8 read ports and 4 write ports. Such a configuration matches the expectations of a 4-issue machine, with 4 pairs of operands (8 reads) and 4 possible results (4 writes). An SSIA will have an identical read-port count (a 4-issue stack provides 8 operands). However, each stack element is capable of retiring an uncommitted value to the SSIA, so it might be thought that there are "*n*" write ports for an *n*-deep discrete logic stack portion. However, there are limitations on practical numbers of retirement buses, and in practice the ability to retire 2, 3, or 4 results to the stack is more realistic and has an impact on the design of the tag match logic (which we have not considered in this analysis) and therefore makes low orders of concurrent writes more desirable.

A further complication is that each issue slot can write an operand to the stack, albeit only to the top element. Do each of these write channels count as write ports in the traditional sense? With the restricted nature of the destination (top of stack only) it seems that it makes sense not to count these as true write ports in their own right. The postulated 12-deep stack with an issue width of four could retire four ALU results and also accept up to four new data values, whilst reading out 8 operands. The extreme interpretation is that this is therefore an 8+8 port operand store. Given that the 8 write channels are actual two groups of four channels with functionally different behaviors, we do not think this is the best way to compare like-for-like (as far as this is possible). We therefore come to a fairly loose conclusion: a suitable SSIA model for comparison would assume the equivalent of 1 write port and two read

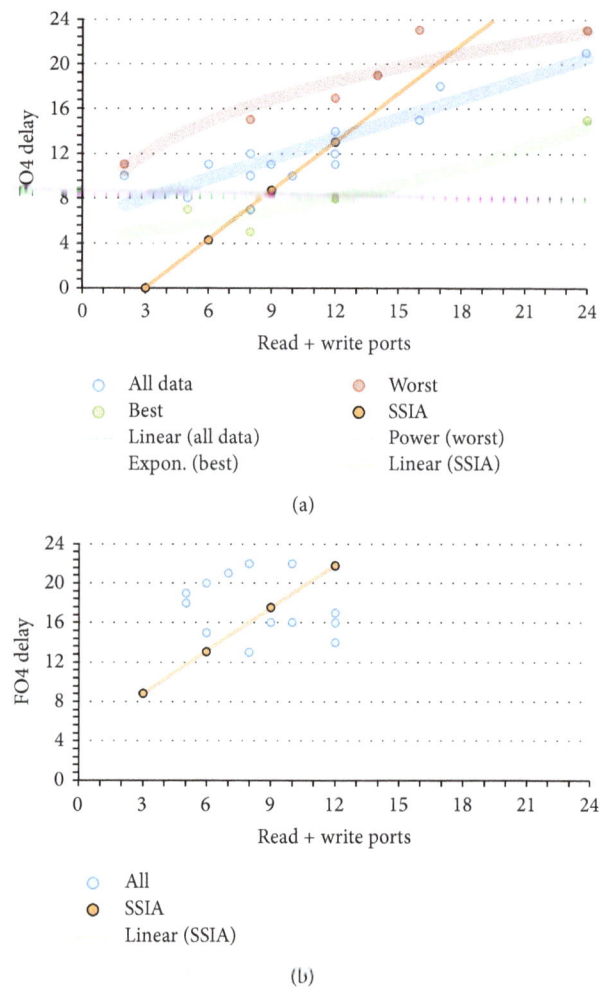

(a)

(b)

FIGURE 17: Comparison between SSIA and register file timing. (a) Access times and (b) cycle times, plotted on same scales.

ports per issue slot. Note that the stack size has no direct impact on the number of write ports assumed as this is largely a function of issue width; however, this is a parameter that could be investigated further.

To perform a comparison of SSIA versus register file performance, we collated a range of over forty FO4 timing data citations and timings for register files as reported in the literature [20–37]. These are detailed in Table 11. Access-time data appears to be more widely reported than cycle times (28 citations versus 13 in our sample). We consider both the access time and the cycle time of our SSIA in comparison to the register file. The population of access-time data for register files (from Table 11) was large enough to derive both best-case and worst-case envelopes and also a general trend for port count versus access delay, based on the whole data set. The data for cycle time was not sufficiently populated to support meaningful trend extrapolation. However, the position of the SSIA cycle time predictions within the group highlights its relative timing behavior. Plotting this analysis gives the graph shown in Figures 17(a) and 17(b), which show the population of register file FO4 characteristics for access time (Figure 17(a)) and cycle time (Figure 17(b)) alongside

TABLE 11: Reported register file delay times.

Reference	Ports[a] Tot (R + W)	T_{READ}[b] (FO4)	T_{CYCLE}[b] (FO4)	Comments
[20, 21]	10 (5 + 5)		22	250 nm CMOS, AMD K7 88 × 90 bit
[20, 22]	8 (4 + 4)	7		130 nmCMOS, 256 × 32 bit
[20, 23]	5 (3 + 2)	7	18	250 nm CMOS, 16 × 64 bit
[20, 24]	5 (3 + 2)	8		250 nm CMOS, 32 × 64 bit, IBM PowerPC
[20, 25]	8	0	13	500 nm CMOS, 32 × 64 bit
[20]	6 (4 + 2)	11	15	32 × 64 bit logical effort model
[20]	9 (6 + 3)	11	16	32 × 64 bit logical effort model
[20]	12	12	17	32 × 128 bit logical effort model
[26]	12 (8 + 4)	17		100 nm CMOS, 16 × 32 bit, low power
[26]	12	8		100 nm CMOS, 16 × 32 bit, high speed
[27]	8 (6 + 2)	5		250 nm SOI, 32 × 64 bits
[28]	12 (4 + 4)	11		100 nm, 160 × 64, two-bank, 440 reg
[28]	12 (4 + 4)	14		100 nm, 160 × 64, two-bank, 80 reg
[28]	24 (16 + 8)	15		100 nm, 160 × 64, two-bank, 40 reg
[28]	24 (16 + 8)	21		100 nm, 160 × 64, two-bank, 80 reg
[29]	17	18		1 um CHMOS, 128 × 64 bit,
[30]	14 (10 + 4)	19		400 nm, 116 × 64 bit
[31]	16 (10 + 6)	23		110 nm CMOS, 34 × 64 bit
[32]	24 (16 + 8)	23		130 nm, 512 reg
[33]	10	10	16	80 × 64 bit, various 250–35 nm
[34]	16 (12 + 4)	15		75 nm CMOS, 128 register,
[35]	2 (1 + 1)	11		180 nm, 160 reg 8 bank
[35]	2 (1 + 1)	10		180 nm, 160 reg 4 bank
[35]	8 (4 + 4)	15		180 nm, 160 reg, 1 bank
[35]	8 (4 + 4)	12		180 nm, 100 reg, 1 bank
[35]	8 (4 + 4)	10		180 nm, 60 reg, 1 bank
[35]	8 (4 + 4)	10		180 nm, 60 reg, 4 bank
[36]	5 (3 + 2)		19	500 nm CMOS, 128 reg
[36]	6 (3 + 3)		20	500 nm CMOS, 128 reg
[36]	7 (4 + 3)		21	500 nm CMOS, 128 reg
[36]	8 (4 + 4)		22	500 nm CMOS, 128 reg
[37]	12 (8 + 4)		14	500 nm CMOS, 48 reg,
[37]	12 (8 + 4)		16	500 nm CMOS, 96 reg

[a]Ports are stated as T (R + W) where T is port total, and bracketed figures (R + W) represent read and write ports where known.
[b]Delays are stated as FO4 delay, assuming 1 FO4 delay equates to an approximation scale of 2 nm per ps [38].

the same timing characteristics for the tristate-multiplexer SSIA model.

The comparative analysis presented in Figure 17 shows that the predicted models for the proposed SSIA configuration have a delay characteristic that is very competitive with register file for both access time and cycle time. This is certainly true for operand pair issue widths of 1, 2, 3, and 4 (equating to port counts of 3, 6, 9, and 12). One can observe that SSIA model appears to have increased delay penalty relative to register file for extreme port counts. For cycle times, where data was only available for register files up to 12 ports, the SSIA model appears competitive across the issue width/port count range plotted. Overall, it can be stated, based upon the data available in this comparison, that SSIA has highly competitive performance across both access and cycle times for up to 8 read ports and issue widths up to 4.

Taking the worst case delays from Table 10 (cycle times), it is possible to make tentative frequency estimates for a pipelined architecture limited by the superscalar store, with frequencies of around 2.5 GHz, 1.7 GHz, 1.3 GHz, and 1.0 GHz for issue widths of 1, 2, 3, and 4, respectively. A complete layout for one bit of an n-wide SSIA is given in Figure 18.

8. Conclusions

In this paper, we have considered a novel approach to operand management, using a stack based approach with a scheme which we believe is a new and novel approach for multiple operand issue, permitting superscalar in-order issue and out-of-order completion. Our methodology has been detailed and a model for building suitable stack structures is demonstrated.

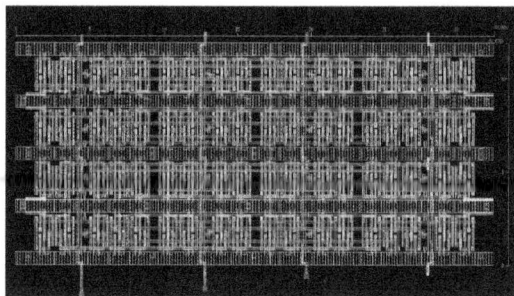

FIGURE 18: Suggested abutted multiplexer row/issue slot structure for 4-wide issue. Rows represent issue slots. Core area is ~32 × 11.9 μm = 378 μm^2 per bit (top four cells).

The cascading nature of the logic structure has two aspects: first of all, it allows a linear growth in logic cost and area, as well as delay characteristics and power consumption, which are potentially advantageous. However, the cascading nature also means that at some point (an issue width of 8, equivalent to a port count of 20, for example) the SSIA as presented here becomes less desirable as those costs accumulate. However, with realistic issue widths the cascading effect does not reach an uncompetitive behavior. Delay evaluations have been made, utilizing industry standard tools for core building block characterization, and making reasonable assumptions for configurations of systems equating to various issue widths. We have compared our projected performance data against a range of existing register file models and found comparable performance is viable. There are a number of possible enhancements that could be made to this base design in order to improve performance. These include

(i) advanced bit-cell design,

(ii) ganged pass gates rather than tristates,

(iii) look-ahead schemes to reduce cascading,

(iv) dynamic precharge of common node,

(v) selective clock-gating of empty cells.

Combining these possibilities, we envisage that cycle delays might be halved and perhaps more. This suggests that a 2 GHz 4-wide issue stack is conceivable with 90 nm CMOS with careful optimization and more innovative design, implying operand issue bandwidth somewhere in the region of 8 to 12 Gigawords per second at 90 nm. Naturally, more advanced processes will offer further performance gains.

One of the most interesting aspects of the superscalar stack is its ability to deliver issue slot operands at different times within the cycle time window, a unique behavior which we believe will allow more freedom in layout floor planning at a higher level of abstraction. This is particularly interesting if wire pipelining is introduced [41]. We also expect to observe (and perhaps to enhance further) a power-spreading effect that reduces peak power over machine cycle time scales due to the ripple-through effects of logic in our structures. This has some potential to reduce power density hot spots in the operand store as well as reducing peak power spikes. When combined with the knowledge that "useful" register lifespans

are often a small fraction of the power-hungry register file [39, 40, 42], SSIA starts to look more interesting as a possible candidate upon which to base superscalar systems. Work on more advanced multiplexer design continues to be a current topic [43], and there is substantial scope to learn from this and improve upon the implementations reported here.

In the wider context, it has been fashionable to consider the stack machine as outdated. However, the potential for such architectures to deliver complex operand and instruction issue models highlights fresh opportunities and offers a new twist in the development of stack machines and related queue machines. Combining this with significantly better stack code optimization frameworks and models highlighted earlier [8–10, 14, 15] suggests that stack machines might be overdue a fresh examination in view of the trend for many simpler cores per chip rather than fewer but more complex ones. We believe that new avenues have been opened up by our initial study, in answering one question we have uncovered many others. A more comprehensive VLSI oriented study is a highly desirable next step in this work. Collectively these objectives will allow a complete design characterization for a prototype superscalar stack processor to be achieved. We therefore expect to continue to evaluate these new and novel SSIA architectures in the future and hope to report further findings in due course.

Conflict of Interests

The authors declare that there is no conflict of interests regarding the publication of this paper.

Acknowledgment

This paper was supported by work undertaken via visiting academic spells hosted by the Circuits and Systems Research Centre, University of Limerick.

References

[1] http://www.itrs.net/links/2011itrs/home2011.htm.

[2] G. Venkatesh, J. Sampson, N. Goulding-Hotta, S. K. Venkata, M. B. Taylor, and S. Swanson, "QsCores: trading dark silicon for scalable energy efficiency with quasi-specific cores," in *Proceedings of the 44th Annual IEEE/ACM Symposium on Microarchitecture (MICRO '44)*, pp. 163–174, ACM, December 2011.

[3] R. J. Ribando and K. Skadron, "Many-core design from a thermal perspective," in *Proceedings of the 45th Design Automation Conference (DAC '08)*, pp. 746–749, Anaheim, Calif, USA, June 2008.

[4] D. Sylvester and H. Kaul, "Future performance challenges in nanometer design," in *Proceedings of the 38th Design Automation Conference*, pp. 3–8, ACM, June 2001.

[5] H. O. Ron, K. W. Mai, and A. Fellow, "The future of wires," *Proceedings of the IEEE*, vol. 89, no. 4, pp. 490–504, 2001.

[6] H. Esmaeilzadeh, E. Blem, R. St. Amant, K. Sankaralingam, and D. Burger, "Dark silicon and the end of multicore scaling," in *Proceedings of the 38th Annual International Symposium on Computer Architecture (ISCA '11)*, pp. 365–376, IEEE, 2011.

[7] I. E. Sutherland, R. F. Sproull, and D. F. Harris, *Logical Effort: Designing Fast CMOS Circuits*, Morgan Kaufmann, 1999.

[8] P. Koopman, "A preliminary exploration of optimized stack code generation," in *Proceedings of the Rochester Forth Conference*, Rochester, NY, USA, 1992.

[9] B. Chris, "Inter-boundary scheduling of stack operands: a preliminary study," in *Proceedings of the EuroForth*, pp. 3–11, 2000.

[10] M. Shannon and C. Bailey, "Global stack allocation: register allocation for stack machines," in *Proceedings of the Euroforth Conference*, 2006.

[11] C. Bailey and M. Weeks, "An experimental investigation of single and multiple issue ILP speedup for stack-based code," in *Proceedings of the EuroForth Conference*, pp. 19–24, 2000.

[12] US Patent 6148391: System for Simultaneously Accessing one or More Stack Elements by multiple functional units, and related US patent 6026485: Instruction Folding for A Stack-Machine.

[13] C. Bailey, "A proposed mechanism for super-pipelined instruction-issue for ILP stack machines," in *Proceedings of the EUROMICRO Systems on Digital System Design (DSD '04)*, pp. 121–129, IEEE, September 2004.

[14] C. Bailey and H. Shi, "Instruction level parallelism of stack-code under varied issue widths, and one-level branch prediction," in *Proceedings of the IADIS International Conference on Applied Computing (AC '05)*, pp. 23–30, Algarve, Portugal, February 2005.

[15] C. Bailey, R. Sotudeh, and M. Ould-Khaoua, "The effects of local variable optimisation in A C-based stack processor environment," in *Proceedings of the 1994 Euroforth Conference*, 1994.

[16] T. J. Stanley and R. G. Wedig, "A performance analysis of automatically managed top of stack buffers," in *Proceedings of the 14th Annual International Symposium on Computer Architecture (ISCA '87)*, pp. 272–281, ACM, 1987.

[17] C. Bailey, "A proposed mechanism for super-pipelined instruction-issue for ILP stack machines," in *Proceedings of the Euromicro Symposium on Digital System Design (DSD '04)*, pp. 121–129, IEEE, 2004.

[18] C. Jesshope, "Microthreading a model for distributed instruction-level concurrency," *Parallel Processing Letters*, vol. 16, no. 2, pp. 209–228, 2006.

[19] S. Galal and M. Horowitz, "Energy-efficient floating-point unit design," *IEEE Transactions on Computers*, vol. 60, no. 7, pp. 913–922, 2011.

[20] N. Burgess, "Logical Effort analysis of multi-port register file architectures," in *Proceedings of the Conference Record of the 37th Asilomar Conference on Signals, Systems and Computers*, vol. 1, pp. 887–891, IEEE, November 2003.

[21] M. Golden and H. Partovi, "500 MHz, write-bypassed, 88-entry, 90-bit register file," in *Proceedings of the Symposium on VLSI Circuits*, pp. 105–108, IEEE, June 1999.

[22] R. K. Krishnamurthy, A. Alvandpour, G. Balamurugan, N. R. Shanbhag, K. Soumyanath, and S. Y. Borkar, "A 130-nm 6-GHz 256 × 32 bit leakage-tolerant register file," *IEEE Journal of Solid-State Circuits*, vol. 37, no. 5, pp. 624–632, 2002.

[23] R. L. Franch, J. Ji, and C. L. Chen, "A 640-ps, 0.25-µm CMOS, 16 × 64-b three-port register file," *IEEE Journal Solid State Circuits*, vol. 32, no. 8, pp. 1288–1292, 1997.

[24] O. Takahashi, J. Silberman, S. Dhong, P. Hofstee, and N. Aoki, "690ps read-access latency register file for a GHz integer microprocessor," in *Proceedings of the 1998 IEEE International Conference on Computer Design*, pp. 6–10, Austin, Tex, USA, October 1998.

[25] W. Hwang, R. V. Joshi, and W. H. Henkels, "A 500-MHz, 32-word x 64 bit, eight-port self-resetting CMOS register file," *IEEE Journal of Solid-State Circuits*, vol. 34, no. 1, pp. 56–67, 1999.

[26] C. H. Hua and W. Hwang, "Low power multiple access port register file design in 100 nm CMOS technology," in *Proceedings of the 14th VLSI/CAD Symposium*, Hualien, Taiwan, August 2003.

[27] R. V. Joshi, W. Hwang, S. C. Wilson, and C. T. Chuang, ""Cool low power" 1 GHz multi-port register file and dynamic latch in 1.8 V, 0.25 µm SOI and bulk technology," in *Proceedings of the International Symposium on Low Power Electronics and Design (ISLPED '00)*, pp. 203–206, July 2000.

[28] M. Kondo and H. Nakamura, "A small, fast and low-power register file by bit-partitioning," in *Proceedings of the 11th International Symposium on High-Performance Computer Architecture (HPCA-11 '05)*, pp. 40–49, IEEE, February 2005.

[29] R. D. Jolly, "A 9-ns, 1.4-gigabyte/s, 17-ported CMOS register file," *IEEE Journal of Solid-State Circuits*, vol. 26, no. 10, pp. 1407–1412, 1991.

[30] C. Asato, "A 14-port 3.8-ns 116-word 64-b read-renaming register file," *IEEE Journal of Solid-State Circuits*, vol. 30, no. 11, pp. 1254–1258, 1995.

[31] N. Tzartzanis, W. W. Walker, H. Nguyen, and A. Inoue, "A 34word × 64b 10R/6W write-through self-timed dual-supply-voltage register file," in *Proceedings of the IEEE International Solid-State Circuits Conference, Digest of Technical Papers (ISSCC '02)*, vol. 2, pp. 338–537, San Francisco, Calif, USA, February 2002.

[32] N. S. Kim and T. Mudge, "The microarchitecture of a low power register file," in *Proceedings of the International Symposium on Low Power Electronics and Design (ISLPED '03)*, pp. 384–389, August 2003.

[33] V. Agarwal, M. S. Hrishikesh, S. W. Keckler, and D. Burger, "Clock rate versus IPC: the end of the road for conventional microarchitectures," in *Proceedings of the 27th Annual International Symposium on Computer Architecture (ISCA '00)*, vol. 28, pp. 248–259, ACM, New York, NY, USA, 2000.

[34] K. Puttaswamy and G. H. Loh, "Implementing register files for high-performance microprocessors in a die-stacked (3D) technology," in *Proceedings of the IEEE Computer Society Annual Symposium on Emerging VLSI Technologies and Architectures*, IEEE, Karlsruhe, Germany, March 2006.

[35] R. Balasubramonian, S. Dwarkadas, and D. H. Albonesi, "Reducing the complexity of the register file in dynamic superscalar processors," in *Proceedings of the 34th ACM/IEEE Annual International Symposium on Microarchitecture (MICRO '01)*, pp. 237–248, December 2001.

[36] J. Curz, A. Gonzalez, M. Valero, and N. P. Tophan, "Multi-banked register file architectures," in *Proceedings of the 27th International Symposium on Computer Architecture (ISCA '00)*, Vancouver, Canada, June 2000.

[37] K. I. Farkas, N. P. Jouppi, and P. Chow, "Register file design considerations in dynamically scheduled processors," in *Proceedings of the 2nd International Symposium on High-Performance Computer Architecture (HPCA '96)*, pp. 40–51, February 1996.

[38] D. Chinnery and K. Keutzer, *Closing the Gap between ASIC & Custom: Tools and Techniques for High-Performance ASIC Design*, Springer, 2002.

[39] P. Montesinos, W. Liu, and J. Torrellas, "Using register lifetime predictions to protect register files against soft errors," in *Proceedings of the 37th Annual IEEE/IFIP International Conference*

on Dependable Systems and Networks (DSN '07), pp. 286–295, IEEE, June 2007.

[40] L. Jin, W. Wu, J. Yang, C. Zhang, and Y. Zhang, "Reduce register files leakage through discharging cells," in *Proceedings of the 24th International Conference on Computer Design (ICCD '06)*, pp. 114–119, IEEE, October 2006.

[41] V. Nookala and S. S. Sapatnekar, "Designing optimized pipelined global interconnects: Algorithms and methodology impact," in *Proceedings of the IEEE International Symposium on Circuits and Systems (ISCAS '05)*, pp. 608–611, IEEE, May 2005.

[42] Z. Hu and M. Martonosi, "Reducing register file power consumption by exploiting value lifetime characteristics," *Proceedings of the Workshop on Complexity-Effective Design (WCED '00)*, vol. 1, pp. 1829–1841, 2000.

[43] R. Singh, G.-M. Hong, M. Kim, J. Park, W.-Y. Shin, and S. Kim, "Static-switching pulse domino: a switching-aware design technique for wide fan-in dynamic multiplexers," *Integration, the VLSI Journal*, vol. 45, no. 3, pp. 253–262, 2012.

## 8

High Throughput Pseudorandom Number Generator Based on Variable Argument Unified Hyperchaos

Kaiyu Wang,[1] **Qingxin Yan,**[1] **Shihua Yu,**[2] **Xianwei Qi,**[1] **Yudi Zhou,**[1] **and Zhenan Tang**[1]

[1] Department of Electronic Engineering, Dalian University of Technology, Gaoxinyuanqu Linggong Road 2, Dalian 116024, China
[2] School of Computer Science and Technology, Hulunbuir College, Xuefu Road, Hulunbuir 021008, China

Correspondence should be addressed to Zhenan Tang; qsyanqingxin@126.com

Academic Editor: Marcelo Lubaszewski

This paper presents a new multioutput and high throughput pseudorandom number generator. The scheme is to make the homogenized Logistic chaotic sequence as unified hyperchaotic system parameter. So the unified hyperchaos can transfer in different chaotic systems and the output can be more complex with the changing of homogenized Logistic chaotic output. Through processing the unified hyperchaotic 4-way outputs, the output will be extended to 26 channels. In addition, the generated pseudorandom sequences have all passed NIST SP800-22 standard test and DIEHARD test. The system is designed in Verilog HDL and experimentally verified on a Xilinx Spartan 6 FPGA for a maximum throughput of 16.91 Gbits/s for the native chaotic output and 13.49 Gbits/s for the resulting pseudorandom number generators.

1. Introduction

Pseudorandom number (PN) is the 01 sequence which has the randomness similar to noise. It has been widely used in digital communication, cryptography, computer games, and numerical computation [1–3]. Chaos is the phenomenon which shows very complex nonlinear dynamic characteristics in a deterministic system. And it has excellent properties such as nonperiodicity, broad bandwidth, and sensitivity to initial value [4, 5]. So Chaos and PN have a natural link. And compared to other PN sequences like m sequences, and so forth, the PN sequence generated by chaotic system has advantages like larger key space, longer cycle, and so forth.

Currently, researches of chaotic pseudorandom number generator (PRNG) are more focused on the digital implementation of low dimensional chaos such as Logistic chaos, Tent chaos, and Lorenz chaos. While these algorithms have significant advantages in some respects, like simpler construction, fewer resources consuming, and faster computing speed, they also have the fatal weakness that cannot be ignored to PRNG like smaller secret key space, periodic problem, and relatively lower throughput. Therefore, implementing a PRNG based on higher-order chaos equations seems more

advantage because the hyperchaos has multiple positive Lyapunov exponent and more controllable parameters and the output of system will have more complex randomness. The hyperchaotic encryption signal is harder to decode than low dimensional encryption signal [6]. And hyperchaos can provide multiple outputs, improve the throughput, and process multiple target signal [7, 8].

In 2002, Lu et al. proposed the unified chaos that can make Lorenz chaos, Lu chaos, and Chen chaos into a unified chaotic system and realize continued transition from one to another [9]. In 2011, Ma and Wang proposed the unified hyperchaos [10]. This algorithm makes the system continued transition from Lorenz hyperchaos through Lu hyperchaos to Chen hyperchaos with one system parameter changing from 0 to 1.

In this paper, we propose a novel variable parameters hyperchaotic PRNG structure which is composed by homogenized Logistic chaos and unified hyperchaos cascade. As [10] proposed the structure that needs to vary the system parameter from 0 to 1 to change chaotic class, and Logistic chaotic output is exactly between 0 and 1, so that they have a natural link. This paper uses the homogenization algorithm proposed in [11] to deal with Logistic output to provide

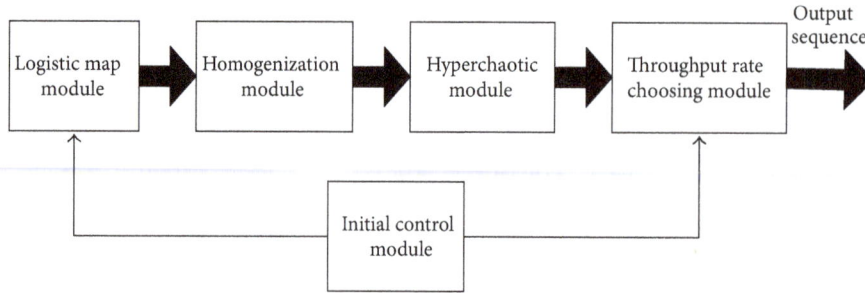

FIGURE 1: Flowchart of chaotic system.

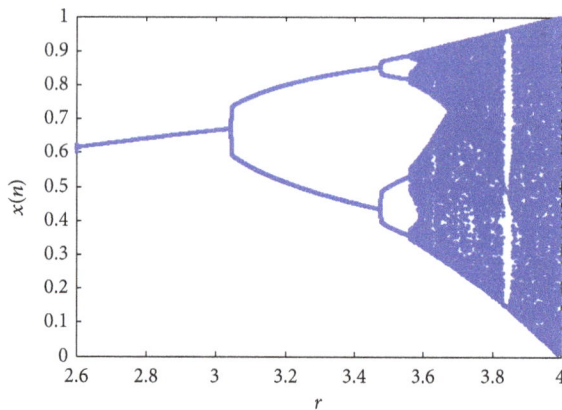

FIGURE 2: Logistic map bifurcation diagram.

variable parameters to the unified hyperchaos in [10]. With this method we can extend cycle of pseudorandom sequence and increase the complexity of system. And through the simple XOR processing to the output of hyperchaos, the system can generate multiple new pseudorandom sequences, greatly improving throughput. The system is designed in Verilog HDL and experimentally verified on a Xilinx Spartan 6 FPGA for a maximum throughput of 16.91 Gbits/s for the native chaotic output and 13.49 Gbits/s for the resulting PRNG output. And the output channel is increased to 26 roads. The output sequence is shown to pass the NIST SP. 800-22 test suite [12] and DIEHARD test suite to indicate statistical randomness.

This paper is organized as follows: Section 2 discusses the algorithms composed of the variable argument hyperchaos and demonstrates chaotic nature; Section 3 describes the details of its implementation in hardware; Section 4 introduces the test results of the output sequence and the resource consuming after FPGA implementation; Section 5 is conclusion.

2. Proposed Variable Argument Unified Hyperchaotic PRNG

As shown in Figure 1, the proposed variable argument unified hyperchaotic PRNG is mainly composed of five modules. They are the Logistic map module which provides

the parameter, homogenization module which homogenizes the output of Logistic map, unified hyperchaotic module, throughput rate choosing module which controls the number of output channel, and initial control module which controls the system. The core algorithms are the Logistic chaos, homogenization algorithm, and unified hyperchaos. Now we will discuss these three algorithms.

2.1. Logistic Chaos. Logistic chaos is one of the most studied chaos systems. It is applied in many chaos systems because of its simple description. The Logistic chaos is described as follows:

$$x_{n+1} = r x_n (1 - x_n)$$
$$0 \le x_n \le 1, \quad 0 \le r \le 4, \quad n = 0, 1, 2, \ldots \tag{1}$$

Iteration of Logistic chaos is affected by parameters r and initial value x_0. Small changes of the two values will lead to significantly different output. When r is in the range $3.569945672 \le r \le 4$, the numbers generated in successive iterations of the mapping become chaotic, and output is always between $[0, 1]$, just as bifurcation diagram Figure 2. We take $r = 4$ to realize in hardware easily.

2.2. Homogenization Algorithm. The Logistic chaotic output in this research is homogenized to make it become a uniform pseudorandom sequence, so that the parameters input into the unified hyperchaos can be more complex and more randomness. Now, we will introduce the transform method.

The IEEE double format consists of three fields: a 52-bit fraction, f; an 11-bit biased exponent, e; and a 1-bit sign, s; then, any real number can be expressed as the following equation:

$$x = ((-1)^s \times 2^{e-1023} \times 1.f)_{10} = (\{s, e, f\})_2. \tag{2}$$

Definition 1. Left-shift b-operation of f, $f_{\leftarrow b}$, is a new fraction obtained by discarding the left-most b bits of f and then padding the result with $b - 1$ bits 0 and 1 bit 1 on the right, if the 51st-bit, 50th-bit, …, $(51 - b + 1)$th-bit in f equals zeroes, while the $(51 - b)$th-bit in f equals one.

Definition 2. The bit-transformation of f, BT$\{f\}$: in (1), the fraction f can be rewritten in the binary-coded form

$f = f_{51}f_{50}\cdots f_1f_0$. Parse f into higher 26-bit block f_H and lower 26-bit block f_L is as follows:

$$f_H = f_{51}f_{50}\cdots f_{27}, \qquad f_L = f_{26}f_{25}\cdots f_0. \qquad (3)$$

Then, reverse f_L into f'_L; that is,

$$f'_L = f_0f_1\cdots f_{26}. \qquad (4)$$

Then

$$\mathrm{BT}\{f\} = \{f_H \oplus f'_L, f_L\}. \qquad (5)$$

Now, one defines (5) as bit-transformation of f, $\mathrm{BT}\{f\}$.

Definition 3. Bit-transformation of real numbers: suppose $x = \{s, e, f\} \in G$, G represent all real numbers. The bit-transformation of $\{s, e, f\}$ is defined by

$$\mathrm{RBT}\{x\} = \{0, 1023 - b, \mathrm{BT}\{f\}_{\leftarrow b}\}. \qquad (6)$$

Note that a bit-transformation of real numbers is composed of a bit-transformation and a left-shift b-operation, so $\mathrm{RBT}\{x\}$ is a multiple-to-one map function.

After the conversion like (6), the pseudorandom sequence can be made uniform. One realizes the homogenization algorithm on FPGA to deal with Logistic chaos. Import the output of Logistic chaos with preprocessing and postprocessing into MATLAB. The result is shown in Figure 3. It has obtained the good effect of homogenization and achieves the goal of the interference transformation, homogenization.

2.3. Unified Hyperchaos. The unified hyperchaotic system is shown as the following equations:

$$\dot{x} = (26a + 10)(y - x) \qquad (7a)$$

$$\dot{y} = (28 - 44a)x - xz + (29a - 1)y - v \qquad (7b)$$

$$\dot{z} = xy - \frac{(8 + a)z}{3} \qquad (7c)$$

$$\dot{v} = 0.1(1 - a)yz + ax + 0.2. \qquad (7d)$$

Obviously, when the parameter a increases from 0 to 1, the systems (7a), (7b), (7c), and (7d) evolve from hyperchaotic Lorenz system to hyperchaotic Chen system. The maximum Lyapunov exponent (MLE) and the Lyapunov dimension (D_L) are often used to measure a chaotic system in a state of chaos case or period orbit case. It is well known that the MLE and D_L satisfy at least one MLE greater than zero and $2 < D_L < 3$ for chaos case, two MLE greater than zero, and $3 < D_L < 4$ for hyperchaos case. For systems (7a), (7b), (7c), and (7d), when $a \in [0, 1]$, the Lyapunov exponent spectrum and the Lyapunov dimension are shown as Figures 4 and 5. As it is shown in Figure 4, all points from 0 to 1 except $a = 0.14$, there are two MLE greater than zero. And as it is shown in Figure 5, all points from 0 to 1 except $a = 0.14$, $3 < D_L < 4$. It means the system is hyperchaotic system only except individual parameter points. And the individual bad

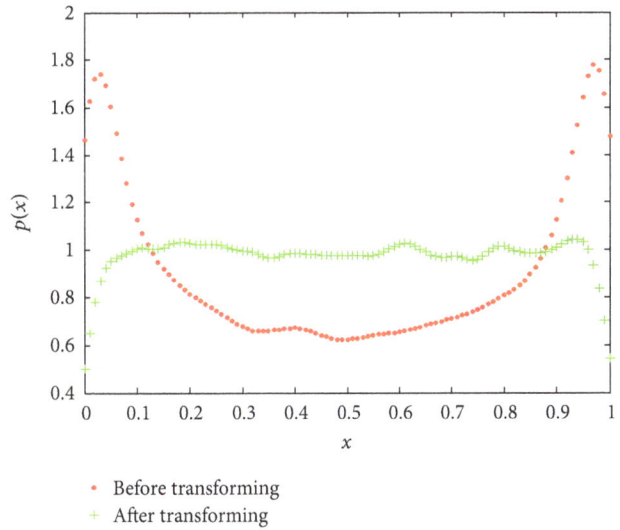

- Before transforming
+ After transforming

FIGURE 3: Homogenization effect.

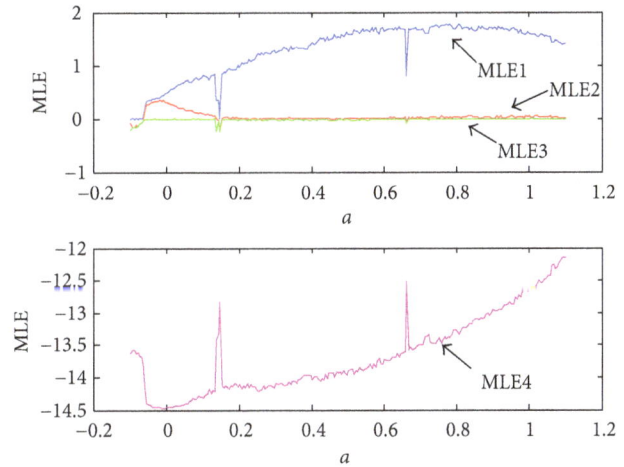

FIGURE 4: Max Lyapunov exponent spectrum.

parameter point can be removed by the means of hardware implementation.

In order to see clearly that, when the parameter a increases from 0 to 1, the systems (7a), (7b), (7c), and (7d) evolve from hyperchaotic Lorenz system to hyperchaotic Chen system, we plot the phase diagram with different parameter as shown in Figures 6 and 7.

2.4. Variable Argument Unified Hyperchaotic Algorithm. The variable argument unified hyperchaotic PRNG we proposed is based on the above three algorithms. Logistic chaos generates sequence between 0 and 1. Then, the sequence is processed by homogenization algorithm to be made uniform. After that, the uniform pseudorandom sequence is introduced as changing parameter to be imported into unified hyperchaotic system to control the output. So that the system varies in different state of hyperchaotic system and increases

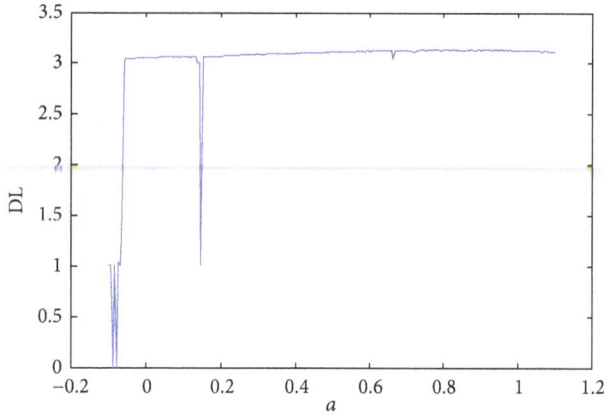

FIGURE 5: Lyapunov dimension.

the output sequence cycle and has more complex dynamic characteristics and optimizes the statistical properties.

3. Hardware Implementation

To implement easily in the hardware, the differential equations (7a), (7b), (7c), and (7d) are discretized. Euler approximation has been shown to provide the best chaotic response, occupy the lowest area, and provide the highest speed compared with Runge-Kutta method and other methods [13]. Therefore, the Euler approximation is applied to discretize the continuous-time systems (7a), (7b), (7c), and (7d) for the digital domain:

$$x_{n+1} = h\left((26a + 10)\left(y_n - x_n\right)\right) + x_n \tag{8a}$$

$$y_{n+1} = h\left((28 - 44a)\, x_n - x_n z_n\, (29a - 1)\, y_n - v_n\right) + y_n \tag{8b}$$

$$z_{n+1} = h\left(x_n y_n - \frac{(8 + a)\, z_n}{3}\right) + z_n \tag{8c}$$

$$v_{n+1} = h\left(0.1\,(1 - a)\, y_n z_n + a x_n + 0.2\right) + v_n. \tag{8d}$$

While the chaotic systems are running in finite precision, the fixed-point arithmetic is preferable over floating point mathematics because it requires less hardware resources and computation time. And under the same word length fixed-point format has a higher accuracy [14, 15]. So we select fixed-point format to represent data. Also due to the limited precision, the digital realization of chaotic systems has degradation dynamics and tends to period orbit case, namely, finite precision. Based on the theory [16] proposed that the cycle of chaotic sequence will grow exponentially with growth of format word length, we use 32-bit fixed-point number format to realize the chaotic system to prevent the finite precision effect. In unified hyperchaotic system, the fixed-point two's complement format is used with the 7 most significant bits for sign and integer part and the remaining for the fractional part. But Logistic chaotic system is used with the 1 most significant bit for integer part and the remaining for the fractional part as its output is always positive number.

As shown in Figure 1, the PRNG we proposed has three core algorithms: Logistic, homogenized, and unified hyperchaotic. And they are cascade structure. Therefore, this work employs a pipelined architecture between the three modules, so that the register between these three modules can be updated in each clock and increase hardware utilization efficiency.

Logistic Module is controlled by control module so that the Logistic Module outputs the same value in m clock cycles. As a result of the pipeline structure, unified hyperchaotic module will read an input as a unified hyperchaotic parameter at each rising edge of the clock, so unified hyperchaotic modules will calculate the output with the same parameter in every m clock cycles. So, if Logistic period is n, then the entire system's period is $m * n$.

As unified hyperchaotic module has four dimensional outputs, it could provide operation space for subsequent processing. We add throughput rate choosing module after unified hyperchaotic module to make bitwise operation among the initial four outputs (x, y, z, v). Based on the conclusion proposed in [7] about the fact that doing bitwise XOR operation on chaotic system output can get better PN sequence, in throughput rate choosing module, we do bitwise XOR operation on two different output sequences $A \oplus B$ (like $x \oplus y$, $x \oplus z$, etc.) or three different output sequences $A \oplus B \oplus C$ (like $x \oplus y \oplus z$, $x \oplus z \oplus v$, etc.) or do bitwise XOR operation on A's higher 16 bits and B's lower 16 bits and then merge these 16 bits with A's lower 16 bits $\{A_{\text{high}} \oplus B_{\text{low}}, A_{\text{low}}\}$ (like $\{x_{[31:16]} \oplus y_{[15:0]}, x_{[15:0]}\}$, etc.) to improve the throughput. After these XOR operations, this system can provide up to 26 channels output. And this module can be configured to decide which channel or which several channels can be output.

4. Result

The proposed variable argument unified hyperchaotic PRNGis designed in Xilinx ISE 12.2 environment using Verilog HDL and experimentally verified on a Xilinx Spartan 6 XC6SLX100 FPGA. In order to fully test the output, through controlling the throughput rate choosing module, make all XOR modules work to output all 26 channels and analyze these output data with below tests.

4.1. Phase Diagram. We import the PRNG's output $\{x, y, z, v\}$ into MATLAB as shown in Figure 8. From Figure 8 we can get the system is switching in different chaotic system with the number of iterations increase and parameter change; it effectively improves the complexity of output.

4.2. Correlation. Among the 26 channels output, there are 22 channels which are produced by XOR operation through 4 original outputs (x, y, z, v). To analyze the cross-correlation among these outputs, we import them into MATLAB. Figure 9 shows cross-correlation result between x related outputs $(x \oplus y, x \oplus z, x \oplus v, x \oplus x \oplus z)$ and original x output. The results show that the XOR operation outputs are still correlated to their original channels, as indicated by a peak at zero-lag.

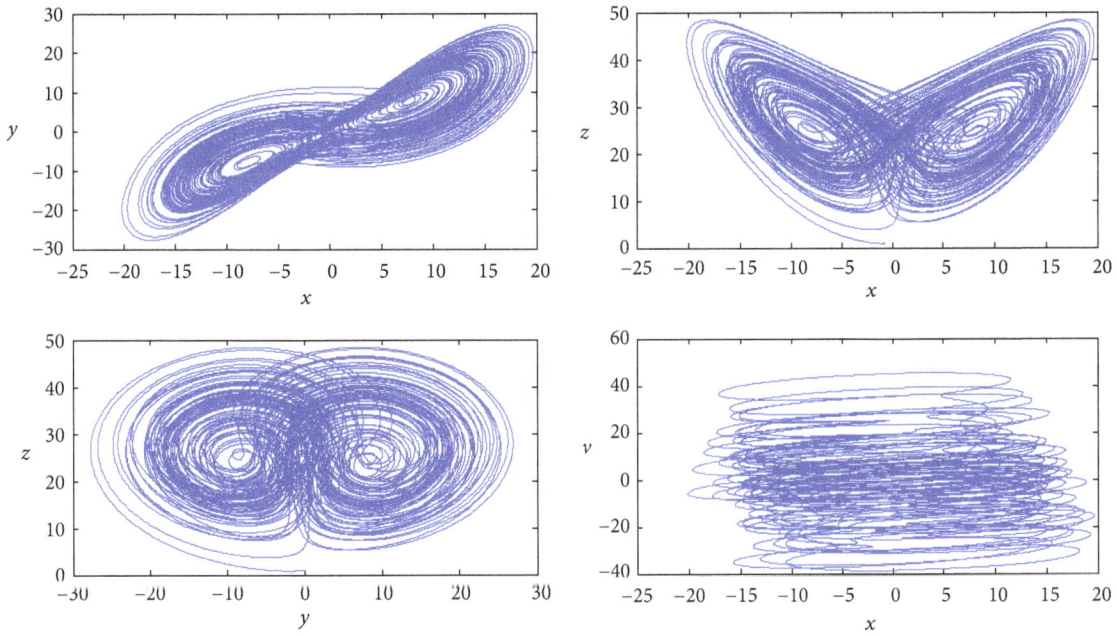

FIGURE 6: Phase diagram when $a = 0.03$.

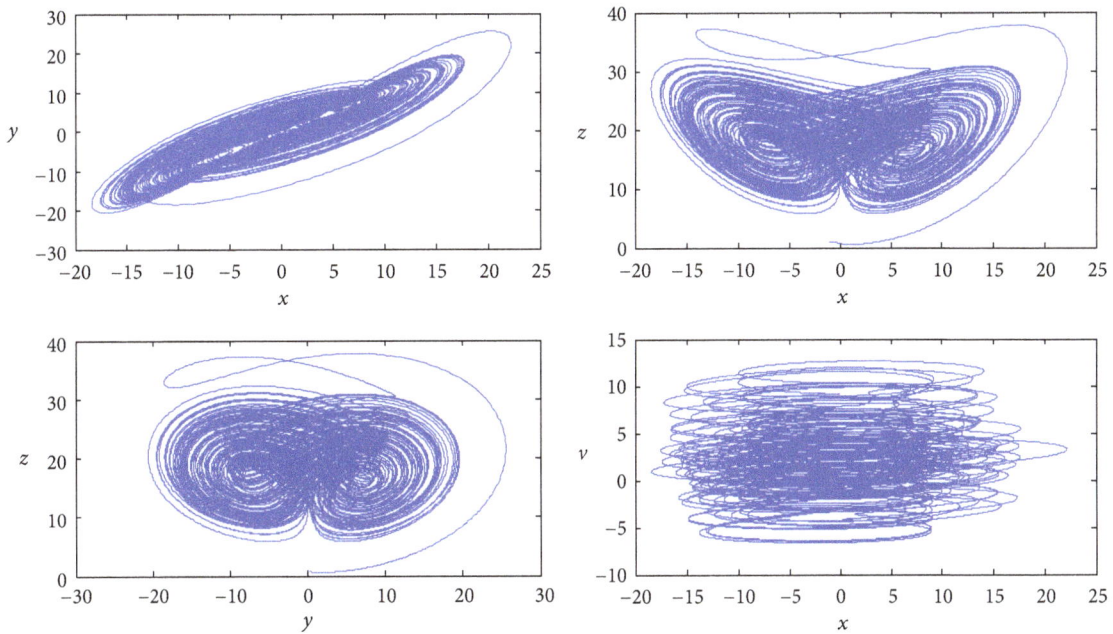

FIGURE 7: Phase diagram when $a = 0.73$.

However, most peaks are below 0.5, and at other delays, cross-correlation coefficients are below 0.3. About other cross-correlation results, they are similar to the above one.

4.3. Pseudorandom Number Test. If the chaotic system is regarded as a PRNG, not all output bits can meet the requirements of randomness. As in the digital context, it creates an uneven distribution of pseudorandomness across the output bits. The MSBs are not only biased but also highly correlated, while the LSBs show desirable statistical randomness [17]. For this kind of situation, we first test all 32 bits. If the sequence cannot pass the tests, we will discard the highest one. Then, we will test the remaining bits. Repeat these actions until the sequence can pass tests. After tested by NIST SP800-22 test suit and DIEHARD test suit, test results show that in the 26 channels of FPGA output, 12 channels

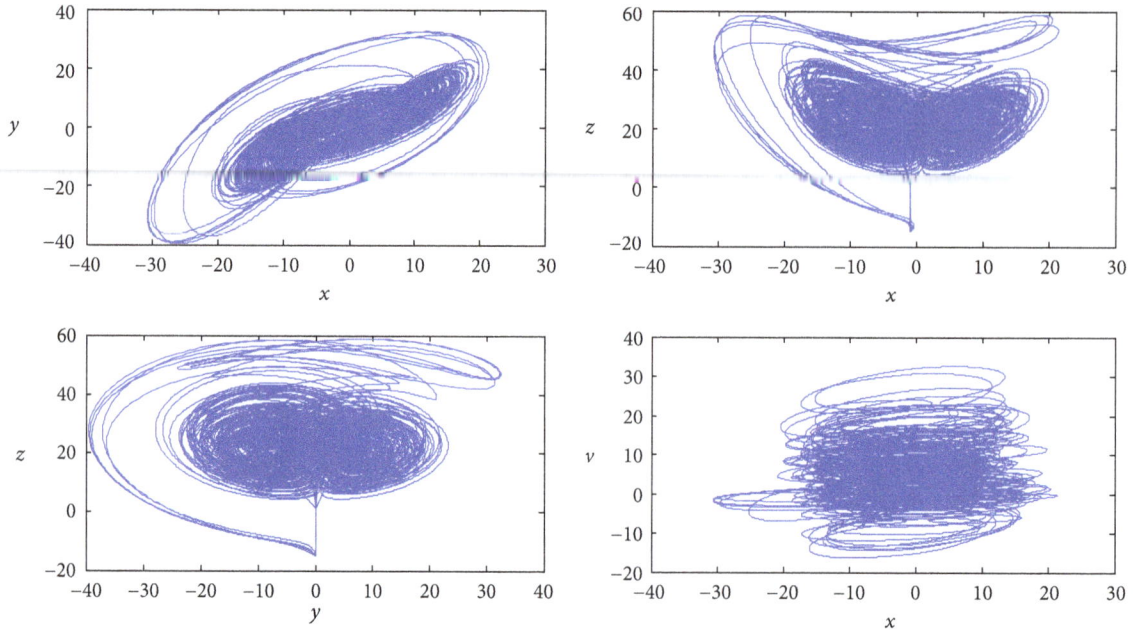

FIGURE 8: System output phase diagram.

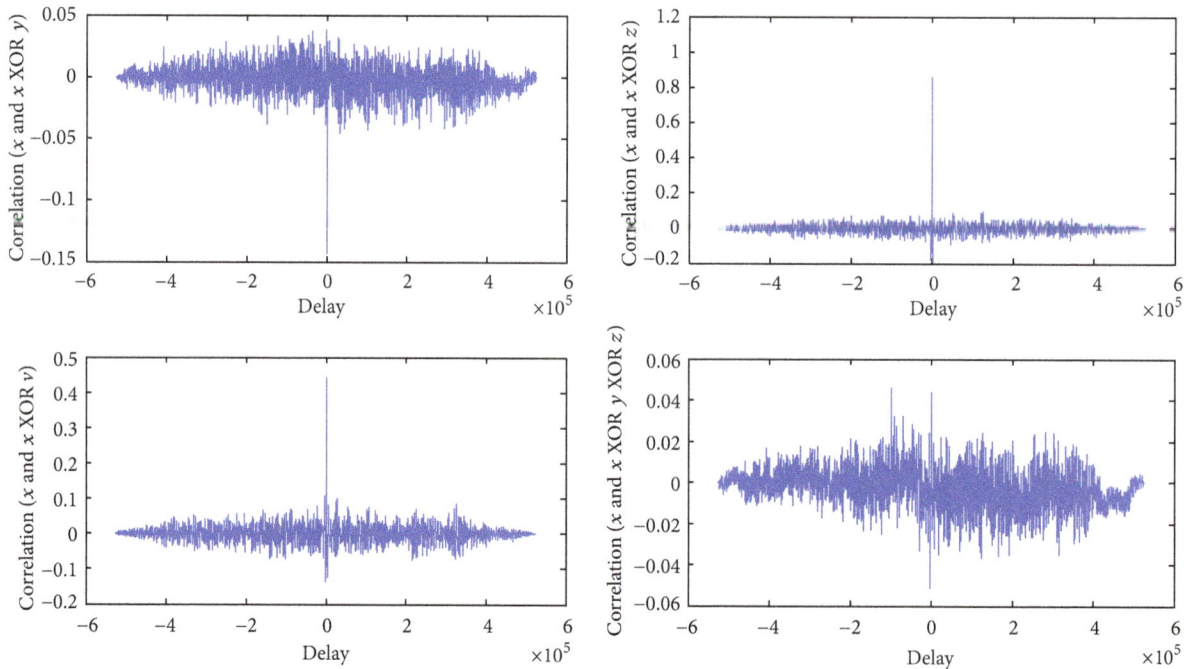

FIGURE 9: X related cross-correlation diagram.

$\{A_{\text{high}} \oplus B_{\text{low}}, A_{\text{low}}\}$ can pass tests with all 32 bits and other 14 channels can pass with lower 20 bits. So, as a PRNG, we make the system output 32 bits in 12 channels $\{A_{\text{high}} \oplus B_{\text{low}}, A_{\text{low}}\}$ and output lower 20 bits in other 14 channels. And from the test result of DIEHARD, we can get the conclusion that the quality from higher to lower is $\{A_{\text{high}} \oplus B_{\text{low}}, A_{\text{low}}\}$, $A \oplus B \oplus C$, $A \oplus B$ and original 4 outputs.

We take the lower 20 bits $x, y, z \oplus v, x \oplus y \oplus z$ NIST test results and DIEHARD test results as the representative list in this paper, in Tables 1 and 2. The remaining 22 roads also pass the tests but not list in this paper. In conclusion, the PRNG we proposed can provide as high as 26 channels output and 13.49 Gbits/s throughput.

4.4. Hardware Resource Utilization. After the FPGA synthesis, Slice Registers resources utilization rate is 1%, Slice LUTs is 5%, and DSP block is 55%. The designed system can

TABLE 1: Result of NIST SP800-22 test.

	X P value	Y P value	Z XOR V P value	X XOR Y XOR Z P value
Frequency	0.791780	0.041550	0.979257	0.961716
Block frequency	0.366274	0.648388	0.367493	0.117584
Cumulative sums	0.744906	0.078225	0.664299	0.985758
Runs	0.383170	0.481335	0.872882	0.017885
Longest run of ones	0.753431	0.746675	0.173653	0.222029
Rank	0.123860	0.050112	0.202915	0.849442
FFT	0.168669	0.868803	0.868803	0.934178
Nonperiodic template matchings	0.988648	0.677613	0.896375	0.857033
Overlapping template matchings	0.545254	0.239949	0.392905	0.672648
Universal statistical	0.303032	0.748731	0.096754	0.951902
Approximate entropy	0.977511	0.223329	0.790074	0.634836
Random excursions	0.379538	0.707823	0.490369	0.981034
Random excursions variant	0.509760	0.958761	0.613635	0.937700
Serial	0.380005	0.582563	0.953135	0.876081
linear complexity	0.418562	0.193589	0.352896	0.198838
TEST status	SUCCESS	SUCCESS	SUCCESS	SUCCESS

Length of bit = 1000000, number of bit streams = 100, and confidence level = 99%.

TABLE 2: Result of DIEHARD test.

	X P value	Y P value	Z XOR V P value	X XOR Y XOR Z P value
BIRTHDAY SPACINGS	0.352976	0.728427	0.767687	0.293804
OVERLAPPING 5-PERMUTATION	1.000000	1.000000	1.000000	0.999995
RANK TEST for 31×31 matrices	0.364438	0.610135	0.417598	0.741397
RANK TEST for 32×32 matrices	0.594228	0.321434	0.431023	0.514951
RANK TEST for 6×8 matrices	0.335467	0.156574	0.227157	0.688417
OVERLAPPING 20-tuples BITSTREAM	0.6421225	0.413186	0.5047115	0.453427
OPSO	0.4453782	0.545113	0.552886	0.470565
OQSO	0.8444143	0.830028	0.720432	0.473571
DNA	0.8535612	0.902319	0.650219	0.597742
COUNT-THE-1's TEST on a stream of bytes	0.3500190	0.210574	0.306527	0.183018
COUNT-THE-1's TEST for specific bytes	0.7102660	0.641749	0.623289	0.490400
PARKING LOT	0.643434	0.068029	0.787648	0.837780
MINIMUM DISTANCE	0.689712	0.467954	0.970672	0.960276
3DSPHERES	0.130624	0.701084	0.231284	0.013265
SQEEZE	0.952062	0.781055	0.454853	0.518254
OVERLAPPING SUMS	0.862138	0.132331	0.913215	0.724329
RUNS	0.578362	0.298342	0.834721	0.302906
CRAPS	1.000000	1.000000	0.679313	0.841946

provide 26-way output with 32 bits. Taking into account the pipeline structure influence, the throughput of system is as high as 16.91 Gbits/s at maximum clock frequency. And its random number throughput rate is as high as 13.49 Gbits/s. Specific numbers of resource consumption and throughput are shown in Table 3. To make the quantitative analysis on the resource consumption and throughput, we adopt the following definition. Gate count is estimated as Gc = 8 × (LUTs + FFs) and the area efficiency is assessed through a figure of merit determined as FOM = $(N_{\mathrm{PRNG}} \times f_{\mathrm{CLK}})/$Gc. The PRNG proposed in this paper is compared with several low dimensional chaotic PRNG in Table 4. From Table 4 we can

TABLE 3: Experimental results on Xilinx Spartan 6 FPGA.

	Experimental results on the Xilinx Spartan 6 FPGA
Registers	1780
LUTs	3068
Fully used LUT-FF pairs	854
DSP48A1s	100
Frequency (MHz)	20.811
Single output bits	32
Number of output channel	26
System throughput (Mb/s)	17314.752
PRNG throughput (Mb/s)	13818.504

TABLE 4: Comparison with previously reported chaos-based PRNGS.

System		Area (Gc)	T.put	FOM	NIST
Chen et al. [18]	LOG.Map	9622	200	0.02	Pass
Li et al. [19]	LOG.Map	9136	200	0.02	Fail
Chen et al. [20]	LOG.Map	31655	3200	0.10	Pass
This one	Hyp.Chaos	31376	13819	0.44	Pass

get that although this work spent higher hardware resource than low dimensional chaotic PRNG, we get much higher throughput and higher FOM.

5. Conclusion

In this paper, we propose a novel variable parameters hyperchaotic PRNG structure which is composed of homogenized Logistic chaos and unified hyperchaos cascade. Take the homogenized Logistic chaotic output as the unified hyperchaotic parameter to make the output sequence in different chaotic system. In this way, system will be more complex and have longer period. At the same time, add a throughput rate control module after the output of the unified hyperchaotic module, through simple XOR processing; the output of the 4 road hyperchaos can be extended to 26 road and greatly improve the throughput of the system. The PRNGis designed in Xilinx ISE 12.2 environment using Verilog HDL and experimentallyverified on a Xilinx Spartan 6 FPGA. The throughput is up to 16.91 Gbits/s for the chaotic system. As a PRNG, it can provide 26 channels output as pseudorandom sequence which all pass NIST SP800-22 test and DIEHARD test. And its random number throughput rate is as high as 13.49 Gbits/s.

Therefore, due to the variable argument unified hyperchaotic PRNG has advantages like high output complexity, multidimensional output, and high throughput rate; it is very suitable for being applied to multiobjective signal processing field like multiobjective control and secure communications, and so forth.

Conflict of Interests

The authors declare that there is no conflict of interests regarding the publication of this paper.

References

[1] Q. Rong and Y. Fang, "Pseudo random sequence generator based on variable structure chaos," *Modern Electronics Technique*, vol. 35, no. 11, pp. 64–67, 2012.

[2] H. P. Ren, *The design of chaotic key system based on FPGA [M.S. dissertation]*, Dalian Maritime University, Ganjingzi, China, 2011.

[3] P. Dabal and R. Pelka, "FPGA implementation of chaotic pseudo-random bit generators," in *Proceedings of the 19th International Conference on Mixed Design of Integrated Circuits and Systems (MIXDES '12)*, pp. 260–264, Warsaw, Poland, May 2012.

[4] P. Li, Z. Li, W. A. Halang, and G. Chen, "A multiple pseudorandom-bit generator based on a spatiotemporal chaotic map," *Physics Letters A*, vol. 349, no. 6, pp. 467–473, 2006.

[5] K. Wang, W. Pei, H. Xia, and Y. Cheung, "Pseudo-random number generator based on asymptotic deterministic randomness," *Physics Letters A*, vol. 372, no. 24, pp. 4388–4394, 2008.

[6] H. Wang, L. Cheng, and J.-H. Peng, "Application of hyperchaos to encrypting digital signals," *Journal of Northeast Normal University (Natural Science Edition)*, vol. 32, no. 2, pp. 31–35, 2000.

[7] A. S. Mansingka, M. Affan Zidan, M. L. Barakat, A. G. Radwan, and K. N. Salama, "Fully digital jerk-based chaotic oscillators for high throughput pseudo-random number generators up to 8.77 Gbits/s," *Microelectronics Journal*, vol. 44, no. 9, pp. 744–752, 2013.

[8] F. Jin-Qing, "Several advances in chaos-based communication and research of information security associated with networks," *Journal of Systems Engineering*, vol. 25, no. 6, pp. 725–741, 2010.

[9] J. H. Lu, G. R. Chen, and D. Z. Cheng, "Bridge the gap between the Lorenz system and the Chen system," *International Journal of Bifurcation and Chaos*, vol. 12, no. 12, pp. 2917–2926, 2002.

[10] C. Ma and X. Wang, "Bridge between the hyperchaotic Lorenz system and the hyperchaotic Chen system," *International Journal of Modern Physics B*, vol. 25, no. 5, pp. 711–721, 2011.

[11] S. Li-Yuan, X. Yan-Yu, and Z. Sheng, "How homogenize Chaos-based Pseudo-random sequences," in *Proceedings of the International Conference on Computer Science and Software Engineering (CSSE '08)*, pp. 793–796, December 2008.

[12] A. Rukhin, J. Soto, M. Smid et al., "A statistical test suite for random and pseudorandom number generators for cryptographic applications," NIST Special Publication 800-22, 2010.

[13] M. A. Zidan, A. G. Radwan, and K. N. Salama, "The effect of numerical techniques on differential equation based chaotic generators," in *Proceedings of the 23rd International Conference on Microelectronics (ICM '11)*, pp. 1–4, Hammamet, Tunisia, December 2011.

[14] A. Pande and J. Zambreno, "A chaotic encryption scheme for real-time embedded systems: design and implementation," *Telecommunication Systems*, vol. 52, no. 2, pp. 551–561, 2013.

[15] L. Wang, W. Liu, H. Shi, and J. M. Zurada, "Cellular neural networks with transient chaos," *IEEE Transactions on Circuits and Systems II: Express Briefs*, vol. 54, no. 5, pp. 440–444, 2007.

[16] B. Zhang, *Performance analysis and optimization of chaotic PN sequence [M.S. thesis]*, Hangzhou Dianzi University, Jianggan, China, 2009.

[17] M. L. Barakat, A. S. Mansingka, A. G. Radwan, and K. N. Salama, "Generalized hardware post-processing technique for chaos-based pseudorandom number generators," *ETRI Journal*, vol. 35, no. 3, pp. 448–458, 2013.

[18] S. Chen, T. Hwang, and W. Lin, "Randomness enhancement using digitalized modified logistic map," *IEEE Transactions on Circuits and Systems II: Express Briefs*, vol. 57, no. 12, pp. 996–1000, 2010.

[19] C.-Y. Li, T.-Y. Chang, and C.-C. Huang, "A nonlinear PRNG using digitized logistic map with self-reseeding method," in *Proceedings of the International Symposium on VLSI Design, Automation and Test (VLSI-DAT '10)*, pp. 108–111, April 2010.

[20] S.- L. Chen, T. Hwang, S.-M. Chang, and W.-W. Lin, "A fast digital chaotic generator for secure communication," *International Journal of Bifurcation and Chaos in Applied Sciences and Engineering*, vol. 20, no. 12, pp. 3969–3987, 2010.

## 9

Parallel Jacobi EVD Methods on Integrated Circuits

Chi-Chia Sun,[1] Jürgen Götze,[2] and Gene Eu Jan[3]

[1] Department of Electrical Engineering, National Formosa University, Wunhua Road 64, Huwei 632, Taiwan
[2] Information Processing Lab, Technology University of Dortmund, Otto-Hahn-Strase 4, 44221 Dortmund, Germany
[3] Institute of Electrical Engineering, National Taipei University, University Road 151, San Shia District, New Taipei City 23741, Taiwan

Correspondence should be addressed to Chi-Chia Sun; ccsun@nfu.edu.tw

Academic Editor: Sungjoo Yoo

Design strategies for parallel iterative algorithms are presented. In order to further study different tradeoff strategies in design criteria for integrated circuits, A 10×10 Jacobi Brent-Luk-EVD array with the simplified μ-CORDIC processor is used as an example. The experimental results show that using the μ-CORDIC processor is beneficial for the design criteria as it yields a smaller area, faster overall computation time, and less energy consumption than the regular CORDIC processor. It is worth to notice that the proposed parallel EVD method can be applied to real-time and low-power array signal processing algorithms performing beamforming or DOA estimation.

1. Introduction

We are on the edge of many important developments which will require parallel data and information processing. The transmission systems are using higher and higher frequencies and the carrier frequencies are increasing to 10 GHz and above. Because of the smaller wavelength more antennas can be implemented on a single device leading to massive MIMO systems. Parallel VLSI architectures will be needed in order to provide the required computational power for 10 GHz and above, massive MIMO, and big data processing [1, 2].

In parallel matrix computation at the circuit level, implementing an iterative algorithm on a multiprocessor array results in a tradeoff between the complexity of an iteration step and the number of required iteration steps. Therefore, as long as the algorithm's convergence properties are guaranteed, it is possible to adjust the architecture, which can significantly reduce the complexity with regard to the implementation. Computing the parallel eigenvalue decomposition (EVD) as a preprocessing step to MUSIC or ESPRIT algorithm with Jacobi's iterative method is used as an important example as the convergence of this method is extremely robust to modifications of the processor elements [3–6].

In [7], it was shown that Brent-Luk-EVD architecture with a modified CORDIC for performing the plane rotation

of the Jacobi algorithm can be realized in advanced VLSI design. Based on it, a Jacobi EVD array is realized by implementing a scaling-free microrotation CORDIC (μ-CORDIC) processor in this paper, which only performs a predefined number of CORDIC iterations. Therefore, the size of the processor array can be reduced for implementing a large-scale EVD array in parallel VLSI architectures. After that, several modifications of the algorithm/processor are studied and their impact on the design criteria is investigated for different sizes of EVD array (10×10 to 80×80). Finally, a strategy to comply with the design criteria is established, especially in terms of balancing the number of microiterations and the computational complexity. The proposed architecture is ideal for real-time antenna array applications, such as a flying object carrying an antenna array for beamforming or DOA estimation that would require a real-time, low-power, and efficient architecture for EVD, or joint time-delay and frequency estimation using a sensor network.

This paper is organized as follows. Serial and parallel Jacobi methods are described in Section 2. In Section 3, the design issues of the parallel Jacobi EVD array are discussed, leading to the simplification from a regular full CORDIC to the μ-CORDIC processor with an adaptive number of iterations. Section 4 shows the implementation results. Section 5 concludes this paper.

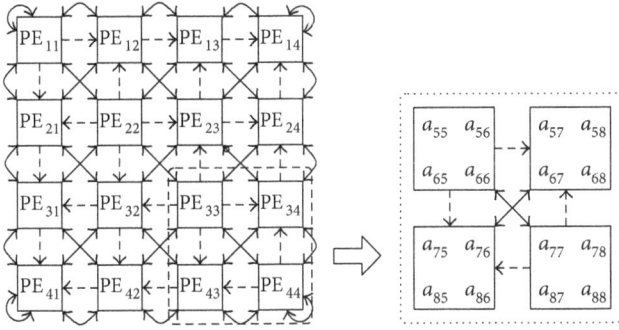

FIGURE 1: A 4×4 Brent-Luk-EVD array, where $n = 8$ for an 8×8 symmetric matrix [3].

2. Parallel Eigenvalue Decomposition

2.1. Jacobi Method. An eigenvalue decomposition of a real symmetric $n \times n$ matrix A is obtained by factorizing A into three matrices $A = Q \wedge Q^T$, where Q is an orthogonal matrix ($QQ^T = I$) and $\wedge$ is a diagonal matrix containing the eigenvalues of A. The Jacobi method approximates the EVD iteratively as follows:

$$A_{k+1} = Q_k A_k Q_k^T, \quad \text{with } k = 0, 1, 2, \ldots, \qquad (1)$$

where Q_k is an orthonormal plane rotation by the angle θ in the (i, j) plane.

The plane rotations Q_k, where $k = 1, 2, 3, \ldots$, can be executed in various orders to obtain the eigenvalues. The most common order of sequential plane rotations $\{Q_k\}$ is called cyclic-by-row, meaning (i, j) is chosen as follows:

$$(i, j) = (1, 2) (1, 3) \cdots (1, n) (2, 3) \cdots (2, n) \cdots (n - 1, n). \qquad (2)$$

The execution of all $N = n(n - 1)/2$ index pairs (i, j) is called a sweep. Matrix A will converge into a diagonal matrix $\wedge$ once k sweeps are applied, where $\wedge$ contains the eigenvalues $\lambda_1, \lambda_2, \ldots, \lambda_n$:

$$\lim_{k \to \infty} A_k = \text{diag} [\lambda_1, \lambda_2, \ldots, \lambda_n] = \begin{bmatrix} \lambda_1 & 0 & \cdots & 0 \\ 0 & \lambda_2 & & \vdots \\ \vdots & & \ddots & 0 \\ 0 & \cdots & 0 & \lambda_n \end{bmatrix}. \qquad (3)$$

2.2. Jacobi EVD Array. Instead of performing the plane rotations Q_k one by one in a cyclic-by-row order, they can be separated into multiple subproblems and executed in parallel on a $\log n$ dimensional multicore platform. Ahmedsaid et al. [3] first presented a parallel array based on Jacobi's method. It consists of $n/2 \times n/2$ PEs and each PE contains a 2×2 subblock of the matrix A. Figure 1 shows a typical 4×4 EVD array

with 16 PEs. This Jacobi array can perform $n/2$ subproblems in parallel. Initially, each PE holds a 2×2 submatrix of A:

$$\text{PE}_{pq} = \begin{pmatrix} a_{2p-1,2q-1} & a_{2p-1,2q} \\ a_{2p,2q-1} & a_{2p,2q} \end{pmatrix}, \qquad (4)$$

where p and $q = 1, 2, \ldots, n/2$.

A rotation angle has to be chosen in order to zero out the off-diagonal elements of the submatrix by solving a 2×2 symmetric EVD subproblem as shown in the following:

$$\begin{bmatrix} a_{ii}' & a_{ij}' \\ a_{ji}' & a_{jj}' \end{bmatrix} = R \cdot \begin{bmatrix} a_{ii} & a_{ij} \\ a_{ji} & a_{jj} \end{bmatrix} \cdot R^T, \qquad (5)$$

where $R = \begin{bmatrix} \cos \theta & -\sin \theta \\ \sin \theta & \cos \theta \end{bmatrix}$.

The maximal reduction $\{a_{ij}', a_{ji}'\} = 0$ can be obtained by applying the optimal angle of rotation θ_{opt}:

$$\theta_{opt} = \frac{1}{2} \arctan \left(\frac{2a_{ij}}{a_{jj} - a_{ii}} \right), \qquad (6)$$

where the range of θ_{opt} is limited to $|\theta_{opt}| \le \pi/4$.

This optimal angle θ_{opt}, which can annihilate the off-diagonal elements ($a_{2p-1,2q}$ and $a_{2p,2q-1}$), is computed using diagonal PEs in (6). Once these rotation angles are computed, they will be sent to the off-diagonal PEs. This transmission is indicated by the dashed lines in the vertical and horizontal direction in Figure 1. All off-diagonal PEs will perform a two-sided rotation with the corresponding rotation angles obtained from the row (θ_r) and column (θ_c), respectively.

Once these rotations are applied, the matrix elements are interchanged between processors as indicated by the diagonal solid lines in Figure 1, for the execution of the next $n/2$ rotations. One sweep needs to perform $n - 1$ of these parallel rotation steps. After several sweeps (iterations) are executed, the eigenvalues will concentrate in the diagonal PEs.

3. CORDIC Approach

3.1. Regular CORDIC. Within each PE, a simple way to solve the subproblem of (5) in VLSI for zeroing out the off-diagonal elements is to use the CORDIC algorithm. An orthogonal CORDIC rotator is defined as [8, 9]

$$x_{i+1} = A_i \left[x_i - y_i \cdot d_i \cdot 2^{-i} \right]$$

$$y_{i+1} = A_i \left[y_i + x_i \cdot d_i \cdot 2^{-i} \right]$$

$$z_{i+1} = z_i - d_i \cdot \tan^{-1} 2^{-i} \qquad (7)$$

$$A_i = \sqrt{1 + 2^{-2i}} \quad i = 1, 2, 3, \ldots, n$$

when $n \to \infty$, $A_n \cong 1.647$.

In the Cartesian coordinate system, the CORDIC orthogonal rotation mode can be used to compute (5) by separating the two-sided rotation into two parts, $G = [G_1^T; G_2^T] = A \cdot R^T$ and $R \cdot G$. $A \cdot R^T$ that is computed by

$$G_1 = \begin{bmatrix} a_{ii}^r, a_{ij}^r \end{bmatrix}^T = \begin{bmatrix} \cos\theta & -\sin\theta \\ \sin\theta & \cos\theta \end{bmatrix} \cdot \begin{bmatrix} a_{ii}, a_{ij} \end{bmatrix}^T,$$

$$G_2 = \begin{bmatrix} a_{ji}^r, a_{jj}^r \end{bmatrix}^T = \begin{bmatrix} \cos\theta & -\sin\theta \\ \sin\theta & \cos\theta \end{bmatrix} \cdot \begin{bmatrix} a_{ji}, a_{jj} \end{bmatrix}^T,$$

$$\tag{8}$$

where the plane rotation with the desired rotation angle θ_{opt} is executed using two CORDIC rotators. The CORDIC processors apply n steps, usually $n = 32$ for single floating precision. A constant scaling value $K = 1/A_n = 0.6073$ is subsequently required to fix the rotated vectors $G_1 = [a_{ii}^r, a_{ji}^r]^T$ and $G_2 = [a_{ij}^r, a_{jj}^r]^T$ in order to retain the orthonormality. Similarly, these two CORDIC rotators can also be applied to compute $R \cdot G$:

$$\begin{bmatrix} a_{ii}', a_{ji}' \end{bmatrix}^T = \begin{bmatrix} \cos\theta & -\sin\theta \\ \sin\theta & \cos\theta \end{bmatrix} \cdot \begin{bmatrix} a_{ii}^r, a_{ji}^r \end{bmatrix}^T,$$

$$\begin{bmatrix} a_{ij}', a_{jj}' \end{bmatrix}^T = \begin{bmatrix} \cos\theta & -\sin\theta \\ \sin\theta & \cos\theta \end{bmatrix} \cdot \begin{bmatrix} a_{ij}^r, a_{jj}^r \end{bmatrix}^T.$$

$$\tag{9}$$

Meanwhile, the angle θ_{opt} can also be determined by using the CORDIC orthogonal vector mode. The CORDIC rotates the input vector through whatever angle is necessary to align the resulting vector with the x-axis:

$$x_n = A_n \sqrt{x_0^2 + y_0^2}$$

$$y_n = 0 \tag{10}$$

$$z_n = z_0 - d_i \cdot \tan^{-1} 2^{-i}.$$

The CORDIC with an orthogonal vector mode can compute the arctangent result iteratively $\theta = \arctan(y/x)$, if the angle accumulator is initialized with zero ($z_0 = 0$).

In the VLSI design, two common approaches can be used to realize the CORDIC dependence flow graph in hardware: the folded (serial) or the parallel (pipelining) [10, 11]. Note that we limit our efforts to the conventional CORDIC iteration scheme, as given in (7). In Figure 2(a), the structure of a folded CORDIC PE is shown, which requires a pair of adders for plane rotation and another adder for steering the next angle direction (computing the following z_i and d_i). All internal variables are buffered in the registers separately until the iteration number is large enough to obtain the result. The signs of all three intermediate variables are fed into a control unit that generates the rotation direction flags d_i to steer the add or suboperations and keep track of the rotation angle z_i. For example, off-diagonal PE$_{43}$ can directly apply the flags d_i from PE$_{33}$ to (8)'s G_1 and PE$_{44}$ to (8)'s G_2. After the rotation, the required scaling procedure can be obtained using the part of Figure 2(b) that fixes A_n, where two multiplexers are required to select the inputs into the barrel shifters. This folded dependence graph is typical for

the orthogonal rotation mode and benefits in a small area in the VLSI design.

In practice, the angle accumulator is not required for the off-diagonal PEs. The d_i from (7) can be used to steer the rotators. Thus, the transmission on the vertical and horizontal dashed lines in Figure 1 will be replaced by a sequence of d_i flags, meaning that the off-diagonal computation efforts for computing the optimal angle θ_{opt} can be omitted.

3.2. Simplified μ-Rotation CORDIC. As the process technologies continue to shrink, it becomes possible to directly implement a Brent-Luk-EVD array with the Jacobi method [12, 13]. However, the size of the EVD array that can be implemented on the current configurable device with the regular CORDIC is still small, say, 4×4. Therefore, we must simplify the architecture in order to integrate more processors. A scaling-free μ-CORDIC for performing the plane rotation in (5) is used [5, 6], where the number of inner iterations is reduced from 32 iterations to only one iteration.

The definition of μ-CORDIC can be developed from (7) as

$$x_{i+1} = \widehat{m}\left[x_i - y_i \cdot d_i \cdot 2^{-i}\right]$$

$$y_{i+1} = \widehat{m}\left[y_i + x_i \cdot d_i \cdot 2^{-i}\right] \tag{11}$$

$$\widehat{m} = \sqrt{\cos^2\theta + \sin^2\theta} = 1 + \epsilon,$$

where $\widehat{m}$ is the required scaling factor per iteration and ϵ is the scaling error. The idea of the μ-CORDIC rotation is to reduce the number of iterations of the full CORDIC to only a few iterations. Meanwhile, the scaling error ϵ will be small enough to be neglected as long as the orthonormality is retained. Figure 3 shows four different methods for different sizes of μ-rotation angles and Table 1 shows a lookup table for the μ-CORDIC, listing 32 approximated rotation angles for each μ-rotation type, the required number of shift-add operations and its computation cycles. Note that the approximated angles are stored as two times of $\tan\theta$. When the rotation angle is very tiny (i.e., ϵ is tiny, too), Type I with only one iteration will comply with the limited working range $1 - 2^{-(n_m+1)} < \widehat{m} < 1 + 2^{-(n_m+1)}$, if the selected n_m ($n_m \in 1 \cdots 32$) is larger than 16. In Figure 3(a), a pair of shift-add operations realizing one iteration step is sufficient. Furthermore, it is scaling free when the angle $2 \times \tan\theta \leq 3.05176 \times 10^{-5}$. These orthonormal μ-rotations are chosen such that they satisfy a predefined accuracy condition in order to approximate the original rotation angles and are constructed by the least computation efforts.

Next, for the Type II rotation (as shown in Figure 3(b)), when n_m is selected from 8 to 15 for small angles, two pairs of shift-add operations are enough to retain the orthonormality. Moreover, when the n_m is selected from 5 to 7, Type III requires three μ-rotations. No scaling is required by Types I through III. Finally, for large rotation angles, the scaling errors cannot be omitted. Figure 3(d) shows the corresponding dependence flow graph for Type IV. Besides the rotation

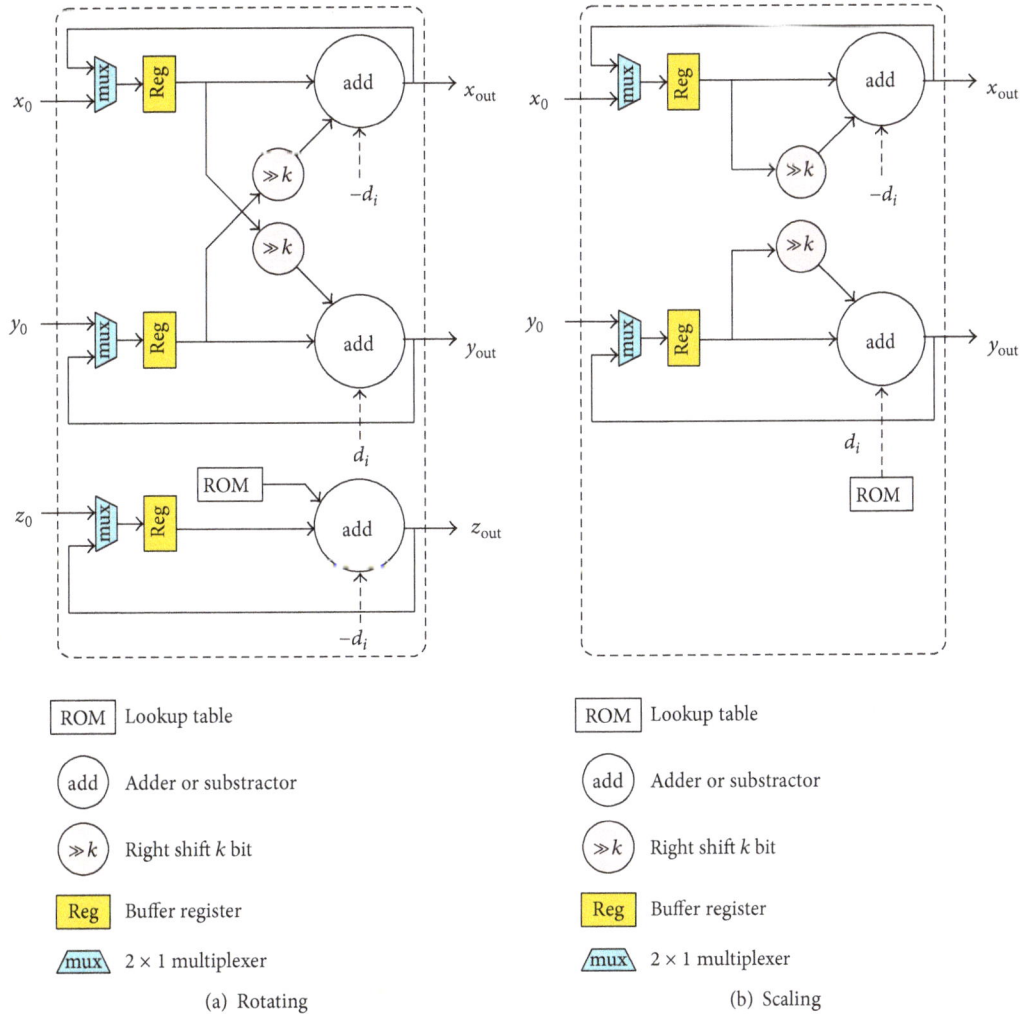

(a) Rotating   (b) Scaling

FIGURE 2: Flow graph of a folded CORDIC (recursive) processor.

itself, it requires two pairs of shift-add operations at the beginning of the flow graph, while 2 to 4 pairs of shift-add operations are required to fix the scaling factor $\widehat{m}$:

$$\widehat{m} = \left(1 + 2^{-2(k+1)}\right)\left(1 + 2^{-4(k+1)}\right)\cdots\left(1 + 2^{-2^{M}(k+1)}\right).$$

(12)

Note that the scaling costs M = 2 to 4 pairs of shift-add operations. In general, the cost of Type IV is bounded by $2 + M$ pairs of shift-add operations. For example, when the index k is 2, the scaling is

$$\widehat{m}_2 = \left(1 + 2^{-6}\right)\left(1 + 2^{-12}\right)\left(1 + 2^{-24}\right).$$

(13)

These four subtypes have three identical parts: Type I with one iteration, the scaling part of Type IV, and the second iteration of Type II. These three parts can be integrated together by using multiplexers to select the data paths, as shown in Figure 4, where 2 adders, 2 shifters, and 4 multiplexers are required [5].

3.3. *Adaptive μ-CORDIC Iterations.* To improve the computational efficiency, the μ-CORDIC has been modified to perform 6 iterations per cycle as CORDIC-6. As the global clock in a synchronous circuit is determined by the critical path, the maximum timing delay per iteration is 6 cycles (when the index k is 1, Type IV). Therefore, the inner iteration steps of the angles are repeated until they are close to the critical one. The required number of repetitions is quoted in Table 1. For example, when the rotation angle index k is 8, it will repeat three times from the index $k = 8$ to the index $k = 10$; when the rotation angle index k is 20, it is repeated six times from the index $k = 20$ to the index $k = 25$. On the other hand, we can adjust the number of iterations by selecting the average angle during the last sweep and name it as CORDIC-mean.

4. Experimental Results

4.1. *Matlab Simulation.* The full CORDIC with 32 iteration steps, the μ-CORDIC with one iteration step, and two different adaptive modes have been tested using numerous

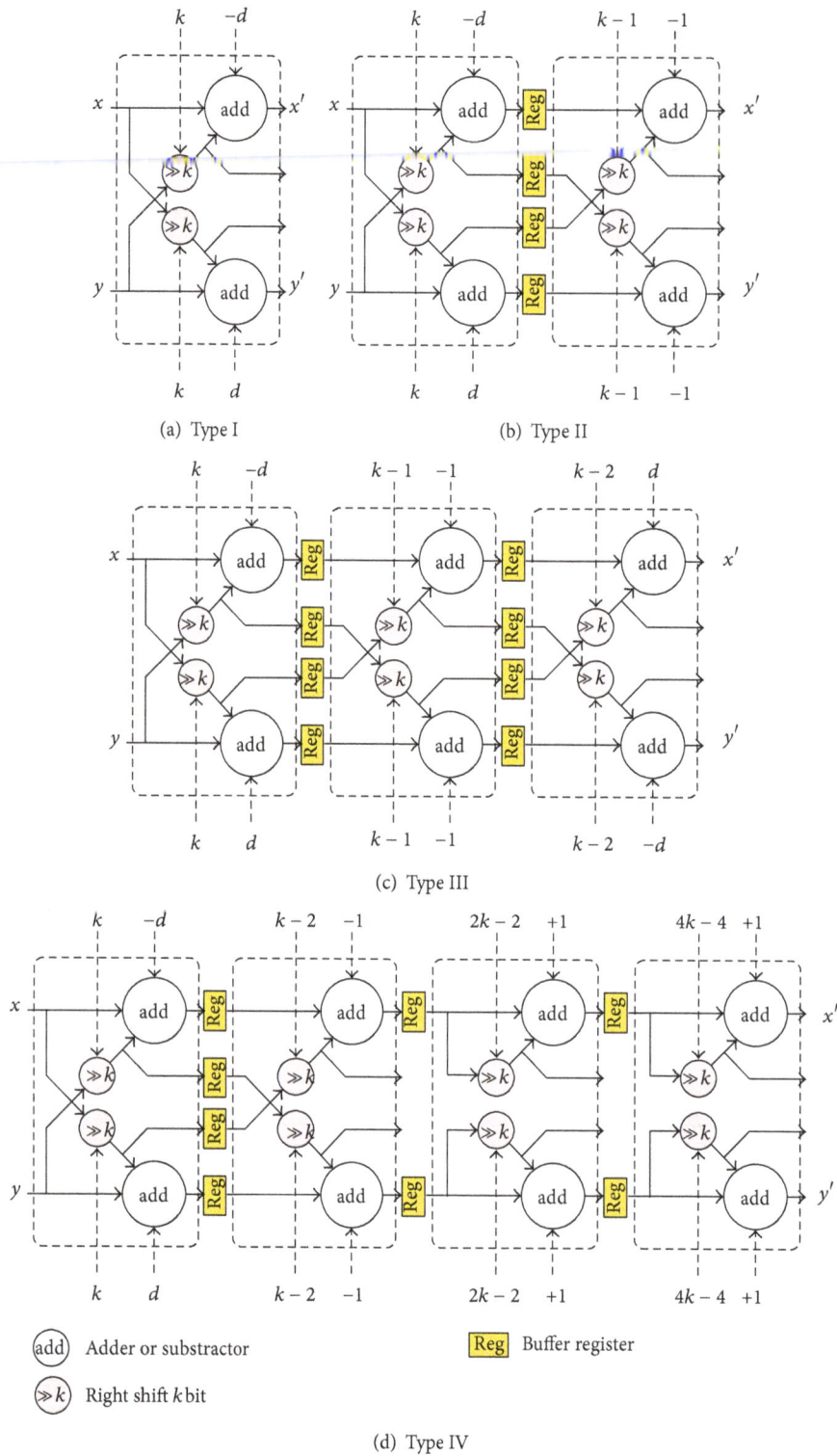

(a) Type I   (b) Type II

(c) Type III

(d) Type IV

add   Adder or substractor

Reg   Buffer register

$\gg k$ Right shift k bit

FIGURE 3: Four μ-CORDIC rotations.

random symmetric matrices A of size 8×8 to 160×160 (i.e., EVD array sizes range from 4×4 to 80×80). Figure 5(a) shows the average number of sweeps needed to compute the eigenvalues/eigenvectors for each size of the EVD array, where the sweep number increases monotonically.

When the Jacobi EVD array size is 10×10, the μ-CORDIC requires 12 sweeps while the full CORDIC only requires 6 sweeps per EVD computation. If we adjust the inner rotations to six times, the sweep number will be 10, smaller than the μ-CORDIC but more than the full CORDIC. Note that using

TABLE 1: The lookup table for μ-rotations CORDIC with 32-bit accuracy, showing the rotation type, the $2 \times \tan\theta$ angle, the required shift-add operations for rotation and scaling, the required cycle delay, and repeat number for CORDIC-6.

Index k	Type	Angle $2 \times \tan\theta$	Shift-add rot.	Shift-add sca.	Cycle cnt.	Cycle re.
1	IV	1.49070	4	8	6	1
2	IV	0.54296	4	6	5	1
3	IV	0.25501	4	6	5	1
4	IV	0.12561	4	4	4	1
5	III	$6.25841\,10^{-2}$	6	0	3	2
6	III	$3.12606\,10^{-2}$	6	0	3	2
7	III	$1.56263\,10^{-2}$	6	0	3	2
8	II	$7.81266\,10^{-3}$	4	0	2	3
9	II	$3.90627\,10^{-3}$	4	0	2	3
10	II	$1.95313\,10^{-3}$	4	0	2	3
11	II	$9.76563\,10^{-4}$	4	0	2	3
12	II	$4.88281\,10^{-4}$	4	0	2	3
13	II	$2.44141\,10^{-4}$	4	0	2	3
14	II	$1.22070\,10^{-4}$	4	0	2	4
15	II	$6.10352\,10^{-5}$	4	0	2	5
16	I	$3.05176\,10^{-5}$	2	0	1	6
17	I	$1.52588\,10^{-5}$	2	0	1	6
18	I	$7.62939\,10^{-6}$	2	0	1	6
19	I	$3.81470\,10^{-6}$	2	0	1	6
20	I	$1.90735\,10^{-6}$	2	0	1	6
21	I	$9.53674\,10^{-7}$	2	0	1	6
22	I	$4.76837\,10^{-7}$	2	0	1	6
23	I	$2.38419\,10^{-7}$	2	0	1	6
24	I	$1.19209\,10^{-7}$	2	0	1	6
25	I	$5.96046\,10^{-8}$	2	0	1	6
26	I	$2.98023\,10^{-8}$	2	0	1	6
27	I	$1.49012\,10^{-8}$	2	0	1	6
28	I	$7.45058\,10^{-9}$	2	0	1	5
29	I	$3.72529\,10^{-9}$	2	0	1	4
30	I	$1.86265\,10^{-9}$	2	0	1	3
31	I	$9.31323\,10^{-10}$	2	0	1	2
32	I	$4.65661\,10^{-10}$	2	0	1	1

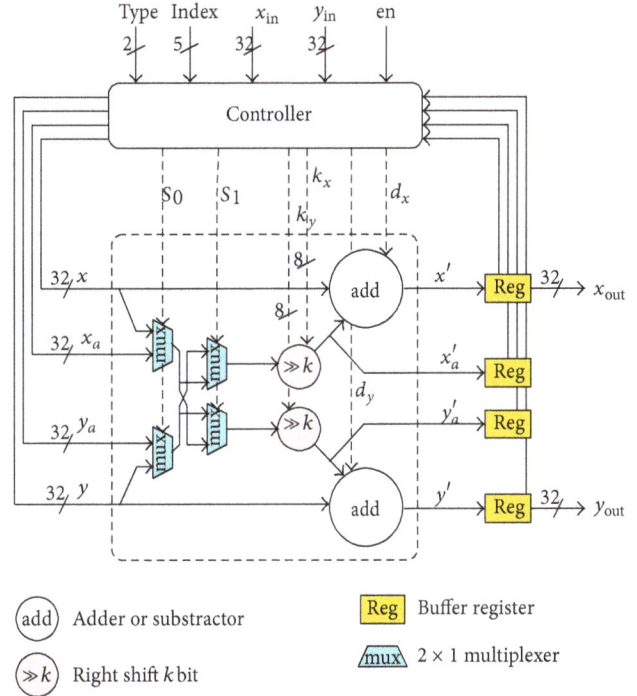

FIGURE 4: The block diagram of a scaling-free μ-CORDIC PE, including 2 adders, 2 shifters, and 4 multiplexers.

shift-add operations than others. The adaptive CORDIC-6 method can offer a compromise between the hardware complexity and the computational effort.

Figure 5(c) shows the off-diagonal Frobenius norm versus the sweep numbers for each array size of 80×80 with double floating precision. Each rotation method converges to the predefined stop criteria: $\|A_{\text{off}}\|_F \times 10^{-8}$. The $\|A_{\text{off}}\|_F$ is the Frobenius norm of the off-diagonal elements of A (i.e., $A_{\text{off}} = A - \text{diag}(\text{diag}(A))$).

Figure 5(d) shows the reduction of the off-diagonal Frobenius norm versus the sweep numbers for single floating precision. It can be noticed that the off-norms do not reach the convergence criteria, and each size of the EVD array has different stop criteria for each rotation method (default IEEE 754 single). Therefore, we can first analyze the Frobenius norm of the off-diagonal elements in Matlab and then observe it until it reaches its maximal reduction. Afterwards, a lookup table can be generated and directly assign these stop criteria to the target hardware circuit or IP component.

the average rotation angle to decide the rotation number as the CORDIC-mean seems to be an unwise method because it requires more sweeps. Although the μ-CORDIC requires double sweeps than the full CORDIC, it actually reduces the number of the inner CORDIC rotations, which results in improved computational complexity. For example, a 10×10 array with the Full CORDIC PE needs 6 sweeps $\times$ 32 inner CORDIC rotations and the CORDIC-6 needs 10 sweeps $\times$ 6 inner CORDIC rotations whereas the μ-CORDIC PE requires only 12 sweeps $\times$ 1 inner CORDIC rotation. In Figure 5(b), the average number of shift-add operations required for each rotation method for different sizes of EVD arrays is demonstrated whereas μ-CORDIC needs significantly fewer

4.2. VLSI Implementation. The μ-CORDIC is modeled and compared to the folded Full CORDIC in VHDL with the resizing feature. These two methods have been integrated into parallel EVD arrays, with sizes 4×4 and 10×10, through a configurable interface separately. After that, they have been synthesized by using the Synopsys Design Compiler with the TSMC 45 nm standard cell library. Note that the word length is 32 bits with the IEEE 754 single floating precision for both CORDIC methods using the same floating point unit from

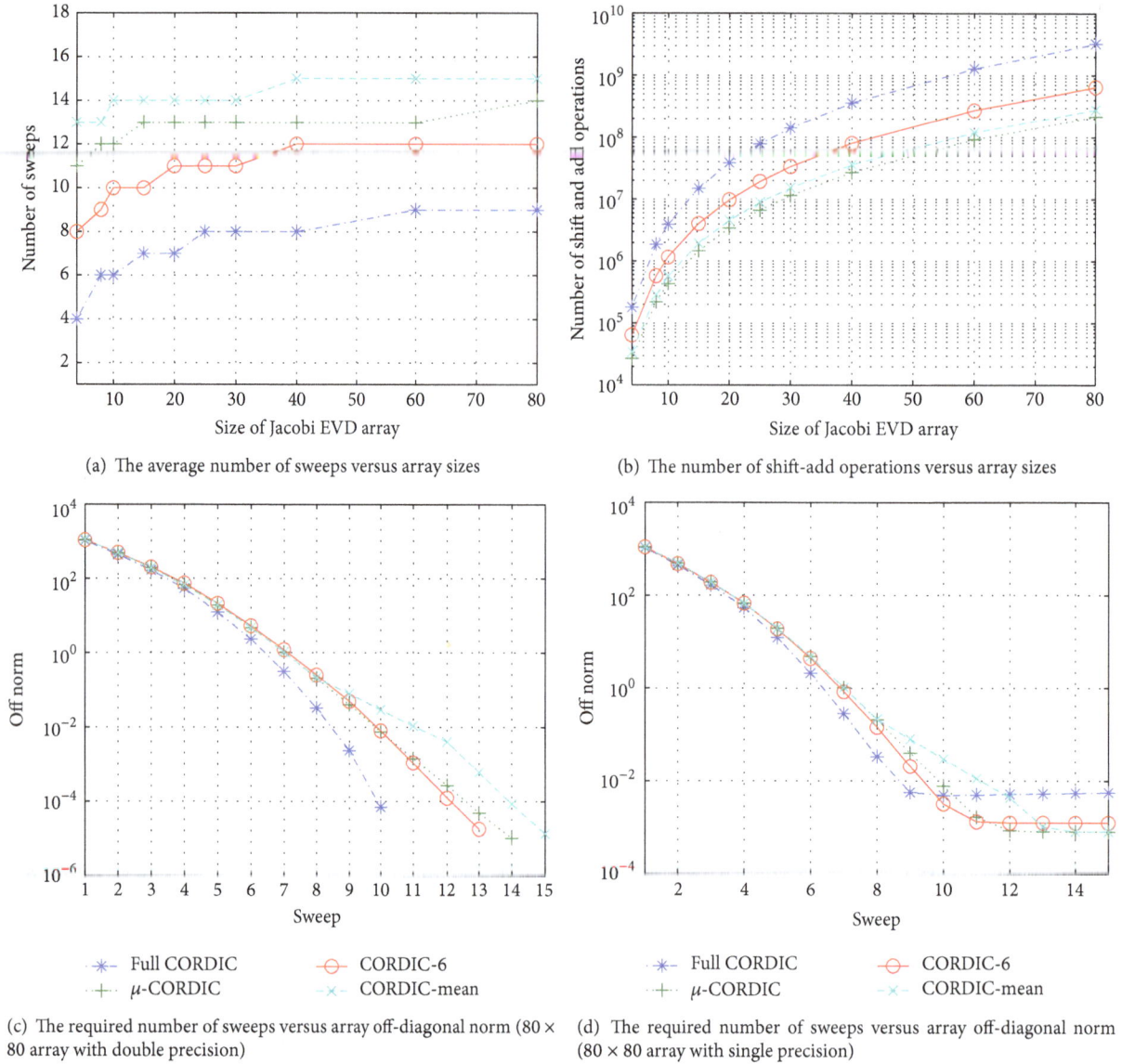

(a) The average number of sweeps versus array sizes

(b) The number of shift-add operations versus array sizes

(c) The required number of sweeps versus array off-diagonal norm (80 × 80 array with double precision)

(d) The required number of sweeps versus array off-diagonal norm (80 × 80 array with single precision)

FIGURE 5: Simulations of four rotation methods.

OpenCORE. Table 2 lists the synthesis results for area, timing delay, and power consumption.

As expected, the combinational logic area and the power consumption of the μ-CORDIC PE are much smaller than the Full CORDIC. Furthermore, in order to determine the time required to compute the EVD of a $n \times n$ symmetric matrix, it can be obtained by

$$T_{\text{total}} = T_{\text{delay}} \times K_{\text{iteration}} \times K_{\text{sweep}} \times [3(n-1) + \Delta + 3],$$
(14)

where $n = 8, 16, 20, 30, \ldots, 160$, $\Delta = n/2 - 1$.

The total timing delay per EVD operation is defined by the critical timing delay × the number of inner CORDIC rotations × average number of outer sweeps × size of the matrix A. It can be observed that the total operation time is dependent on the relationship between the inner CORDIC rotations

and the outer sweeps. Therefore, one obtains a speedup by a factor of 21.4 by reducing the number of inner CORDIC rotations. Although the reduction of power consumption is less significant due to an extra μ-CORDIC's controller and multiplexers, it actually 6 consumes much less energy per EVD computation due to the shorter computation time. Note that the μ-CORDIC PE requires two inner iterations on average due to the different rotation cycles, from six to one inner iteration, as shown in Table 1. Figure 6 shows the energy consumption for sizes of the array from 4×4 to 80×80. Both rotation methods consume much less energy than the Full CORDIC, where the 6-CORDIC can obtain a factor of 40.9 and the μ-CORDIC can obtain a factor of 104.3 on average for energy reduction compared to the Full CORDIC.

In [14], a Jacobi single cycle-by-row EVD algorithm [15] has been implemented with a single CORDIC processor.

TABLE 2: Comparison of 4×4 and 10×10 Jacobi EVD arrays.

	PE array size	Full CORDIC 4×4	μ-CORDIC 4×4	Full CORDIC 10×10	μ-CORDIC 10×10
	Combinational	$0.847 \, \text{mm}^2$	$0.296 \, \text{mm}^2$	$5.143 \, \text{mm}^2$	$1.829 \, \text{mm}^2$
Area	Noncombinational	$0.390 \, \text{mm}^2$	$0.123 \, \text{mm}^2$	$2.306 \, \text{mm}^2$	$0.833 \, \text{mm}^2$
	Total	$\mathbf{1.237} \, \text{mm}^2$	$\mathbf{0.419} \, \text{mm}^2$	$\mathbf{7.449} \, \text{mm}^2$	$\mathbf{2.662} \, \text{mm}^2$
	Cell	62.283 mW	18.239 mW	388.379 mW	123.215 mW
Power	Net	0.465 mW	0.433 mW	2.993 mW	2.678 mW
	Leakage	11.909 mW	3.765 mW	86.136 mW	23.966 mW
	Total	**74.657** mW	**22.437** mW	**477.508** mW	**149.859** mW
Timing	Critical	4.454 ns	1.213 ns	4.286 ns	2.247 ns
	Frequency	224.5 MHz	824.4 MHz	233.3 MHz	445 MHz

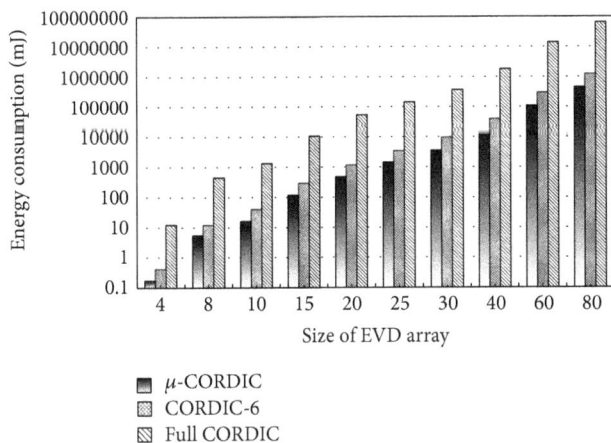

FIGURE 6: The energy consumption per EVD operation with each size of EVD array (operating at 100 MHz).

Since it requires a very complex controller and lookup tables, the throughput is not comparable with a real Brent-Luk's parallel EVD array [13]. In comparison to [13], Full CORDIC for Jacobi Brent-Luk-EVD parallel architecture is implemented in FPGA; however, current configurable device can only perform 4×4 EVD array. The experimental results show that performing the unitary rotation in CORDIC processor is a good solution. It required smaller area size, improved the overall computation time, and reduced the energy consumption. Furthermore, the unitary-rotation method can be also applied to other more efficient CORIDC architectures as long as the orthogonality is obtained during CORDIC iterations, such as pipeline CORDIC [16, 17], or implementing the rotators with better adder structures [18, 19].

As the process technologies continue to shrink, it becomes possible to directly implement a Brent-Luk-EVD array with the Jacobi method [12, 13]. However, the size of the EVD array that can be implemented on the current configurable device with the regular CORDIC is still small, say, 4×4. Therefore, we must simplify the architecture in order to integrate more processors. A scaling-free μ-CORDIC for performing the plane rotation in (5) is used [5, 6], where the number of inner iterations is reduced from 32 iterations to only one iteration.

5. Conclusions

The EVD was computed by the parallel Jacobi method, which was selected as an example for a typical iterative algorithm which exhibits very robust convergence properties. A configurable Jacobi EVD array with both Full CORDIC and μ-CORDIC is implemented in order to further study the tradeoff strategies in design criteria for parallel integrated circuits. The experimental results indicate that the presented μ-CORDIC method can reduce the size of the combinational logic, speed up the overall computation time, and improve the energy consumption.

Conflict of Interests

The authors declare that there is no conflict of interests regarding the publication of this paper.

Acknowledgment

The authors would like to thank the National Science Council of Taiwan for the support under the contract of NSC 101-2218-E-150-001.

References

[1] S. Aggarwal and K. Khare, "CORDIC-based window implementation to minimise area and pipeline depth," *IET Signal Processing*, vol. 7, no. 5, pp. 427–435, 2013.

[2] H. M. Ahmed, J. Delosme, and M. Morf, "Highly concurrent computing structure for matrix arithmetic and signal processing," *IEEE Computer Magazine*, vol. 15, no. 1, pp. 65–82, 1982.

[3] A. Ahmedsaid, A. Amira, and A. Bouridane, "Improved SVD systolic array and implementation on FPGA," in *Proceedings of the IEEE International Conference on Field-Programmable Technology*, pp. 3–42, 2003.

[4] R. Andraka, "Survey of CORDIC algorithms for FPGA based computers," in *Proceedings of the ACM/SIGDA 6th International Symposium on Field Programmable Gate Arrays (FPGA '98)*, pp. 191–200, February 1998.

[5] I. Bravo, P. Jiménez, M. Mazo, J. L. Lázaro, and A. Gardel, "Implementation in FPGAS of Jacobi method to solve the eigenvalue and eigenvector problem," in *Proceedings of the International Conference on Field Programmable Logic and Applications (FPL '06)*, pp. 1–4, Madrid, Spain, August 2006.

[6] R. P. Brent and F. T. Luk, "The solution of singular-value and symmetric eigenvalue problems on multiprocessor arrays," *SIAM Journal on Scientific and Statistical Computing*, vol. 6, no. 1, pp. 69–84, 1985.

[7] G. H. Golub and C. F. van Loan, *Matrix Computations*, Johns Hopkins University Press, Baltimore, Md, USA, 3rd edition, 1996.

[8] J. Götze and G. J. Hekstra, "An algorithm and architecture based on orthonormal μ-rotations for computing the symmetric EVD," *Integration, the VLSI Journal*, vol. 20, no. 1, pp. 21–39, 1995.

[9] J. Götze, S. Paul, and M. Sauer, "An efficient Jacobi-like algorithm for parallel eigenvalue computation," *IEEE Transactions on Computers*, vol. 42, no. 9, pp. 1058–1065, 1993.

[10] A. Hakkarainen, J. Werner, K. R. Dandekar, and M. Valkama, "Widely-linear beamforming and RF impairment suppression in massive antenna arrays," *Journal of Communications and Networks*, vol. 15, no. 4, pp. 383–397, 2013.

[11] S. Klauke and J. Götze, "Low power enhancements for parallel algorithms," in *Proceedings of the IEEE International Symopsium on Circuits and Systems*, pp. 234–237, 2001.

[12] Y. Liu, C.-S. Bouganis, and P. Y. K. Cheung, "Hardware architectures for eigenvalue computation of real symmetric matrices," *IET Computers and Digital Techniques*, vol. 3, no. 1, pp. 72–84, 2009.

[13] P. K. Meher and S. Y. Park, "CORDIC designs for fixed angle of rotation," *IEEE Transactions on Very Large Scale Integration (VLSI) Systems*, vol. 21, no. 2, pp. 217–228, 2013.

[14] K. K. Parhi and T. Nishitani, *Digial Signal Processing for Multimedia Systems*, Marcel Dekker, 1999.

[15] S. Purohit and M. Margala, "Investigating the impact of logic and circuit implementation on full adder performance," *IEEE Transactions on Very Large Scale Integration (VLSI) Systems*, vol. 20, no. 7, pp. 1327–1331, 2012.

[16] B. Ramkumar and H. M. Kittur, "Low-power and area-efficient carry select adder," *IEEE Transactions on Very Large Scale Integration (VLSI) Systems*, vol. 20, no. 2, pp. 371–375, 2012.

[17] F. Rusek, D. Persson, B. K. Lau et al., "Scaling up MIMO: opportunities and challenges with very large arrays," *IEEE Signal Processing Magazine*, vol. 30, no. 1, pp. 40–46, 2013.

[18] C.-C. Sun and J. Götze, "VLSI circuit design concept for parallel iterative algorithms in nanoscale," in *Proceedings of the 9th International Symposium on Communications and Information Technology (ISCIT '09)*, pp. 688–692, Icheon, Republic of Korea, September 2009.

[19] J. S. Walther, "A unified algorithm for elementary functions," in *Proceedings of the Spring Joint Computer Conference*, pp. 379–385, 1971.

## 10

Analysis and Implementation of Kidney Stone Detection by Reaction Diffusion Level Set Segmentation Using Xilinx System Generator on FPGA

Kalannagari Viswanath[1] and Ramalingam Gunasundari[2]

[1]Pondicherry Engineering College, Puducherry 605 014, India
[2]Department of ECE, Pondicherry Engineering College, Puducherry 605 014, India

Correspondence should be addressed to Kalannagari Viswanath; viswa_kv@pec.edu

Academic Editor: Mohamed Masmoudi

Ultrasound imaging is one of the available imaging techniques used for diagnosis of kidney abnormalities, which may be like change in shape and position and swelling of limb; there are also other Kidney abnormalities such as formation of stones, cysts, blockage of urine, congenital anomalies, and cancerous cells. During surgical processes it is vital to recognize the true and precise location of kidney stone. The detection of kidney stones using ultrasound imaging is a highly challenging task as they are of low contrast and contain speckle noise. This challenge is overcome by employing suitable image processing techniques. The ultrasound image is first preprocessed to get rid of speckle noise using the image restoration process. The restored image is smoothened using Gabor filter and the subsequent image is enhanced by histogram equalization. The preprocessed image is achieved with level set segmentation to detect the stone region. Segmentation process is employed twice for getting better results; first to segment kidney portion and then to segment the stone portion, respectively. In this work, the level set segmentation uses two terms, namely, momentum and resilient propagation (R_{prop}) to detect the stone portion. After segmentation, the extracted region of the kidney stone is given to Symlets, Biorthogonal (bio3.7, bio3.9, and bio4.4), and Daubechies lifting scheme wavelet subbands to extract energy levels. These energy levels provide evidence about presence of stone, by comparing them with that of the normal energy levels. They are trained by multilayer perceptron (MLP) and back propagation (BP) ANN to classify and its type of stone with an accuracy of 98.8%. The prosed work is designed and real time is implemented on both Filed Programmable Gate Array Vertex-2Pro FPGA using Xilinx System Generator (XSG) Verilog and Matlab 2012a.

1. Introduction

Kidney stone disease is one of the major life threatening ailments persisting worldwide. The stone diseases remain unnoticed in the initial stage, which in turn damages the kidney as they develop. A majority of people are affected by kidney failure due to diabetes mellitus, hypertension, glomerulonephritis, and so forth. Since kidney malfunctioning can be menacing, diagnosis of the problem in the initial stages is advisable. Ultrasound (US) image is one of the currently available methods with noninvasive low cost and widely used imaging techniques for analyzing kidney diseases [1]. Shock wave lithotripsy (SWL), percutaneous nephrolithotomy (PCNL),

and relative super saturation (RSS) are the available practices to test urine. The Robertson Risk Factor Algorithms (RRFA) are open and are used for laparoscopic surgery; these algorithms are assigned for exceptional [2] special cases. Hyaluronan is a large (>106 Da) linear glycosaminoglycan composed of repeating units of glucuronic acid (GlcUA) and N-acetyl glucosamine (GlcNAc) disaccharides [3]. It has a significant role in a number of processes that can eventually lead to renal stone disease, including urine concentration, uric acid, salt form crystal, crystallization inhibition, crystal retention, magnesium ammonium phosphate, and amino acid.

Rahman and Uddin have proposed diminution of speckle noise and segmentation from US image. It not only detects problem in the kidney region but also provides image quality enhancement [1]. Hafizah proposed kidney US images and divides them into four dissimilar categories: normal, bacterial infection, cyotic disease, and kidney stones, using gray level cooccurrence matrix (GLCM). This categorization helps doctors to identify the abnormalities in kidney [4]. Rathi and Palani have proposed a Hierarchical Self-Organizing Map (HSOM) for brain tumours using the segmentation technique and wavelets packets. Accuracy of the results was found to be correct up to 97% [5]. Norihiro Koizumi has proposed high intensity focused ultrasound (HIFU) technique for terminating tumours and stones [6, 7]. Viswanath and Gunasundari propose content descriptive multiple stones detection using level set segmentation, wavelets processing for identification of kidney stone, and artificial neural network (ANN) for classification. The results yielded a maximum accuracy of 98.66% [8]. The MLP-BP ANN is found to perform better in terms of accuracy 92% with a speed of 0.44 sec and it is found to be very sensitive [9, 10]. The noninvasive combination of renal using pulsed cavitation US therapy proposed that shock wave lithotripsy (ESWL) has become a customary for the treatment of calculi located in the kidney and ureter [11]. Tamilselvi and Thangaraj have proposed seeded region growing based on segmentation and classification of kidney images with stone sizes using CAD system [12]. Bagley et al. estimate location of urinary stones with unenhanced computed tomography (CT) using half-radiation (low) dose compared with the standard dose. Out of the 50 patients examined, 35 patients were found to have a single stone while the rest had multiple stones [13]. The solution for local minima and segmentation problem was proposed by Thord Andersson, Gunnar Lathen, with modified gradient search and level set segmentation technique [2]. Templates based technique was proposed by Emmanouil Skounakis for 3D detection of kidneys and their pathology in real time. Its accuracy was found to be 97.2% and abnormalities in kidneys had an accuracy of 96.1% [7]. Gabor function is used for achievement of optimal sharpening and smoothening of 2D image in both time and frequency resolutions [14]. Chen et al. have proposed the finite element method based 3D tumor growth prediction using longitudinal kidney tumor images [15]. Using the linear elastic theory Owen et al. proposed pressure finding in fluid for calculating the depth of shock wave scattering by kidney stone in water [16]. The pH value based prediction of stone formation epidemiologically has been proposed by Kok [17]. Datar has proposed the segmentation of the desired portion using initial seed selection, growing, and region merging without any edge detection [18]. Multilayer perceptron and back propagation implementation on FPGA and ASIC design were carried out by Raj and Pinjare [19].

This research paper ensues as follows. In Section 2 problem statement is demarcated, Section 3 defines the proposed method, in Section 4 image segmentation to locate the kidney stone, in Section 5 optimized energy calculation of segmented portion is discussed, in Section 6 wavelets based energy extraction, Section 7 explains the artificial neural networks classifiers used, in Section 8 experiments results are discussed, and in the last section conclusion of the paper with scope for future work is given.

2. Problem Statement

The kidney malfunctioning can be life intimidating. Hence early detection of kidney stone is essential. Precise identification of kidney stone is vital in order to ensure surgical operations success. The ultrasound images of kidney comprise speckle noise and are of low contrast which makes the identification of kidney abnormalities a difficult task. As a result, the doctors may find identification of small stones and the type is difficult and challenging for identify the small kidney stones and their type appropriately. To address this issue, a reaction diffusion level set segmentation is proposed to identify location of the stone; it is implemented in real time on Vertex-2Pro FPGA with Verilog HDL using Xilinx System Generator blocks from Matlab 2012a which is compatible with xilinx13.4 ISE and lifting scheme wavelets subbands are employed for extraction of the energy levels of the stone. The results are analyzed using MLP-BP ANN algorithms for classification and its type of stone [20].

3. Methodology

Figure 1 shows the overall block diagram of the proposed method. It consists of the following blocks via kidney image database, image preprocessing, image segmentation, wavelet processing, and ANN classification.

3.1. Kidney Image Database. Kidney image database consists of nearly 500 US kidney images collected from different individuals of various hospitals. It consists of both normal and abnormal images stored in the database. One of the images is selected from the database and subjected to stone detection process.

3.2. Image Preprocessing. The aim of preprocessing is to improve the acquired low contrast ultrasound image with speckle noise. It suppresses the undesired distortions and enhances certain image features significant for further processing and stone detection. Without preprocessing, the US image quality may not be good for analyzing. For surgical operations, it is essential to identify the location of kidney stone accurately. Preprocessing helps to overcome this issue of low contrast and speckle noise reduction. Figure 2 shows the steps involved in preprocessing of US image, which are as follows:

(1) image restoration,

(2) smoothing and sharpening,

(3) contrast enhancement.

3.2.1. Image Restoration. Image restoration is meant to mitigate the degradation of the US image. Degradation may be due to motion blur, noise, and camera misfocus. The main purpose of image restoration is to reduce the degradations

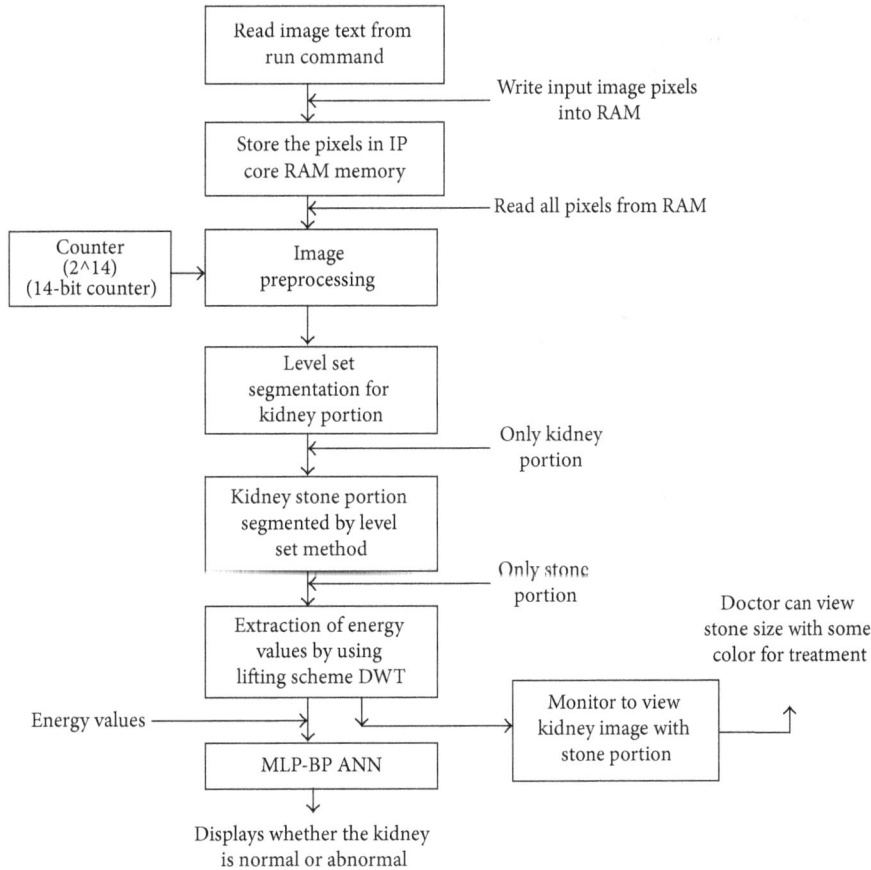

FIGURE 1: Proposed block diagram for kidney stone detection.

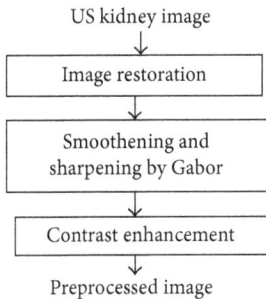

FIGURE 2: Preprocessing of kidney image.

3.2.2. Smoothing and Sharpening. The restored image is enriched with optimal resolution in both spatial and frequency domains using Gabor filter. This filter acts as a band pass filter with local spatial frequency distribution [6]. Image smoothing and removal of noise is performed using convolution operator. The standard deviation of the Gaussian function can be modified to tune the degree of smoothening and its hardware results are shown in Figure 3.

3.2.3. Contrast Enhancement. Histogram equalization is employed for improvement of the low contrast US image and achievement of the uniform intensity. This approach can be used on the image as a whole or to a part of an image. In this system, contrast enhancement of the images is executed by transforming the image intensity values, such that the histogram of the output image approximately matches a specified histogram and its results are shown in Figure 4. The input and output signal are of same data type.

4. Image Segmentation

In the segmentation process, five level set methods in all are discussed, all implemented and compared.

4.1. Conventional LSS. In the conventional level set method, consider a closed parameterized planar curve or surface,

that are caused during acquisition of US scanning. In this system, level set function is used for proper orientation. Using plane curve motion, curve smoothers, shrinks are eventually removed [1]. Thus Merriman and Sethian proposed evolution between $\max(k, 0)$ and $\min(k, 0)$:

$$f(x) = \begin{cases} \max(k, 0), & \text{if } a(x, y) < G(x, y) \\ \min(k, 0), & \text{otherwise}, \end{cases} \quad (1)$$

where $a(x, y)$ is average intensity small neighborhood and $G(x, y)$ is median in the same neighborhood.

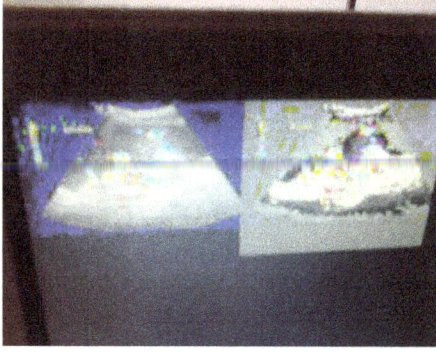

FIGURE 3: Hardware results of smoothening filter.

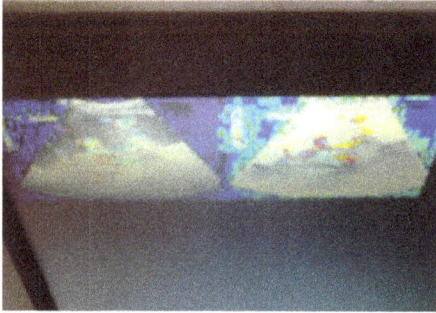

FIGURE 4: Hardware result of contrast enhancement.

denoted by $C(y,t) : [0,1] x R^+ \rightarrow R^n$, where $n = 2$ is for planar curve and $n = 3$ is for surface, and t is the artificial time generated by the movement of the initial curve or surface $C_0(y)$ in its inward normal direction $\tilde{N}$. The curve or surface evolution equation is as follows [21]:

$$C(y, t = 0) = C_0(y),$$
$$C_t = F\overline{N}, \tag{2}$$

where F is the force function.

In the above equation intrinsic drawback of interactively solving (2) lies in its difficulty to handle topological changes of the moving front, such as splitting and merging [21]. This problem can be eliminated by using the level set method (LSM) by modifying the above equation (2) by taking the derivative with respect to time t on the both sides, yielding the following equation:

$$\phi_t + \nabla_\phi \cdot C_t = \phi_t + \Delta_\phi \cdot F\overline{N} = 0,$$
$$\phi(x, t = 0) = \phi_0(x), \tag{3}$$

where gradient operator $\nabla(\cdot) = \partial(\cdot)/\partial x_1, \partial(\cdot)/\partial x_2, \ldots, \partial(\cdot)/\partial x_n$ and $\phi_0(x)$ is the initial LSF $C_0(y) = \{x \mid \phi_0(x) = 0\}$.

But (3) fails for too flat or too steep near the zero level set to address this issue re-initialization is introduced.

4.2. Reinitialization LSS. But, during evolution, the level set function (LSF) fails for too flat or too steep near the zero level set, causing serious numerical errors. Therefore, a procedure

called reinitialization is periodically employed to reshape it to be a signed distance function (SDF). In reinitialization the distance signed function is $\phi(x) = 1 \pm \text{dist}^2(x)$ where $\text{dist}(\cdot)$ is a distance function and $\pm$ denotes the sign inside and outside the contour [21]. But it has many problems, such as expensive computational cost, blocking the emerging of new contours [21], failures when the LSF deviates much from an SDF, and inconsistency between theory and implementation using XSG shown in Figure 5. Therefore, some formulation has been proposed to regularize the variational LSF to eliminate the reinitialization and computational cost. The following reinitialization equation is given by

$$\phi_t + S(\phi_0)(|\nabla\phi|) = 0, \tag{4}$$

where $\phi_0 = \phi_0/\sqrt{\phi_0^2 + \Delta x^2}$, ϕ_0 is the initial LSF, and Δx is the spatial step.

4.3. Distance Regularized Level Set Evolution (DRLSE). Li et al., proposed a signed distance penalizing energy functional is given by

$$P(\phi) \frac{1}{2} \int_\Omega (|\nabla\phi| - 1)^2 dx. \tag{5}$$

Equation (5) measures the closeness between an LSF (ϕ) and an SDF in the domain $\Omega \subset R^n$, $n = 2$ or 3. By calculus of variation [21], the gradient flow of $P(\phi)$ is obtained as

$$\phi_t = -P_\phi(\phi) = \text{div}(r_1(\phi)\nabla\phi). \tag{6}$$

Equation (6) is a diffusion equation with rates

$$r_1(\phi) = 1 - \frac{1}{|\nabla\phi|}. \tag{7}$$

However, $r_1(\varphi) \rightarrow -\infty$ when $|\nabla\varphi| \rightarrow 0$, which may cause oscillation in the final LSF φ. This problem is solved by applying a new diffusion rate

$$r_2(\phi) = \begin{cases} \dfrac{\sin(2\prod|\nabla\phi|)}{2\prod|\nabla\phi|}, & \text{if } |\nabla\phi| \leq 1 \\ 1 - \dfrac{1}{|\nabla\phi|}, & \text{if } |\nabla\phi| \geq 1 \end{cases} \tag{8}$$

and a constrained level set diffusion rate as

$$r_3(\phi) = H_p(|\nabla\phi| - 1), \tag{9}$$

where $H_p(z) = (1/2)[1 + (2/\prod)\arctan(z/\rho)]$.

And ρ is a fixed parameter. The DRLSE methods using $r_1(\varphi), r_2(\varphi)$, and $r_3(\varphi)$ are called generalized DRLSE such as GDRLSE1, GDRLSE2, and GDRLSE3, respectively.

4.4. Reaction Diffusion LSS. The RD equation is constructed by adding a diffusion term into the conventional LSE equation. Such an introduction of diffusion to LSE makes LSE stable without reinitialization. The diffusion term "$\varepsilon\Delta\phi$" was

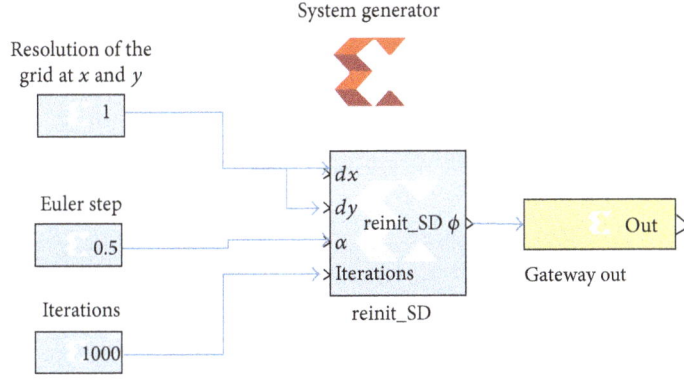

FIGURE 5: Hardware implementation of signed distance function using XSG.

added to the LSE equation (3); we get the following equation for RD:

$$\phi_t = \varepsilon\Delta\phi - \frac{1}{\varepsilon}L(\phi), \quad x\varepsilon\Omega \subset R^n$$

$$\text{Subject} \longrightarrow \phi(x, t = 0, \varepsilon) = \phi_0(x),$$

(10)

where ε is a small positive constant, $L(\varphi)$ for PDE-based LSM or $L(\varphi) = -F\delta(\varphi)$ for variational LSM, Δ is the Laplacian operator defined by $\Delta(\cdot) = \sum_{i=1}^{n}(\partial^2(\cdot)/\partial x_i^2)$, and $\phi_0(x)$ is the initial LSF. Equation (10) has two dynamic processes, the diffusion term $\varepsilon\Delta\phi$ gradually regularizes the LSF to be piecewise constant in each segment domain Ω_i and the reaction term "$-\varepsilon^{-1}L(\varphi)$" forces the final stable solution of (10) to $L(\varphi) = 0$, which determines Ω_i.

RD Level Set Segmentation Algorithm

(1) Initialization is as follows: $\phi^n = \phi_0$, $n = 0$.

(2) Compute $\emptyset^{n+1/2}$ as $\phi^{n+1/2=\phi^n=\Delta t_2 L(\phi^n)}$.

(3) Compute $\emptyset^{n+1}$ as $\phi^{n+1/2=\phi^n+\Delta t_2 L(\phi^n)}$.

(4) If $\emptyset^{n+1}$ satisfies stationary condition, stop; otherwise, $n = n + 1$ and return to step (2): $\phi^n = \phi^{n+1/2}$.

From the analysis in DRLSE and RD, the equilibrium solution of (10) is seen t as $\varepsilon \rightarrow 0^+$, which is the characteristic of phase transition. On the other hand, the said equation intrinsic problem of phase transition, that is, the stiff parameter ε^{-1}, makes (10) difficult to implement. In TSSM section, we propose a splitting method to implement (10) to reduce the side effect of stiff parameter ε^{-1}.

4.4.1. Two-Step Splitting Method (TSSM) for RD. A TSSM algorithm to implement RD has been proposed in [21] to generate the curvature-dependent motion. In [21] the reaction function is first forced to generate a binary function with values 0 and 1, and then the diffusion function is applied to the binary function to generate curvature-dependent motion, different from [21], where the diffusion function is used to generate curvature-dependent motion, in our proposed RD based LSM; the LSE is driven by the reaction function, that

is, the LSE equation. Therefore, use of the diffusion function to regularize the LSF generated by the reaction function is proposed, following TSSM to solve the RD. The steps are as follows:

(1) Solve the reaction term $\varphi_t = -\varepsilon^{-1}L(\varphi)$ with $\varphi(x, t = 0) = \varphi^n$ till some time T_r to obtain the intermediate solution, denoted by $\varphi^{n+1/2} = \varphi(x, T_r)$.

(2) Solve the diffusion term $\varphi_t = \varepsilon\Delta\varphi$, $\varphi(x, t = 0) = \varphi^{n+1/2}$ till some time T_d, and then the final level set is $\varphi^{n+1} = \varphi(x, T_d)$.

Although the second step faces the rush of moving the zero level set away from its original position. But this can be eliminated through the choice of a td small enough compared to the spatial resolutions.

In the above two terms, by choosing small T_r and T_d, we can discretely approximate $\varphi^{n+1/2}$ and φ^{n+1} as $\varphi^{n+1/2} = \varphi^n + \Delta t_1(-\varepsilon^{-1}L(\varphi^n))$ and $\varphi^{n+1} = \varphi^{n+1/2} + \Delta t_2(\varepsilon\Delta\varphi^{n+1/2})$, respectively, where the time steps Δt_1 and Δt_2 represent the times T_r and T_d, respectively. Obviously, we can integrate the parameter ε into the time steps Δt_1 and Δt_2 as $\Delta t_1 \leftarrow \Delta t_1(-\varepsilon^{-1})$ and $\Delta t_2 \leftarrow \Delta t_2\varepsilon$, and hence similar to the diffusion-generated or convolution-generated curvature motion, we only need to consider the two time steps Δt_1 and Δt_2 to keep numerical stability.

Speed of Segmentation. For the reinitialization methods, (4) should be iterated several times to make the LSF be an SDF while keeping the zero level set stationary. This is highly time-consuming for the reinitialization methods [21]. The GDRLSE methods are computationally much more efficient than reinitialization method. Equation (8) in each iteration of the computation of GDRLSE includes two components: the regularization term and LSE term driven by force F, in each iteration of RD method; the computation also includes two similar components. The only difference is that we split the computation into two steps: first compute the LSE term, and then compute the diffusion term. Therefore, the computation complexity of RD is similar to that of GDRLSE methods and, using Xilinx system generator implemented on hardware, its XSG are shown in the Figure 6. The proposed design is implemented using Verilog HDL, configured on Vertex-2Pro

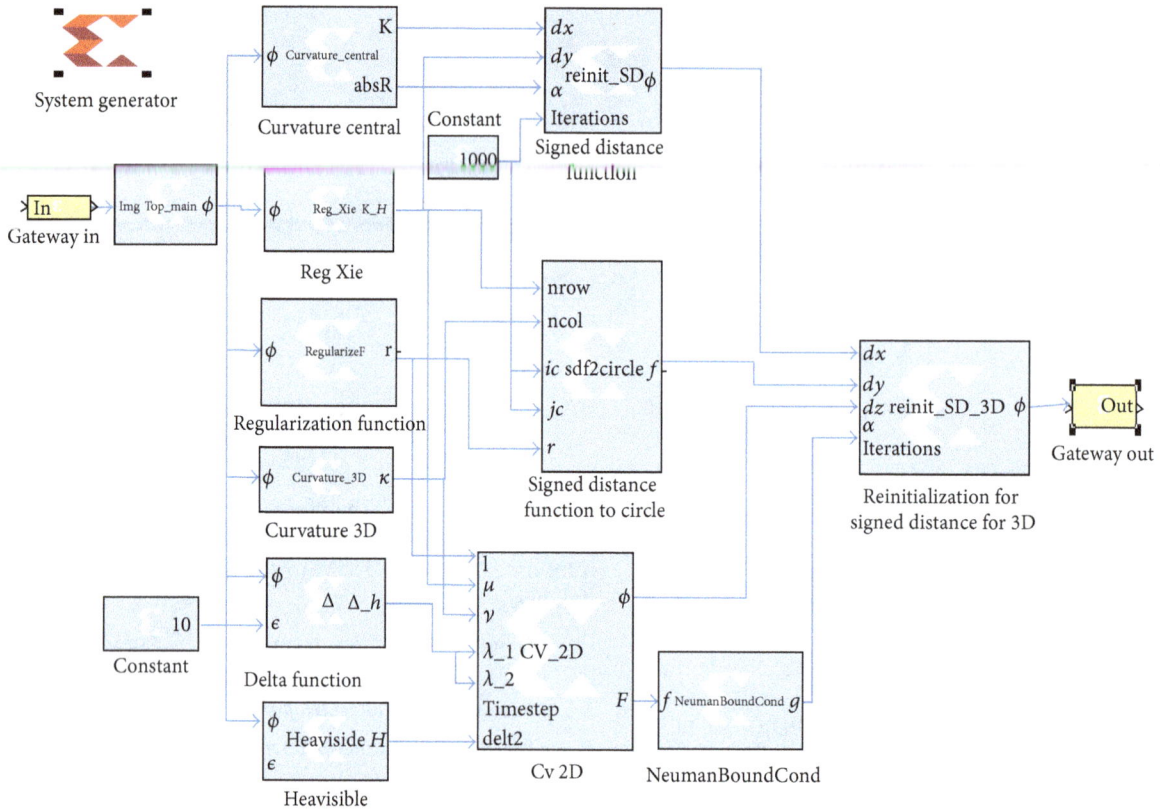

FIGURE 6: Curvature central, regularization, delta, signed distance, signed distance function to circle and reinitialization of signed distance of reaction, and diffusion level set function segmentation.

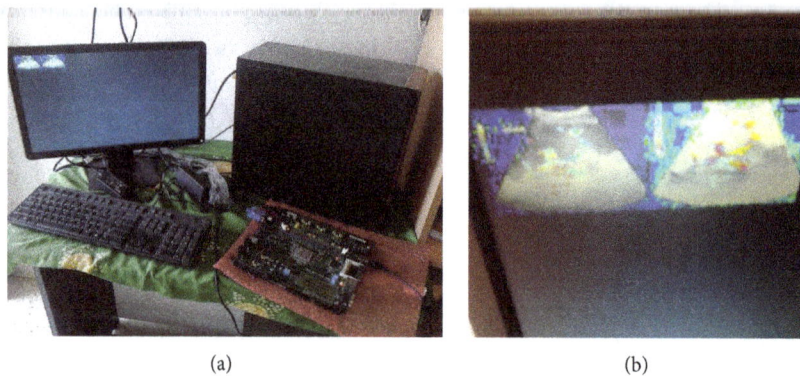

| (a) | (b) |

FIGURE 7: The FPGA output in the monitor through VGA. The first image shows the kidney portion segmented from the US image and second image witnesses the detected stone in that image indicated with red color.

FPGA and output of FPGA is applied to monitor through VGA for displaying input image and processed image as shown Figure 7.

5. Lifting Scheme Wavelets Processing

The segmented image (only stone) obtained from the previous block is applied to the lifting scheme wavelet processing block. This block consists of Daubechies filter (Db12), Symlets filter (sym12), and Biorthogonal filter (bio3.7, bio3.9, and bio4.3). In *Daubechies filter (Db12)* the number 12 denotes the number of vanishing moments. The higher the number of vanishing moments, the smoother the wavelet (and longer the wavelet filter) and the length of the wavelet (and scaling) filter should be twice that of the number [5]. *Symlets filter (sym12)* extracts the kidney image features and analyses the discontinuities and abrupt changes contained in the signals. One of the 12th-order Symlets wavelets is used for feature extraction. *Biorthogonal filter (bio3.7, bio3.9, and bio4.4)* filter's wavelet energy signatures were considered and averages of horizontal and vertical coefficients details were

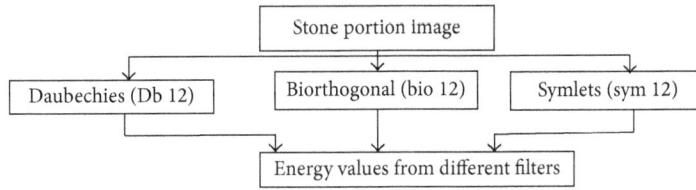

FIGURE 8: Wavelets filters to extract energy features.

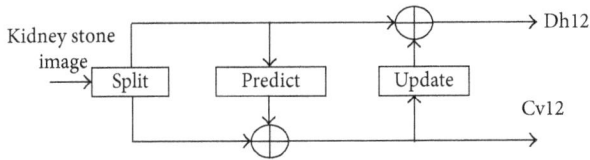

FIGURE 9: 2D lifting scheme DWT.

calculated. Figure 8 shows different filter used in the lifting scheme that gives different energy levels or energy features. These energy features will have significant difference if there is any stone present in the particular region or location. The identification of type of stone is explained in next section.

In 2D lifting scheme wavelets transformation consists of update and predictor to get Db12, sym12, bio3.7, bio3.9, and bio4.3 as shown in Figure 9.

The equations of predict and update are given by

$$d_i^1 = \alpha \left(x_{2i} + x_{2i+2} \right) + x_{2i+1},$$

$$a_i^1 = \beta \left(d_i^1 + d_{i-1}^1 \right) + x_{2i},$$

$$d_i^2 = \gamma \left(a_i^1 + a_{i+1}^1 \right) + d_i^1,$$

$$a_i^2 = \delta \left(d_i^2 + d_{i-1}^2 \right) + a_i^1,$$

$$d_i = \frac{d_i^2}{\varepsilon},$$

(11)

where x_{2i} and x_{2i+2} are even pixels, x_{2i+1} is odd pixels of stone image, and $\alpha, \beta, \gamma, \delta, \varepsilon$ are the constants.

6. ANN Classification

Two architectures are used in the ANN classification, namely, multilayer perceptron and back propagation, which are described in detail in the following sections.

6.1. Multilayer Perceptron (MLP). A multilayer perceptron is a feedforward artificial neural network algorithm that helps in the mapping of different sets of energy and average values obtained from the wavelets subbands energy extraction shown in Table 1. These energy values are given to the input layer and multiplied with initial weights. The back propagation is the modified version of linear perceptron which uses three or more hidden layers with the nonlinear activation function. The back propagation is the most extensively used learning algorithm for multilayer perceptron in

TABLE 1: Min and max features of kidney images database which are enlarged table of the table in the GUI of lifting scheme wavelets.

	Min-max
Db12 Dh1 average	0.0026–0.0169
Db12 cV energy	0,0011–9.0990e − 04
Sym12 Dh1 average	0.0026–0.0169
Sym12 cV energy	0.0011–9.0990e − 04
rbio3.7 Dh1 average	0.0052–0.0255
rbio3.7 cD energy	0.0010–8.8080e − 04
rbio3.7 cV energy	0.0010–9.5594e − 04
rbio3.7 Dh1 average	0.0066–0.0272
rbio3.9 cD energy	0.0010–8.0706e − 04
rbio3.9 cV energy	0.0014–8.9330e − 04
rbio3.9 cV average	0.0039–0.0061
rbio4.4 Dh1 average	0.0011–8.1203e − 04
rbio4.4 cH energy	1.4336e − 04–8.7336e − 04
rbio4.4 cV energy	0.0010–9.7450e − 04

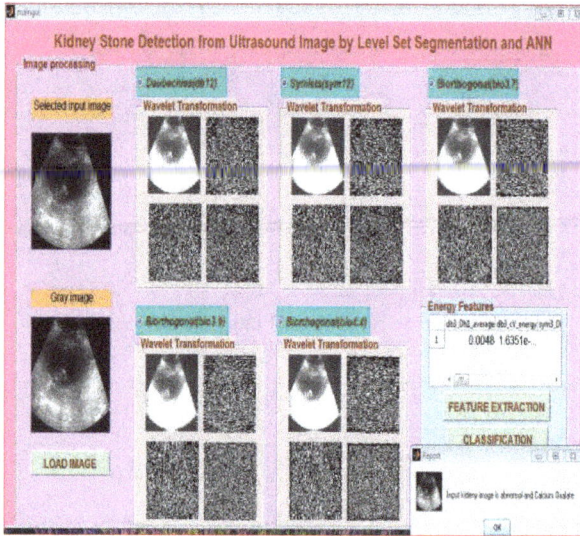

FIGURE 10: Lifting scheme wavelets subband decomposition, energy feature extraction, and classifications.

Compass plot for the all energy and average features of database kidney stone detection

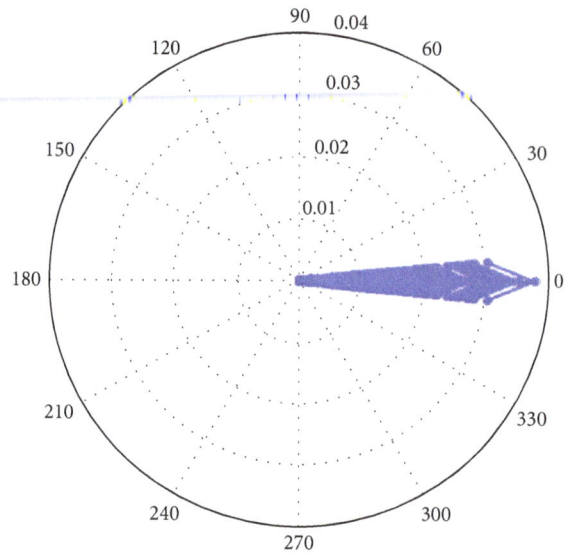

FIGURE 11: Energy and average feature values are within 0 to 1 range.

neural networks and it employs gradient descent to minimize the mean squared error between the network output value and the desired output value. These error signals are taken for completion of the weight updates which represent the power of knowledge learnt by the back propagation [22]. Multilayer perceptron with back propagation (MLP-BP) is the core algorithm. Based on the literature survey, MLP-BP algorithm was found to be better than the other algorithms in terms of accuracy, speed, and performance.

7. Implementation and Results

The implementation of the proposed work is completed using Verilog on Vertex-2Pro FPGA and Matlab 2012a. The Graphical User Interface (GUI) is created for the developed system as shown in Figure 10. From the database of the US kidney images, one kidney image is uploaded through the GUI. The uploaded image is preprocessed and is shown in the GUI. The image segmentation option present in the GUI is executed as the next step to segment the kidney stone portion from the image. The segmented image is performed with wavelet processing by picking one of the lifting wavelet filters shown in the GUI. After selecting the filter for energy level extraction, a specific wavelet code will be invoked to get the subsequent image. Then the feature extraction option is selected to get list of energy levels extracted from the segmented image.

In the GUI shown in Figure 10 there is another table which lists energy levels of all the kidney images present in the database. This is performed for testing the accuracy of MLP-BP ANN system in identifying the kidney images as normal or abnormal and to mention the type of the detected stone. Essentially the database should have both normal and abnormal images in which we should have knowledge of the number of normal and abnormal image count. During the test, it is found that our system can classify the kidney images

TABLE 2: Design summary of proposed hardware implementation.

Parameters	Used	Available	Utilization
Number of slices Register	377	28,800	1%
Number of flip flops used	376		
Number of latches used	507	28,200	1%
Mo of slice LUTs	424	28,200	1%
Dynamic power	0.488 W		
Gate delay	16.5 ns		

as normal and abnormal almost with accuracy of 98.8%. The energy levels of all kidney images from database are extracted and plotted and are shown in Figure 12. The maximum peak plot indicates presence of stone. All extracted features are from range of 0 to 1 as shown in Figure 11.

Table 1 shows the lists of energy levels extracted from the segmented image. This table is the enlarged version of the table shown in the GUI. The rows of the table give the individual energy levels of each kidney image in the database. The columns of the table give the energy level extracted from the images in the database with respect to each wavelet filter. The first two columns correspond to Daubechies filter; the third and the fourth columns correspond to symlets12. The fifth, sixth, and seventh columns correspond to Biorthogonal filter (Bio3.7). The eighth, ninth, and tenth columns correspond to Biorthogonal filter (Bio3.9). The eleventh, twelfth, and thirteenth columns correspond to Biorthogonal filter (Bio4.4).

The proposed work is implemented on Vertex-2Pro FPGA and the device speed is increased by 35% with a gate delay of 3.765 ns. The number of slices LUTs are deceased by 23% in comparison with the existing methods. The design summary table of the proposed hardware model is shown in Table 2.

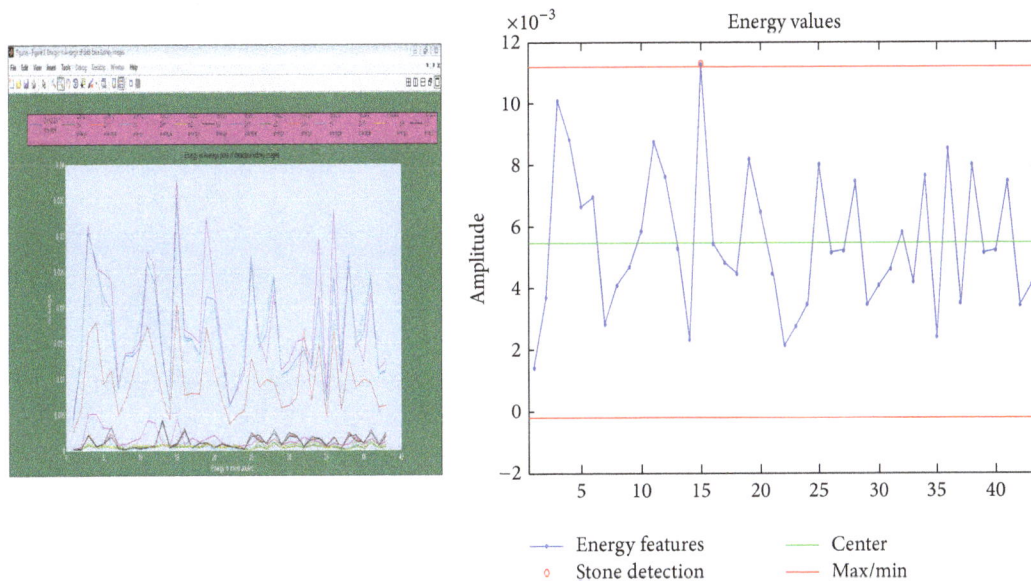

FIGURE 12: Lifting scheme filters energy and average features of database kidney images.

8. Conclusion and Scope for Future Work

The proposed work is implemented on Vertex-2Pro FPGA employing level set segmentation with momentum and resilient propagation parameters. It is found to be capable of any satisfactory achievement in identifying the stones in the US kidney image. The device performance is also pleasing with very low utilization of resources. The energy levels extracted from the lifting scheme wavelet subbands, that is, Daubechies (Db12), Symlets (sym12), and Biorthogonal filterers (bio3.7, bio3.9, and bio4.4), give the perfect indication of the difference in the energy levels of the stone portion compared to that of normal kidney region. The ANN is trained with normal kidney image and classified image input for normal or abnormal conditions by considering extracted energy levels from wavelets filters. The developed system is examined for different kidney images from the database and the results are effective in classifying the types of stone successfully with the accuracy of 98.8% [23]. Thus this system can be readily utilized in the hospitals for patients with abnormality in kidney. This work proves that the combination of level set segmentation, lifting scheme wavelet filters, and multilayer perceptron with back propagation means a better approach for the detection of stones in the kidney. In the future work, the system will be designed for real time implementation by placing biomedical sensors in the abdomen region to capture kidney portion. The captured kidney image is subjected to the proposed algorithm to process and detect stone on FPGA using hardware description language (HDL). The identified kidney stone in the image is displayed with colour for easy identification and visibility of stone in monitor.

Conflict of Interests

The authors declare that there is no conflict of interests regarding the publication of this paper.

References

[1] T. Rahman and M. S. Uddin, "Speckle noise reduction and segmentation of kidney regions from ultrasound image," in *Proceedings of the 2nd International Conference on Informatics, Electronics and Vision (ICIEV '13)*, pp. 1–5, IEEE, Dhaka, Bangladesh, May 2013.

[2] W. G. Robertson, "Methods for diagnosing the risk factors of stone formation," *Arab Journal of Urology*, vol. 10, no. 3, pp. 250–257, 2012.

[3] B. Hess, "Metabolic syndrome, obesity and kidney stones," *Arab Journal of Urology*, vol. 10, no. 3, pp. 258–264, 2012.

[4] W. M. Hafizah, "Feature extraction of kidney ultrasound images based on intensity histogram and gray level co-occurrence matrix," in *Proceedings of the 6th Asia Modeling Symposium (AMS '12)*, pp. 115–120, IEEE, May 2012.

[5] V. P. G. P. Rathi and S. Palani, "Detection and characterization of brain tumor using segmentation based on HSOM, wavelet packet feature spaces and ANN," in *Proceedings of the 3rd International Conference on Electronics Computer Technology (ICECT '11)*, vol. 6, pp. 274–277, IEEE, Kanyakumari, India, April 2011.

[6] N. Koizumi, J. Seo, D. Lee et al., "Robust kidney stone tracking for a non-invasive ultrasound theragnostic system—servoing performance and safety enhancement," in *Proceedings of the IEEE International Conference on Robotics and Automation (ICRA '11)*, pp. 2443–2450, Shanghai, China, May 2011.

[7] M. E. Abou El-Ghar, A. A. Shokeir, H. F. Refaie, and A. R. El-Nahas, "Low-dose unenhanced computed tomography for diagnosing stone disease in obese patients," *Arab Journal of Urology*, vol. 10, no. 3, pp. 279–283, 2012.

[8] K. Viswanath and R. Gunasundari, "Kidney stone detection from ultrasound images by Level Set Segmentation and multilayer perceptron ANN," in *Proceedings of the International Conference on Communication and Comuting (IMCIET-ICCE '14)*, pp. 38–48, Elsevier, 2014.

[9] N. Dheepa, "Automatic seizure detection using higher order moments & ANN," in *Proceedings of the 1st International Conference on Advances in Engineering, Science and Management (ICAESM '12)*, pp. 601–605, March 2012.

[10] K. Kumar, "Artificial neural network for diagnosis of kidney stone disease," *International Journal of Information Technology and Computer Science*, vol. 7, pp. 20–25, 2012.

[11] P. M. Morse and H. Feshbach, "The variational integral and the Euler equations," in *Methods of Theoretical Physics Part I*, pp. 276–280, 1953.

[12] P. R. Tamilselvi and P. Thangaraj, "Computer aided diagnosis system for stone detection and early detection of kidney stones," *Journal of Computer Science*, vol. 7, no. 2, pp. 250–254, 2011.

[13] D. H. Bagley, K. A. Healy, and N. Kleinmann, "Ureteroscopic treatment of larger renal calculi (>2 cm)," *Arab Journal of Urology*, vol. 10, no. 3, pp. 296–300, 2012.

[14] L. Shen and S. Jia, "Three-dimensional gabor wavelets for pixel-based hyperspectral imagery classification," *IEEE Transactions on Geoscience and Remote Sensing*, vol. 49, no. 12, pp. 5039–5046, 2011.

[15] X. Chen, R. Summers, and J. Yao, "FEM-based 3-D tumor growth prediction for kidney tumor," *IEEE Transactions on Biomedical Engineering*, vol. 58, no. 3, pp. 463–467, 2011.

[16] N. R. Owen, O. A. Sapozhnikov, M. R. Bailey, L. Trusov, and L. A. Crum, "Use of acoustic scattering to monitor kidney stone fragmentation during shock wave lithotripsy," in *Proceedings of the IEEE Ultrasonics Symposium*, pp. 736–739, Vancouver, Canada, October 2006.

[17] D. J. Kok, *Metaphylaxis, Diet and Lifestyle in Stone Disease*, vol. 10, Arab Association of Urology, Production and Hosting by Elsevier B.V., 2012.

[18] D. S. Datar, "Color image segmentation based on Initial seed selection, seeded region growing and region merging," *International Journal of Electronics, Communication & Soft Computing Science and Engineering*, vol. 2, no. 1, pp. 13–16, 2012.

[19] P. C. P. Raj and S. L. Pinjare, "Design and analog VLSI implementation of neural network architecture for signal processing," *European Journal of Scientific Research*, vol. 27, no. 2, pp. 199–216, 2009.

[20] J. Martínez-Carballido, C. Rosas-Huerta, and J. M. Ramírez-Cortés, "Metamyelocyte nucleus classification using a set of morphologic templates," in *Proceedings of the Electronics, Robotics and Automotive Mechanics Conference (CERMA '10)*, pp. 343–346, IEEE, Morelos, Mexico, September-October 2010.

[21] C. Li, C. Xu, C. Gui, and M. D. Fox, "Distance regularized level set evolution and its application to image segmentation," *IEEE Transactions on Image Processing*, vol. 19, no. 12, pp. 3243–3254, 2010.

[22] M. Stevenson, R. Winter, and B. Widrow, "Sensitivity of feedforward neural networks to weight errors," *IEEE Transactions on Neural Networks*, vol. 1, no. 1, pp. 71–80, 1990.

[23] K. Viswanath and R. Gunasundari, "Design and analysis performance of kidney stone detection from ultrasound image by level set segmentation and ANN classification," in *Proceedings of the International Conference on Advances in Computing, Communications and Informatics (ICACCI '14)*, pp. 407–414, IEEE, New Delhi, India, September 2014.

## 11

Process Variation Aware Wide Tuning Band Pass Filter for Steep Roll-Off High Rejection

Jian Chen and Chien-In Henry Chen

Department of Electrical Engineering, Wright State University, Dayton, OH 45435, USA

Correspondence should be addressed to Jian Chen; chen.92@wright.edu

Academic Editor: Mohamed Masmoudi

A wide tuning band pass filter (BPF) with steep roll-off high rejection and low noise figure is presented. The design feature of steep roll-off high stopband rejection (>20 dB) and low noise figure (<6 dB) provides a wide tuning frequency span (1–2.04 GHz) to accept desirable signals and reject close interfering signals. The process variation aware design approach demonstrates robustness of the BPF after calibration from process variations, operating in 1.04 GHz tuning frequency span: almost zero deviation on center frequency, an average maximum deviation 1.16 dB on a nominal pass band gain of 55.6 dB, and an average maximum deviation 1.06 MHz on a nominal bandwidth of 12.3 MHz.

1. Introduction

When RF devices are upgraded to support future standards, IF amplifiers, mixers, band pass filters (BPFs), modulators, and demodulators are preferable to be tunable and cost-effective to meet different frequency band standards [1]. Reconfigurable RF minimizes duplicated RF front-end components and hence reduces energy and cost. Considering cognitive radio in the TV bands, there are narrow spectral holes between strong TV transmitter signals. To avoid blocking of the receiver, strong TV signals have to be rejected by tunable BPF of high linearity, high Q, and high stopband rejection. A typical example is $Q = 50$ for 10 MHz bandwidth in a wide tuning frequency range of 500 MHz in the TV bands. Reconfigurable RF devices include power amplifiers, antennas, band pass filters, and matching networks with high tuning speed and high linearity for reasons of size and cost.

Tunable BPF features include tunable frequency range, pass band gain, tuning center frequency span, bandwidth, noise figure, and stopband rejection. Several state-of-the-art CMOS design techniques such as gm-C filters, Q enhanced LC filters, N-path, and pseudo-N-path filters have been presented. Most of them have either low band pass gain, high noise figure, or low tuning frequency range [2–15].

The gm-C filters have the advantage of low power and high frequency. Their weak linearity is improved by applying linearization techniques such as resistive source degeneration, dynamic source degenerated differential pair, tunable feedback, and adaptive feedback. However, these techniques have low pass band gain and narrow tuning range [7, 10]. The Q-enhanced LC band pass filter through an adjustable negative-conductance generator is capable of operating in low-voltage supply but it has low pass band gain (−5 dB) and high noise figure (26.8 dB) [8]. Other Q-enhanced LC filter based designs were proposed in [9, 11]. The BPF in [9] consumes less power and has a tuning frequency range (400 MHz) with a pass band gain (23 dB). A low pass band distortion BPF proposed in [11] has a low pass band gain (0 dB). The high-Q integrated switched capacitor band pass filter [12] has a broad frequency band but a low pass band gain (−2 dB) and high power. A design using high-Q N-path band pass filter is proposed [13], which has a tuning frequency range (0.1–1 GHz) but low band pass gain (−2 dB). A novel inductorless tunable switched capacitor band pass filter based on N-path periodically time-variant circuit operates in a high frequency range (4–4.44 GHz) but has low pass band gain (−12 dB) and high noise figure (14 dB) [14].

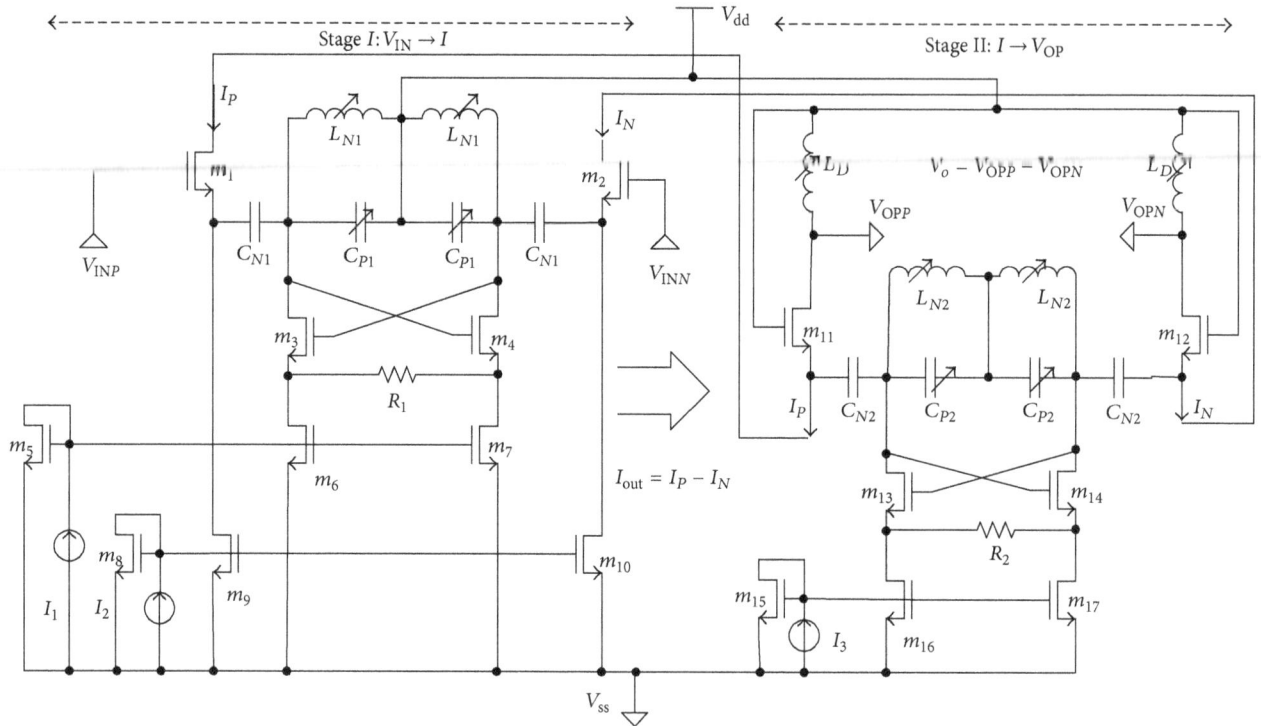

FIGURE 1: Schematic of band pass filter.

In this paper, a tunable BPF is presented, which has a wide tuning range, high band pass gain, high stopband rejection, low noise figure, and low power. The paper is organized as follows. Section 2 presents the architecture and operation principle of the tunable BPF. Section 3 presents design approach to BPF for specified center frequency and bandwidth. Section 4 presents process variation aware design approach to BPF for tunable center frequency and bandwidth. BPF design calibration after process variation is also presented. Section 5 presents measurements and performance analysis of BPF after design calibration. Section 6 is the conclusion.

2. BPF Architecture and Principle Operation

The tunable BPF with differential cascade architecture is divided into two stages. Stage I is the transconductance stage. Stage II is the transimpedance stage. The transconductance stage converts input voltages V_{INP} and V_{INN} to output currents I_P and I_N while the transimpedance stage converts input currents I_P and I_N to output voltages V_{OPP} and V_{OPN}, as shown in Figure 1. The BPF is a type of double notch filter [2], which has two single notch filters with LC parallel series resonant combination, as depicted in Figure 2.

The input LC impedance for parallel series resonant combination is expressed as Z_N. Its small equivalent signal model is depicted in Figure 3, where the drain current I_D of transistor m_9 (Figure 1) serves as driver of the amplifier and I_N is the current feeding the LC parallel series resonant combination. I_L is the current through transistor m_1. The transconductance of m_1 is g_{m1} and the load resistance seen

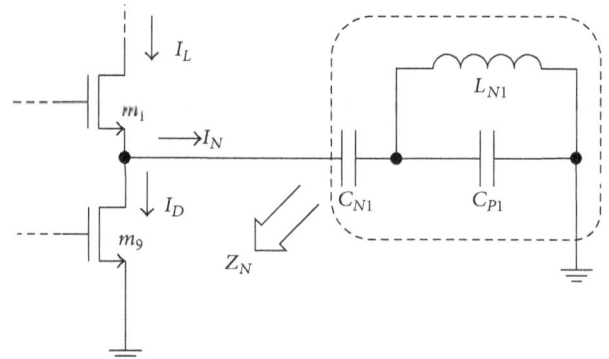

FIGURE 2: Schematic of single notch filter.

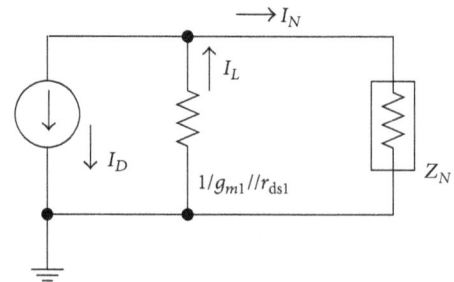

FIGURE 3: Small equivalent signal model for single notch filter.

from the source of m_1 is $1/g_{m1}$. The input impedance Z_{N1} is expressed in (1). Consider

$$Z_{N1} = \frac{1 + s^2 L_{N1}\left(C_{N1} + C_{P1}\right)}{SC_{N1}\left(1 + s^2 L_{N1} C_{P1}\right)}. \tag{1}$$

Considering the parallel series LC network, its parallel resonant frequency f_{P1} is higher than the zero frequency f_{Z1}, as shown below:

$$f_{P1} = \frac{1}{2\pi\sqrt{L_{N1}C_{P1}}}, \tag{2}$$

$$f_{Z1} = \frac{1}{2\pi\sqrt{L_{N1}\left(C_{N1} + C_{P1}\right)}}. \tag{3}$$

As shown in Figure 3, the equivalent resistance of $1/g_{m1}$ and r_{ds1} in parallel is approximately $1/g_{m1}$ for $r_{ds1} \gg 1/g_{m1}$. The amplitude I_N is then expressed as

$$I_N = I_D \frac{1}{1 + g_{m1}Z_N} = \frac{g_{m1}V_{INN}}{1 + g_{m1}Z_N}, \tag{4}$$

where I_D is equal to $g_{m1}V_{INN}$. Similarly, the current I_P can be expressed by replacing v_{INN} with v_{INP}. The differential output current I_{out} ($= I_P - I_N$) of the transconductance stage is simply obtained after replacing v_{INP} with v_{IN} ($= v_{INP} - v_{INN}$). The same LC parallel series combination with different impedance L_{N2}, C_{N2}, C_{P2} is applied in the transimpedance stage with v_{out} ($= v_{o1} - v_{o2}$). Hence, the input impedance of the second LC stage is given by

$$Z_{N2} = \frac{1 + s^2 L_{N2}\left(C_{N2} + C_{P2}\right)}{S\left(C_{N2} + s^2 L_{N2}C_{P2}\right)}. \tag{5}$$

The resonant frequencies are expressed by

$$f_{P2} = \frac{1}{2\pi\sqrt{L_{N2}C_{P2}}}, \tag{6}$$

$$f_{Z2} = \frac{1}{2\pi\sqrt{L_{N2}\left(C_{N2} + C_{P2}\right)}}. \tag{7}$$

The differential output voltage of the transimpedance filter stage is then expressed by

$$v_{out} = sL_D \frac{g_{m1}Z_{N2}}{1 + g_{m1}Z_{N2}} I_{out}. \tag{8}$$

When $f = f_{Z1}$ then $Z_{N1} \approx 0$. And when $f = f_{P1}$ then $Z_{N1} \approx \infty$. Therefore, the output current $I_{out} = I_P - I_N$ in (4) can be expressed as

$$I_{out} \approx \begin{cases} g_{m1}v_{IN} & \text{when } f = f_{Z1}, \\ 0 & \text{when } f = f_{P1}. \end{cases} \tag{9}$$

Similarly, $Z_{N2} \approx 0$ when $f = f_{Z2}$. And $Z_{N2} \approx \infty$ when $f = f_{P2}$. Hence the output voltage in (6) is given by

$$v_{out} \approx \begin{cases} 0 & \text{when } f = f_{Z2}, \\ sL_D I_{out} & \text{when } f = f_{P2}. \end{cases} \tag{10}$$

It is observed from (9) and (10) that the transmission zero happens at f_{P1} in transconductance stage and happens

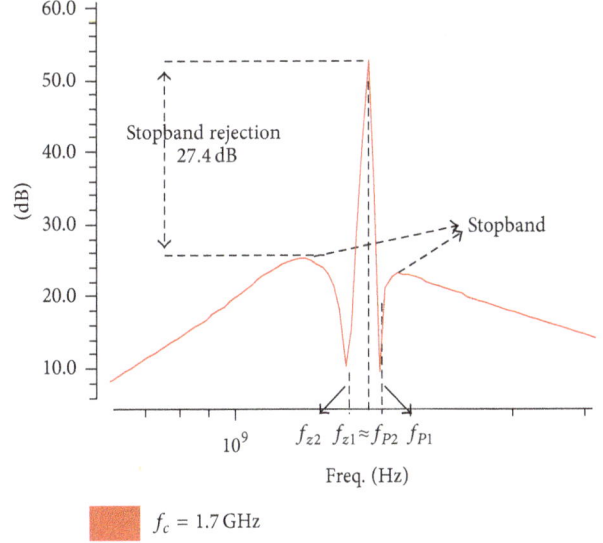

FIGURE 4: Example of frequency response of the proposed band pass filter (f_c = 1.7 GHz, BW = 12.4 MHz, and stopband rejection = 27.4 dB).

at f_{Z2} in transimpedance stage. Meanwhile, the peak value happens at f_{Z1} in transconductance stage and happens at f_{P2} in transimpedance stage. Correspondingly, the peak conductance g_{max} and impedance Z_{max} are obtained at f_{Z1} and f_{P2}. Finally, the overall transfer function of the BPF is obtained from (3) to (9). Consider

$$v_{out} = v_{IN} \frac{g_{m1}^2 L_D Z_{N2} s}{\left(1 + g_{m1}Z_{N1}\right)\left(1 + g_{m1}Z_{N2}\right)}. \tag{11}$$

The peak output can be expressed in (12) when $f = f_{Z1}$, and also $f_{Z1} \approx f_{P2}$. Consider

$$v_{out} \approx \begin{cases} 0 & \text{when } f = f_{P1} \text{ or } f_{Z2}, \\ g_{max}Z_{max}v_{IN} & \text{when } f = f_{Z1} \approx f_{P2}, \end{cases} \tag{12}$$

where $g_{max} = I_{out}/v_{IN}$ and $Z_{max} = v_{out}/I_{out} = sL_D$.

From (12) the peak gain happens at center frequency (the pole f_{P2} and the zero f_{Z1}). A wide tuning range is realized by changing L_{N1} and L_{N2}. A high base band gain is achieved by maximizing g_{max} and Z_{max}, which are determined by the Q factor of the inductors and by location of the poles and zeros. The Q-factor of inductors is improved by using a cross-coupled transistor pair, creating negative impedance in parallel with the inductors. The location of poles and zeros can be adjusted by varying C_N and C_P.

The pole frequency can be adjusted to be slightly greater than the zero frequency by making C_{N1} much less than C_{P1} and making C_{N2} much smaller than C_{P2}. A steep roll-off can be achieved. Figure 4 demonstrates the frequency response of the proposed BPF at f_c = 1.7 GHz and BW = 12.4 MHz. Its stopband rejection is 27.4 dB. Meanwhile, the peak center frequency gain can be increased when $f_{P2} \approx f_{Z1}$ by making $C_{N1} + C_{P1} \approx C_{P2}$.

Given: BPF (Figure 1) with NMOS (m_1 to m_{17}), inductors (L_{N1}, L_{N2}, L_D), capacitors (C_N, C_{P1}, C_{N2}, and C_{P2}), current sources (I_1, I_2, I_3), power (pw)

Input: V_{inn} and V_{inp}

Objective: BPF to meet a specified center frequency and bandwidth

Output: NMOS sizes, capacitors and inductors value

Design approach: //BPF design to meet the specified center frequency and bandwidth//
(1) Specify DC power (pw_0), center frequency (f_{c0}), bandwidth (BW_0);
(2) Calculate I_1, I_2 and I_3 based on pw_0;
(3) Set w_3, w_4, w_5 based on I_1, I_2 and I_3;
(4) Set initial value $k_3(0)$, $k_4(0)$ and $k_5(0)$;
(5)   Given C_{N1}, C_{P1}, C_{N2}, C_{P2}, L_{N1}, L_{N2} and L_D,
        set $C_{N1} \ll C_{P1}$ and $C_{N2} \ll C_{P2}$ to make f_{p1} a little greater than f_{z1} and make f_{p2} a little greater than f_{z2};
        set $C_{p2} \approx C_{p1} + C_{n1}$ to make $f_{p2} \approx f_{z1}$;
(6) $i = 0$
(7)   while ($i < N$) //Optimize k_3, k_4, k_5 in N iterations
(8)   {$k_3(i) = k_3(0) + \left(k_{3\,max} - k_3(0)\right) * i/N$;
          $k_4(i) = k_5(i) = k_4(0) + \left(k_{4\,max} - k_4(0)\right) * i/N$;
(9)   do DC simulation;
(10)   if $|pw - pw_0|/pw > 0.01$ then $i = i + 1$ and go to Step (7);
(11)   else exit;
(12)   end if}
(13)   adjust C_{N1}, C_{P1}, C_{N2}, C_{P2}, L_{N1}, L_{N2};
          do AC simulation; //find f_c
(14)   If $|f_c - f_{c0}|/f_{c0} < 0.01$ then go to Step (17)
(15)   else go to Step (13)
(16)   end if;
(17)   adjust L_D;
(18)   do AC simulation; //find BW
(19)   If $|BW - BW_0|/BW_0 < 0.01$ then exit;
(20)   else go to Step (17)
(21)   end if;

ALGORITHM 1: A BPF design to meet the specified center frequency and bandwidth.

TABLE 1: Transistor size information.

Transistor	m_1	m_2	m_3	m_4	m_5	m_6	m_7	m_8	m_9	m_{10}	m_{11}	m_{12}	m_{13}	m_{14}	m_{15}	m_{16}	m_{17}
Width	w_1	w_1	w_6	w_6	w_4	$k_4 w_4$	$k_4 w_4$	w_3	$k_3 w_3$	$k_3 w_3$	w_2	w_2	w_7	w_7	w_5	$k_5 w_5$	$k_5 w_5$
Length	L	L	L	L	L	L	L	L	L	L	L	L	L	L	L	L	L

3. BPF Design with Fixed Center Frequency and Bandwidth

Algorithm 1 presents sizing approach of BPF (Figure 1) in 180 nm CMOS process. The power supply is 1.8 V. The desirable power consumption is denoted as pw_0 while the desirable center frequency and bandwidth are denoted as f_{c0} and BW_0, respectively. With the relationship of current mirror circuit, the current I_1 drives branch currents I_{m6} and I_{m7}; the current I_2 drives branch currents I_{m9} and I_{m10}; the current I_3 drives branch currents I_{m16} and I_{m17}. Note that I_{mxx} is the current through the transistor m_{xx}. Normally, the driver current (I_1, I_2, I_3) is made by the tenth of its load current through transistor (m_6, m_7, m_9, m_{10}, m_{16}, m_{17}) based on the power consumption pw.

The next step involves setting width and length of transistor m_1 to m_{17} and multiplicand k_3, k_4, k_5 for current mirror load current through m_3, m_4, m_5, which is precalculated and summarized in Table 1. The initial value for k_3, k_4, k_5 is 10. It is then optimized to have the load current through transistor (m_6, m_7, m_9, m_{10}, m_{16}, m_{17}) as 1/10 of its driver current (I_1, I_2, I_3).

Using 180 nm CMOS process, the value of C_1 to C_4 and L_1 to L_4 is calculated based on (2), (3), (6), and (7). Given $C_{N1} \ll C_{P1}$ and $C_{N2} \ll C_{P2}$, f_{P1} and f_{P2} are set to be a little greater than f_{Z1} and f_{Z2}. Also, $f_{Z1} \approx f_{P2}$ if $C_{p2} \approx C_{p1} + C_{n1}$. The initial parameter setting is completed in steps 1–5 in Algorithm 1. As shown in Algorithm 1, k_3, k_4, and k_5 are optimized in step 8 till power consumption is met. Next, adjust C_{N1} to C_{P2}, L_{N1}, L_{N2} and do AC simulation until the center frequency f_c is close to f_{c0} (<1% error). Finally, adjust L_D and do AC simulation until the bandwidth BW is close to BW_0 (<1% error).

Given: f_c, BW_0, and initial values of L_{N1}, L_{N2}, L_D.
Input: V_{inn} and V_{inp}
Objective: tunable BPF to meet f_c and BW_0
Output: L_{N1}, L_{N2}, C_{P2}

Design approach: //Stage A to meet the specified f_c (Steps (1)–(10))//
(1) Find the range of f_c ($f_{c1} < f_c < f_{c2}$) in Table 2;
(2) Set initial values $L_D = L_{D2}$, $L_{N1} = L_{N2} = L_{N12}$, and set C_{N1}, C_{N2}, C_{P1}, C_{P2}, δ_1, δ_2, δ_3, M, N, P, Q;
(3) Repeat
(4)   do AC simulation;
(5)   decrease $L_{N1} = L_{N2}$ by $\Delta L_{N1} = (L_{N12} - L_{N11})/M$;
(6) Until $|f - f_c| < \delta_1$;
(7) Repeat
(8)   do AC simulation;
(9)   decrease C_{P2} by $\Delta C = (C_{P22} - C_{P21})/N$;
(10) Until $|f - f_c| < \delta_2$;

Design approach: //Stage B to meet the specified BW_0 (Steps (11)–(17))//
(11) Find f_c and its corresponding BW_{min} in Table 2;
(12) If $BW_0 < BW_{min}$ then break;
(13) Else
(14) Do {
(15)   decreases inductor L_D by $\Delta L = (L_{D2} - L_{D1})/P$;
(16)   do AC simulation to find the bandwidth BW;
(17) } While $|BW - BW_0| < \delta_3$;

Design approach: //Stage C to calibrate BPF design after considering process variations (Steps (18)–(32))//
(18) Select N (i.e., $N = 100$) cases of BPF designs with different f_{c0}, BW_0 and A_V after Stage A and B;
(19) Perform Monte Carlo analysis to obtain μ and σ for BW;
(20) Compute $(BW - (\mu - \sigma))$; //calculate the BW deviation
(21) Choose top 30% cases having the worst BW deviation and find their center frequency (f_c) and corresponding gain (A_v);
(22) Compute the average of center frequency (f_{avg}) and gain deviation (ΔA_v)
(23) Compute ΔL_D from (17);
(24) For $i \leftarrow 1$ to 30 //consider the top 30 cases
(25) $L_{D\_new}(i) = L_{D\_old}(i) + \Delta L_D$;
(26)   do AC simulation;
(27)   if $BW_{new}(i) > BW_{old}(i)$ then set $BW(i) = BW_{new}(i)$;
(28)   else $\{L_{D\_new}(i) = L_{D\_old}(i) - \Delta L_D$;
(29)   do AC simulation;
(30)   set $BW(i) = BW_{new}(i)\}$;
(31)   end if;
(32) End for;

ALGORITHM 2: Process variation aware tunable BPF design approach.

4. Process Variation Aware Tunable BPF Design

4.1. Tunable BPF with Specified Center Frequency and BW. The design approach described in Section 3 is sequential and iterative loops for parameter setting and sequential results are affecting one another. The final design meets unique specifications and its design parameters are difficult to be changed to become tunable for different center frequency or bandwidth. In this regard, a design approach to make BPF tunable is proposed, which stores BPF design parameters as design reference and increases design space for wide tuning frequency range.

The design approach for tunable BPF is divided into three stages. Stage A is to meet the center frequency f_c. Stage B is to meet the BW. Stage C is to calibrate the BPF design to meet f_c and BW after process variations. They are shown in Figure 5. Algorithm 2 depicts detailed design approach in

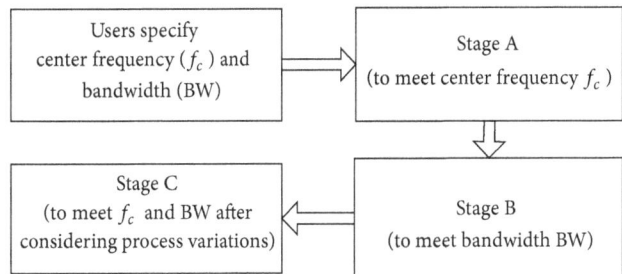

FIGURE 5: Block diagram of process variation aware tunable BPF design.

the three stages and explains how to obtain design parameters for the tunable BPF. In stage A, it is desirable to narrow the range of design parameters, which are primarily related to the center frequency. By referring to a look-up table of

TABLE 2: Initial look-up table of inductors obtaining peak gain and minimum BW for different center frequencies.

L_{N1} (nH)	L_{N2} (nH)	L_D (nH)	f_c (GHz)	BW_{min} (MHz)	Peak gain (dB)
1.30	1.30	281.0	1.00	5.34	65.3
1.10	1.10	269.4	1.07	5.47	69.3
1.00	1.00	210.0	1.15	5.85	68.3
0.90	0.90	196.0	1.20	6.33	64.7
0.80	0.80	183.0	1.26	6.14	67.8
0.70	0.70	169.0	1.35	6.87	66.2
0.60	0.60	164.0	1.45	7.17	66.3
0.50	0.50	136.0	1.58	7.79	66.1
0.40	0.40	118.0	1.78	8.92	65.0
0.30	0.30	93.00	2.04	9.91	64.9
0.20	0.20	55.00	2.51	12.4	63.6
0.10	0.10	27.50	3.55	16.6	63.1
0.05	0.05	14.20	5.01	21.4	62.9

inductors (Table 2), peak pass band gain and minimum BW for different center frequencies in a wide tuning frequency range are obtained. These parameters are obtained by design approach in Algorithm 1. Given f_c to tunable BPF design, the first step is to find the adjacent center frequency (i.e., upper f_{c2} and lower f_{c1}) from Table 2. Next, find the upper L_{N12} and lower inductor L_{N11} from f_{c2} and f_{c1}, respectively. For example, if $f_c = 1.7$ GHz, then f_{c1} is 1.58 GHz and f_{c2} is 1.78 GHz. Also, $L_{N12} = 0.5$ nH and $L_{N11} = 0.4$ nH. Thereafter, set the initial value $L_D = L_{D2}, L_{N1} = L_{N2} = L_{N12}$. Calculate C_{N1} to C_{P2} value from equations in Section 2. Make $C_{P22} = 1.1C_{P2}, C_{P21} = 0.9C_{P2}$ and do AC simulation. L_{N1} and L_{N2} are the primary impact factors to determine the center frequency. If $|f - f_c| > \delta_1$ (i.e., 15 MHz), decrease L_1 and L_2 by $\Delta L_{N1} = (L_{N12} - L_{N11})/M$ and then conduct AC simulation. Otherwise, decrease the second primary impact factor C_{P2} by $\Delta C_{P2} = (C_{P22} - C_{P21})/N$ until $|f - f_c| < \delta_2$ (i.e., 1 MHz). Then, go to stage B to meet the specified bandwidth, as shown in steps 11–17 in Algorithm 2. If $BW < BW_{min}$, then BW is out of design range. Otherwise, decrease L_D by $\Delta L = (L_{D2} - L_{D1})/P$ and do AC simulation till $|BW - BW_0| < \delta_3$ (i.e., 0.1 MHz).

Then, the specified BW is met. M, N, and P are the number of repeats determined by the accuracy ($\delta_1, \delta_2, \delta_3$) in simulation. Consider

$$\Delta L_{N1} = \frac{L_{larger} - L_{smaller}}{\text{number of repeats}} = \frac{L_{N12} - L_{N11}}{M}, \quad (13)$$

$$\Delta C_{P2} = \frac{C_{larger} - C_{smaller}}{\text{number of repeats}} = \frac{C_{p22} - C_{p21}}{N}, \quad (14)$$

$$\Delta L_D = \frac{L_{larger} - L_{smaller}}{\text{number of repeats}} = \frac{L_{D2} - L_{D1}}{P}, \quad (15)$$

$$A_v = \frac{v_{out}}{v_{IN}} = g_{max}Z_{max} = sg_{max}L_D, \quad (16)$$

$$|\Delta A_v| = 2\pi f_{avg}g_{max}\Delta L_D. \quad (17)$$

4.2. Design Example of a Tunable BPF for f_c (1.7 GHz) and BW (12.4 MHz).

Table 3 summarizes transistor sizes and parameter settings based on the design approach in Section 3. Prior to implementing the tunable BPF design, the desirable center frequency f_c (1.7 GHz) and BW (12.4 MHz) are specified. The number of repeats depends on the design accuracies δ_1, δ_2, and δ_3. For example, δ_1, δ_2, and δ_3 are given 15 MHz, 1 MHz, and 0.1 MHz and the number of repeats M, N, and P is 20, 40, and 25. As shown in Table 2, the center frequency f_c (1.7 GHz) lies between f_{c1} (1.58 GHz) and f_{c2} (1.78 GHz), the corresponding inductor L_D lies between L_{D1} (136 nH) and L_{D2} (118 nH), and the inductors L_{N1} and L_{N2} lie between L_{N11} (0.4 nH) and L_{N12} (0.5 nH). Therefore, set the initial value $L_D = L_{D2} = 136$ nH, $L_{N1} = L_{N2} = L_{N12} = 0.5$ nH, and $C_{N1} = 2$ pF, $C_{P1} = 18$ pF, $C_{N2} = 3$ pF, $C_{P2} = C_{N1} + C_{P1} = 20$ pF, $C_{P22} = 1.1C_{P2} = 22$ pF, $C_{P21} = 0.9C_{P2} = 18$ pF and do AC simulation. If $|f - f_0| > \delta_1$ (>15 MHz), decrease L_{N1} and L_{N2} by $\Delta L_{N1} = (L_{N12} - L_{N11})/M = 0.005$ nH and then do AC simulation. Otherwise, decrease the second primary impact factor C_{P2} by $\Delta C_{P2} = (C_{P22} - C_{P21})/N = 0.1$ pF until $|f - f_c| < \delta_2$ (<1 MHz). After the specified f_c is met, then go to stage B to meet the specified BW = 12.4 MHz.

As shown in steps 11–17 in Algorithm 2, the first step is to find f_c and its corresponding bandwidth (BW_{min}). If $BW < BW_{min}$, then BW is out of range. If $BW > BW_{min}$, decrease L_D by $\Delta L = (L_{D2} - L_{D1})/P = 0.84$ nH and run AC simulation. Repeat until $|BW - BW_0| < \delta_3$ (<0.1 MHz). Then, f_c (1.7 GHz) and BW (12.4 MHz) are met. The proposed BPF provides a high pass band gain between 45.2 and 65.1 dB and tunable pass band between 5.5 and 51.2 MHz, while the center frequency is varied from 1.0 to 2.04 GHz, which is depicted in Figure 6.

4.3. BPF Calibration after Process Variations.

The BPF design calibration to meet the specified BW after process variation is presented in stage C in Algorithm 2. The initial step in stage C is to select N (i.e., $N = 100$) cases of different f_c, BW, and A_V and conduct Monte Carlo simulation to obtain BW's mean (μ) and standard deviation (σ). The top 30% design cases contributing to the worst deviation of (BW − ($\mu - \sigma$)) are selected to calculate the average center frequency (f_{avg}) and the corresponding gain deviation (ΔA_v). Then, ΔL_D can be calculated from (17). Note that ΔL_D is used to calibrate L_D, which in turn calibrates BW. Finally, calibrated L_D and calibrated BW for each case are obtained. Table 4 shows the tunable BPF BW comparison with and without calibration after process variations. Ten design cases of center frequency which varied from 1 to 2.04 GHz are compared. Their BW is close to 12.3 MHz. Considering the case example of $f_c = 1.35$ GHz and BW = 12.2 MHz in Table 4, the bandwidth deviation BW − ($\mu - \sigma$) before calibration is 0.88 MHz and after calibration is 0.42 MHz, which accounts for 52.3% improvement. Consider all 10 case examples, the average bandwidth deviation BW−($\mu-\sigma$) before calibration is

TABLE 3: Transistor size and parameter setting.

m_1 (um)	m_2 (um)	m_3 (um)	m_4 (um)	m_5 (um)	m_6 (um)	m_7 (um)	m_8 (um)	m_9 (um)
150	150	2	2	2	19	19	4	46.4
m_{10} (um)	m_{11} (um)	m_{12} (um)	m_{13} (um)	m_{14} (um)	m_{15} (um)	m_{16} (um)	m_{17} (um)	L (um)
46.4	90	90	2	2	2	19	19	180
M	N	P	δ_1 (MHz)	δ_2 (MHz)	δ_3 (MHz)	I_1 (mA)	I_2 (mA)	I_3 (mA)
20	40	25	15	1	0.1	50	300	50

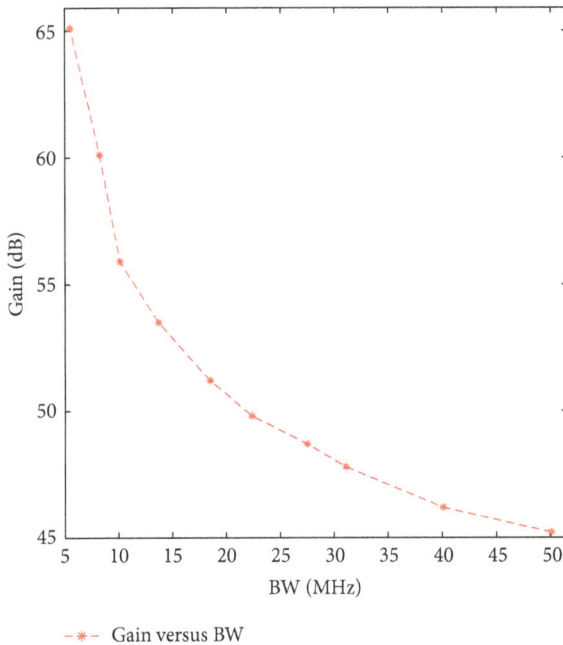

−*− Gain versus BW

FIGURE 6: Pass band gain versus tunable bandwidth ($45.2\,\text{dB} \leq A_V \leq 65.1\,\text{dB}$, $5.5\,\text{MHz} \leq \text{BW} \leq 51.2\,\text{MHz}$) for $1.0\,\text{GHz} \leq f_c \leq 2.04\,\text{GHz}$.

TABLE 4: Tunable BPF BW comparison with and without calibration after process variations.

f_c (GHz)	BW (MHz)	$\mu - \sigma$		$\text{BW} - (\mu - \sigma)$	
		w/o calibration	w calibration	w/o calibration	w calibration
1.00	12.1	11.15	11.56	0.95	0.54
1.07	12.3	10.96	11.52	1.34	0.78
1.15	12.1	12.51	12.34	−0.41	−0.24
1.20	12.2	11.24	11.62	0.96	0.58
1.26	12.3	11.25	11.85	1.05	0.45
1.35	12.2	11.32	11.78	0.88	0.42
1.45	12.4	11.83	12.09	0.57	0.31
1.58	12.5	12.15	12.39	0.35	0.11
1.78	12.6	12.81	12.74	−0.21	−0.14
2.04	12.6	12.27	12.49	0.33	0.11

0.58 MHz and after calibration is 0.292 MHz, which accounts for 49.6% improvement.

5. Measurement and Performance Analysis

The performance of ten tunable BPF designs with the pass band nearly constant (12.1–12.6 MHz) is shown in Table 5. The bandwidth is 12.1 MHz, 12.3 MHz, 12.1 MHz, 12.1 MHz, 12.2 MHz, 12.3 MHz, 12.2 MHz, 12.4 MHz, 12.5 MHz, 12.6 MHz, and 12.6 MHz, respectively. The center frequency is varied from 1.0 to 2.04 GHz: 1.0 GHz, 1.07 GHz, 1.15 GHz, 1.20 GHz, 1.26 GHz, 1.35 GHz, 1.45 GHz, 1.58 GHz, 1.78 GHz, and 2.04 GHz. The nominal pass band gain before considering process variations is between 53.4 and 59.9 dB. Considering process variations on t_{ox}, L_{eff}, and v_t with a maximum 10% variation on their nominal values, the Monte Carlo simulation results show robustness of the BPF:

zero deviation on center frequency. The average maximum deviation on pass band gain is 1.16 dB on a nominal pass band gain of 55.6 dB. And the average maximum deviation on bandwidth is 1.06 MHz on a nominal bandwidth of 12.3 MHz.

Table 6 summarizes the calibrated performance of the ten tunable BPF designs after process variations. Figure 7 shows a low noise figure (<6 dB), while the center frequency is varied from 1.0 to 2.04 GHz.

6. Conclusion

This paper presented an effective design approach to optimize design parameters of a BPF to achieve tunable center frequency and bandwidth in a wide frequency span, for example, 1.04 GHz. Process variations on channel length, physical oxide thickness, and threshold voltage were considered in Monte Carlo simulation. The BPF design calibration to compensate bandwidth deviation from process variations was presented and evaluated. Considering process variations in 180 nm CMOS process and the central frequency which varied from 1.0 to 2.04 GHz, it was shown in Table 6 that the BW deviation (BW_σ) is 0.292 MHz on the pass band mean (BW_μ) of 12.01 MHz. The pass band gain variation ($A_{v(\sigma)}$) is

TABLE 5: BPF f_c, A_v, and BW considering process variations.

Nominal (GHz)	f_c after process variation				Nominal (dB)	A_v after process variation				Nominal (MHz)	BW after process variation			
	t_{ox} [min, max] (GHz)	L_{eff} [min, max] (GHz)	t_{ox} [min, max] (GHz)	All [min, max] (GHz)		t_{ox} [min, max] (dB)	L_{eff} [min, max] (dB)	t_{ox} [min, max] (dB)	All [min, max] (dB)		t_{ox} [min, max] (MHz)	L_{eff} [min, max] (MHz)	t_{ox} [min, max] (MHz)	All [min, max] (MHz)
1.00	1.00, 1.00	1.00, 1.00	1.00, 1.00	1.00, 1.00	53.4	52.19, 53.21	52.75, 53.82	52.51, 53.02	53.64, 53.91	12.1	10.95, 13.91	10.65, 12.09	11.82, 12.46	10.24, 11.49
1.07	1.07, 1.07	1.07, 1.07	1.07, 1.07	1.07, 1.07	54.7	54.26, 55.25	54.77, 54.82	54.65, 54.91	54.14, 55.45	12.3	11.65, 12.81	11.36, 12.07	11.69, 12.39	10.93, 11.06
1.15	1.15, 1.15	1.15, 1.15	1.15, 1.15	1.15, 1.15	53.5	54.98, 55.68	55.59, 55.94	55.42, 55.87	54.22, 56.88	12.1	12.16, 12.30	11.81, 12.25	11.95, 12.35	11.51, 13.27
1.20	1.20, 1.20	1.20, 1.20	1.20, 1.20	1.20, 1.20	53.7	53.19, 54.20	53.77, 54.99	53.48, 54.08	54.89, 54.92	12.2	11.20, 13.78	10.91, 12.29	11.98, 12.59	10.47, 11.54
1.26	1.26, 1.26	1.26, 1.26	1.26, 1.26	1.26, 1.26	54.9	54.63, 55.40	55.03, 55.11	54.88, 55.17	55.03, 55.68	12.3	11.76, 12.78	11.49, 12.14	11.92, 12.42	11.20, 11.34
1.35	1.35, 1.35	1.35, 1.35	1.35, 1.35	1.35, 1.35	55.2	54.98, 55.44	55.24, 55.39	55.07, 55.40	54.89, 55.88	12.2	11.59, 12.73	11.39, 12.08	11.87, 12.29	11.29, 11.45
1.45	1.45, 1.45	1.45, 1.45	1.45, 1.45	1.45, 1.45	56.8	56.10, 57.67	56.44, 56.87	56.78, 56.96	55.57, 57.41	12.4	12.10, 12.19	11.87, 12.12	11.97, 12.32	11.64, 12.03
1.58	1.58, 1.58	1.58, 1.58	1.58, 1.58	1.58, 1.58	57.1	56.46, 57.92	56.77, 57.19	57.08, 57.29	55.94, 57.69	12.5	12.39, 12.63	12.03, 12.46	12.36, 12.62	11.87, 12.32
1.78	1.78, 1.78	1.78, 1.78	1.78, 1.78	1.78, 1.78	57.2	56.00, 58.40	56.05, 57.15	56.98, 57.31	55.21, 57.00	12.6	12.41, 12.94	12.67, 12.89	12.59, 12.67	12.74, 13.04
2.04	2.04, 2.04	2.04, 2.04	2.04, 2.04	2.04, 2.04	59.9	59.43, 60.20	59.34, 59.87	59.79, 59.93	58.93, 59.75	12.6	12.15, 12.67	12.27, 12.31	12.32, 12.33	12.10, 12.44
AMD	0.00	0.00	0.00	0.00		0.62	0.56	0.21	1.16		0.66	0.76	0.33	1.06

Note: * AMD denotes average maximum deviation in comparison with its nominal value.

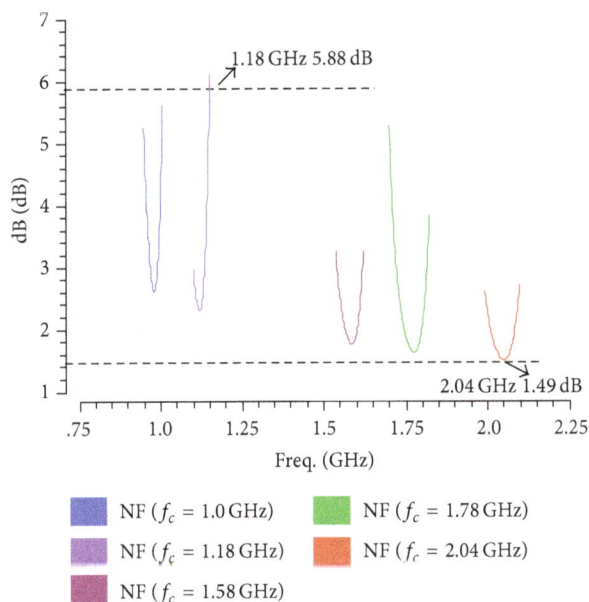

FIGURE 7: Noise figure measurement for different center frequencies.

TABLE 6: Calibrated performance after process variation for the ten tunable BPF designs in Table 5.

Parameter	Our work
CMOS technology (nm)	180
Filter order	4
Supply voltage (V)	1.8
Frequency tuning range (GHz)	1.0–2.04
BW mean (BW_μ) (MHz)	12.01
BW deviation (BW_σ) (MHz)	0.292
Pass band gain mean ($A_{v(\mu)}$) (dB)	55.02
Pass band gain deviation ($A_{v(\sigma)}$) (dB)	0.58
NF (dB)	<6
Power (mW)	15

0.58 dB on the pass band gain mean ($A_{v(\mu)}$) of 55.02 dB. Low noise figure (<6 dB) and steep roll-off high stopband rejection (>20 dB) make the tunable BPF attractive in reconfigurable RF and cognitive radio in TV band applications.

Conflict of Interests

The authors declare that there is no conflict of interests regarding the publication of this paper.

References

[1] A. S. Hussaini, R. Abd-Alhameed, and J. Rodriguez, "Tunable RF filters: survey and beyond," in *Proceedings of the 18th IEEE International Conference on Electronics, Circuits and Systems (ICECS '11)*, pp. 512–515, December 2011.

[2] C. Barth, I. R. Linscott, and U. S. Inan, "A double notch RF filter architecture for SAW-less GPS receivers," in *Proceedings of the IEEE International Symposium of Circuits and Systems (ISCAS '11)*, pp. 1804–1807, May 2011.

[3] S. A. Vallese, A. Bevilacqua, C. Sandner, M. Tiebout, A. Gerosa, and A. Neviani, "Analysis and design of an integrated notch filter for the rejection of interference in UWB systems," *IEEE Journal of Solid-State Circuits*, vol. 44, no. 2, pp. 331–343, 2009.

[4] H. Lee, H. Seo, I. Choi, T. Chung, D. Jeong, and B. Kim, "A RF CMOS band-pass tracking filter with enhanced Q and high linearity," in *Proceedings of the Asia-Pacific Microwave Conference (APMC '11)*, pp. 1901–1904, December 2011.

[5] J. Robert Tourret, S. Amiot, M. Bernard et al., "SiP tuner with integrated LC tracking filter for both cable and terrestrial TV reception," *IEEE Journal of Solid-State Circuits*, vol. 42, no. 12, pp. 2809–2821, 2007.

[6] A. N. Mohieldin, E. Sánchez-Sinencio, and J. Silva-Martínez, "A 2.7-V 1.8-GHz fourth-order tunable LC bandpass filter based on emulation of magnetically coupled resonators," *IEEE Journal of Solid-State Circuits*, vol. 38, no. 7, pp. 1172–1181, 2003.

[7] H. Le-Thai, H. H. Nguyen, H. N. Nguyen, H. S. Cho, J. S. Lee, and S. G. Lee, "An IF bandpass filter based on a low distortion transconductor," *IEEE Journal of Solid-State Circuits*, vol. 45, no. 11, pp. 2250–2261, 2010.

[8] F. Dülger, E. Sánchez-Sinencio, and J. Silva-Martínez, "A 1.3-V 5-mW fully integrated tunable bandpass filter at 2.1 GHz in 0.35-μm CMOS," *IEEE Journal of Solid-State Circuits*, vol. 38, no. 6, pp. 918–928, 2003.

[9] A. X. He and W. B. Kuhn, "A 2.5-GHz low-power, high dynamic range, self-tuned Q-enhanced LC filter in SOI," *IEEE Journal of Solid-State Circuits*, vol. 40, no. 8, pp. 1618–1628, 2005.

[10] Y. Sun, C. J. Jeong, I. Y. Lee, J. S. Lee, and S. G. Lee, "A 50-300-MHz low power and high linear active RF tracking filter for digital TV tuner ICs," in *Proceedings of the IEEE Custom Integrated Circuits Conference (CICC '10)*, pp. 1–4, San Jose, Calif, USA, September 2010.

[11] B. Georgescu, I. G. Finvers, and F. Ghannouchi, "2 GHz Q-enhanced active filter with low passband distortion and high dynamic range," *IEEE Journal of Solid-State Circuits*, vol. 41, no. 9, pp. 2029–2039, 2006.

[12] A. El Oualkadi, M. El Kaamouchi, J.-M. Paillot, D. Vanhoenacker-Janvier, and D. Flandre, "Fully integrated high-Q switched capacitor bandpass filter with center frequency and bandwidth tuning," in *Proceedings of the IEEE Radio Frequency Integrated Circuits Symposium (RFIC '07)*, pp. 681–684, June 2007.

[13] A. Ghaffari, E. A. M. Klumperink, M. C. M. Soer, and B. Nauta, "Tunable high-Q N-path band-pass filters: modeling and verification," *IEEE Journal of Solid-State Circuits*, vol. 46, no. 5, pp. 998–1010, 2011.

[14] T. A. Vu, S. Sudalaiyandi, H. A. Hjortland, O. Nass, and T. S. Lande, "An inductorless 3–5 GHz band-pass filter with tunable center frequency in 90 nm CMOS," in *Proceedings of the IEEE International Symposium on Circuits and Systems (ISCAS '13)*, pp. 1284–1287, IEEE, Beijing, China, May 2013.

[15] M. Darvishi, R. van der Zee, E. A. M. Klumperink, and B. Nauta, "Widely tunable 4th order switched gm-C band-pass filter based on N-path filters," *IEEE Journal of Solid-State Circuits*, vol. 47, no. 12, pp. 3105–3119, 2012.

## 12

VLSI Architectures for Image Interpolation: A Survey

C. John Moses,[1] D. Selvathi,[2] and V. M. Anne Sophia[1]

[1] Department of ECE, St. Xavier's Catholic College of Engineering, Nagercoil 629003, India
[2] Department of ECE, Mepco Schlenk Engineering College, Sivakasi 626005, India

Correspondence should be addressed to C. John Moses; jofjef@yahoo.com

Academic Editor: Marcelo Lubaszewski

Image interpolation is a method of estimating the values at unknown points using the known data points. This procedure is used in expanding and contrasting digital images. In this survey, different types of interpolation algorithm and their hardware architecture have been analyzed and compared. They are bilinear, winscale, bi-cubic, linear convolution, extended linear, piecewise linear, adaptive bilinear, first order polynomial, and edge enhanced interpolation architectures. The algorithms are implemented for different types of field programmable gate array (FPGA) and/or by different types of complementary metal oxide semiconductor (CMOS) technologies like TSMC 0.18 and TSMC 0.13. These interpolation algorithms are compared based on different types of optimization such as gate count, frequency, power, and memory buffer. The goal of this work is to analyze the different very large scale integration (VLSI) parameters like area, speed, and power of various implementations for image interpolation. From the survey followed by analysis, it is observed that the performance of hardware architecture of image interpolation can be improved by minimising number of line buffer memory and removing superfluous arithmetic elements on generating weighting coefficient.

1. Introduction

In digital image scaling, image interpolation algorithms are used to convert an image from one resolution to another resolution without losing the visual content in the image. In the colour, image interpolation is the process of estimating the missing colour samples to reconstruct a full colour image [1]. Image scaling is widely used in many fields, ranging from consumer electronics, such as digital camera, mobile phone, tablet, display devices and medical imaging like computer assisted surgery (CAS) and digital radiographs [2]. In many applications, from consumer electronics to medical imaging, it is desirable to improve the restructured image quality and processing performance of hardware implementation [3]. For example, a video source with a 640 × 480 video graphics arrays (VGA) resolution may need to fit the 1920 × 1080 resolution of a high definition multimedia interface (HDMI).

Image up scaling [4] methods are implemented for a variety of computer equipments like printers, digital television, media players, image processing systems, graphics renderers, and so on. On the other hand, high resolution image may need to be scaled down to a small size in order to fit the lower resolution of small liquid crystal display panels. That is, the image scaling is a challenging and very significant issue in digital image processing [5]. The problem is to attain a digital image to be displayed on a large bitmap from unique data sample in a smaller grid, and this image should appear like it had been attained with a sensor having the resolution of the up-scaled image or, as a minimum, present a "natural" texture. Methods that are normally used to solve the problem (i.e., pixel replication, bilinear, or bi-cubic interpolation) do not realize these requirements, producing images with visual artifacts like pixelization, jagged contours, and over smoothing. Therefore, a set of advanced adaptive methods have been presented [4].

The interpolation algorithms can be classified as adaptive and nonadaptive algorithms [3]. Nonadaptive algorithms perform interpolation in a fixed pattern for every pixel, that is, by averaging neighbouring pixels. This causes an artifact, the zipper effect in the interpolated image. This leads to a blur across the edges. This technique is fixed irrespective of the input image features. Adaptive algorithms use both spectral and spatial features present in the neighbourhood pixel in order to interpolate the missed pixel as close to the original as possible [6]. To restore more accurate and visually pleasing results, image spatial or spectral correlation or both have been exploited. Adaptive algorithms can detect local spatial features present in the pixel neighbourhood and then make effective choices as to which predictor to use for that neighbourhood. The result is a reduction or elimination of zipper-type artifacts.

Further, the interpolation methods are generally classified into two classes as [7] spatial domain and frequency domain. Spatial domain technique is adopted frequently for real-time applications because it has low complexity. The other method uses proper transform such as discrete cosine transform (DCT), discrete fourier transform (DFT), or wavelet transform to scale images in the frequency domain. Due to its higher computational complexity and memory requirements, the frequency domain technique is not appropriate for real-time or low cost applications.

In many practical real-time applications, the scaling process is included in end user equipment and it can be implemented by VLSI. VLSI architecture is implemented using either FPGA or application specific integrated circuit (ASIC). FPGA is designed to be configured by a customer or a designer often manufacturing. Thus, FPGAs are used to construct reconfigurable computing systems with high performance and low cost. But ASIC is not reconfigurable. Recent FPGA platforms have millions of gates, up to 500 MHz clock frequency, reasonably large on chip memory and fast input output interface [8]. FPGAs are widely used for real-time implementation of various image processing algorithms such as motion detection, image enhancement, image correlation, and image compression. In this paper, different hardware architectures of interpolation methods are discussed and analysed their performance.

Section 2 explains the hardware architecture of image interpolation, Section 3 provides the important characteristics of different hardware implementations of image interpolation algorithms using FPGA/TSMC. Section 4 deals with a detailed analysis of different optimization parameters of VLSI architecture like gate count, speed, and power. Finally, Section 5 concludes the analysis.

2. Hardware Architecture of Image Interpolation

The hardware architecture of image interpolation includes the coordinate accumulator, line buffer, weighting coefficients generator, vertical interpolator, and horizontal interpolator. Each of the units performs specified operations for the interpolation [14]. The hardware architecture with its different units is shown in Figure 1.

2.1. Co-Ordinate Accumulator.
The coordinate accumulator calculates the corresponding coordinate of the current target pixel to be interpolated. Its inputs are two scaling factors: one is for the horizontal direction scale_x and the other is for the vertical direction scale_y. These factors provide the scaling

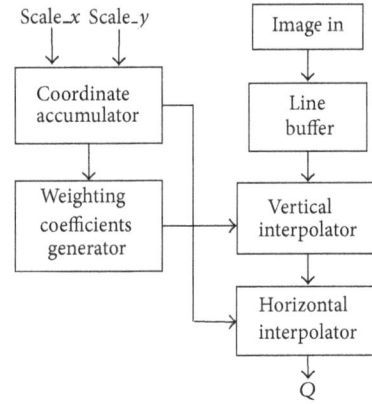

FIGURE 1: Block diagram of the hardware architecture of image interpolation.

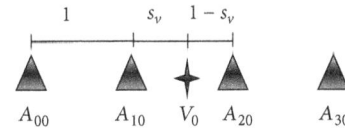

FIGURE 2: Vertical interpolation.

ratio of each direction. The current target pixel's coordinate (x_c, y_c) is calculated as [7]

$$x_c = x_{c-1} + \text{scale\_}x,$$
$$y_c = y_{c-1} + \text{scale\_}y. \tag{1}$$

Equation (1) represents the coordinate of the previous target pixel.

2.2. Line Buffers (LB).
The line buffers are used to store source image data.

2.3. Weighting Coefficient Generator.
In 1D interpolation, vertical and horizontal weighting coefficients have to be determined. The method of calculating vertical weighting coefficients is identical to the way of obtaining horizontal weighting coefficients. The way of calculating vertical and horizontal weighting coefficients is shown in Figures 2 and 3, respectively [14].

According to Figure 2 the vertical weighting coefficients can be calculated as

$$vw_i = -(1 - s_v)^2 \times s_v,$$
$$vw_{i+1} = (1 - s_v) + 2(1 - s_v)^2 s - (1 - s_v)s_v^2,$$
$$vw_{i+2} = s_v + 2(1 - s_v)s_v^2 - (1 - s_v)^2 s_v, \tag{2}$$
$$vw_{i+3} = -(1 - s_v) \times s_v^2,$$

where vw_i, vw_{i+1}, vw_{i+2} and vw_{i+3} are the vertical coefficients of the corresponding source pixels A_{00}, A_{10}, A_{20}, and A_{30}. s_v is the distance between the source pixel $A_{i+1,j}$ and the vertical interpolation pixel v_0.

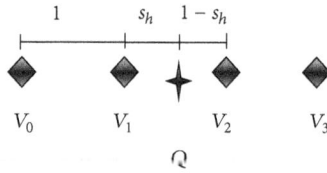

FIGURE 3: Horizontal interpolation.

Similarly, based on Figure 3, all the horizontal weighting coefficients can be found by

$$hw_j = -\left(1 - s_h\right)^2 \times s_h,$$

$$hw_{j+1} = \left(1 - s_h\right) + 2\left(1 - s_h\right)^2 s - \left(1 - s_h\right) s_h^2,$$

$$hw_{j+2} = s_h + 2\left(1 - s_h\right) s_h^2 - \left(1 - s_h\right)^2 s_h,$$

$$hw_{j+3} = -\left(1 - s_h\right) \times s_h^2,$$

$$(3)$$

where hw_j, hw_{j+1}, hw_{j+2}, and hw_{j+3} are the horizontal coefficients of the corresponding virtual pixels V_0, V_1, V_2, and V_3. S_h is the distance between the vertical interpolation pixel V_0 and target pixel Q.

2.4. Vertical Interpolator. After the vertical weighting coefficients are obtained, the vertical interpolation operation is performed by using (2). For the 16 source image pixels, four new pixel values are obtained from vertical interpolation. These pixels are called "virtual pixels."

2.5. Horizontal Interpolator. The vertical interpolated pixels are used to interpolate the target pixel through the horizontal direction. The horizontal weighting coefficients are processed by using (3) to obtain the target pixel Q.

3. Interpolation Methods

This chapter briefs a review of different image interpolation algorithms such as winscale, bi-cubic, linear, polynomial convolution, bilinear, and adaptive scaling algorithms. Various algorithms are discussed by their hardware characteristics and visual quality. The hardware characteristics such as area utilization and speed and power consumptions are mainly focused upon. The quality of the interpolation algorithms can be expressed in dB (decibel) with peak signal-to-noise ratio (PSNR) of scaled image and the original image.

3.1. Digital Image Scaling Algorithm. Andreadis and Amantiadis proposed an image interpolation algorithm for both grayscale and colour images of any resolution in any scaling factor [5]. This algorithm uses a mask of maximum four pixels and calculates the final luminosity of each pixel combining two factors such as the percentage of area that mask covers from each source pixel and the difference in luminosity between the source pixels. This interpolation operates on linear area domain and uses continuous area filtering. It can perform both upscale and downscale processes for fast

real-time implementation. This method is implemented on Quartus II FPGA with the operating frequency of 55 MHz. The hardware architecture of this interpolation uses 20 additions and 13 multiplications. The root mean square error (RMSE) of this method is very less than the RMSE of other interpolation scheme like nearest neighbour, bilinear, and winscale.

3.2. Winscale Image Interpolation. Image scaling using winscale [9] is proposed by Kim et al. The aim of this scaling algorithm is to improve the quality and to reduce the computational complexity by reducing number of operations per pixel. This algorithm can perform both scale up and scale down transform using an area pixel model rather than a point pixel model. To provide low complexity, it uses a maximum of four pixels of an original image to calculate one pixel of a scaled image. Also, it can provide better quality by the characteristics such as fine-edge and changeable smoothness. The hardware of winscale is implemented using an FPGA for displaying scenes in a liquid crystal display panel. winscale has good scale property with low complexity. It preserves the edge characteristic of an image well and handles streaming data directly and requires only small amount of memory. Thus, image interpolation can be done by four-line buffers in spite of up/down ratio. Further, this work tells that the winscale algorithm has low RMSE and it has better image quality than the bilinear algorithm. This implementation utilizes 29000 gates at operating frequency of 65 MHz. winscale can be used in various digital display devices that need image scaling, especially in applications that require good image quality with low hardware cost.

3.3. VLSI Design of Winscale for Digital Image Scaling. Lin et al. proposed a VLSI design of winscale algorithm [10] for digital image scaling. This winscale technique uses an area pixels model to evaluate a scaled pixel. The original image and the scaled image are treated as rectangular and the intensity pixel is evenly scattered in rectangular area. The scaled image could be obtained by source image with scaling up/down in various ratios. The scale ratio that comprises horizontal scale ratio (HSR) and vertical scale ratio (VSR) is greater than 1.0 for scale up and less than 1.0 for scale down. The VLSI architecture of this scaling algorithm is implemented by UMC 0.18 μm CMOS standard cell library. The total gate count is 17414 with the operating frequency of 130.24 MHz. This architecture occupies $450 \times 450 \, \mu m^2$ core area of chip and total power consumption is 19.41 mW. Further, this can be used for processing image interpolation for high definition television (HDTV) in real time.

3.4. Bi-Cubic Convolution Interpolation. An efficient VLSI design of bi-cubic convolution interpolation [11] for digital image processing algorithm is presented by Lin et al. The architecture of reducing the computational complexity of generating coefficients and decreasing the number of memory access times is proposed. The hardware architecture of this method is based on Figure 1. In this method, the number of memory access time, for interpolating a row is fixed and does not vary with the scale ratio. This method generates weighting coefficients as shown in Figures 2 and 3. The number of multipliers and adders used to generate coefficients is less than the original bi-cubic algorithm. It provides a simple hardware architecture design and low computation cost and it is also easy to implement. Based on this technique, the high-speed VLSI architecture has been successfully designed and implemented with TSMC 0.13 μm standard cell library. The simulation results demonstrate that the high performance architecture of bi-cubic convolution interpolation at 279 MHz with 30643 gates in a $498 \times 498 \ \mu m^2$ chip is able to process digital image scaling for HDTV, in real time. The hardware of this algorithm is also analyzed by using FPGA. The FPGA implementation demonstrates that the algorithm requires only about 437 logic blocks but the bi-cubic algorithm which is proposed by Nuño-Maganda and Arias-Estrada [16] requires about 890 logic blocks.

3.5. Real-Time FPGA Linear Convolution Interpolation. A linear interpolation [12] is presented by Lin et al., which is low-cost hardware architecture with digital image scaling for real-time requirements. This architecture is also based on Figure 1 and this generates weighting coefficients based on Figures 2 and 3. The scheme has the advantage of low operation complexity which reduces the coefficient generating effort and hardware cost with the interpolation quality, compatible to that of bi-cubic convolution interpolation [16]. The operation of linear convolution interpolation requires 16 weighting coefficients, generated from 16 neighbouring pixels of source image. Therefore, the number of adders and subtractors used to generate weighting coefficients in this method are much less than the bi-cubic algorithm [16]. The hardware architecture of this algorithm is implemented on Virtex-II FPGA. Consequently, this architecture has solved the problem of computational complexity of interpolation and simplified the hardware circuit. The high performance architecture of extended linear interpolation can be simulated at 104 MHz with 379 LBs (Logic Blocks) which is able to process digital image scaling. This architecture requires about 26200 gates but the architecture which is proposed by Kim et al. [9] requires about 29000 gates. In addition, this method achieves that higher PSNR (33.12). But the winscale and bi-cubic achieve lower PSNR as 28.1 and 33.05, respectively for the 3/2 upscaling image after 2/3 downscaling.

3.6. Extended Linear Interpolation. An extended linear interpolation for real-time digital image processing is presented by Lin et al. [13]. This image interpolation is a low cost architecture with the interpolation quality compatible to that of bi-cubic convolution interpolation [22]. This method has the similar hardware architecture as shown in Figure 1. Further, this method has similar visual quality (PSNR) as the bi-cubic interpolation algorithm and higher PSNR than the winscale algorithm [9]. The architecture is capable of reducing the computational complexity of generating weighting coefficients. A 2-dimensional interpolation method is decomposed as two 1-dimensional operations (vertical interpolation and horizontal interpolation) as shown in Figures 2 and 3, respectively. This is to solve the problem of computational complexity of interpolation and furthermore to simplify the circuit, and to reduce chip area. This architecture is implemented on the Virtex-II FPGA, and the VLSI architecture has been successfully designed and implemented with TSMC 0.13 μm standard cell library. The high performance architecture of extended linear interpolation that can be simulated at 267 MHz with 26200 gates in a $452 \times 452 \ \mu m^2$ chip is able to process digital image scaling for HDTV in real time.

3.7. Efficient Architecture of Extended Linear Interpolation (EAELI). An efficient extended linear image interpolation is proposed by Lin et al. [14]. The hardware architecture of this interpolation is shown in Figure 1. The extended linear interpolation requires 16 weighting coefficients produced from 16 neighbouring pixels of source image. This method reduces computational complexity of generating weighting coefficients. The principles of generating coefficients are shown in Figures 1 and 2 and by (2) and (3). The number of arithmetic elements, such as adders and subtractors for generating weighting coefficients, are less than the bi-cubic [16] and winscale [9] image interpolations. This architecture is designed on Virtex-II FPGA and with TSMC 0.13 technologies. This utilizes 25980 gates at 267 MHz to process image interpolation on HDTV in real time. The FPGA implementation tells that this architecture utilizes only about 379 configurable logic blocks (CLBs) but the bi-cubic architecture [16] requires about 890 CLBs. Furthermore, this method provides higher PSNR than winscale [9] and bi-cubic [16] algorithms. The achieved PSNR of this method is 35.29 for the test image (tank) for 3/2 upscaling after 2/3 downscaling.

3.8. Piecewise Linear Convolution Interpolation. A piecewise linear convolution interpolation, with third-order approximation, for real-time image processing, is presented by Lin et al. [20]. This work presents a high-performance architecture of a piecewise linear convolution interpolation for digital image. The hardware architecture of this interpolation is based on Figure 1. The kernel of this method, built up of piecewise linear polynomial, approximates the ideal sinc-function in interval $[-2, 2]$. The number of arithmetic elements like adders and subtractors used for generating weighting coefficients are very less than the bi-cubic algorithm. The number of CLBs utilized by Virtex-II FPGA for various algorithms such as bi-cubic [11, 16] and this algorithm are 890, 437, and 393, respectively. This is able to process various-range image scaling for HDTV in real time. Thus, the architecture reduces the computational complexity of generating weighting coefficients and provides a simple hardware architecture design and low computation cost and it is easy to meet real-time requirement. The architecture is implemented on the Virtex-II FPGA, and the VLSI architecture has been designed and implemented with TSMC 0.13 μm standard cell library. The simulation results indicate that the interpolation quality of the proposed architecture is better than cubic convolution interpolations [11, 16]. This implementation has higher PSNR (49.53) than the keys (49.29) and bi-cubic (46.38) for the test image (airplane) of scaled from 3/4 downscaling after 4/3 upscaling.

3.9. An Edge-Oriented Image Scaling Algorithm. An edge-oriented area-pixel scaling algorithm [15] has been presented by Chen et al. This algorithm aims to achieve low cost; the area-pixel scaling technique is implemented with low-complexity VLSI architecture in this design. A simple edge catching technique is adopted to preserve the image edge features effectively so as to achieve better image quality. This scaling processor can support floating-point magnification factor and preserve the edge features efficiently by taking into account the local characteristic that existed in those available source pixels around the target pixel. Furthermore, it handles streaming data directly and requires only small amount of memory like one-line buffer rather than a full frame buffer. The FPGA implantation of this method is obtained by Xilinx Virtex-II XC2VP50. The FPGA implementation utilizes only about 581 CLBs but the bi-cubic [16] algorithm utilizes about 890 CLBs. Thus, the seven stage pipelined VLSI architecture of this image scaling processor contains 10.4 K gate counts and yields a processing rate of about 200 MHz by using TSMC 0.18 μm technology, but the bi-cubic architecture [16] utilizes about 29 K gates. The quality comparison shows that this method has higher PSNR as 38.16 than the other methods such as nearest neighbour, bilinear, bi-cubic [16], area-pixel scaling [9], and winscale [5] algorithms. The PSNR is obtained for the test image (crowd) for image reduction from the size of 600 × 600 to 512 × 512.

3.10. A Novel Interpolation Algorithm (NICRMA). Huang et al. presented a novel scaling algorithm for the implementation of (2D) image scalar [7]. This interpolation method is based on the interpolation error theorem. It solves the blurring and checkerboard effects caused by linear interpolation, and efficiently enhance the edge features of scaled images. This method can be conveniently applied to image zooming using both integer and noninteger magnification factors. The hardware architecture of this method is based on Figure 1 and it uses four-line-memory buffer. In the VLSI architecture, the cooperation and hardware sharing techniques have been used, which greatly reduce hardware cost requirements. The number of arithmetic components like adders and subtractors is very less than winscale [9, 15] algorithms. Using a nine-stage pipeline, the scaling circuit contains 13 K gate counts and yields a processing rate of approximately 278 MHz using TSMC 0.13 μm technology. Simulation results indicate a clock period of 3.6 ns and a processing rate of 278 megapixels per second. This implementation can also provide better PSNR than the winscale algorithms.

3.11. Bicubic Interpolation. One of the most extended algorithms for image scaling is bi-cubic interpolation [16]. In this work, hardware architecture for bi-cubic interpolation (HABI) is proposed. The HABI is integrated by three main blocks: the first block generates the coefficients, which implements the bi-cubic function to be used in HABI; the second

block performs the interpolation process and the third one is a control unit that synchronizes the processing and the pipeline stages. The architecture works with monochromatic image, but it can be extended for working with RGB colour images. This design description is implemented on Xilinx Virtex-II pro FPGA. The system runs 10 times faster than an Intel Pentium 4-based PC at 2.4 GHz. The hardware architecture requires about 890 CLBs, 28 BlockRAMs at 100 MHz operating frequency, and 3.50 ms processing time. The architecture is a compact and efficient module that can be combined with other high performance architecture to create more robust application like, tracking system, robotic, and mobile applications. This architecture can also be updated to more parallelism without changing control and memory.

3.12. First-Order Polynomial Convolution Interpolation. Fast first-order polynomial convolution interpolation (FFOPCI) for real-time digital image reconstruction [19] is given by Lin et al. This work presents high-performance architecture of a novel first-order polynomial convolution interpolation for digital image scaling. The architecture is based on Figure 1. Generally, a better quality of image interpolation is achieved by using higher order model but that requires complex computations. This work tries to reduce the computational effort by using first-order polynomial convolution with compatible quality. The kernel of the proposed method is built up of first-order polynomials and it approximates the ideal sinc-funtion in the interval [−2, 2]. This method reduces the computational complexity of generating weighting coefficients and provides a simple hardware architecture design and low computation cost, and it easily meets real-time requirements. The architecture is implemented on the Virtex-II FPGA, and the VLSI architecture has been designed and implemented with the TSMC 0.13 μm standard cell library. The number of adders and subtractors used to generate weighting coefficients in the architecture is much less than that of bi-cubic interpolation [11]. Consequently, this architecture has solved the problem of computational complexity of interpolation and furthermore simplified the circuit, and reduced chip area. The Virtex-II FPGA utilizes 414 CLBs for this algorithm, but the Virtex-II utilizes 483 CLBs for bi-cubic [11] algorithm.

3.13. Bilinear Interpolation. A novel approach to real-time bilinear interpolation is proposed by Gribbon and Bailey [22]. This work presents a real-time bilinear interpolation, which is useful in applications, such as lens distortion correction where the input coordinates follow a curved path that spans multiple rows. Gribbon and Bailey aim to improve the quality of interpolated image with high speed. To improve the speed of operation a special caching is proposed to retrieve the required pixels in a single clock cycle. Also, they use the three point interpolation instead of using the normal four point interpolation. The hardware architecture of this interpolation is implemented on Spartan-II FPGA. The FPGA implementation utilizes about 329 CLBs out of 1172 CLBs and three numbers of BlockRAM out of 14 BlockRAMs. The major drawback of this system is that the use of the three-point interpolation instead of using the normal four-point interpolation generates additional errors into the measured pixel values. Bilinear interpolation can be used for lens distortion correction, but it is complicated by the truth that input coordinates do not emerge on straight lines.

3.14. Parallel Bilinear Interpolation.

Generalised parallel bilinear interpolation architecture for vision systems [21] is given by Fahmy. Bilinear image interpolation is extensively used in computer vision for generating pixel values for locations that lie off the pixel grid in an image. For every sub-pixel, the values of four neighbours are utilized to calculate the interpolated value. This work improves memory access by parallelism in embedded memories and increases the speed of interpolation. This method attains performance of 250 M samples per second in a modern device. The memory requirements of this method differ linearly with image size. For larger images that would need external memory, the resulting speed would depend on the greatest access speeds for the external memory. The hardware architecture of this method is implemented on Xilinx Virtex 4 XC4VSX55 FPGA. This architecture utilizes 362 Slices, 256 numbers of 18 K bit Block memories, and three numbers of DSP48 blocks. This system can operate at 250 MHz and can process 512×512 pixel images. This system is above 30 times faster as compared with the software implementation. This also suggested that Virtex-5 device can be used for processing 1024×1024 pixel images. This system can be used for computer vision. The major drawback of this system is that it requires a large memory.

3.15. Adaptive Scalar Bilinear Interpolation.

To implement a 2D image scalar, a novel scaling algorithm [3] has been proposed by Chen et al. Bilinear interpolation, a clamp filter, and a sharpening spatial filter are involved in this algorithm. For its low complexity and high quality, the bilinear interpolation algorithm is used. Generally, bilinear interpolation produces blurring and aliasing effects. To reduce these effects, prefilters such as clamp and sharpening spatial filter are added. By using 5×5 combined convolution filter instead of 3×3 matrix coefficients, the memory requirement and computation complexity are reduced. The VLSI architecture is optimized by cooperation and hardware sharing. The hardware architecture of this interpolation scheme is implemented on FPGA and designed by using TSMC $0.18\,\mu m$ and TSMC $0.13\,\mu m$ CMOS libraries. The FPGA implementation of this scheme utilizes about 9.28 K gates, but the winscale [9, 10] algorithms utilize about 29 K gates and 17.4 K gates, respectively. The bi-cubic algorithms [11] and real-time FPGA [12] also utilize a large amount of gate as 30.6 K and 26.2 K, respectively, by comparing this method. This VLSI architecture of bilinear interpolation algorithm can achieve 280 MHz and its chip area is $46418\,\mu m^2$ synthesized by a $0.13\,\mu m$ CMOS process. Furthermore, this method has higher PSNR (32.07) for the test image (splash) than the winscale [5, 9, 10], bi-cubic [11, 12] and other bilinear algorithms. This higher PSNR is achieved by using combined filter, adaptive technology, and bilinear interpolation. Thus, this algorithm provides a low cost, low power consumption, low memory demand, high performance, and good quality VLSI architecture for many video and image scaling applications.

3.16. SL Chen's Image Scaling Algorithm.

Low cost, high quality image scaling algorithm [17] is proposed by Chen. This image scaling algorithm consists of a sharpening spatial filter, a clamp filter, and a bilinear interpolation. Bilinear interpolation is used to reduce the blurring and aliasing artifacts. The sharpening spatial and clamp filters are added as prefilters to minimize the memory buffers and computing resources. A T-model and inversed T-model convolution kernels are produced for realizing the sharpening spatial and clamp filters. In addition, the combined filter is designed by combining two T-models or inversed T-model filters which requires only a one-line buffer memory. Chen introduces sharing of computing resources and hardware to reduce the computational effort and hardware cost. The number of multiplication for this scheme is about four but the other algorithms [12, 17, 19] require 13, 10, and 32 multiplications, respectively. Also, this work requires only one-line buffer but winscale [19] and bi-cubic [12] require four- and six-line buffers, respectively. The VLSI architecture of this method contains 6.08 K gate counts, and the chip area is $30378\,\mu m^2$ synthesized by TSMC $0.13\,\mu m$ CMOS process. It consumes 6.9 mW at 280 MHz operating frequency with a 1.1 V supply voltage. The PSNR of this method is 28.78 for the test image (Doub.) by using combined filter and bilinear interpolation. This PSNR is higher than the PSNR of other methods such as bilinear, winscale [19], bi-cubic [12], edge-enhanced [17], and other adaptive [5] image interpolation algorithms.

3.17. Edge-Enhanced Adaptive Scalar.

A less computation complexity adaptive edge-enhanced scalar has been proposed [18] by Chen for 2D image interpolation applications. In this work, Chen aims to increase the interpolation quality, to reduce the computational complexity, and to reduce the hardware cost. This method includes a linear space-variant edge detector, a less complexity spatial filter, and a bilinear interpolation. The edge detector is used to find the image edges. The spatial filter is used as a prefilter which reduces the blurring effect of bilinear interpolation. An adaptive method enhances the edge detector by choosing pixels of the bilinear interpolation. Furthermore, an algebraic manipulation and a technique of hardware sharing are utilized to optimize bilinear interpolation, which reduce the computation complexity and chip area. This algorithm uses eight 8-bit registers and it can process streaming data directly and needs a one-line-buffer memory. The hardware of this work is analyzed by using Altera Cyclone II FPGA. The FPGA implementation of this method utilizes 6.65 K gates, but the other algorithms utilize a large number of gates. For example, winscale [10], edge-oriented [15], and an adaptive [3] algorithms utilize 17.4 K, 10.4 K, and 9.27 K gates, respectively. The VLSI circuit of this method contains 6.67 K gate count and operates at 280 MHz by utilizing TSMC $0.13\,\mu m$ CMOS technology. This method achieves higher PSNR as 42.08 by using sharp spatial filter and 41.91 by using bilinear interpolation for the test image (splash) for image scaling down from the size of

TABLE 1: Comparison of computing resources.

Interpolation	Multiplication (numbers)	Addition (numbers)
Digital image scaling [5]	13	20
Winscale [9]	10	11
VLSI design of Winscale [10]	9	12
Bi-cubic [11]	32	53
Real time FPGA [12]	32	53
Extended linear [13]	4	3
EAELI [14]	8	12
Edge-oriented [15]	13	14
NICRMA [7]	11	14
HABI [16]	20	N/A
Adaptive bilinear [3]	3	50
SL Chen's image scaling [17]	4	36
Edge-enhanced adaptive scalar [18]	3	19

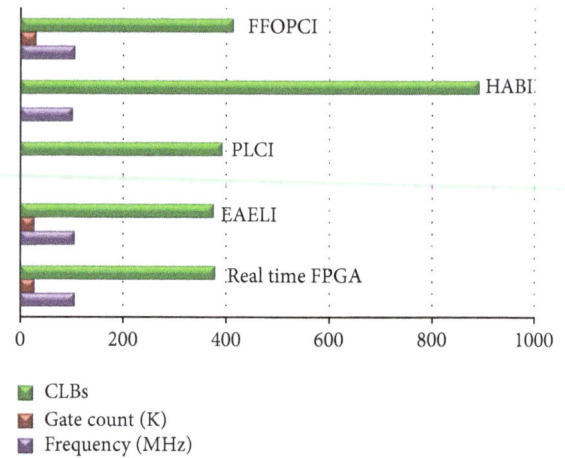

FIGURE 4: CLBs, gate count, and frequency of FPGA (Virtex-II) based interpolation methods.

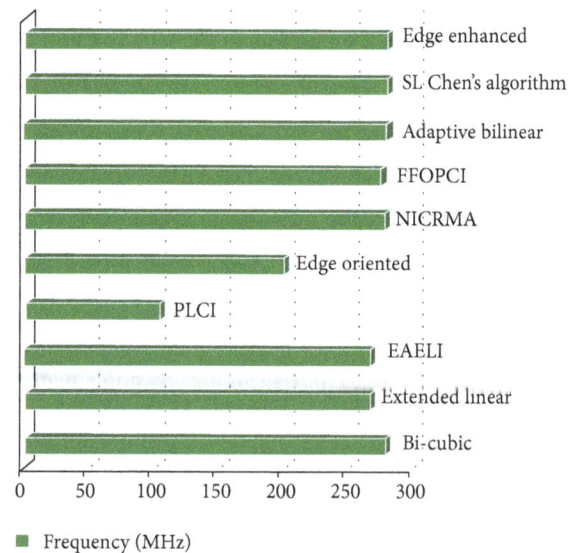

FIGURE 5: Frequency of TSMC 0.13 μm based interpolation methods.

512 × 512 to 720 × 480. Based on PSNR, this method can give better image quality than other methods such as bilinear, winscale [6], bi-cubic [5], adaptive [3], and edge-enhanced [15] algorithms.

4. Performance Analysis

The performance analysis can be given in terms of parameters such as line buffer, gate count, frequency, and power. Figures 4, 5, and 6 and Tables 1, 2, 3, and 4 show the comparison of those parameters for various interpolation algorithms using VLSI architectures. These architectures are implemented by using different FPGA devices or by using TSMC 0.18 μm/TSMC 0.13 μm CMOS libraries for different image processing purposes.

Table 1 describes that the number of arithmetic elements such as multiplication and addition are used for various interpolation techniques. Generally, the number of arithmetic element is a major factor that affects the computational complexity of VLSI circuits. Table 1 demonstrates two interpolation algorithms such as bi-cubic [11] and real-time FPGA [12] that require a large number of multiplications and additions as 32 and 53, respectively. Also, it shows that the other two interpolation algorithms such as edge-enhanced adaptive scalar [18] and an adaptive bilinear [3] require less number of multiplications as three but the adaptive linear requires a large number of addition as 50. Furthermore, it describes that the extended linear [13] requires less number of multiplications as four and number of additions are only three. Therefore, the computation complexity of extended linear interpolation [13] is very less as compared with other interpolation algorithms.

Table 2 gives the comparison of different interpolation architectures in terms of number of memory buffer, gate count, frequency, and number of configurable logic blocks. The different types of FPGA used for designing these interpolation schemes are Virtex II, Virtex IV, Quartus II, and Spartan II devices. Among the various FPGA devices,

Xilinx Virtex-II is mostly used. By comparing Virtex-II based interpolation methods, the HABI [16] system requires a large number of CLBs as 890. But the EAELI [14] requires a less number of CLBs as 376 and the gate count of this method is only 25.98 K. Also, based on Table 1, EAELI [14] requires a less number of multiplications as eight and a less number of additions as 12 as compared with HABI [16]. Therefore, among the various implementations on Virtex-II, the EAELI [14] is an area efficient method. Further, by comparing adaptive bilinear [3] and edge-enhanced adaptive scalar [18], the work in [18] uses less number of line buffer memory as only one and it requires less gate count as 6.65 K. But, [3] requires four-line buffer memory and 9.28 K. Therefore, [18] is an area efficient implementation on Quartus-II FPGA.

Figure 4 tells that the EAELI [14] uses less number of CLBs than other methods [12, 16, 19, 20]. Further, it illustrates that the operating frequency is almost similar for all the methods except PLCI [20].

TABLE 2: Comparison of FPGA based interpolation method.

Interpolation algorithms		Parameters			
	FPGA	Memory buffer (numbers)	Gate count (K)	Frequency (MHz)	CLBs (numbers)
Winscale [9]	FPGA	4 line	29	65	N/A
Bi-cubic [11]	FPGA	4 line	N/A	N/A	437
HABI [16]	Virtex-II	N/A	N/A	100	890
FFOPCI [19]	Virtex-II	N/A	28.9	104.3	414
PLCI [20]	Virtex-II	N/A	N/A	N/A	393
Real time FPGA [12]	Virtex-II	4 line	26.2	104	379
EAELI [14]	Virtex-II	N/A	25.98	104.3	376
Parallel bilinear [21]	Virtex-IV	N/A	N/A	250	362
Adaptive bilinear [3]	Quartus-II	4 line	9.28	67	N/A
Edge-enhanced adaptive scalar [18]	Quartus-II	1 line	6.65	72.57	N/A
Digital image scaling [5]	Quartus-II	4 line	N/A	55	N/A
Bilinear [22]	Spartan-II	N/A	N/A	N/A	329

TABLE 3: Comparison of TSMC 0.18 μm based interpolation methods.

Interpolation algorithms	Parameters				
	Memory buffer (numbers)	Gate count (K)	Frequency (MHz)	Throughput (pixels/second) Frames/sec	Power (mW)
NICRMA [7]	4 line	12.9	200	N/A	11.69
Adaptive bilinear [3]	4 line	9.27	200	200 M/30 f/sec	14.7
SL Chen's image scaling [17]	1 line	6.08	200	200 M/30 f/sec	6.45
Edge-enhanced adaptive scalar [18]	1 line	6.65	200	200 M/30 f/sec	10.23

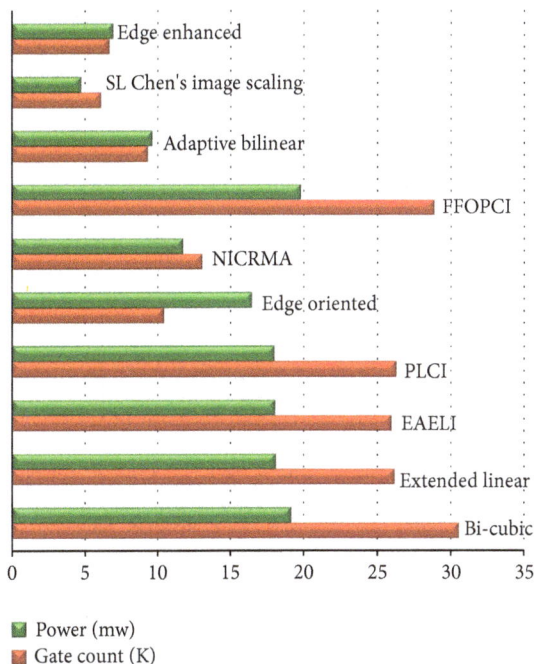

FIGURE 6: Power and gate count of TSMC 0.13 μm based interpolation methods.

Figure 4 also shows the comparison of various interpolation methods such as FFOPCI [19], HABI [16], PLCI [20], EAELI [14], and real-time FPGA [12] in terms of their number of CLBs used, gate count, and frequency on Virtex-II based implementation.

Table 3 illustrates the different VLSI architectures of image interpolation in terms of line-buffer memory, gate count, throughput, and power consumption on TSMC 0.18 μm technology. Based on Tables 1 and 3, as adaptive bilinear [3] requires three multiplications and a large number of additions as 50, this method utilizes 9.27 K gate count with four-line-memory buffer on TSMC 0.18 μm technology. But, the SL Chen's image scaling [17] consumes a low power as 6.45 mW and this method requires only a less amount of gate count as 6.08 K with only one-line-memory buffer and four multiplications and 36 additions. Table 3 also illustrates the interpolation methods such as adaptive bilinear [3], SL Chen's image scaling [17], and edge-enhanced adaptive scalar [18] provide the same throughput as 200 Mega pixels per second to operate a HDMI video resolution of wide quad super extended graphics array (WQSXGA) (3200 × 2048) at 30 frames per second in real-time multimedia image processing.

Table 4 lists the different parameters like memory buffer, gate count, frequency, throughput, and power of different

TABLE 4: Comparison of TSMC 0.13 μm based interpolation methods.

Interpolation algorithms	Memory buffer (numbers)	Gate count (K)	Frequency (MHz)	Throughput (pixels/sec) Frames/sec	Power (mW)
FFOPCI [19]	NA	28.9	275	NA	19.80
Bi-cubic [11]	4 line	30.6	279	NA	19.17
Extended linear [13]	4 line	26.2	267	NA	18.14
EAELI [14]	NA	25.98	267	NA	18.07
PLCI [20]	NA	26.3	104.3	NA	18.02
Edge-oriented [15]	1 line	10.4	200	NA	16.47
NICRMA [7]	4 line	13	278	278 M/30 f/sec	11.69
Adaptive bilinear [3]	4 line	9.28	280	280 M/30 f/sec	9.6
Edge-enhanced adaptive scalar [18]	1 line	6.67	280	280 M/30 f/sec	6.9
SL Chen's image scaling [17]	1 line	6.08	280	280 M/30 f/sec	4.68

image interpolations on TSMC 0.13 μm VLSI technology. Based on Tables 1 and 4, the SL Chen's image scaling [17] requires a less number of arithmetic elements and therefore this method utilizes less power as 4.68 mW and less gate count as 6.08 K on TSMC 0.13 μm technology with only one-line-memory buffer. But, the bi-cubic algorithm [11] uses a large number of arithmetic elements (32 multiplications and 53 additions) and therefore this method requires a large amount of gate as 30.6 K and consumes a large amount of power as 19.17 mW with four-line-memory buffer. Therefore, the SL Chen's image scaling [17] can be implemented as an area efficient interpolation. By comparing the throughput of different interpolation methods on TSMC 0.13 μm technology, the methods such as adaptive bi-linear [3], SL Chen's image scaling [17], and edge-enhanced adaptive scalar [18] achieve the 280 Mega pixels per second or 30 frames per second. But the method NICRMA [7] achieves only 27 Mega pixels per second with four-line-memory buffer.

Figure 5 compares the frequency of various interpolation methods on TSMC 0.13 μm technology. This shows that except for PLCI [20] and edge oriented [15], all other methods operate on similar range of frequency between 260 MHz and 280 MHz.

Figure 6 demonstrates power utilization and gate count of various interpolation methods on TSMC 0.13 μm technologies. This shows the various interpolation methods such as PLCI [20], EAELI [14]. Extended-linear [13] and bi-cubic architecture [11] utilize a power ranges between 18 mW and 20 mW. Also, they require a large gate count ranges between 25 K and 31 K. The methods such as NICFMA [7] and edge oriented [15] utilize a moderate range of power and gates. But, the methods such as edge enhanced [18], SL Chen's image scaling [17], and adaptive bilinear [3] utilize a less amount of power ranges between four mW and 10 mW. Also, they require a less amount of gate ranges between 6 K and 10 K.

5. Conclusion

Digital image interpolation algorithms are widely used in many fields of digital image and video applications. These applications require improved image quality and high processing performance of hardware requirements. This survey focuses on different interpolation algorithms and their hardware performance. The algorithms such as Adaptive bilinear, SL Chen's image scaling algorithm, Extended linear interpolation algorithm provide low cost, low power consumption, low memory demand, and high performance. Among the various interpolation algorithms, SL Chen's image scaling consumes less power as 4.68 mW and requires a less gate count as 6.08 K with higher operating frequency of 280 MHz and one-line-memory buffer by TSMC 0.13 μm VLSI technologies, whereas the NICRMA algorithm which requires moderate power of 11.69 mW contains 13 K gate counts, yields a processing rate of approximately 278 MHz using TSMC-0.13 μm technology, and provides better image quality with four-line-memory buffer. Based on the survey, it can be concluded that the performance of hardware architecture of image interpolation can be improved by reducing number line buffer memory and by removing redundant arithmetic elements on generating weighting coefficient.

Conflict of Interests

The authors declare that there is no conflict of interests regarding the publication of this paper.

References

[1] I. Pekkucuksen and Y. Altunbasak, "Multiscale gradients-based colour filter array interpolation," *IEEE Transactions on Image Processing*, vol. 22, no. 1, pp. 157–165, 2013.

[2] T. M. Lehmann, C. Gönner, and K. Spitzer, "Survey: interpolation methods in medical image processing," *IEEE Transactions on Medical Imaging*, vol. 18, no. 11, pp. 1049–1075, 1999.

[3] S.-L. Chen, H.-Y. Huang, and C.-H. Luo, "A low-cost high-quality adaptive scalar for real-time multimedia applications," *IEEE Transactions on Circuits and Systems for Video Technology*, vol. 21, no. 11, pp. 1600–1611, 2011.

[4] A. Giachetti and N. Asuni, "Real-time artifact-free image upscaling," *IEEE Transactions on Image Processing*, vol. 20, no. 10, pp. 2760–2768, 2011.

[5] I. Andreadis and A. Amanatiadis, "Digital image scaling," in *Proceedings of the IEEE Instrumentation and Measurement Technology Conference (IMTC '05)*, pp. 2028–2032, Ottawa, Canada, May 2005.

[6] S. P. Jaiswal, V. Jakhetiya, A. Kumar, and A. K. Tiwari, "A low complex context adaptive image interpolation algorithm for real time applications," in *Proceedings of the IEEE International Conference on Instrumentation and Measurement Technology (MTC '12)*, pp. 969–972, 2012.

[7] C.-C. Huang, P.-Y. Chen, and C.-H. Ma, "A novel interpolation chip for real-time multimedia applications," *IEEE Transactions on Circuits and Systems for Video Technology*, vol. 22, no. 10, pp. 1512–1525, 2012.

[8] L. Deng, K. Sobti, Y. Zhang, and C. Chakrabarti, "Accurate area, time and power models for FPGA-based implementations," *Journal of Signal Processing Systems*, vol. 63, no. 1, pp. 39–50, 2011.

[9] C.-H. Kim, S.-M. Seong, J.-A. Lee, and L.-S. Kim, "Winscale: an image-scaling algorithm using an area pixel model," *IEEE Transactions on Circuits and Systems for Video Technology*, vol. 13, no 6, pp. 549–553, 2003.

[10] C.-C. Lin, Z.-C. Wu, W.-K. Tsai, M.-H. Sheu, and H.-K. Chiang, "The VLSI design of winscale for digital image scaling," in *Proceedings of the 3rd International Conference on Intelligent Information Hiding and Multimedia Signal Processing (IIHMSP '07)*, vol. 2, pp. 511–514, November 2007.

[11] C.-C. Lin, M.-H. Sheu, H.-K. Chiang, C. Liaw, and Z.-C. Wu, "The efficient VLSI design of BI-CUBIC convolution interpolation for digital image processing," in *Proceedings of the IEEE International Symposium on Circuits and Systems (ISCAS '08)*, pp. 480–483, May 2008.

[12] C.-C. Lin, M.-H. Sheu, H.-K. Chiang, W.-K. Tsai, and Z.-C. Wu, "Real-time FPGA architecture of extended linear convolution for digital image scaling," in *Proceedings of the International Conference on Field-Programmable Technology (FPT '08)*, pp. 381–384, Taipei, Taiwan, December 2008.

[13] C.-C. Lin, M.-H. Sheu, H.-K. Chiang, Z.-C. Wu, J.-Y. Tu, and C.-H. Chen, "A low-cost VLSI design of extended linear interpolation for real time digital image processing," in *Proceedings of the International Conference on Embedded Software and Systems (ICESS '08)*, pp. 196–202, July 2008.

[14] C.-C. Lin, M.-H. Sheu, H.-K. Chiang, C. Liaw, Z.-C. Wu, and W.-K. Tsai, "An efficient architecture of extended linear interpolation for image processing," *Journal of Information Science and Engineering*, vol. 26, no. 2, pp. 631–648, 2010.

[15] P.-Y. Chen, C.-Y. Lien, and C.-P. Lu, "VLSI implementation of an edge-oriented image scaling processor," *IEEE Transactions on Very Large Scale Integration (VLSI) Systems*, vol. 17, no. 9, pp. 1275–1284, 2009.

[16] M. A. Nuño-Maganda and M. O. Arias-Estrada, "Real-time FPGA-based architecture for bicubic interpolation: an application for digital image scaling," in *Proceedings of the International Conference on Reconfigurable Computing and FPGAs (ReConFig '05)*, pp. 1–8, September 2005.

[17] S.-L. Chen, "VLSI implementation of a low cost high quality image scaling processer," *IEEE Transactions on Circuit and System: Express Briefs*, vol. 60, no. 1, pp. 31–35, 2013.

[18] S.-L. Chen, "VLSI implementation of an adaptive edge-enhanced image scalar for real-time multimedia applications," *IEEE Transactions on Circuits and Systems for Video Technology*, vol. 23, no. 9, pp. 1510–1522, 2013.

[19] C.-C. Lin, M.-H. Sheu, C. Liaw, and H.-K. Chiang, "Fast first-order polynomials convolution interpolation for real-time digital image reconstruction," *IEEE Transactions on Circuits and Systems for Video Technology*, vol. 20, no. 9, pp. 1260–1264, 2010.

[20] C.-C. Lin, C. Liaw, and C.-T. Tsai, "A piecewise linear convolution interpolation with third-order approximation for real-time image processing," in *Proceedings of the IEEE International Conference on Systems, Man and Cybernetics (SMC '10)*, pp. 3632–3637, October 2010.

[21] S. A. Fahmy, "Generalised parallel bilinear interpolation architecture for vision systems," in *Proceedings of the International Conference on Reconfigurable Computing and FPGAs (ReConFig '08)*, pp. 331–336, December 2008.

[22] K. T. Gribbon and D. G. Bailey, "A novel approach to real-time bilinear interpolation," in *Proceedings of the 2nd IEEE International Workshop on Electronic Design, Test and Applications (DELTA '04)*, pp. 126–131, January 2004.

## 13

A New CDS Structure for High Density FPA with Low Power

Xiao Wang[1,2,3] **and Zelin Shi**[1,3]

[1]*Shenyang Institute of Automation, Chinese Academy of Sciences, Shenyang 110016, China*
[2]*University of the Chinese Academy of Sciences, Beijing 100049, China*
[3]*Key Laboratory of Opto-Electronic Information Processing, Chinese Academy of Sciences, Shenyang 110016, China*

Correspondence should be addressed to Xiao Wang; wangxiao@sia.cn

Academic Editor: Jose Silva-Martinez

Being an essential part of infrared readout integrated circuit, correlated double sampling (CDS) circuits play important roles in both depressing reset noise and conditioning integration signals. To adapt applications for focal planes of large format and high density, a new structure of CDS circuit occupying small layout area is proposed, whose power dissipation has been optimized by using MOSFETs in operation of subthreshold region, which leads to 720 nW. Then the noise calculation model is established, based on which the noise analysis has been carried out by the approaches of transfer function and numerical simulations using SIMULINK and Verilog-A. The results are in good agreement, demonstrating the validity of the present noise calculation model. Thermal noise plays a dominant role in the long wave situation while $1/f$ noise is the majority in the medium wave situation. The total noise of long wave is smaller than medium wave, both of which increase with the integration capacitor and integration time increasing.

1. Introduction

Infrared detectors have a wide range of applications in areas of military, research, and manufacture, whose core part is an infrared focal plane assembly. The assembly mainly consists of two parts: focal plane arrays (FPAs) that function to convert radiation to current signal and readout integrated circuits (ROIC) that are responsible for realization of serial read and processing of signals sampled by the FPA.

Being an essential part of infrared readout integrated circuit, correlated double sampling (CDS) circuits play important roles in both depressing reset noise and conditioning integration signals [1–3]. Applications of focal planes of large format and high density put forward more harsh demand on low power dissipation and small layout area of a ROIC unit cell. Based on the theory that MOSFETs operating in the subthreshold region consume much less dissipation than those in the depletion region, this paper proposed a low power CDS structure that contains only one sampling capacitor, two switches, and two operation amplifiers (OPs), which saves the layout area [4, 5]. Then the noise calculation model is established, based on which noise analysis has been carried out by the approaches of transfer function and numerical

simulation using SIMULINK and Verilog-A, whose results are in good agreement.

2. Circuit Design

2.1. Operating Principle. The proposed CDS circuit is shown in Figure 1. It comprises two Ops A_1 and A_2 that are connected as buffers, a sampling capacitor Csh and two complementary switches S_1 and S_2. A_1 and A_2 are standard two stage OPs, which are shown in Figure 2. They can provide high gain in order to reduce the error caused by the transmission process of the signals, in the meanwhile guarantee low noise.

The clock timing waveforms of the CDS circuit are also illustrated in Figure 1. After the integrator resets, S_1 and S_2 are both switched on at "t_0" and the reset voltage of the integrator is coupled on the "V_1" node, which is the first sample; then the two switches are off at "t_1," so the charge of Csh remains unchanged until "t_2"; after the integration duration, S_1 is turned on while S_2 remains off at "t_2," and the second sampled signal is stored on "V_1" node. Because of the law of conservation of charge, the voltage of "V_2" node jumps by the difference value between the two sampling processes, which cuts off the error that resulted by the reset process

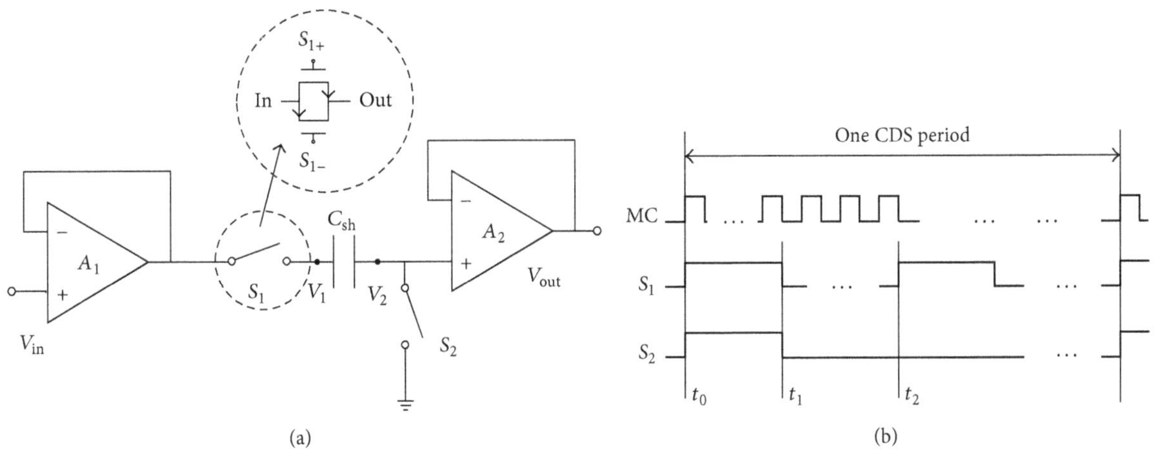

FIGURE 1: Operating principle: (a) structure and (b) operating timing.

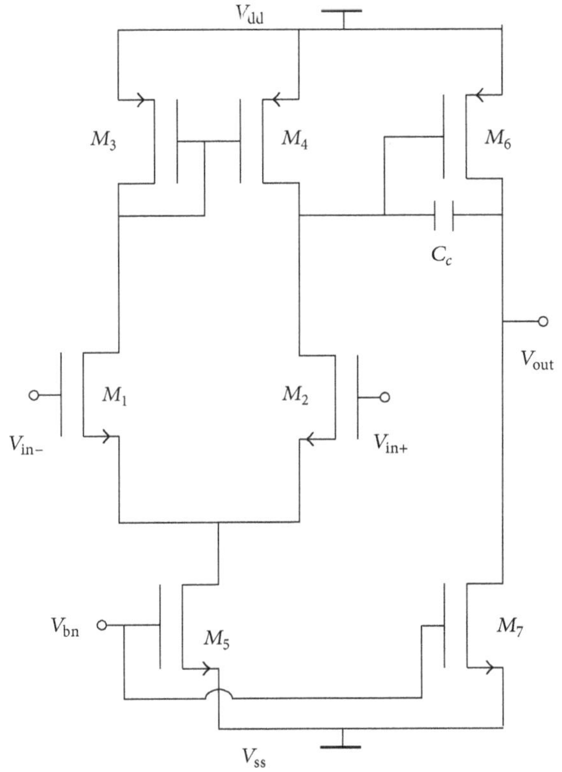

FIGURE 2: Structure of the OPs used in the proposed CDS circuit.

TABLE 1: Performance comparison.

Working region	Gain (dB)	GBW (Hz)	SR (V/us)	Power (W)
Saturation	78	50.5 M	10.2	622 u
Saturation	72.5	1.7 M	0.62	5.5 u
Subthreshold	75.6	1.58 M	0.57	720 n

Ideally, when V_{gs} is lower than V_{th}, the channel between source and drain is shut down. Nevertheless, some electrons still flow across the two ports, known as subthreshold current. Research demonstrates that the subthreshold current is increasing exponentially with the V_{gs} increasing, like the current in BJT. The relationship can be expressed as

$$I_{sub} = I_0 e^{(V_{gs}-V_{th})/nV_T} \left(1 - e^{-V_{ds}/V_T}\right), \tag{1}$$

where I_0 is the drain current when $V_{gs} = V_{th}$, V_T is the thermal voltage, and V_{ds} represents the drain-to-source voltage [6]. It is worth noting that the behavioural model of MOSFETs in subthreshold region is not accurate enough when the process of ICs goes into deep submicron, like 0.18 um. To do the calculation precisely, all the parameters adopted should be those obtained through simulations.

MOSFETs operating in subthreshold region have larger gm-to-channel current ratio than those in saturation region, which implies that subthreshold technology can be applied to optimization power dissipation of analog ICs with the guarantee for sufficient gain [7]. Table 1 is the comparison of the performance for the OPs consisting of MOSFETs working in different regions, where the supply voltages are 0 Volts to 3 Volts. It can be seen that the dissipation of the OP with the design of subthreshold technology succeeds in reducing at least one order of magnitude at the price of tradeoff with frequency character like small SR and GBW.

Figure 3 shows the transient response of the proposed CDS circuit with the two OPs A_1 and A_2 designed by the subthreshold technology, in which the reset voltage is 1 Volt, the output of the integrator is 3 Volts, and S_1 is switched on at "t_2." After slight oscillation the output of the CDS circuit reaches 2 Volts, which is the integration voltage.

of the integrator and suppresses low frequency noise. The proposed circuit structure is easy to implement, where using OP provides the conditions that no extra bias voltage is needed. As is known, capacitors occupy the most layout area; thus the design of only one capacitor saves much area. Besides subthreshold technology applied makes the proposed CDS circuit suitable for ROIC unit cells of the large format FPAs.

2.2. Power Optimization. Subthreshold technology is operating transistors in subthreshold region by providing gate-to-source voltage lower than threshold voltage ($V_{gs} < V_{th}$).

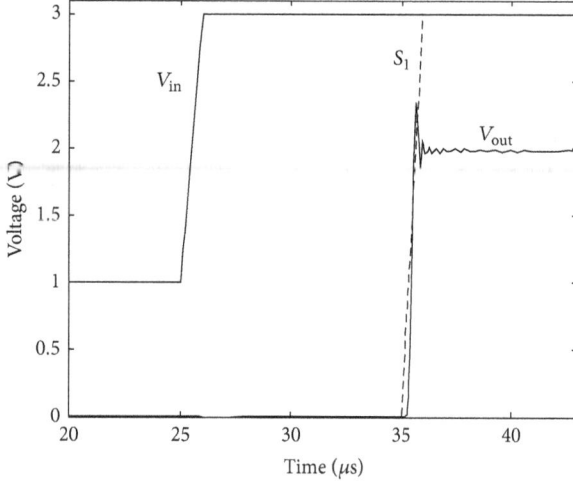

FIGURE 3: The transient response of the CDS circuit.

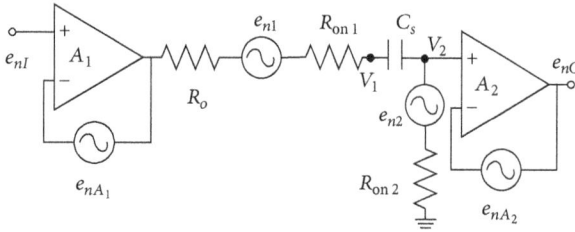

FIGURE 4: Calculation model of noise.

TABLE 2: Parameters of noise calculation.

$R_{\mathrm{on}1}, R_{\mathrm{on}2}$	$2.2\,\mathrm{k\Omega}$
K_{1e1}, K_{1e2}	$4.0 \times 10^{-17}\,\mathrm{V^2/Hz}$
K_{2e1}, K_{2e2}	$4.2 \times 10^{-11}\,\mathrm{V^2/Hz}$
$\mathrm{GBW}_1, \mathrm{GBW}_2$	$1.58\,\mathrm{MHz}$
R_o	$560\,\mathrm{k\Omega}$

3. Noise Analysis

3.1. Noise Calculation Model. The calculation model of noise for the proposed CDS circuit is illustrated in Figure 4, involving four noise sources, which are from the two OPs and the two switches, respectively. The noise sources of the switches are thermal noise and the noise of the OPs is composed of thermal noise and $1/f$ noise.

Referred to the noise, voltage difference appears on C_s at "t_1." Based on the law of charge conservation, the voltage across C_s maintains the same, so the noise voltage of the "V_2" node can be derived from

$$v_2\left(nT_S + T\right) = v_1\left(nT_S + T\right) - \left(v_1\left(nT_S\right) - v_2\left(nT_S\right)\right). \quad (2)$$

For the form of integration of frequency spectrum,

$$v_{n2}^2 = \int_0^\infty e_{n2}^2\left(f\right)df = \int_0^\infty e_{n,v_1}^2\left(f\right)H_{\mathrm{CDS}}^2\left(f\right) + e_{n,v_2}^2\left(f\right)df. \quad (3)$$

The transfer function of $H_{\mathrm{CDS}}(f)$ is given by

$$H_{\mathrm{CDS}}\left(f\right) = 1 - \exp\left(-2\pi jfT\right), \quad (4)$$

where $e_{n,v_1}^2(f)$ and $e_{n,v_2}^2(f)$ are noise power spectrum density (PSD) of "V_1" and "V_2" nodes at "nT_S," respectively [8, 9], which can be found by (5), due to the independence of each noise source. Here $H_{A_1,i}(f)$, $H_{1,i}(f)$, and $H_{2,i}(f)$ are transfer functions to the "V_i" node where i can be 1 or 2 for the following three noise sources: $e_{nA_1}^2(f)$, $e_{nS_1}^2(f)$, and $e_{nS_2}^2(f)$, respectively, which are the reference noise of A_1, the resistance-on noise of S_1, and the resistance-on noise of S_2, respectively. They are described in detail in the appendix:

$$\begin{bmatrix} e_{n,v_1}^2\left(f\right) \\ e_{n,v_2}^2\left(f\right) \end{bmatrix} = \begin{bmatrix} H_{A_1,1}^2\left(f\right), H_{1,1}^2\left(f\right), H_{2,1}^2\left(f\right) \\ H_{A_1,2}^2\left(f\right), H_{1,2}^2\left(f\right), H_{2,2}^2\left(f\right) \end{bmatrix} \begin{bmatrix} e_{nA_1}^2\left(f\right) \\ e_{nS_1}^2\left(f\right) \\ e_{nS_2}^2\left(f\right) \end{bmatrix}. \quad (5)$$

The noise of the end "V_2" goes through A_2 with finite GBW, added by the noise source of A_2. The output noise of the CDS circuit is given by

$$v_{no}^2 = \int_0^\infty \left(e_{n2}^2\left(f\right) + e_{nA_2}^2\left(f\right)\right)H_{\mathrm{UG}}^2\left(f\right)df, \quad (6)$$

where $H_{\mathrm{UG}}(f)$ is the transfer function of the buffers constituted by A_1 or A_2, and GBW represents gain-bandwidth product as seen in a number of textbooks for CMOS design:

$$H_{\mathrm{UG}}\left(f\right) = \frac{1}{1 + f/\mathrm{GBW}}. \quad (7)$$

3.2. Noise Calculation. The noise of the OP and the switch is introduced by MOSFETs. With regard to S_1 and S_2, MOSFETs in linear region produce thermal noise similar to resistance [6], which is given by

$$e_{nS_i}^2 = 4kTR_{\mathrm{on}i}, \quad (8)$$

where k is the Boltzmann constant, T is the absolute temperature, and the MOSFETs of the OP are in the subthreshold region that generates not only thermal noise but also $1/f$ noise. The whole noise of the OP can be modeled as the reference input noise at the input port of the OP, which can be described by

$$e_{nA_i}^2 = K_{1ei} + \frac{K_{2ei}}{f}, \quad (9)$$

where $R_{\mathrm{on}i}$ represents the on-resistance of the switches and K_{1ei} and K_{2ei} represent thermal noise factor and $1/f$ noise factor for the OPs, respectively [7]. Here i is equal to 1 or 2. The calculation of noise adopts the parameters in Table 2, where R_o is the output resistance of OP.

Detectors capable of different wavebands produce a variety of densities of photocurrent, which leads to different integration time needed at certain ability of charge processing.

(a)

(b)

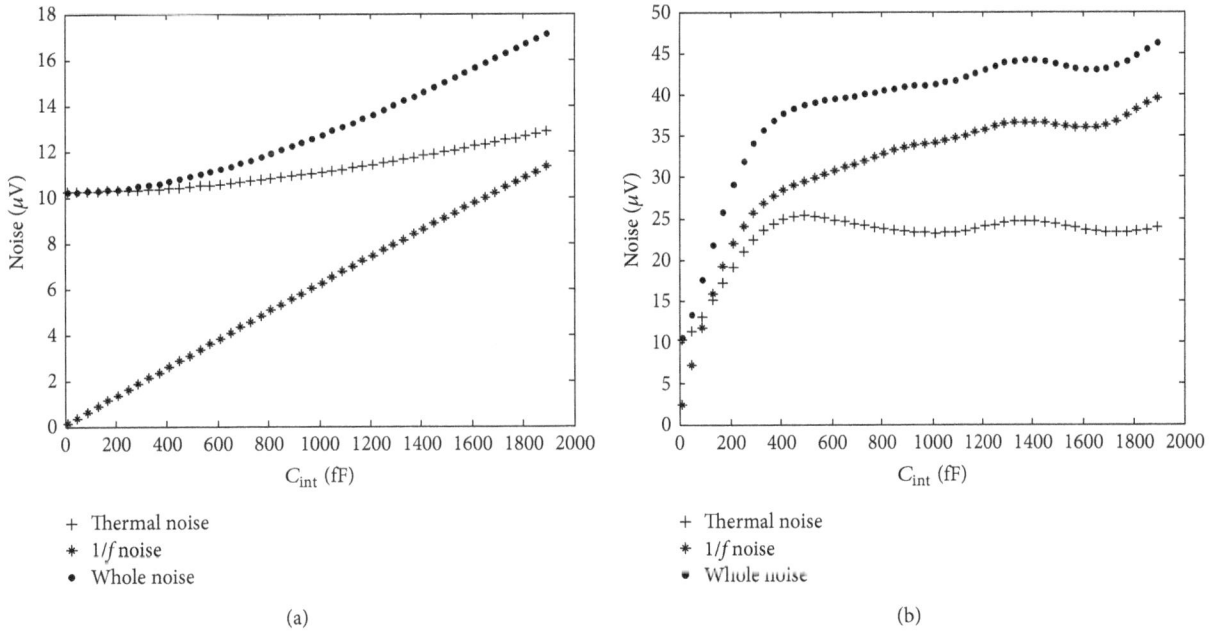

FIGURE 5: Noise versus C_{int}: (a) long wave and (b) medium wave.

We define $K_{int} = T_{int}/C_{int}$ as the integration factor, determined by the value of photocurrent and output swing of the ROIC. T_{int} means integration time, that is, the time interval between the two sampling processes of CDS, and C_{int} is integration capacitor. Figure 5 gives the output noise and its component noise of the proposed CDS circuit as functions of C_{int} under the situations of applications of long wave and medium wave, respectively. K_{int} is 50 K for typical long wave and 1 MEG for medium wave. As can be seen, noise increases with C_{int} increasing for long wave detection, in which $1/f$ noise has greater growth than thermal noise; when $C_{int} <$ 2pF, thermal noise dominates. For medium wave situation, owing to the radiation weaker than that for long wave, larger T_{int} is needed at the same C_{int}. The fact mentioned above results in longer interval between the two sampling processes of CDS, thus causing inferiority of suppressing $1/f$ noise. We can see that under medium wave situation $1/f$ noise is larger than thermal noise and rises with C_{int} increasing. By comparison of the two situations, we conclude that the noise that CDS circuit brings into the signal chain is larger for medium wave application than that for long wave.

Figure 6 shows the noise varying as functions of T_{int} at fixed C_{int}, in which we can see $1/f$ noise is increasing with T_{int} increasing; while thermal noise nearly remains the same, the reason can be concluded through analysis that the increasing of T_{int} results in the increasing of $H^2_{CDS}(f)$ in its low frequency area; thus more noise of low frequency is transmitted to the output node of the CDS circuit.

4. Simulation Experiment

The transient analysis model of the proposed CDS circuit is constructed in SIMULINK, which is shown in Figure 7.

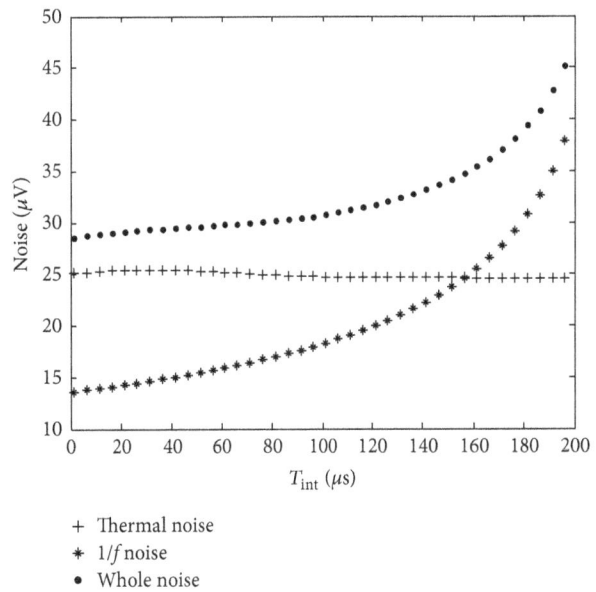

FIGURE 6: Noise versus T_{int}.

Thermal noise can be presented directly using the module in the simulator and $1/f$ noise is modeled by the approach brought out in [10]. HSPICE provides the possibility to simulate circuit noise in AC response by computing the PSD but cannot give the waveform of noise in transient response directly, whereas we carried out an approach to model time domain noise source using Verilog-A in this paper [11], as is shown in Figure 8(a). The noise is filtered by an RC filter, which settles its bandwidth. The waveform can be seen in Figure 8(b).

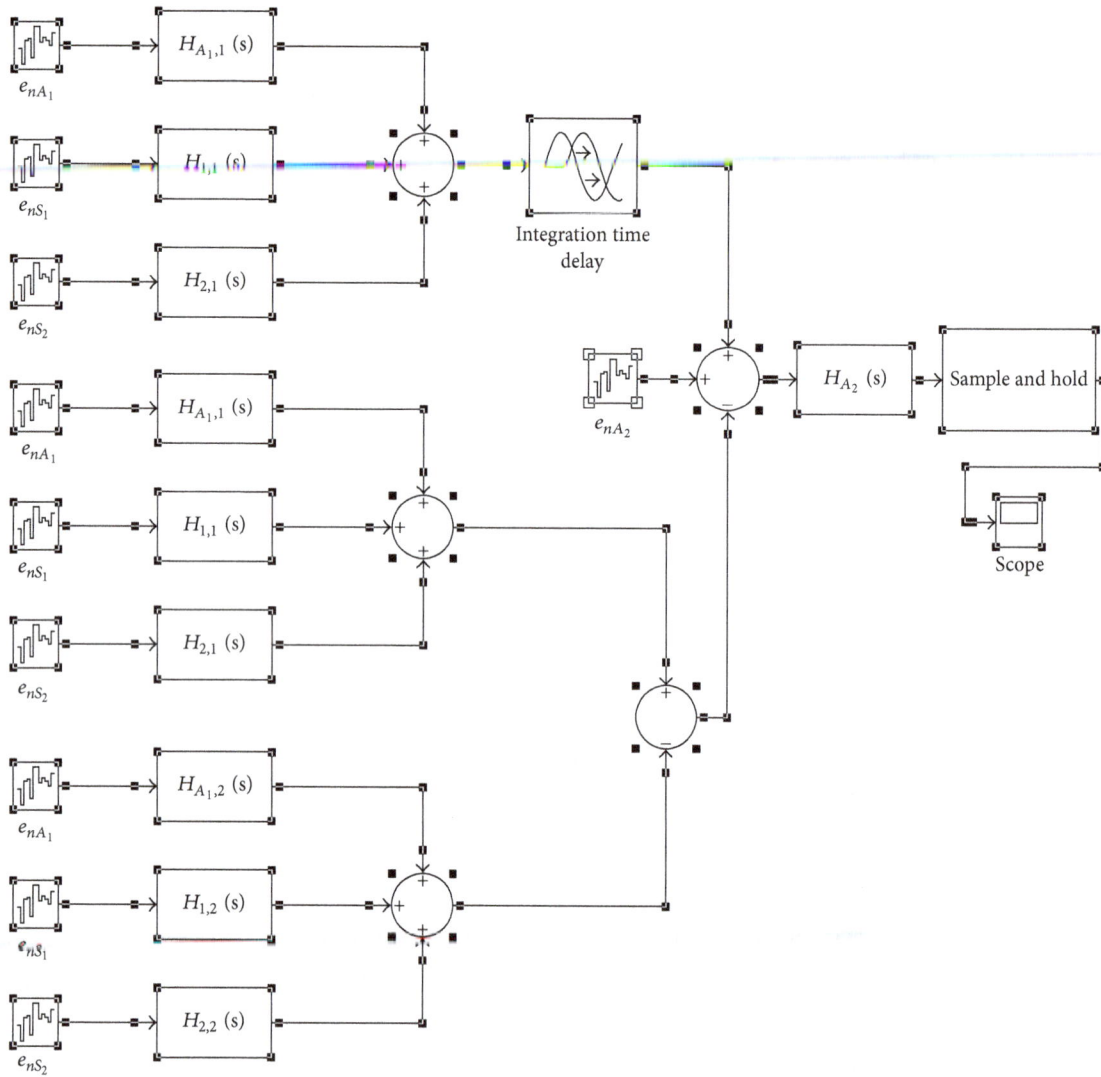

FIGURE 7: Noise model in SIMULINK.

At last we averaged the RMS of the output noise from the scope in Figure 7 and output noise obtained by Verilog-A method to compare with those that were calculated by the equations in Section 2, and the unit of noise is uV. The three sets of results, which are given in Table 3, are in good agreement, therefore proving the feasibility of the method of transfer function noise analysis.

5. Conclusion

For the applications of FPAs of large format and high density, a new structure of CDS circuit is proposed, whose power dissipation has been optimized by subthreshold technology, which leads to 720 nW. Because of using only one sampling capacitor, the proposed CDS circuit occupies small layout area. Then the noise calculation model is established, based on which the noise analysis has been carried out by the approaches of transfer function and numerical simulation using SIMULINK and Verilog-A. The results are in good

agreement, demonstrating the validity of the present noise calculation model. Thermal noise plays a dominant role in the long wave situation while $1/f$ noise is the majority in the medium wave situation. The total noise of long wave is smaller than medium wave, both of which increase with the integration capacitor and integration time increasing.

Appendix

Explanation of (5)

The model of noise source of A_1 transmitting to the nodes across C_s is shown in Figure 9, whose transfer functions that appear in (5) are given by (A.1) and (A.2), respectively:

$$H_{A_1,1}(f) = \frac{1}{1 + f/\text{GBW}} \cdot \frac{1 + sC_{\text{SH}}R_{\text{on}\,2}}{1 + sC_{\text{SH}}(R_o + R_{\text{on}\,1} + R_{\text{on}\,2})},$$
$$(\text{A.1})$$

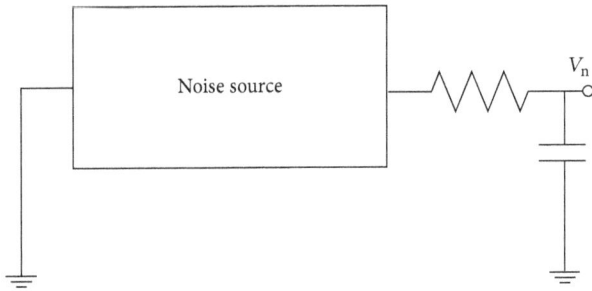

```
//NOISE SOUREC Verilog-A
module Noise_Soure (out);
output out;
electrical out;
Parameter period = 1.0;
Parameter Fall_time = 100;
Parameter vmod = 1;
integer x;
  analog begin
    @(timer (0, period))
    x = $random;
    V(out)<+transition(x, 0, period/Fall_time)/vmod;
  end
endmodule
```

(a)

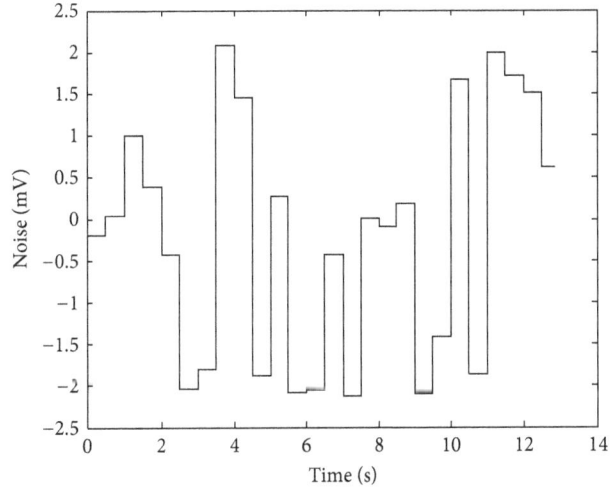

(b)

FIGURE 8: Noise source modelling in Verilog-A: (a) noise source and (b) waveform.

TABLE 3: Comparison of the two methods of noise analysis.

T_{int}	100 ns	1 us	10 us	100 us	150 us	200 us
Theoretical results	28.03	29.39	29.41	31.20	34.02	46.62
SIMULINK results	27.44	28.57	28.82	30.71	34.05	47.52
Verilog-A results	27.19	29.11	28.23	30.26	32.80	49.57

$$H_{A_{1,2}}(f) = \frac{1}{1 + f/\mathrm{GBW}} \cdot \frac{sC_{\mathrm{SH}}R_{\mathrm{on}\,2}}{1 + sC_{\mathrm{SH}}(R_o + R_{\mathrm{on}\,1} + R_{\mathrm{on}\,2})}. \tag{A.2}$$

The model of noise source of S_1 transmitting to the nodes across C_s is shown in Figure 10, whose transfer functions that appear in (5) are given by (A.3) and (A.4), respectively:

$$H_{1,1}(f) = \frac{1 + sC_{\mathrm{SH}}R_{\mathrm{on}\,2}}{1 + sC_{\mathrm{SH}}(R_o + R_{\mathrm{on}\,1} + R_{\mathrm{on}\,2})}, \tag{A.3}$$

$$H_{1,2}(f) = \frac{sC_{\mathrm{SH}}R_{\mathrm{on}\,2}}{1 + sC_{\mathrm{SH}}(R_o + R_{\mathrm{on}\,1} + R_{\mathrm{on}\,2})}. \tag{A.4}$$

The model of noise source of S_1 transmitting to the nodes across C_s is shown in Figure 11, whose transfer functions that appear in (5) are given by (A.5) and (A.6), respectively:

$$H_{2,1}(f) = \frac{sC_{\mathrm{SH}}(R_o + R_{\mathrm{on}\,1})}{1 + sC_{\mathrm{SH}}(R_o + R_{\mathrm{on}\,1} + R_{\mathrm{on}\,2})}, \tag{A.5}$$

$$H_{2,2}(f) = \frac{1 + sC_{\mathrm{SH}}(R_o + R_{\mathrm{on}\,1})}{1 + sC_{\mathrm{SH}}(R_o + R_{\mathrm{on}\,1} + R_{\mathrm{on}\,2})}. \tag{A.6}$$

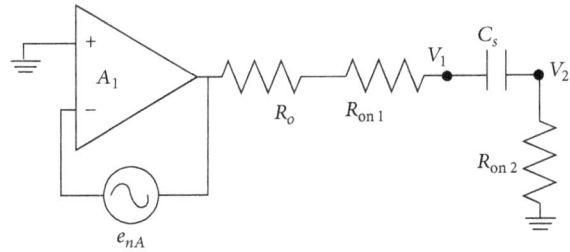

FIGURE 9: Transfer function of noise source of A_1.

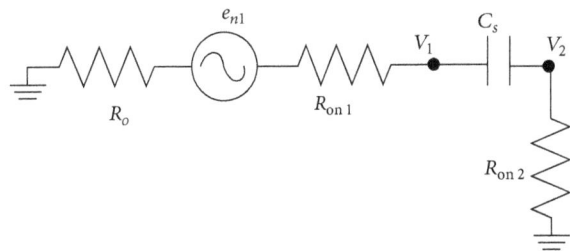

FIGURE 10: Transfer function of noise source of S_1.

Conflict of Interests

The authors declare that there is no conflict of interests regarding the publication of this paper.

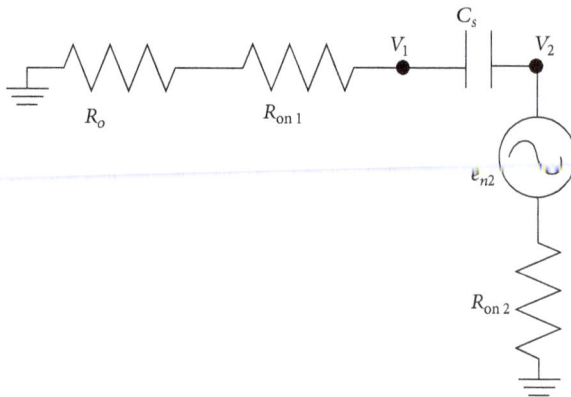

FIGURE 11: Transfer function of noise source of S_2.

Acknowledgment

The authors thank the Key Laboratory of Opto-Electronic Information Processing for the continuous supporting of the research on noise of readout integrated circuits.

References

[1] E. R. Fossum and B. Pain, "Infrared readout electronics for space-science sensors: state of the art and future directions," in *Infrared Technology*, vol. 2020 of *Proceedings of SPIE*, pp. 262–285, November 1993.

[2] R. Richwine, R. Balcerak, C. Rapach, K. Freyvogel, and A. Sood, "A comprehensive model for bolometer element and uncooled array design and imaging sensor performance prediction," in *Infrared and Photoelectronic Imagers and Detector Devices II*, vol. 6294 of *Proceedings of SPIE*, August 2006.

[3] C.-C. Hsieh, C.-Y. Wu, F.-W. Jih, and T.-P. Sun, "Focal-plane-arrays and CMOS readout techniques of infrared imaging systems," *IEEE Transactions on Circuits and Systems for Video Technology*, vol. 7, no. 4, pp. 594–605, 1997.

[4] B. H. Calhoun, A. Wang, and A. Chandrakasan, "Device sizing for minimum energy operation in subthreshold circuits," in *Proceedings of the IEEE Custom Integrated Circuits Conference (CICC '04)*, pp. 95–98, October 2004.

[5] Y. Tsividis, *Operation and Modeling of the MOS Transistor*, McGraw-Hill, New York, NY, USA, 2nd edition, 1999.

[6] B. Razavid, *Design of Analog CMOS Integrated Circuits*, McGraw, New York, NY, USA, 2000.

[7] P. R. Gray, P. J. Hurst, S. H. Lewis, and R. G. Meyer, *Analysis and Design of Analog Integrated Circuits*, Wiley, 2008.

[8] R. Schreier, J. Silva, J. Steensgaard, and G. C. Temes, "Design-oriented estimation of thermal noise in switched-capacitor circuits," *IEEE Transactions on Circuits and Systems I: Regular Papers*, vol. 52, no. 11, pp. 2358–2368, 2005.

[9] J. F. Johnson and T. S. Lomheim, "Hybrid infrared focal-plane signal and noise modeling," in *Infrared Sensors: Detectors, Electronics, and Signal Processing*, vol. 1541 of *Proceedings of SPIE*, pp. 110–126, July 1991.

[10] Z. C. Butler and N. V. Amarasinghe, "Random telegraph signals in deep submicron metal-oxide-semiconductor field-effect transistors," in *Noise and Fluctuations Control in Electronic Devices*, pp. 187–199, American Scientific, 2002.

[11] N. Kawai and S. Kawahito, "Noise analysis of high-gain, low-noise column readout circuits for CMOS image sensors," *IEEE Transactions on Electron Devices*, vol. 51, no. 2, pp. 185–194, 2004.

## 14

Engineering Change Orders Design Using Multiple Variables Linear Programming for VLSI Design

Yu-Cheng Fan, Chih-Kang Lin, Shih-Ying Chou, Chun-Hung Wang, Shu-Hsien Wu, and Hung-Kuan Liu

Department of Electronic Engineering, National Taipei University of Technology, Taipei 10608, Taiwan

Correspondence should be addressed to Yu-Cheng Fan; skystar@ntut.edu.tw

Academic Editor: Chih-Cheng Lu

An engineering change orders design using multiple variable linear programming for VLSI design is presented in this paper. This approach addresses the main issues of resource between spare cells and target cells. We adopt linear programming technique to plan and balance the spare cells and target cells to meet the new specification according to logic transformation. The proposed method solves the related problem of resource for ECO problems and provides a well solution. The scheme shows new concept to manage the spare cells to meet possible target cells for ECO research.

1. Introduction

Engineering change orders (ECO) are important technologies used for changes in integrated circuit (IC) layout and compensate for design problems. Traditionally, when chip shows errors, it often requires new photomasks for all layers. However, photomasks of deep-submicron semiconductor fabrication process are very expensive. In order to save money, ECO technology modifies only a few of the metal layers (metal-mask ECO) to reduce the cost of photomasks for all layers [1].

To perform the ECO, IC designers adopt sprinkling many unused logic gates during IC design flow. When chip is manufactured and shows design errors, IC designers modify the gate-level net-list using the presprinkling unused logic gates. At the same time, the designers track and verify the modification to check formal equivalence after ECO process. The designers must guarantee the revised design matching the revised specification.

How to achieve ECO efficiently? There are some literatures that address this problem and provide related solution. In literature [2], Tan and Jiang describe a typical metal-only ECO flow with four steps that include placement and spare cell distribution, logic difference extraction, metal-only

ECO synthesis, and ECO routing [2]. Kuo et al. insert spare cells with constant insertion for engineering change and describe an iterative method to determine feasible mapping solutions for an EC problem [3]. Besides, in order to perform ECO efficiently, literature [4–9] adopt minimal change EC equations automatically. Brand proposed incremental synthesis method [4]. Huang presented a hybrid tool for automatic logic rectification [5]. Lin et al. addressed logic synthesis techniques for engineering change problems [6]. Shinsha et al. performed incremental logic synthesis through gate logic structure identification [7]. Swamy et al. achieved minimal logic resynthesis for engineering change [8]. Watanabe and Brayton presented another kind of incremental synthesis technique for engineering changes [9]. However, few researchers discuss the resource between spare cells and target cells. Therefore, in order to solve the problems, we adopt linear programming technique to plan and balance the spare cells and target cells in this paper. The proposed scheme meets the new specification according to logic transformation and overcomes the related problems of resource for ECO research.

This paper is organized as follows. In Section 2, we address typical ECO design flow. In Section 3, logic transformation is discussed. In Section 4, multiple variables linear

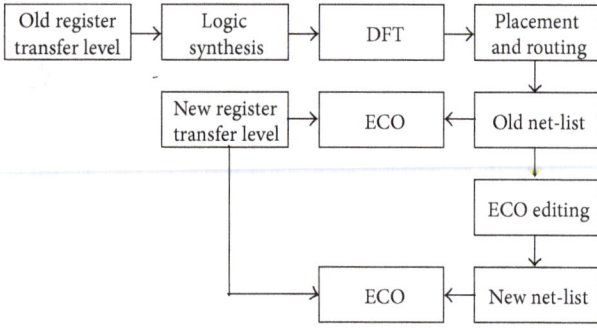

FIGURE 1: A typical ECO design flow.

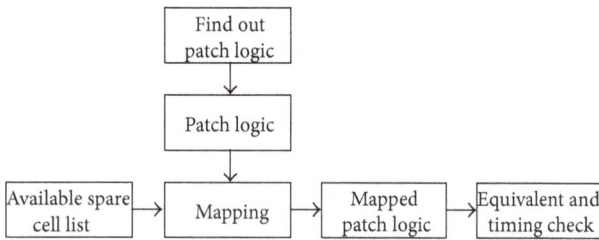

FIGURE 2: Two-phase ECO design flow.

programming for VLSI design is presented. In Section 5, we discuss the advantage and disadvantage of the related works. Finally, we conclude this paper in Section 6.

2. Typical ECO Design Flow

Before describing the proposed method, we address a typical manual ECO design flow in Figure 1. IC designers perform the change in register transfer level and verify fixed code matching the new specification at first. Then, old net-list is scanned to search the possible fix points. After the possible fix points are searched, IC designers modify the net-list and check the functionally equivalent between new net-list and new register transfer level [10–14].

Next, we describe two-phase ECO design flow in Figure 2. To patch the logics of the modified circuit, we prepare available spare cell list. According to logic function, the modified circuit is mapped to specified logics. After patching logic, equivalent check and timing check are performed to make sure that the new function met the new specification.

However, there are some important problems that appear during patching logic. Are there enough spare cells and types to satisfy the consumption of patch logic? How to estimate the quantity and logic types of ECO procedure? In order to solve this problem, we proposed an engineering change orders design using multiple variables linear programming for VLSI design in this paper.

3. Logic Transformation

Before discussing the engineering change orders design using multiple variables linear programming, we addressed ECO logic transformation. Figure 3 describes an ECO problem with an equation out $= (A + (BC)')'$. Figure 3(b) lists the available spare cells. According to the list, we discover the available spare cells are not enough. In order to solve the problem, we adopt another mapping solution with an equation out $= (A'BC)$ instead of the original equation in Figure 3(c). It requires one AND and one INV gate. The mapping solution in Figure 3(c) requires gates fewer than the available spare cells and is constructed with the available spare cells.

However, most of spare cells only provide basic logical functions that include AND, OR, NOT, NAND, and NOR. Half Adder (HA), Full Adder (FA), And-Or-Inverter (AOI), and Or-And-Inverter (OAI) can provide complex logical functions. We can adopt these logical cells to perform ECO function. For example, AOI22 can be implemented by two NAND and one AND cells in Figure 4. According to the existing resources of spare cells, we can resynthesize the changed function lists.

4. Multiple Variables Linear Programming for VLSI Design

Although logic transformation skill makes the ECO technology come true, a chip often does not own enough spare cells to modify the function to meet a new specification. How to allocate limited resource? We should estimate quantity of spare cells and logic transformation rule to perform optimal engineering charge orders.

In Figure 5, it describes the engineering change orders design using multiple variables linear programming for VLSI design and relation of logic transformation. "Logic A" is one kind of spare cells that can be transformed into "Logic a" or "Logic b." Similarly, "Logic B" can be transformed into "Logic a," "Logic b," or "Logic c." "Logic C" performs ECO function instead of "Logic c" or "Logic d." Besides, Logic D is transformed into "Logic c" or "Logic d." Equivalently, "Logic E" is transformed to "Logic d" or "Logic e" to achieve ECO function.

We assume X_1, X_2, X_3, X_4, and X_5 are the number of spare cells, Logic A, Logic B, Logic C, Logic D, and Logic E. Let Y_1, Y_2, Y_3, Y_4, and Y_5 be the desired number of target cells, Logic a, Logic b, Logic c, Logic d, and Logic e.

Besides, Aa is the number of spare cells (Logic A) to be transformed into Logic a and Ab is the number of spare cells (Logic A) to be transformed into Logic b. Similarly, Ba, Bb, and Bc are the number of spare cells (Logic B) to be transformed into Logic a, Logic b, and Logic c. In a similar way, Cc and Cd are the number of spare cells (Logic C) to be transformed into Logic c and Logic d, Dc and Dd are the number of spare cells (Logic D) to be transformed into Logic c and Logic d, and Ed and Ee are the number of spare cells (Logic E) to be transformed into Logic d and Logic e.

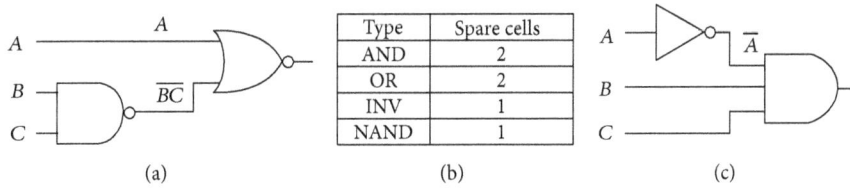

FIGURE 3: Example of an ECO problem. (a) EC equation: output $= (A + (BC)')'$. (b) Spare cells. (c) Mapping: output $= (A'BC)$.

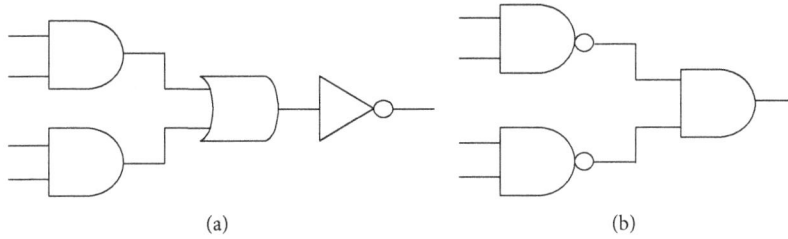

FIGURE 4: AOI22 can be implemented by two NAND and one AND cells.

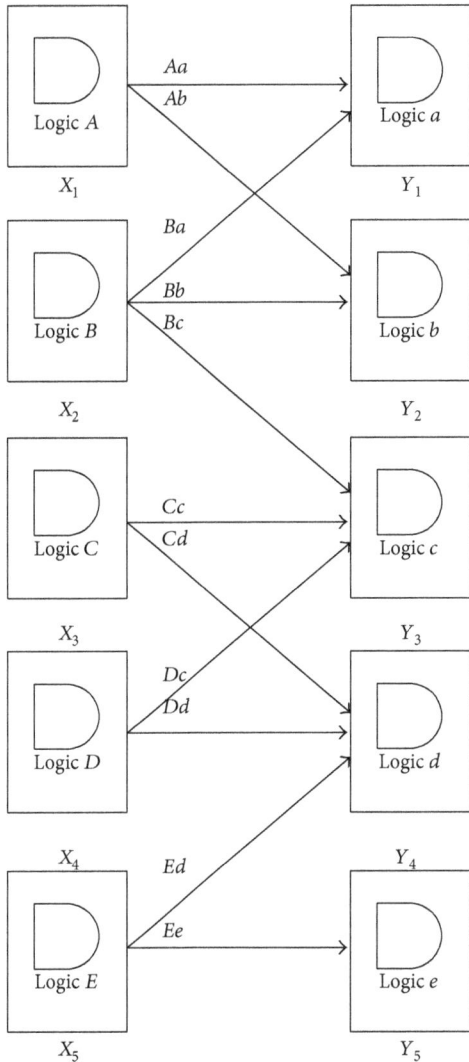

FIGURE 5: ECO design using multiple variables linear programming for VLSI design and relation of logic transformation.

Therefore, the restriction rule of the number of spare cells and transformed target cells in Figure 5 is written as follows:

$$X_1 = Aa + Ab;$$
$$X_2 = Ba + Bb + Bc;$$
$$X_3 = Cc + Cd; \tag{1}$$
$$X_4 = Dc + Dd;$$
$$X_5 = Ed + Ee.$$

Besides, the restriction rule of the engineering change orders design using multiple variables linear programming in Figure 5 is written as follows:

$$Aa + Ba \geq Y_1; \tag{2}$$
$$Ab + Bb \geq Y_2; \tag{3}$$
$$Bc + Cc + Dc \geq Y_3; \tag{4}$$
$$Cd + Dd + Ed \geq Y_4; \tag{5}$$
$$Ed + Ee \geq Y_5. \tag{6}$$

However, spare cells are not often enough; designer should balance the spare cell allocation to meet all requirements of desirable cells.

We assume one case when $Bb \leq Y_2$. In order to provide enough spare cells, we should increase the number of Ab to achieve $Ab + Bb \geq Y_2$.

Similarly, when $Cc + Dc \leq Y_3$, we should increase Bc number to meet $Bc + Cc + Dc \geq Y_3$.

Therefore, we define another restriction rule of the engineering change orders design which is written as follows:

$$Aa = X_1 - Ab;$$
$$Ba = X_2 - Bb - Bc. \tag{7}$$

TABLE 1: ECO methods comparison.

Method	Traditional ECO	Proposed method
Cell resource prediction	Constraint based	Multiple variables linear programming
Predictive precision of patching logic number	Normal precision	High precision
Balance between spare cells and target cells	Low balance	High balance
Restriction rule	Not define	Define
Resource optimization	Not define	Define
Solution boundary	Not define	Define

According to formulas (2) and (7), we can balance the number of Ab, Bb, and Bc to achieve the target number Y_1. Consider the following:

$$X_1 - Ab + X_2 - Bb - Bc \geqq Y_1. \tag{8}$$

In a similar way, we define the restriction rule of the engineering change orders design which is written as follows:

$$Bc = X_2 - Ba - Bb;$$
$$Cc = X_3 - Cd; \tag{9}$$
$$Dc = X_4 - Dd.$$

According to formulas (4) and (9), we can balance the number of Bc, Cc, and Dc to achieve the target number Y_3. Consider

$$X_2 - Ba - Bb + X_3 - Cd + X_4 - Dd \geqq Y_3. \tag{10}$$

We model the engineering change orders problems using multiple variables linear programming. According to the functions, we can understand the engineering change orders relation between supply and requirement. Then, designer can estimate and perform ECO using spare cell efficiently.

5. Discussion

In this Section, we discuss the advantage and disadvantage of the related works. Table 1 shows ECO method comparison. The proposed approach designs a multiple variable linear programming ECO for VLSI design. Our method can predict cell resource accurately using multiple variable linear programming techniques. Traditional ECO is not to predict it well. Besides, our scheme provides a high accurate prediction of patching logic number to balance between spare cells and target cells. It is hard for traditional ECO method to do these. Moreover, we define restriction rule, resource optimization, and solution boundary of ECO problem to increase the efficiency of the proposed ECO method and provide a well solution.

6. Conclusion

In this paper, we proposed an engineering change orders design using multiple variables linear programming for VLSI design. The paper discusses typical ECO design flow, logic transformation, and multiple variables linear programming for VLSI design. The presented scheme estimates the resource of spare cells and provides a well solution of ECO problems.

Conflict of Interests

The authors declare that there is no conflict of interests regarding the publication of this paper.

Acknowledgments

This work was supported by the National Science Council of Taiwan under Grant nos. NSC 101-2221-E-027-135-MY2 and 102-2622-E-027-008-CC3. The authors gratefully acknowledge the Chip Implementation Center (CIC), for supplying the technology models used in IC design.

References

[1] J. A. Roy and I. L. Markov, "ECO-system: embracing the change in placement," IEEE Transactions on Computer-Aided Design of Integrated Circuits and Systems, vol. 26, no. 12, pp. 2173–2185, 2007.

[2] C. Tan and I. H. Jiang, "Recent research development in metal-only ECO," in Proceedings of the 54th IEEE International Midwest Symposium on Circuits and Systems (MWSCAS '11), pp. 1–4, August 2011.

[3] Y. M. Kuo, Y. T. Chang, S. C. Chang, and M. Marek-Sadowska, "Spare cells with constant insertion for engineering change," IEEE Transactions on Computer-Aided Design of Integrated Circuits and Systems, vol. 28, no. 3, pp. 456–460, 2009.

[4] D. Brand, A. Drumm, S. Kundu, and P. Narain, "Incremental synthesis," in Proceedings of the IEEE/ACM International Conference on Computer-Aided Design, pp. 14–18, 1994.

[5] S. Huang, K. Chen, and K. Cheng, "AutoFix: A hybrid tool for automatic logic rectification," IEEE Transactions on Computer-Aided Design of Integrated Circuits and Systems, vol. 18, no. 9, pp. 1376–1384, 1999.

[6] C. Lin, K. Chen, and M. Marek-Sadowska, "Logic synthesis for engineering change," IEEE Transactions on Computer-Aided Design of Integrated Circuits and Systems, vol. 18, no. 2-3, pp. 282–292, 1999.

[7] T. Shinsha, T. Kubo, Y. Sakataya, J. Koshishita, and K. Ishihara, "Incremental logic synthesis through gate logic structure identification," in Proceedings of the IEEE/ACM Conference on Design Automation, pp. 391–397, Jun 1986.

[8] G. Swamy, S. Rajamani, C. Lennard, and R. K. Brayton, "Minimal logic re-synthesis for engineering change," in Proceedings of

the IEEE International Symposium on Circuits and Systems, pp. 1596–1599, 1997.

[9] Y. Watanabe and R. K. Brayton, "Incremental synthesis for engineering changes," in *Proceedings of the IEEE International Conference on Computer Design: VLSI in Computers and Processors (ICCD '91)*, pp. 40–43, Cambridge, Mass, USA, October 1991.

[10] J. Wang, *Finding the Minimal Logic Difference for Functional ECO*, Taiwan Cadence Design Systems, 2012.

[11] Y. C. Fan and H. W. Tsao, "Watermarking for intellectual property protection," *IEE Electronics Letters*, vol. 39, no. 18, pp. 1316–1318, 2003.

[12] Y. Fan and H. Tsao, "Boundary scan test scheme for IP core identification via watermarking," *IEICE Transactions on Information and Systems*, vol. E88-D, no. 7, pp. 1397–1400, 2005.

[13] Y. Fan, "Testing-based watermarking techniques for intellectual -property identification in SOC design," *IEEE Transactions on Instrumentation and Measurement*, vol. 57, no. 3, pp. 467–479, 2008.

[14] Y. Fan and Y. Chiang, "Discrete wavelet transform on color picture interpolation of digital still camera," *VLSI Design*, vol. 2013, Article ID 738057, 9 pages, 2013.

## 15

Novel Receiver Architecture for LTE-A Downlink Physical Control Format Indicator Channel with Diversity

S. Syed Ameer Abbas,[1] S. J. Thiruvengadam,[2] and S. Susithra[1]

[1] Department of Electronics and Communication Engineering, Mepco Schlenk Engineering College, Sivakasi 626 005, India
[2] Department of Electronics and Communication Engineering, Thiagarajar College of Engineering, Madurai 625 015, India

Correspondence should be addressed to S. Syed Ameer Abbas; abbas_mepco@yahoo.com

Academic Editor: Chien-In Henry Chen

Physical control format indicator channel (PCFICH) carries the control information about the number of orthogonal frequency division multiplexing (OFDM) symbols used for transmission of control information in long term evolution-advanced (LTE-A) downlink system. In this paper, two novel low complexity receiver architectures are proposed to implement the maximum likelihood- (ML-) based algorithm which decodes the CFI value in field programmable gate array (FPGA) at user equipment (UE). The performance of the proposed architectures is analyzed in terms of the timing cycles, operational resource requirement, and resource complexity. In LTE-A, base station and UE have multiple antenna ports to provide transmit and receive diversities. The proposed architectures are implemented in Virtex-6 xc6vlx240tffl156-1 FPGA device for various antenna configurations at base station and UE. When multiple antenna ports are used at base station, transmit diversity is obtained by applying the concept of space frequency block code (SFBC). It is shown that the proposed architectures use minimum number of operational units in FPGA compared to the traditional direct method of implementation.

1. Introduction

The goal of third generation partnership project (3GPP) long term evolution-advanced (LTE-A) wireless standard is to increase the capacity and speed of wireless data communication. The LTE-A physical layer is a highly efficient means of conveying both data and control information between an enhanced base station, popularly known as eNodeB, and mobile user equipment (UE). It supports both frequency division duplex (FDD) and time division duplex (TDD) configurations in uplink and downlink operations. Further, it provides a wide range of system bandwidths in order to operate in a large number of different spectrum allocations [1].

LTE-A standard has six physical channels for downlink. They are physical broadcast channel (PBCH), physical downlink shared channel (PDSCH), physical multicast channel (PMCH), physical downlink control channel (PDCCH), physical hybrid automatic repeat request (ARQ) indicator channel (PHICH), and physical control format indicator channel (PCFICH). PBCH carries the basic system information for the other channels to be configured and operated in the LTE-A grid. The PDSCH is the main data-bearing channel. PMCH is defined for future use. In LTE-A, the control signals are transmitted at the start of each subframe in the LTE-A grid. PDCCH is used to carry the scheduling information of different types such as downlink resource scheduling and uplink power control instructions. PHICH is used to send the acknowledgement/negative acknowledgement bit to UEs to indicate whether the uplink user data is correctly received or not. PCFICH carries the control information about the number of orthogonal frequency division multiplexing (OFDM) symbols used for transmission of downlink control information. The high data rate in LTE-A requires high processing demands on all layers of the system which includes high digital signal processing (DSP) hardware processing in the physical layer. Further, the hardware implementation of receiver structures of various physical channels in LTE-A becomes a challenging task as the computational complexity increases.

In [2], receivers were designed for a 2×2 antenna system and for quadrature phase shift keying (QPSK) modulation and quadrature amplitude modulation (16-QAM and 64-QAM). Though successive interference cancellation (SIC) receiver meets the timing requirements in the LTE system, it is complex and the K-best list sphere detector (K-LSD) receiver has high latency. In [3], field programmable gate array (FPGA) and application specific integrated circuit (ASIC) implementations of receivers based on the linear minimum mean-square error (LMMSE), the K-LSD, iterative successive interference cancellation (SIC) detector, and the iterative K-LSD algorithms are carried out for spatial multiplexing based LTE-A system. The SIC algorithm is found to perform worse than the K-LSD when the MIMO channels are highly correlated, while the performance difference diminishes when the correlation decreases. The ASIC receivers are designed to meet the decoding throughput requirements in LTE and the K-LSD is found to be the most complex receiver although it gives the best reliable data transmission throughput. It is shown that the receiver architecture which could be reconfigured to use a simple or a more complex detector as the channel conditions change would achieve the best performance while consuming the least amount of power in the receiver. FPGA implementation of MIMO detector based on two typical sphere decoding algorithms, namely, the Viterbo-Boutros (VB) algorithm and the Schnorr-Euchner (SE) algorithm, is carried out in [4]. In this implementation method, three levels of parallelism are explored to improve the decoding rate: the concurrent execution of the channel matrix preprocessing on an embedded processor and the decoding functions on customized hardware modules, the parallel decoding of real/imaginary parts for complex constellation, and the concurrent execution of multiple steps during the closest lattice point search. The implementation of low-complexity codebook searching engine is proposed to support both LTE and LTE-A operations [5]. In [6], VLSI implementation of a low-complexity multiple input multiple output (MIMO) symbol detector based on a novel MIMO detection algorithm called modified fixed-complexity soft-output (MFCSO) detection is presented. It includes a microcode-controlled channel preprocessing unit, separate channel memory, and a pipelined detection unit. MATLAB-based downlink physical-layer simulator for LTE only for research applications is presented [7]. In [8], maximum likelihood- (ML-) based receiver structures are developed for decoding the downlink control channels PCFICH and PHICH in LTE wireless standard and the performance of the receivers has been analyzed for various configurations. The analytical results were validated against computer simulations but hardware implementation of the structures was not coded or synthesized. In [9], direct implementation of receive algorithms was carried out in FPGA for downlink control channels in LTE. However, most of these works either propose architectures for FPGA implementation or analyze the performance of various receiver structures in a generalized manner. The objective of this paper is to propose novel architectures for FPGA implementation of transmit and receive processing of downlink PCFICH channel in LTE-A standard in particular.

TABLE 1: CFI 32-bit block code.

CFI	$\langle b_{31}, \ldots, b_0 \rangle$
1	01101101101101101101101101101101
2	10110110110110110110110110110110
3	11011011011011011011011011011011
4	00000000000000000000000000000000

1.1. Transmit and Receive Processing of PCFICH. In PCFICH, the control format indicator (CFI) contains a 32-bit code word that represents the value of CFI as 1, 2, 3, or 4. The CFI informs the UE about the number of OFDM symbols used for the transmission of PDCCH information in a subframe. The 32-bit code word corresponding to the value of CFI is scrambled and QPSK modulated. The resultant 16 QPSK complex symbols are mapped to the resource elements of the first OFDM symbol of every subframe after layer mapping and precoding to obtain transmit diversity when two or more antenna ports are used at eNodeB [10]. The 32-bit code words for the four possible values of CFI are given in Table 1. A general block diagram of the transmitter and receiver processing of PCFICH is shown in Figure 1.

The OFDM signal is transmitted through a frequency selective fading channel. It is assumed that the number of receive antenna ports at UE is K. At each receive antenna port of the UE resource-element demapping follows the cyclic prefix removal and fast fourier transformation (FFT). The 16×1 receive signal vector at each antenna port is equalized in frequency domain at each subcarrier using the corresponding 16×1 channel frequency response vector. The outputs of frequency domain equalizer from each antenna port are summed up. The resultant 16×1 complex vector is applied to the maximum likelihood (ML) detector for detecting the CFI value. The objective of this paper is to synthesize and implement the receiver architecture for PCFICH.

The paper is structured as follows. Section 2 explains the system model and basic implementation architectures for single input single output (SISO) and single input multiple output (SIMO) configurations. The system model and basic implementation architecture for multiple input single output (MISO) and multiple input multiple output (MIMO) configurations are described in Sections 3 and 4, respectively. The proposed implementation architectures using folding and superscalar methods are given in Section 5 for SISO, SIMO, MISO, and MIMO configurations. Section 6 analyzes the performance of the proposed architectures and Section 7 concludes the paper with remarks on future work.

2. System Model and Implementation Architecture for SISO and SIMO Configurations

The received signal model for SISO configuration of PCFICH is given by

$$\mathbf{y} = \mathbf{h} \circ \mathbf{d}^{(m)} + \mathbf{w}, \tag{1}$$

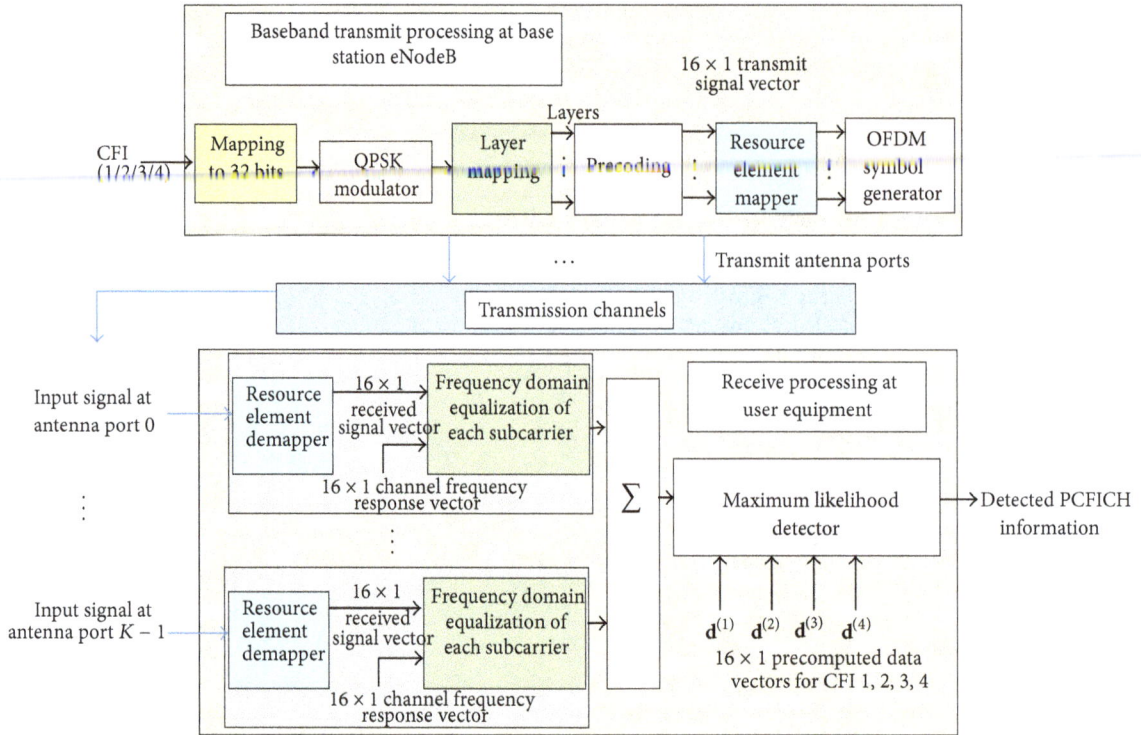

FIGURE 1: Block diagram of transmitter and receiver processing.

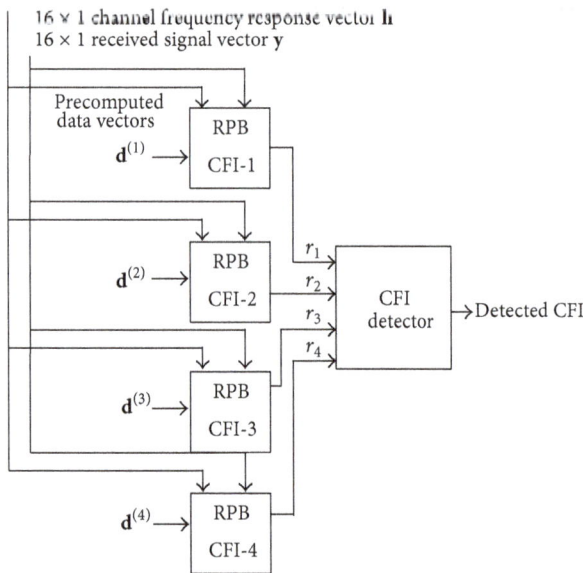

FIGURE 2: Basic architecture for SISO configuration.

where $\mathbf{y} = [y_0, y_1, \ldots, y_{15}]^T$ is a 16 × 1 received signal vector, $\mathbf{h} = [h_0, h_1, \ldots, h_{15}]^T$ is a 16 × 1 channel frequency response vector, $\mathbf{d}^{(m)} = [d_0^{(m)}, d_1^{(m)}, \ldots, d_{15}^{(m)}]^T$ is a 16 × 1 complex QPSK symbol vector corresponding to CFI value

from the set $\{1, 2, 3, 4\}$, "∘" represents the element by element multiplication, and $\mathbf{w}$ is a 16 × 1 additive white noise vector and its elements are zero mean Gaussian random numbers with unit variance. The objective is to detect the value of CFI from the received signal vector $\mathbf{y}$ assuming the channel frequency response vector $\mathbf{h}$ to be known. Using maximum likelihood (ML) principle, CFI is detected as

$$\widehat{\text{CFI}} = \underset{m}{\arg\min} \left\| \mathbf{y} - \mathbf{h} \circ \mathbf{d}^{(m)} \right\|^2. \qquad (2)$$

Figure 2 shows the basic architecture for estimating CFI using (2), in SISO configuration. The received signal vector $\mathbf{y}$ and the channel frequency response vector $\mathbf{h}$ are provided as input to the four receiver processing blocks (RPB) along with precomputed data vectors $\mathbf{d}^{(1)}, \mathbf{d}^{(2)}, \mathbf{d}^{(3)}$, and $\mathbf{d}^{(4)}$. The internal diagram for RPB CFI-1 is shown in Figure 3. It computes the expression $\left\| \mathbf{y} - \mathbf{h} \circ \mathbf{d}^{(1)} \right\|^2$ assuming the CFI = 1. In RPB-m, the precomputed data vector $\mathbf{d}^{(m)}$ is multiplied element by element with the channel frequency response vector. The resultant (16 × 1) vector is subtracted from the (16 × 1) received signal vector $\mathbf{y}$. The sum of squared magnitude of each element in the resultant vector is the output of RPB.

The inputs to the CFI detector are the 16-bit outputs of RPBs r_1, r_2, r_3, and r_4. The CFI detector determines which RPB output has minimum value. The internal diagram for CFI detector circuit which has 4 comparator modules (CM) is shown in Figure 4. In CM-1, input r_2 and one's complement

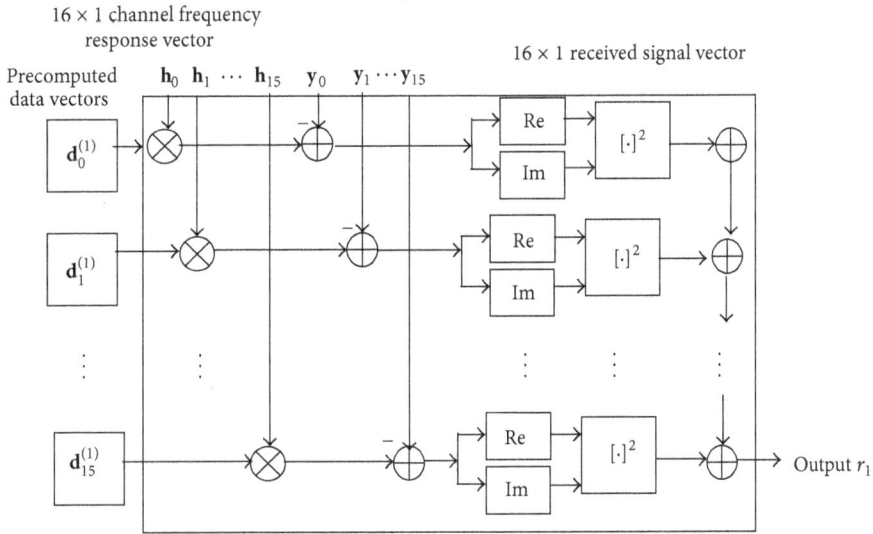

FIGURE 3: Internal architecture of receiver processing block (RPB CFI-1).

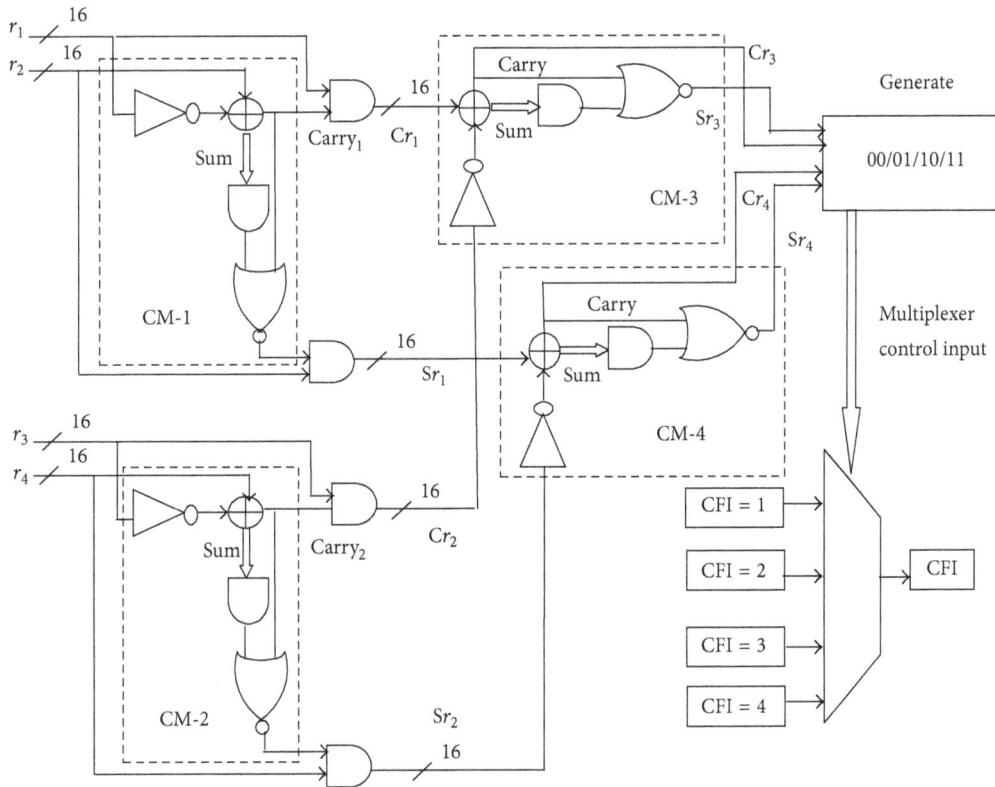

FIGURE 4: Internal architecture of CFI detector.

of input r_1 are added. If carry is generated, then r_1 is less than r_2. The outputs Cr_1 and Sr_1 of the CM-1 are defined as

$$Cr_1 = \begin{cases} r_1 & \text{if carry}_1 = \text{``1''} \, (r_1 < r_2) \\ 0 & \text{if carry}_1 = \text{``0''} \, (r_1 > r_2), \end{cases}$$

$$Sr_1 = \begin{cases} r_2 & \text{if carry}_1 = \text{``0''} \, (r_1 > r_2) \\ 0 & \text{if carry}_1 = \text{``1''} \, (r_1 < r_2). \end{cases} \tag{3}$$

In CM-2, input r_4 and one's complement of input r_3 are added. If carry is generated, then r_3 is less than r_4. The outputs Cr_2 and Sr_2 of CM-2 are defined as

$$Cr_2 = \begin{cases} r_3 & \text{if carry}_2 = \text{``1''} \, (r_3 < r_4) \\ 0 & \text{if carry}_2 = \text{``0''} \, (r_3 > r_4), \end{cases}$$

$$Sr_2 = \begin{cases} r_4 & \text{if carry}_2 = \text{``0''} \, (r_3 > r_4) \\ 0 & \text{if carry}_2 = \text{``1''} \, (r_3 < r_4). \end{cases} \tag{4}$$

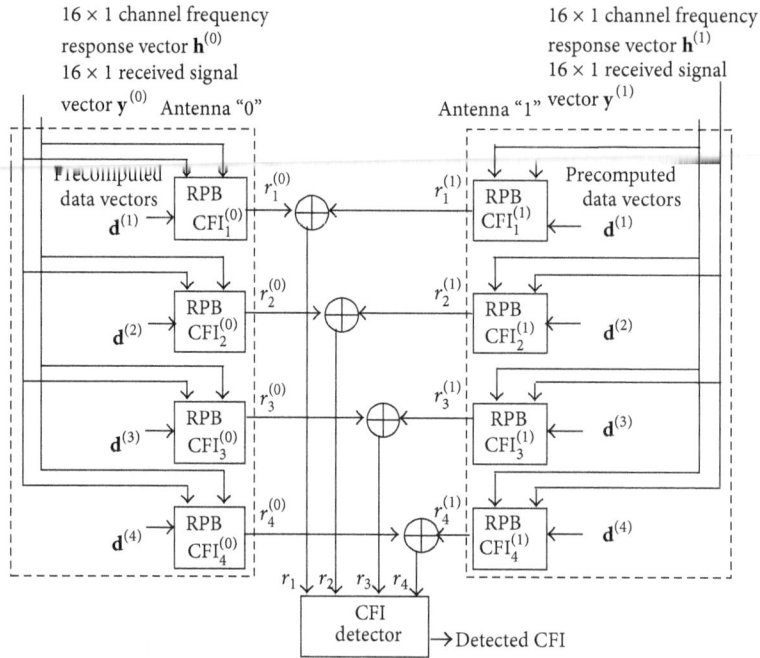

FIGURE 5: Basic architecture for SIMO configuration.

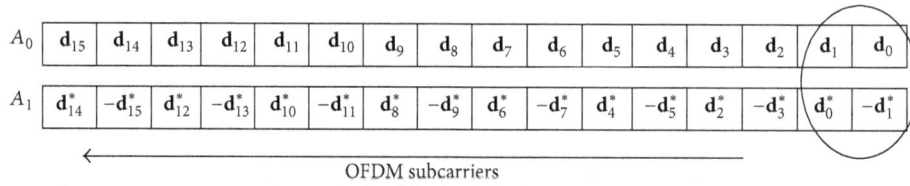

FIGURE 6: Subcarrier mapping for SFBC for 2-antenna system.

The multiplexer control input is activated based on the outputs from CM-3 and CM-4. One of the four outputs Cr_3, Sr_3, Cr_4, and Sr_4 would be "1" based on the minimum value of four inputs r_1, r_2, r_3, and r_4, respectively. Based on this 00, 01, 10, or 11 in the multiplexer control unit would be activated to obtain the detected CFI value.

In SIMO, the 16×1 receive signal vector at the kth receive antenna is modeled as

$$\mathbf{y}^{(k)} = \mathbf{h}^{(k)} \circ \mathbf{d}^{(m)} + \mathbf{w}_k, \quad k = 0, 1, 2, \ldots, K-1, \quad (5)$$

where "K" represents the number of receive antennas at UE, $\mathbf{h}^{(k)}$ is 16×1 channel frequency response vector between the transmit antenna and kth receive antenna, and $\mathbf{w}_k$ is 16×1 noise vector at kth receive antenna. Now, the objective is to detect the value of CFI from the received signal vectors at each receive antenna, assuming the channel frequency response vectors at each receive antenna are known. The maximal ratio combining is carried out at the receiver. Using maximum likelihood (ML) principle, CFI is estimated as [9]

$$\widehat{\mathrm{CFI}} = \min_{m=1,2,3,4} \sum_{k=1}^{K} \left\| \mathbf{y}^{(k)} - \left(\mathbf{h}^{(k)} \circ \mathbf{d}^{(m)} \right) \right\|^2. \quad (6)$$

The basic architecture for estimating CFI using (6) in 1×2 SIMO configuration shown in Figure 5 is similar to the basic architecture of SISO configuration. The received signal vector $\mathbf{y}^{(k)}$ and the channel frequency response vector $\mathbf{h}^{(k)}$ are provided as input to the four receiver processing blocks (RPB-CFI$_m^{(k)}$) at kth receive antenna, along with precomputed data vectors $\mathbf{d}^{(1)}$, $\mathbf{d}^{(2)}$, $\mathbf{d}^{(3)}$, and $\mathbf{d}^{(4)}$. The outputs from the mth RPB at 0th receive antenna $r_m^{(0)}$ and 1st receive antenna $r_m^{(1)}$ are added to get the mth input r_m of the CFI detector circuit.

3. System Model and Implementation Architecture for MISO Configuration

In MISO and MIMO configurations, space frequency block code (SFBC) based layer mapping and precoding are carried out to obtain transmit diversity when two or more antenna ports are used at eNodeB as per the 3GPP LTE wireless standard [1, 11]. It is assumed that 2 antenna ports are used at eNodeB. The 16×1 complex symbol vector output of the modulation mapper is applied to the layer mapper. The 8×1 symbol vectors at layer 0 and layer 1 are given by $[d_0, d_2, d_4, d_6, d_8, d_{10}, d_{12}, \text{ and } d_{14}]$ and $[d_1, d_3, d_5, d_7, d_9, d_{11}, d_{13}, \text{ and } d_{15}]$. The precoding is carried out using

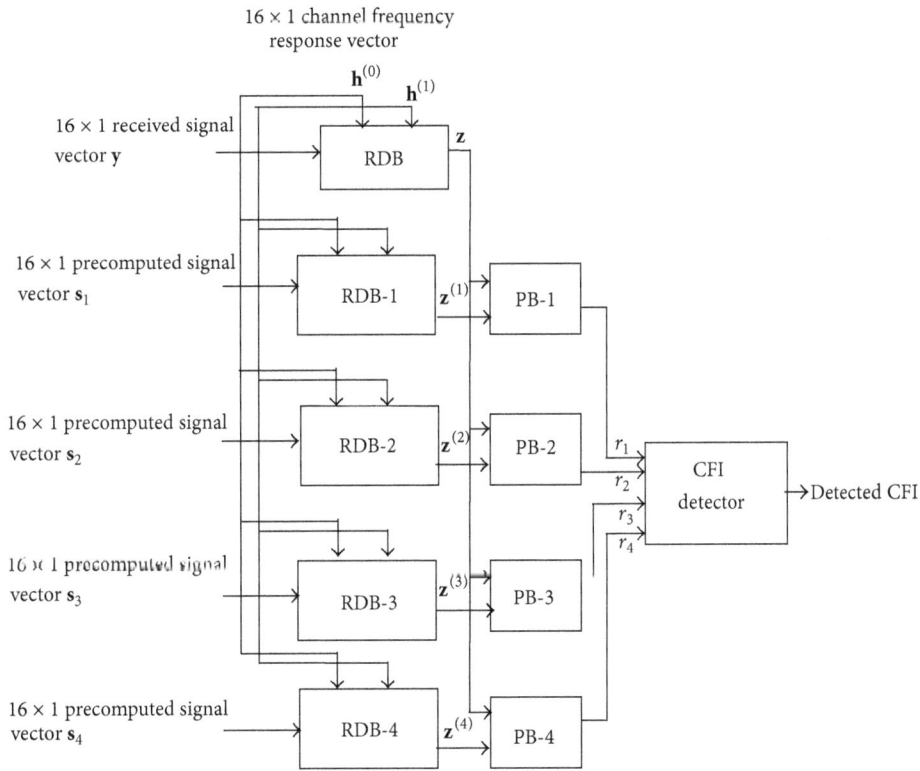

FIGURE 7: Proposed MISO receiver architecture for PCFICH.

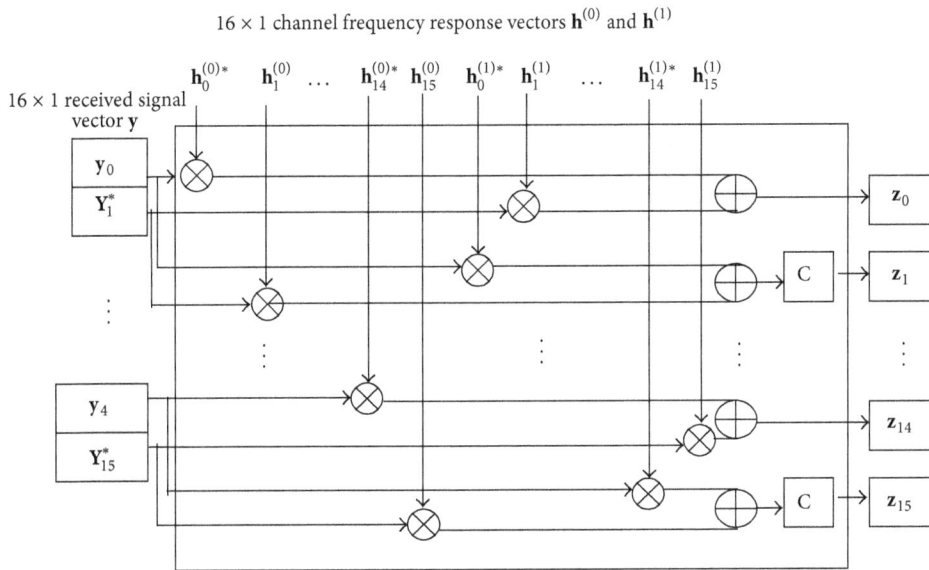

FIGURE 8: Internal architecture of MISO receiver decoding block (RDB).

the SFBC in the LTE-A standard. The precoder output at antenna port 0 (A_0) and antenna port 1 (A_1) is shown in Figure 6.

The notation "*" represents the complex conjugate of the symbol. Basically, in precoding, a symbol d_0 from layer 0 and a symbol d_1 from layer 1 are encoded such that the antenna output is formulated using the orthogonal matrix given by

$$A = \begin{bmatrix} d_0 & d_1 \\ -d_1^* & d_0^* \end{bmatrix}. \tag{7}$$

Precomputed data

$$\mathbf{z}_0 \mathbf{z}_1 \cdots \mathbf{z}_{14} \mathbf{z}_{15}$$

Output vector from RDB1

$\mathbf{z}_0^{(1)}$

$\mathbf{z}_1^{(1)}$

$\mathbf{z}_{14}^{(1)}$

$\mathbf{z}_{15}^{(1)}$

Im

Re

Im

Re

Im

Re

Im

Re

$[\cdot]^2$

$[\cdot]^2$

$[\cdot]^2$

$[\cdot]^2$

Output r_1

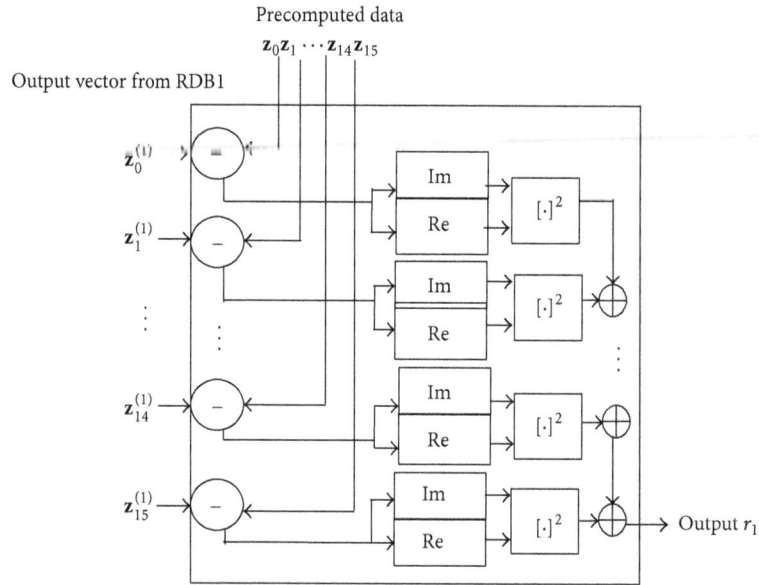

FIGURE 9: Internal architecture of MISO processing block (PB-1).

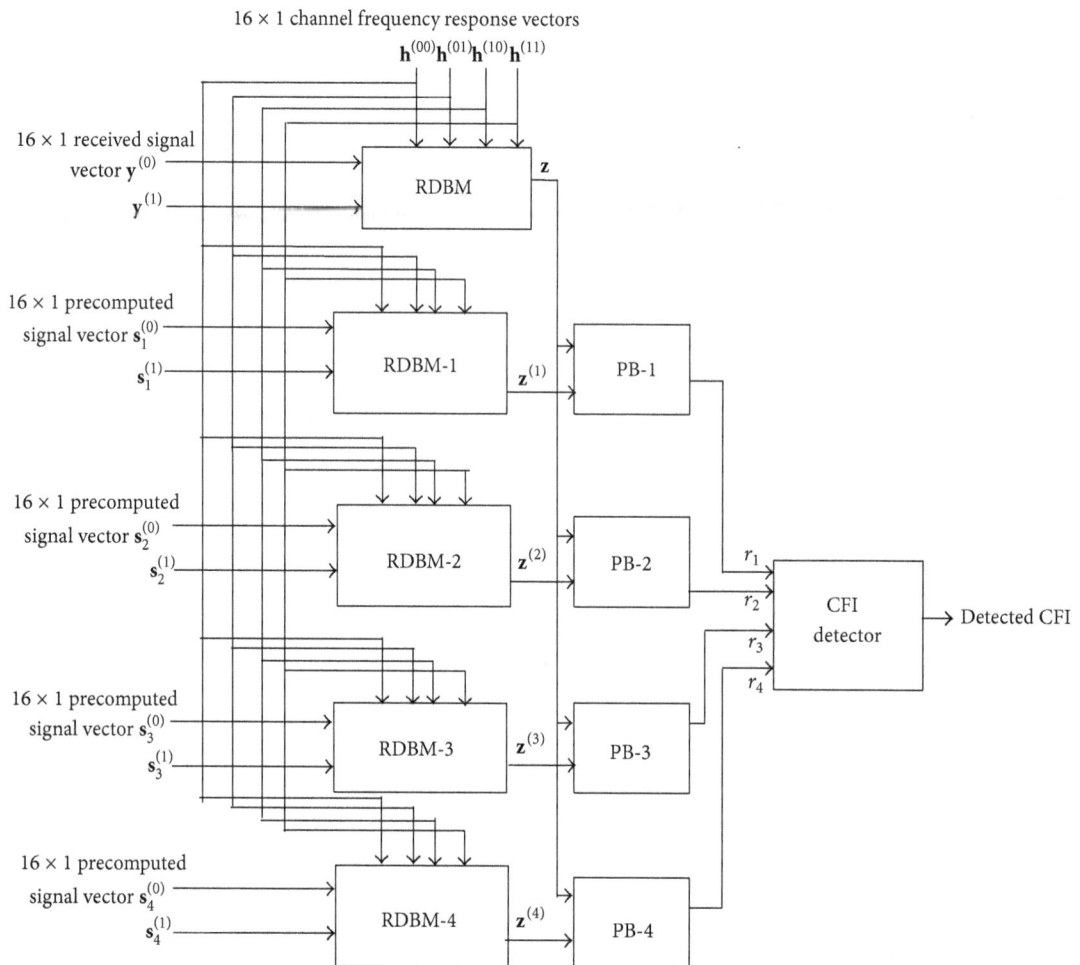

16×1 channel frequency response vectors

$$\mathbf{h}^{(00)} \mathbf{h}^{(01)} \mathbf{h}^{(10)} \mathbf{h}^{(11)}$$

16×1 received signal vector $\mathbf{y}^{(0)}$

$\mathbf{y}^{(1)}$

RDBM

$\mathbf{z}$

16×1 precomputed signal vector $\mathbf{s}_1^{(0)}$

$\mathbf{s}_1^{(1)}$

RDBM-1

$\mathbf{z}^{(1)}$

PB-1

16×1 precomputed signal vector $\mathbf{s}_2^{(0)}$

$\mathbf{s}_2^{(1)}$

RDBM-2

$\mathbf{z}^{(2)}$

PB-2

r_1

r_2

r_3

r_4

CFI detector

Detected CFI

16×1 precomputed signal vector $\mathbf{s}_3^{(0)}$

$\mathbf{s}_3^{(1)}$

RDBM-3

$\mathbf{z}^{(3)}$

PB-3

16×1 precomputed signal vector $\mathbf{s}_4^{(0)}$

$\mathbf{s}_4^{(1)}$

RDBM-4

$\mathbf{z}^{(4)}$

PB-4

FIGURE 10: Proposed MIMO receiver architecture for PCFICH.

16 × 1 channel frequency response vectors $\mathbf{h}^{(00)}, \mathbf{h}^{(01)}, \mathbf{h}^{(10)}$, and $\mathbf{h}^{(11)}$

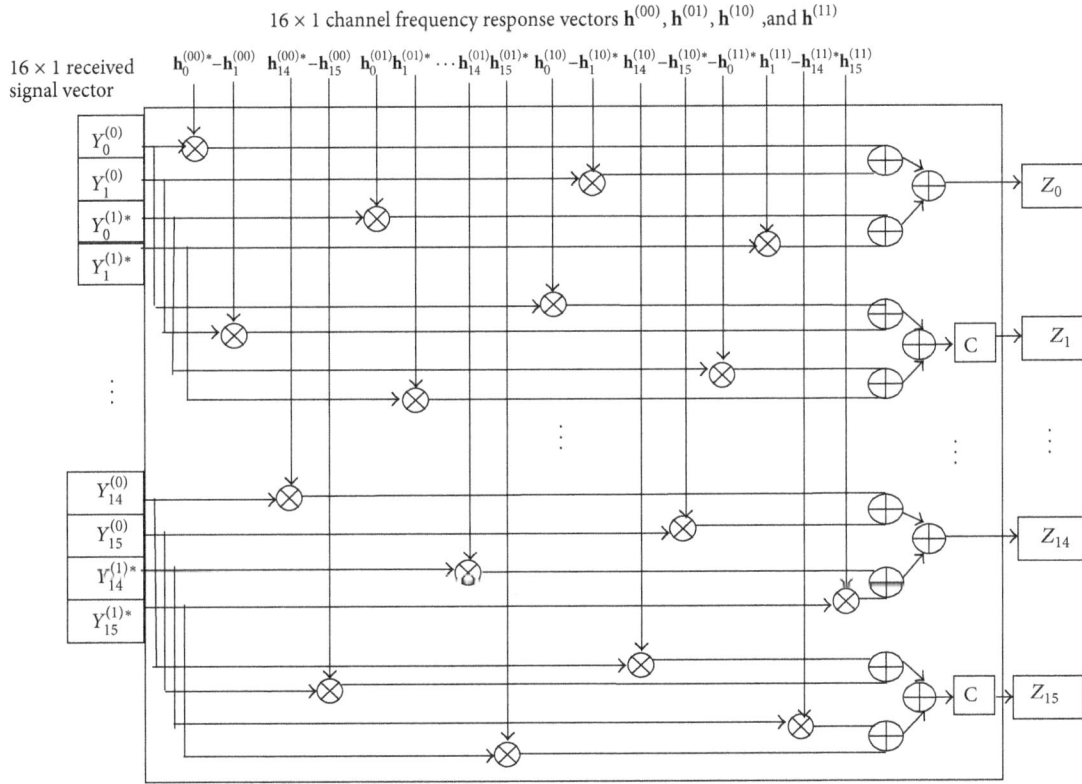

FIGURE 11: Internal architecture receiver decoding block-MIMO (RDBM).

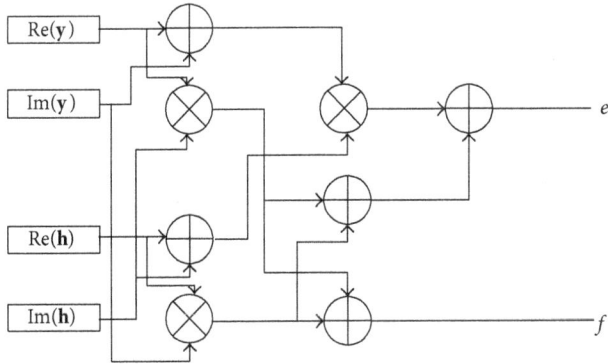

FIGURE 12: Multiplicands rearrangement for a single complex multiplication block.

This is repeated for all the 8 symbols in layer 0 and layer 1. Equation (7) defines the transmission format with the row index indicating the antenna port number and the column index indicating the subcarrier index. In 2 × 1 MISO configuration, the receive signals at ith and $(i+1)$th subcarrier are given in matrix form as

$$\begin{bmatrix} y_i \\ y_{i+1}^* \end{bmatrix} = \begin{bmatrix} h_i^{(0)} & -h_i^{(1)} \\ h_{i+1}^{*(1)} & h_{i+1}^{*(0)} \end{bmatrix} \begin{bmatrix} d_i \\ d_{i+1}^* \end{bmatrix} + \begin{bmatrix} n_i \\ n_{i+1}^* \end{bmatrix},$$

$$(8)$$

for $i = 0, 2, 4, 6, 8, 10, 12, 14,$

where $h_i^{(0)}$ represents the channel frequency response of ith subcarrier between 0th transmit antenna port and receive

antenna, d_i is data symbol at ith subcarrier, and n_i is the noise at ith subcarrier at the receive antenna. Equation (8) can simply be represented as

$$\mathbf{y}_i = \mathbf{H}_{\text{eff},i}\mathbf{d}_i + \mathbf{n}_i, \quad \text{for } i = 0, 2, 4, 6, 8, 10, 12, 14, \quad (9)$$

where $\mathbf{y}_i$ is 2×1 receive signal vector, $\mathbf{H}_{\text{eff},i}$ is the 2×2 channel matrix, $\mathbf{d}_i$ is 2×1 complex signal vector, and $\mathbf{n}_i$ is 2×1 noise vector. The objective is to detect the elements d_i and d_{i+1} of the data vector $\mathbf{d}_i$. Assuming that the elements of channel frequency response matrix $\mathbf{H}_{\text{eff},i}$ are perfectly known at the receiver, the decoder output vector $\mathbf{z}_i$ is given by

$$\mathbf{z}_i = \mathbf{H}_{\text{eff},i}^H \mathbf{y}_i, \quad \text{for } i = 0, 2, 4, 6, 8, 10, 12, 14, \quad (10)$$

where $\mathbf{H}_{\text{eff},i}^H$ is the Hermitian of the 2×2 channel transmission matrix. Equation (10) is expanded as

$$\begin{bmatrix} z_i \\ z_{i+1}^* \end{bmatrix} = \begin{bmatrix} h_i^{*(0)} & h_{i+1}^{(1)} \\ -h_i^{*(1)} & h_{i+1}^{(0)} \end{bmatrix} \begin{bmatrix} y_i \\ y_{i+1}^* \end{bmatrix},$$

$$(11)$$

for $i = 0, 2, 4, 6, 8, 10, 12, 14.$

The elements of decoder output are calculated as

$$z_i = h_i^{*(0)} y_i + h_{i+1}^{(1)} y_{i+1}^*, \quad \text{for } i = 0, 2, 4, 6, 8, 10, 12, 14,$$

$$z_{i+1}^* = -h_i^{*(1)} y_i + h_{i+1}^{(0)} y_{i+1}^*, \quad \text{for } i = 0, 2, 4, 6, 8, 10, 12, 14.$$

$$(12)$$

The PCFICH receive architecture for 2 × 1 MISO configuration is shown in Figure 7. Receiver decoding block (RDB)

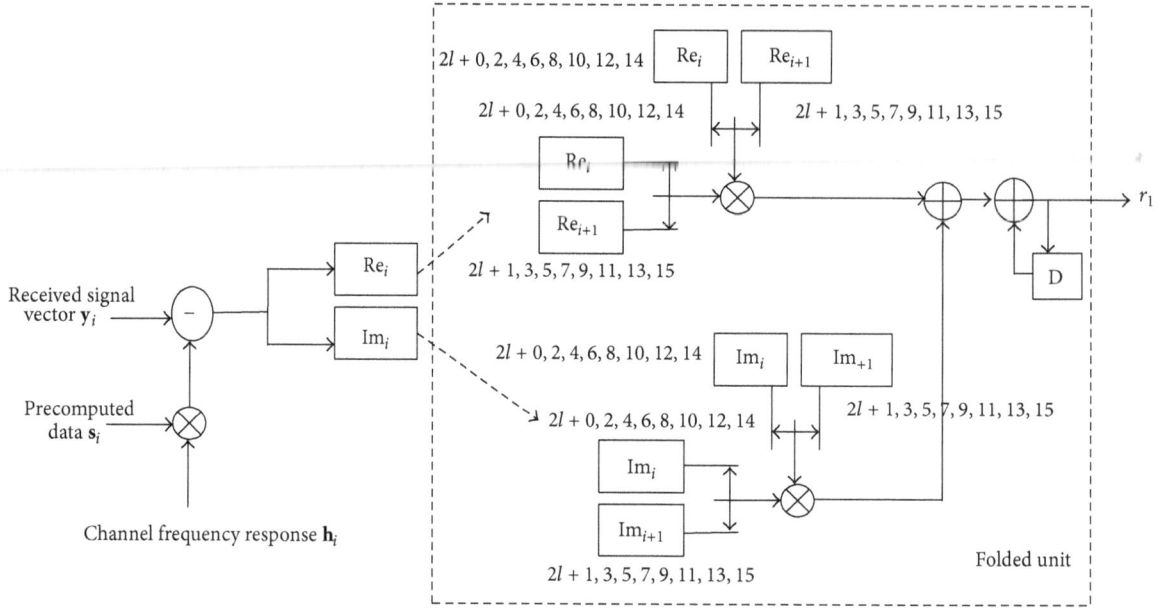

FIGURE 13: Illustration of folded architecture of RPB in SISO and SIMO.

gets the 16×1 received signal vector $\mathbf{y}$ and computes the decoder output vector using (10), assuming that the channel frequency response vectors $\mathbf{h}^{(0)}$ and $\mathbf{h}^{(1)}$ are known. The detailed internal architecture of RDBM is shown in Figure 11. The decoder output vectors $\mathbf{z}_i, i = 0, 2, 4, \ldots, 14$ are stacked as 16×1 vector $\mathbf{z} = [z_0^T, z_2^T, \ldots, z_{14}^T]^T$. The 16×1 precomputed data vectors for CFI $= 1,2,3,4$ are represented as $\mathbf{s}_1, \mathbf{s}_2, \mathbf{s}_3,$ and $\mathbf{s}_4$ respectively.

The detailed structure of receiver decoding blocks (RDB) is shown in Figure 8. The output vectors $\mathbf{z}^{(1)}, \mathbf{z}^{(2)}, \mathbf{z}^{(3)}, \mathbf{z}^{(4)}$ from RDB-1 to RDB-4 are fed to the processing blocks (PB-1 to PB-4). The detailed architecture of PB-1 is shown in Figure 9. The sum of the square magnitude of the elements of difference vector between decoded output vector $\mathbf{z}$ and the precomputed data vector $\mathbf{s}_1$ is the output r_1 of PB-1. Similarly $r_2, r_3,$ and r_4 are computed for CFI $= 2, 3,$ and 4 using PB-2, PB-3, and PB-4, respectively. The processing block outputs $r_1, r_2, r_3,$ and r_4 are applied to the CFI determination circuit shown in Figure 4 to detect the CFI value.

4. System Model and Implementation Architecture for MIMO Configuration

In MIMO system, the signals at ith and $(i + 1)$th subcarrier in the receive array are given by

$$\begin{bmatrix} y_i^{(0)} \\ y_{i+1}^{(0)} \\ y_i^{(1)*} \\ y_{i+1}^{(1)*} \end{bmatrix} = \begin{bmatrix} h_{00} & h_{01} \\ h_{10} & h_{11} \\ h_{01}^* & -h_{00}^* \\ h_{11}^* & -h_{10}^* \end{bmatrix} \begin{bmatrix} d_i \\ d_{i+1}^* \end{bmatrix} + \begin{bmatrix} n_i^{(0)} \\ n_{i+1}^{(0)*} \\ n_i^{(1)} \\ n_{i+1}^{(1)*} \end{bmatrix},$$

$$\text{for } i = 0, 2, 4, 6, 8, 10, 12, 14, \quad (13)$$

where h_{ab} represents the channel frequency response vector between bth transmit antenna and ath receive antenna and $n_i^{(j)}$ represents the noise in ith subcarrier in jth receive antenna. In vector form, it is written as

$$\mathbf{y}_i = \mathbf{H}_{\text{eff},i}\mathbf{d}_i + \mathbf{n}_i, \quad \text{for } i = 0, 2, 4, 6, 8, 10, 12, 14, \quad (14)$$

where $\mathbf{y}_i$ is 4×1 receive signal vector, $\mathbf{H}_{\text{eff},i}$ is the 4×2 channel frequency response vector at ith and $(i + 1)$th subcarrier, $\mathbf{d}_i$ is 2×1 data vector at ith and $(i + 1)$th subcarrier, and $\mathbf{n}_i$ is 4×1 noise vector. The objective is to detect the elements d_i and d_{i+1} of the data vector $\mathbf{d}_i$. Assuming that the elements of channel frequency response matrix $\mathbf{H}_{\text{eff},i}$ are perfectly known at the receiver, the decoder output vector $\mathbf{z}$ is given by

$$\mathbf{z}_i = \mathbf{H}_{\text{eff},i}^H \mathbf{y}_i, \quad \text{for } i = 0, 2, 4, 6, 8, 10, 12, 14, \quad (15)$$

where $\mathbf{H}_{\text{eff},i}^H$ is the Hermitian of the 4×2 channel transmission matrix. This can be expanded as

$$\begin{bmatrix} z_i \\ z_{i+1}^* \end{bmatrix} = \begin{bmatrix} h_{00}^* & -h_{10}^* & h_{01} & h_{11} \\ h_{01}^* & -h_{11}^* & -h_{00} & -h_{10} \end{bmatrix} \begin{bmatrix} y_i^{(0)} \\ y_{i+1}^{(0)} \\ y_i^{(1)*} \\ y_{i+1}^{(1)*} \end{bmatrix}, \quad (16)$$

$$\text{for } i = 0, 2, 4, 6, 8, 10, 12, 14.$$

The decoder outputs are given by

$$z_i = h_{00}^* y_i^{(0)} - h_{10}^* y_{i+1}^{(0)} + h_{01} y_i^{(1)*} + h_{11} y_{i+1}^{(1)*},$$

$$\text{for } i = 0, 2, 4, 6, 8, 10, 12, 14.$$

$$z_{i+1}^* = h_{01}^* y_i^{(0)} - h_{11}^* y_{i+1}^{(0)} - h_{00} y_i^{(1)*} - h_{10} y_{i+1}^{(1)*},$$

$$\text{for } i = 0, 2, 4, 6, 8, 10, 12, 14.$$

$$(17)$$

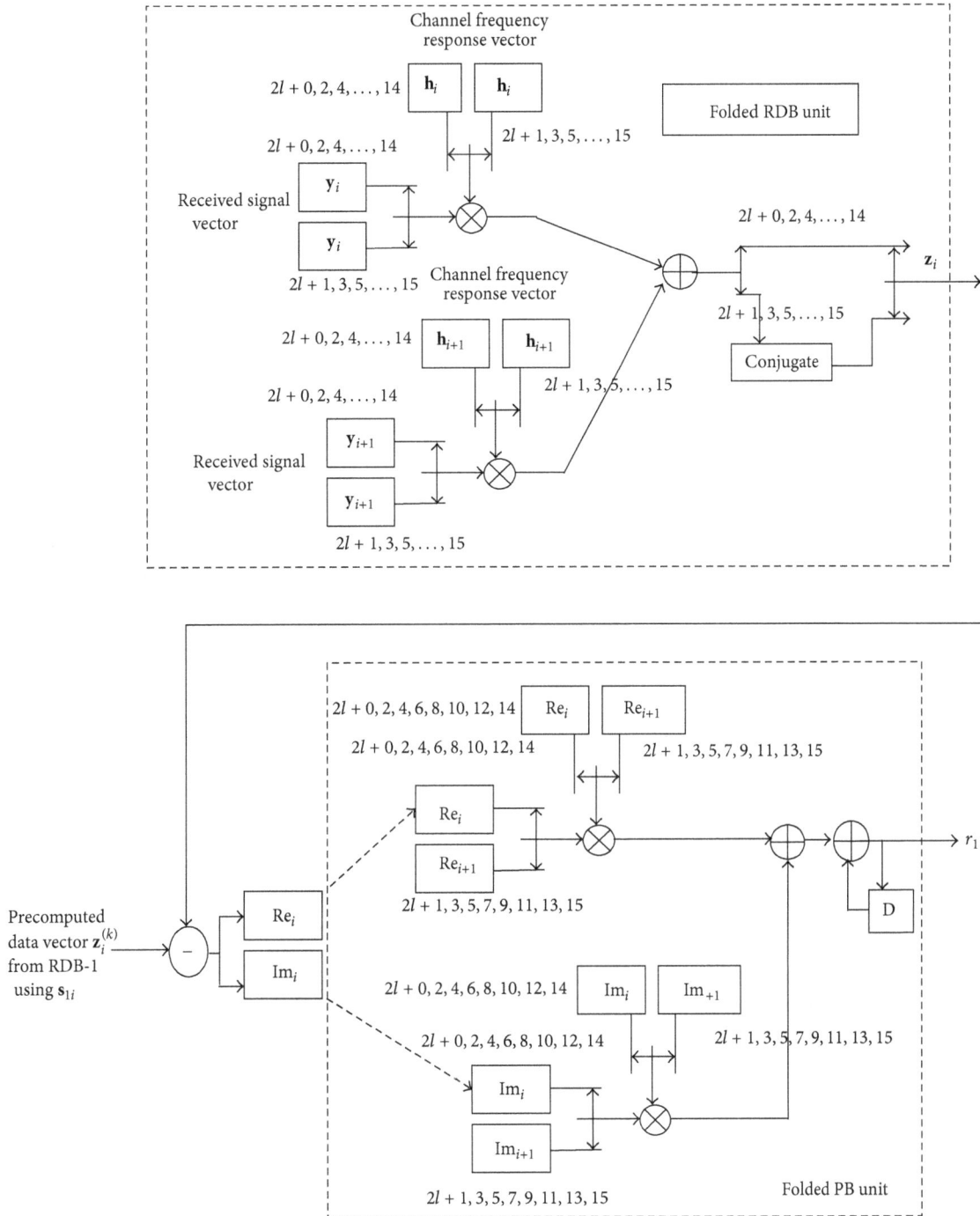

FIGURE 14: Illustration of proposed architecture for RDB and PB in MISO and MIMO. Note: receiver decoding block (RDB) in MISO is termed as RDBM in MIMO.

The PCFICH receiver architecture of 2×2 MIMO configurations is shown in Figure 10.

Receiver decoding block (RDBM) gets the 16×1 received signal vector $\mathbf{y}$ and computes the decoder output vector using (14), assuming that the channel frequency response vectors $\mathbf{h}^{(00)}, \mathbf{h}^{(01)}, \mathbf{h}^{(10)},$ and $\mathbf{h}^{(11)}$ are known. The 16×1 precomputed data vectors for CFI = 1, 2, 3, and 4 are represented as $\mathbf{s}_1^{(0)}$, $\mathbf{s}_2^{(0)}, \mathbf{s}_3^{(0)},$ and $\mathbf{s}_4^{(0)},$ respectively, for antenna 0, and as $\mathbf{s}_1^{(1)}$,

$\mathbf{s}_2^{(1)}, \mathbf{s}_3^{(1)},$ and $\mathbf{s}_4^{(1)},$ respectively, for antenna 1. The received signal vectors $\mathbf{y}_i^{(0)}$ and $\mathbf{y}_i^{(1)}$ multiply with the four channel estimation vectors to give decoded output vector $\mathbf{z}$ that is sent to the processing block (PB) which is shown in Figure 9. The decoder outputs $z_i, i = 0, 2, 4, \dots, 14$ are stacked as 16×1 vector $\mathbf{z} = [z_0^T, z_2^T, \dots, z_{14}^T]^T$. Similarly, RDBM1 gives output vector $\mathbf{z}^{(1)}$ using the precomputed data vectors $\mathbf{y}_1^{(0)}$ and $\mathbf{y}_1^{(1)}$ and channel estimation vectors. The architecture of PBs and

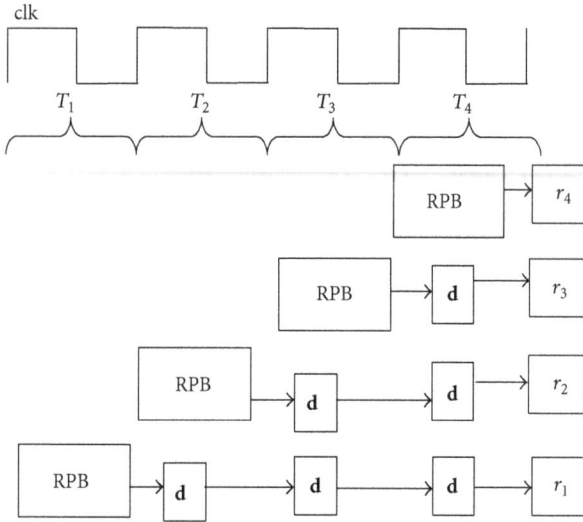

FIGURE 15: Illustration of superscalar method for SISO and SIMO (with no complex multiplications and operating from T_1 to T_4).

the CFI detection architecture are similar to that of the MISO system. The sum of the squared magnitude of the difference between each element in the decoded output vector $\mathbf{z}$ and its precomputed data in the vector $\mathbf{z}^{(1)}$ is the output r_1 of PB1. Similarly r_2, r_3, and r_4 are computed for other CFI. The r_1, r_2, r_3, and r_4 are compared to determine the minimum value by the CFI detector shown in Figure 4.

5. PCFICH Receiver Implementation Methods

The PCFICH receiver architectures can be implemented directly based on the basic architectures developed in Sections 3 and 4. But, in order to effectively utilize the resources in FPGA, the implementation of basic architectures is done using the modified novel architectures based on VLSI DSP techniques, namely, folding and superscalar processing approach.

5.1. Direct Implementation with Multiplicands Rearranged Method. In the receiver architecture for SISO and SIMO, the 16×1 received signal vector is directly subtracted from the precomputed data vector for a given CFI. This requires lesser number of multipliers and adders when compared to MISO and MIMO. In MISO and MIMO configurations, complex multiplications are necessary for the multiplication of $\mathbf{H}_{eff}^H$ with the received signal vector. It increases the number of multiplications in the CFI detection process. Hence, optimum rearrangement of the terms is carried out to minimize the number of multiplications. Further, the intermediate products are reused in the calculation of real and imaginary parts. Consider the multiplication of two complex numbers $\mathrm{Re}\{h\} + j\,\mathrm{Im}\{h\}$ and $\mathrm{Re}\{y\} + j\,\mathrm{Im}\{y\}$. The output real part ($e$) and imaginary part ($f$) terms are given by

$$e = \mathrm{Re}\{h\}\,\mathrm{Re}\{y\} - \mathrm{Im}\{h\}\,\mathrm{Im}\{y\},$$
$$f = \mathrm{Re}\{h\}\,\mathrm{Im}\{y\} + \mathrm{Im}\{h\}\,\mathrm{Re}\{y\}. \tag{18}$$

It requires four multiplications and two additions. To reduce the number of multiplications, the terms in (18) are rearranged as

$$e = [\mathrm{Re}\{y\} - \mathrm{Im}\{y\}]\,[\mathrm{Re}\{h\} - \mathrm{Im}\{h\}]$$
$$\mathrm{Re}\{y\}\,\mathrm{Im}\{h\} + \mathrm{Im}\{y\}\,\mathrm{Re}\{h\}, \tag{19}$$
$$f = \mathrm{Re}\{y\}\,\mathrm{Im}\{h\} + \mathrm{Im}\{y\}\,\mathrm{Re}\{h\}.$$

Since the terms $\mathrm{Re}\{y\}\,\mathrm{Im}\{h\}$ and $\mathrm{Im}\{y\}\,\mathrm{Re}\{h\}$ are in (19), it requires only three multiplications but five additions. This kind of rearrangement of the multiplicands is employed in the processing blocks at the cost of increased additions as shown in Figure 12.

5.2. Proposed Architecture Using Folding Method. Folding architecture systematically determines the control circuits in DSP architectures where multiple algorithmic operations are time-multiplexed to a single functional unit [12]. It is used for synthesis of DSP architectures that can be operated at single or multiple clocks. It reduces the number of hardware functional units (FUs) by a factor of N at the expense of increased computation time.

The folding architecture is introduced in the receiver structure of RPB in SISO and SIMO configurations and of RPB and PB in MISO and MIMO configurations as shown in Figures 13 and 14, respectively. For SISO RPB, there are 16 hardware lines to calculate the value of r_1 each requiring two multipliers. Hence the number of multipliers used in one RPB is 32. In order to reduce the number of multipliers and adders, folding architecture is proposed. This architecture uses only two multipliers and performs the operation of a single hardware line 16 times in sequential way. The difference between the product of channel frequency response vector with the precomputed data vector and the received signal vector is stored in registers. At a time, one resultant signal pair involves in computation using two multipliers to get the value of z_i. Four switches operating in system clock speed are involved in the architecture where two switches are used to pass the real part of the signal to one multiplier, while the other two switches are used to pass the imaginary part of the signal to another multiplier. The multipliers pass the products to the first adder for z_i. The output of the first adder is passed to the second adder with a delay to accumulate the values z_0 to z_{15} into a register in subsequent clock cycles. This process requires 16 clock cycles and the CFI is detected at the 17th clock cycle. Though it takes longer time for the clock cycles to get the output, the resources are minimized in this method.

The folded architecture of decoding block of MISO and MIMO involving complex multiplication of the channel frequency response vector and the receive signal vector is shown in Figure 14. There are 2 complex multiplications and one addition in each of the 16 hardware lines. Hence total resource elements used are 32 complex multiplications and 16 additions. The folded architecture which reduces to just 2 complex multiplications and one addition requires five switches. Two switches are used to pass the first element of the receive signal vector and its corresponding channel

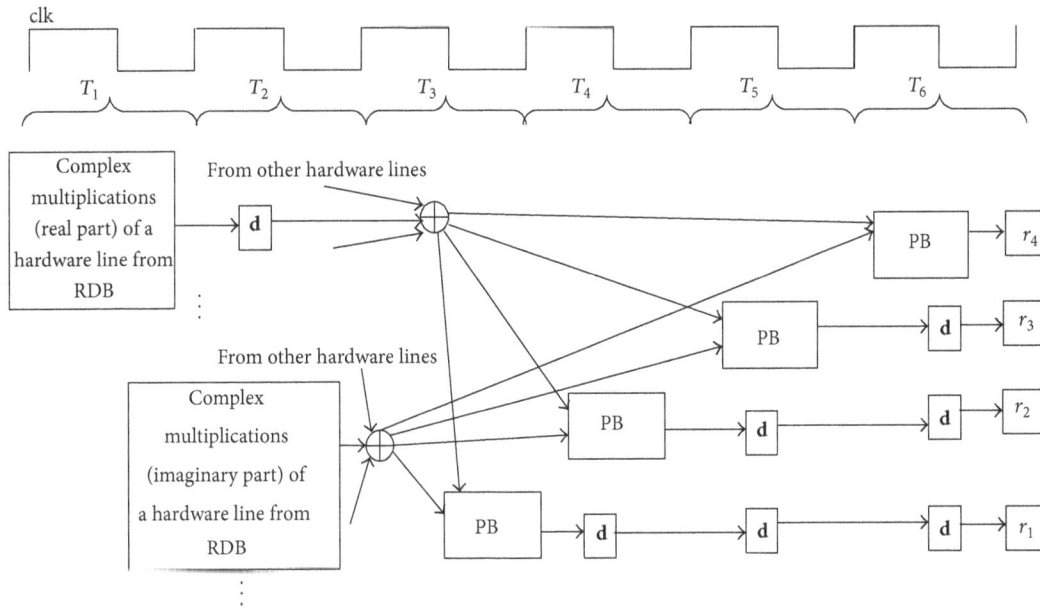

FIGURE 16: Illustration of superscalar method for MISO and MIMO (with complex multiplications and operating from T_1 to T_6).

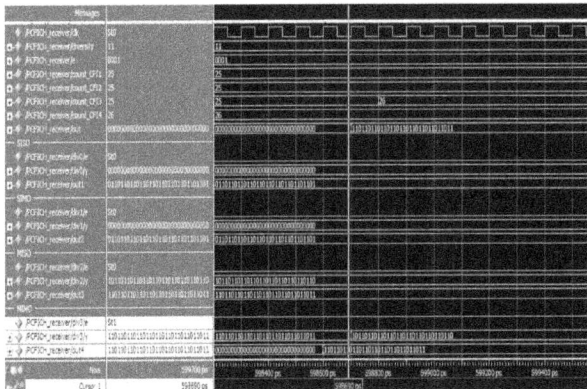

FIGURE 17: Simulation waveform for PCFICH receiver.

frequency response vector to one multiplier and other two switches are used to pass the second element of receive signal vector and its channel frequency response vector to another multiplier. These four switches operate in system clock speed. The multipliers pass their products to the adder through the fifth switch before moving to PB. This process requires 16 clock cycles and the CFI is detected at the 17th clock cycle.

5.3. Proposed Architecture Using Superscalar Method. Superscalar approach is another low resource utilizing VLSI DSP technique. The superscalar processing method includes parallel processing and pipelining strategies. In this case, parallel operation for the 16 pairs of hardware lines is arranged with pipelining of the subtraction and square magnitude operations for each CFI. SISO configuration does not have complex multiplications and it has only square magnitude operations. Hence the RPB of SISO has 16 hardware lines each having 2 multipliers which results to a total of 32 multipliers.

This setup requires more hardware resources than folding, but the output is obtained at every 4th clock cycle as shown in Figure 15. SIMO configuration which involves two receive antenna signal processing, requires twice the number of multiplications as that of SISO and the output is obtained at every 4th clock cycle. The block "*d*" represents the delay element introduced to buffer the values and produce the outputs at the same time instant.

For MISO configuration the RDB has 16 hardware lines, with 2 complex multiplications each. Since each complex multiplication requires four real multiplications, RDB can be executed in two clock cycles by reusing 64 multipliers. 32 multipliers are required for PB taking 4 clock cycles. Hence 96 multipliers are required in MISO configuration. For MIMO configuration, the RDB requires reuse of 128 multipliers taking 2 clock cycles and an additional 32 multipliers are required for the PB taking 4 clock cycles. Hence 160 multipliers are required for MISO configuration and the output is obtained at every 6th clock cycle as shown in the Figure 16. The block "*d*" represents the delay element introduced to buffer the values and produce the outputs at the same time instant.

6. Results and Discussion

The proposed receiver architectures for PCFICH in SISO, SIMO, MISO, and MIMO configurations are implemented using the Xilinx PlanAhead tool on the Virtex-6 FPGA xc6vlx240tffl156-1 device board. The target device Virtex-6 has only 768 DSP elements. Table 2 shows the performance of the proposed architectures using folding and superscalar methods being compared with the direct implementation of PCFICH receiver, in terms of resource utilisation, speed, and power for all the SISO, SIMO, MISO, and MIMO

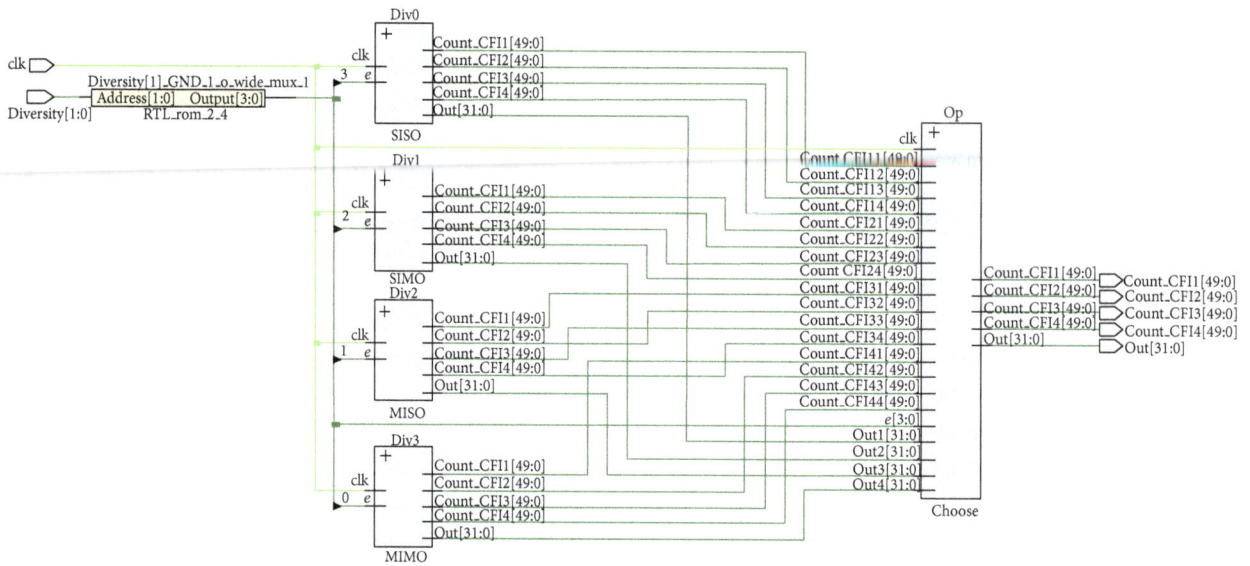

FIGURE 18: RTL schematic for combined PCFICH architecture with diversity.

TABLE 2: Performance of proposed architectures based on folding and superscalar method.

Diversity	Method	Multipliers	Adders	DSP elements	LUTs	Total delay (ns)	Speed (MHz)	Dynamic power
SISO	Direct	125	245	125	5479	39.081	25.587	—
	Folding (16T)	8	81	16	2561	114.448	8.737	84
	Superscalar (4T)	32	182	66	2731	69.333	14.423	93
SIMO	Direct	250	494	250	10942	40.278	24.827	—
	Folding (16T)	16	165	32	5117	130.704	7.651	159
	Superscalar (4T)	64	318	130	5623	53.873	18.562	170
MISO	Direct	224	580	594	14880	43.023	23.243	—
	Folding (16T)	14	101	39	3950	255.264	3.917	173
	Superscalar (6T)	96	338	196	6156	80.495	12.423	208
MIMO	Direct	320	844	675	17380	56.962	17.555	—
	Folding (16T)	20	155	46	4395	256.528	3.898	374
	Superscalar (6T)	160	465	262	6932	85.822	11.652	382

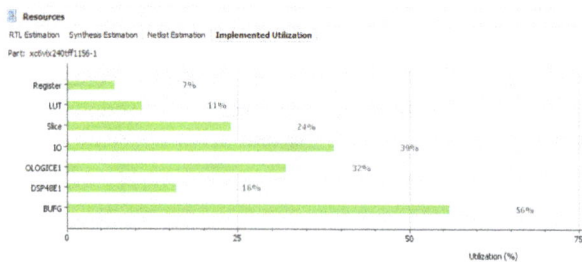

FIGURE 19: Resource utilization graph for generalized architecture.

configurations. The proposed architectures based on folding and superscalar processing methods require less number of resource elements.

In the folding approach, resource utilization is less compared to the direct and superscalar approach at the cost of reduced speed of operation but it is suitable for real-time frame timings. When the LTE-A system operates at 1.4 MHz bandwidth, maximum time available for detection at each subcarrier is 992.063 ns since each slot of 0.5 ms duration in a frame (10 ms radio frame duration) consists of 7 OFDM symbols and there are 72 subcarriers along one OFDM symbol. The total delay in the receiver architecture is within the LTE time constraint. The dynamic power consumption is less in the folding method compared to superscalar method due to decrease in block arithmetic. Direct method does not require sequential execution and clocking and hence total power consumption is due to static power. Hence, it is inferred that the proposed architecture based on folding method is more suitable for CFI detection. The simulation waveform of the proposed architecture based on folding method is shown in Figure 17 for SISO, SIMO, MISO, and MIMO configurations.

A general architecture based on folding method which operates at all the four SISO, SIMO, MISO, and MIMO configurations has also been developed. In this architecture,

TABLE 3: Resource requirements of proposed architecture using folding method.

Parameter	Multipliers	Adders	Minimum clock period (ns)	Total delay (ns)	Speed (MHz)	Total dynamic power (mW)
Value	58	518	16.033	256.528	3.898	1019

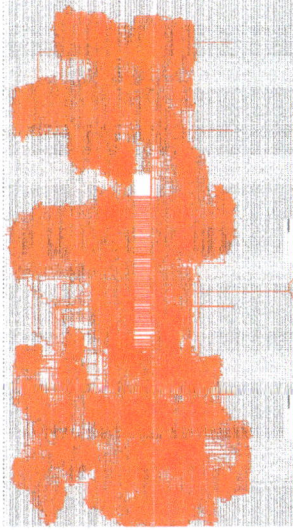

FIGURE 20: Implemented device in FPGA editor.

a control variable "e" is used to enable or disable the submodules SISO, SIMO, MISO, or MIMO according to the selection input "diversity." CFI is detected at every 17th clock cycle. The synthesis results of a general architecture based on folding show that it utilizes minimum resources in XC6VLX240TFF1156-1 Virtex 6 device (768 DSPs). This is summarized in Table 3. Dynamic power consumption is due to internal switching contributed by the clock (246 mW), logic (670 mW), and the block arithmetic (103 mW).

Figure 18 shows the RTL schematic of 4 diversity blocks "div0," "div1," "div2," and "div3" corresponding to SISO, SIMO, MISO, and MIMO controlled by wires named "e." Power consumed includes both static power and dynamic power due to internal switching.

Figure 19 shows the resource utilization graph which shows the percentage of registers, lookup tables (LUTs), slices, DSP elements, and buffers used.

Figure 20 shows the implemented device in FPGA editor with the implemented components and interconnections between the components configured into the FPGA device.

7. Conclusion

In this paper, low complexity, low resource single, or multi-antenna CFI detection at the receiver system has been proposed and analyzed using modelsim and implementation in the Virtex-6 device in Xilinx PlanAhead tool. In the receiver, computational complexity and the resource utilization are minimized by employing arithmetic operational rearrangement and suboptimal sequential DSP algorithm called the folding approach. The proposed architecture using folding

method complies with the LTE frame timing constraint in SISO, SIMO, MISO and MIMO configurations. It is a suitable solution for the area optimized hardware implementation of receiver structures for PCFICH. In future, a total hardware accommodating all the physical downlink control channels of the 3GPP-LTE-A with low resource utilization could be synthesized and implemented.

Conflict of Interests

The authors do not have direct financial relation with any commercial identity mentioned in the paper or any other conflict of interests.

Acknowledgments

The authors wish to express their sincere thanks to All India Council for Technical Education, New Delhi, for the Grant to do the Project titled Design of Testbed for the Development of Optimized Architectures of MIMO Signal Processing (no. 8023/RID/RPS/039/11/12). They are also thankful to the Managements of Mepco Schlenk Engineering College, Sivakasi, and Thiagarajar College of Engineering, Madurai, for their constant support and encouragement to carry out this research work successfully.

References

[1] 3GPP TS 36. 211 Version 11. 0. 0 Release 11 (2012-10) Evolved Universal Terrestrial Radio Access (E-UTRA), Physical Channels and Modulation, 2012.

[2] J. Ketonen and M. Juntti, "SIC and K-best LSD receiver implementation for a MIMO-OFDM system," in Proceedings of the 16th European Signal Processing Conference (EUSIPCO '08), Lausanne, Switzerland, August 2008.

[3] J. Ketonen, M. Juntti, and J. R. Cavallaro, "Performance—complexity comparison of receivers for a LTE MIMOOFDM system," IEEE Transactions on Signal Processing, vol. 58, no. 6, pp. 3360–3372, 2010.

[4] X. Huang, C. Liang, and J. Ma, "System architecture and implementation of MIMO sphere decoders on FPGA," IEEE Transactions on Very Large Scale Integration Systems, vol. 16, no. 2, pp. 188–196, 2008.

[5] Y. Lin, Y. Chen, C. Chu, C. Zhan, and A. Wu, "Dual-mode low-complexity codebook searching algorithm and VLSI architecture for LTE/LTE-advanced systems," IEEE Transactions on Signal Processing, vol. 61, no. 14, pp. 3545–3562, 2013.

[6] D. Wu, J. Eilert, R. Asghar, and D. Liu, "VLSI implementation of a fixed-complexity soft-output MIMO detector for high-speed wireless," EURASIP Journal on Wireless Communications and Networking, vol. 2010, Article ID 893184, 13 pages, 2010.

[7] C. Mehlführer, M. Wrulich, J. C. Ikuno, D. Bosanska, and M. Rupp, "Simulating the long term evolution physical layer," in

Proceedings of the 17th European Signal Processing Conference (EUSIPCO '09), pp. 1471–1478, Glasgow, UK, August 2009.

[8] S. J. Thiruvengadam and L. M. A. Jalloul, "Performance analysisof the 3GPP-LTE physical control channels," *EURASIP Journalon Wireless Communications and Networking*, vol. 2010, Article ID 914934, 10 pages, 2010.

[9] S. M. Alamouti, "A simple transmit diversity technique for wireless communications," *IEEE Journal on Selected Areas in Communications*, vol. 16, no. 8, pp. 1451–1458, 1998.

[10] S. S. A. Abbas and S. J. Thiruvengadam, "FPGA implementation of 3GPP-LTE-A physical downlink control channel using diversity techniques," *International Journal of Wireless and Mobile Computing*, vol. 9, no. 2, p. 84, 2013.

[11] S. Ahmadi, *LTE-Advanced: A Practical Systems Approach to Understanding 3GPP LTE Releases 10 and 11 Radio Access Technologies*, Sassan Ahmadi Academic Press, 2013.

[12] K. K. Parhi, *VLSI Digital Signal Processing Systems—Design and Implementation*, Wiley-Interscience, 1999.

## 16

A Novel Scan Architecture for Low Power Scan-Based Testing

Mahshid Mojtabavi Naeini and Chia Yee Ooi

Department of Electronic Systems Engineering, Malaysia-Japan International Institute of Technology, Universiti Teknologi Malaysia, Jalan Sultan Yahya Petra, 54100 Kuala Lumpur, Malaysia

Correspondence should be addressed to Chia Yee Ooi; ooichiayee@utm.my

Academic Editor: Jose Carlos Monteiro

Test power has been turned to a bottleneck for test considerations as the excessive power dissipation has serious negative effects on chip reliability. In scan-based designs, rippling transitions caused by test patterns shifting along the scan chain not only elevate power consumption but also introduce spurious switching activities in the combinational logic. In this paper, we propose a novel area-efficient gating scan architecture that offers an integrated solution for reducing total average power in both scan cells and combinational part during shift mode. In the proposed gating scan structure, conventional master/slave scan flip-flop has been modified into a new gating scan cell augmented with state preserving and gating logic that enables average power reduction in combinational logic during shift mode. The new gating scan cells also mitigate the number of transitions during shift and capture cycles. Thus, it contributes to average power reduction inside the scan cell during scan shifting with low impact on peak power during capture cycle. Simulation results have shown that the proposed gating scan cell saves 28.17% total average power compared to conventional scan cell that has no gating logic and up to 44.79% compared to one of the most common existing gating architectures.

1. Introduction

Nowadays, design-for-testability (DFT) techniques have become an inseparable consideration for testing modern microelectronic designs as they play an important role in the improvement of the test quality and reducing the test application time in the VLSI digital circuits. The next generation of deep submicron circuits will rely not only on the low power VLSI design but also on the DFT methods targeting low power testing.

Full scan design has become one of the most popular structured DFT methods and is widely used in testing of sequential circuit since it has solved the difficulties in control and observation of the internal nodes of circuit by providing external access to all storage elements of the design. However, from the view point of power dissipation, scan-based architectures are very expensive as each scan test pattern contributes to a shift operation with high power consumption [1]. Moreover, since there is less correlation among scan test patterns generated by an Automatic Test Pattern Generation (ATPG) tool compared to the data during normal mode, high switching activities incurred in capture

mode have increased test power drastically over chip power limitations. The elevated power consumption during test application can cause severe problems in the Circuit-Under-Test (CUT). The problem of excessive power consumption during test application can mainly fall into two subproblems: (1) excessive average power consumption and (2) excessive peak power consumption.

Problems due to Excessive Average Power Consumption. Average power consumption is the total distribution of power over time period and is calculated using the ratio of consumed energy to test time [1]. This excessive average power consumption due to testing, in the first place, produces extra heat in CUT which has inevitable role in appearing hot spots, circuit premature destruction, degradation of performance, functional failures, and, as a result, circuit reliability degradation. The main mechanisms which lead to these structural degradations are *corrosion* (oxidizing of conductors), *electromigration* (molecular migration of the conductor structure toward the electronic flow), *hot-carrier-induced* defects, or *dielectric breakdown* (loss of insulation of the dielectric barrier) [2]. Excessive average power affects not only temperature increase but also temperature variations. These temperature variations may induce timing variations during test and, in some cases, may lead to test-induced yield loss [3]. In addition, intensive heat generated by high switching activity during the test process has negative influences on the circuit packaging cost to make the CUT tolerable to higher level temperature.

Problems due to Excessive Peak Power Consumption. Peak power is the highest power value at any given time instant [1]. Often, the time window to define peak power is restricted to one clock period. However, in practice, it was outlined in [1] that restricting the time window to just one clock cycle is not realistic enough since the power consumption within one clock cycle may not be large enough to elevate the temperature over the thermal capacity limit of the chip. As pointed in [3, 4], excessive peak power dissipation comes with a high instantaneous current demand due to *high switching activity* during test application time which may cause power supply noise (PSN). The excessive noise may induce several phenomena as outlined below.

(1) Changing the logic value at some internal nodes of the circuit leads to failing of good dies and consequently unnecessary yield loss.

(2) Ground bounce or voltage droop: by increasing switching activities during test time, voltage glitches may be observed at some signal lines which can change rise/fall time of the gates (timing performance degradation) causing good dies to be declared "fail." Thus, unwanted yield loss happens.

(3) IR-drop is referred to decrease (increase) in the power (ground) rail voltage and is linked to the existence of a nonnegligible resistance between the rail and each node in the CUT.

(4) Cross talk is referred to capacitive coupling between neighboring nets within an IC. By increasing PSN, the voltage at some gates in the circuit is reduced (voltage drop) causing these gate to show higher delays (performance degradation), possibly leading to test fail and yield loss.

Therefore, reducing the switching activities at any instant time mitigates the average power and hence the peak power of the chip [5]. Moreover, peak power and average power reduction during test contribute to enhanced reliability of the test and improvement of yield [6].

For scan-based design, test power concerns include shift power reduction and capture power. Shift power consumption is due to the transitions occurring in scan cells when the adjacent bits in test vector have different values. These transitions not only cause switching activity in scan cells but they are also propagated to the combinational logic through scan cells outputs. Capture power consumption is referred to transitions that happen within scan cells when they have different values before and after capture. However, in some literatures, transitions in combinational part in launch cycles also have been considered as capture power. In general, average power reduction can be accomplished via shift power

reduction while peak power reduction can be achieved by reducing capture power since the logic values in many scan cells changes simultaneously during this mode. In scan-based testing, the major portion of the power and energy is dissipated during shift process as reported in [7] since a large portion of the test application time includes shift cycles especially for large industrial designs with long scan chains. So, in our research we have mainly focused on the problem of shift power reduction while capture power also has been taken into account. Although power issue is one of the major concerns in testing of modern VLSI circuits, other parameters such as test application time, area overhead, and fault coverage should not be ignored.

The remainder of the paper is organized as follows. In Section 2, the existing related methods for reducing switching activity in combinational logic and scan chain are reviewed. Section 3 elaborates the proposed gating scan architecture. Simulation results have been presented in Section 4. Finally, Section 5 draws the conclusion.

2. Previous Works

For scan-based designs in particular, there are two sources of power consumption during shift mode. The switching activities happen in the scan chain caused by scan-ripple and the switching activities in the combinational logic due to propagation of rippling transitions in the scan chain. Switching activity in the combinational part contributes to a large portion of the total switching activity in the circuit [11]. One of the most straight-forward ways for shift power reduction is to reduce switching activity in the combinational logic by isolating stimulus path of scan cells from combinational logic during shift cycle because the major source of dynamic power in CUT is the propagation of ripple transitions from the scan cells to the combinational logic during scan shifting. Some authors [7, 12] have tried to gate the stimulus paths of flip-flops to a constant logic "1" or "0" by utilizing gating logic (NOR [7], transmission gate (TG) together with a pull-up or a pull-down transistor [12]) at the scan cells output, thus eliminating spurious switching in logic gates. However, they are able to block only one transient (from "1" to "0" or vice versa) at the first level of the combinational circuit when the circuit mode changes from normal/capture to shift mode. Thus, the unblocked transient still can propagate to the deeper level of the combinational logic, causing many transitions at circuit internal lines before reaching the steady state. Other approaches [13, 14] have experimented the gating logic (MUX [13], extra inverter-base latch [14]) to hold the scan output at the previous logic and completely block any redundant switching activity in the combinational logic during shift mode. An enhanced scan structure has been reported in [11] for delay fault testing. In this method, each scan cell has been augmented into an AND-NOR-based hold latch at the scan cell output to ensure that the scan stimulus paths remain unchanged during shift session. The hold latch has been implemented by a cross-coupled gate of NOR and two AND gates to hold the scan cells' stimulus paths. However, these approaches suffer from large overhead in terms of area and propagation delay.

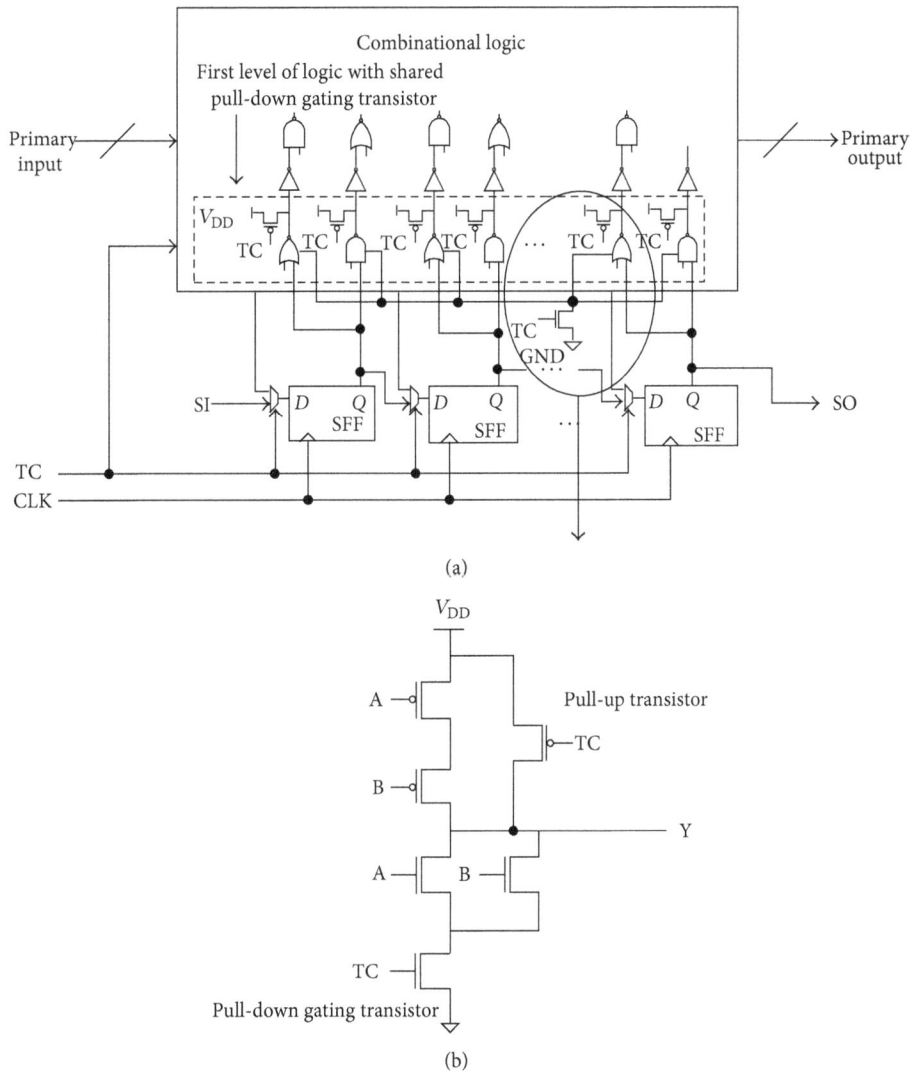

FIGURE 1: (a) FLS structure [8] and (b) transistor level schema of first level gating in FLS.

The first level supply (FLS) gating scheme proposed by Bhunia et al. [8] has inserted a common pull-down gating transistor to the gates at the first level of logic which are typically connected to the scan chain outputs. In order to prevent the outputs of the first level gates from being floating, they have added another pull-up transistor to force the output to V_{DD}. Both the gating and pull-up transistors are controlled by test control signal (TC). TC is assigned "0" in shift mode such that the first level NOR gate is disabled since it is cut off from the supply. Figure 1 shows FLS structure.

Although FLS scheme has less overhead in terms of area, propagation delay, or even switching activity in gating logic compared with other gating schemes, it is unable to block all transients in the combinational logic because the output of the first level gates is forced to a fixed value, which is similar to NOR and TG gating. Another approach by Bhunia et al. [15, 16] gives an alternative solution for Enhanced Scan in delay fault testing named First Level Hold (FLH) which gates the rippling transition to the combinational logic during shift mode. In addition to two gating pull-up and pull-down

transistors, a cross-coupled inverter latch controlled by TC signal has been employed to hold the output of the first level gates during pattern shifting until the next pattern is launched. A local TC signal has been generated for each first level gate to drive the related gating and holding logic. Gating and holding logics inserted to all first level gates in addition to hardware for local TC generation applied to individual gates have large overhead in terms of area and power consumption. All of the mentioned techniques can reduce shift power in combinational logic effectively. However, they suffer from large overhead in terms of area and propagation delay.

In order to eliminate performance degradation due to gating logic, Suhag et al. [17] have reported a modified transistor level design of scan flip-flop for critical paths. The modified scan cell has been implemented with nine extra transistors compared to conventional scan cell to tie the output of scan cell to combinational logic to the constant value "1" and improve the performance on this path. However, nine extra transistors in the scan cell structure result in significant switching activity and peak power elevating with

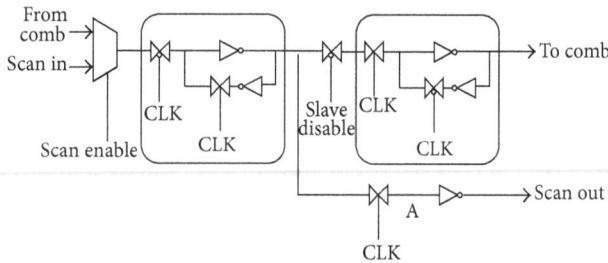

FIGURE 2: Modified scan flip-flop for low power delay fault testing [9].

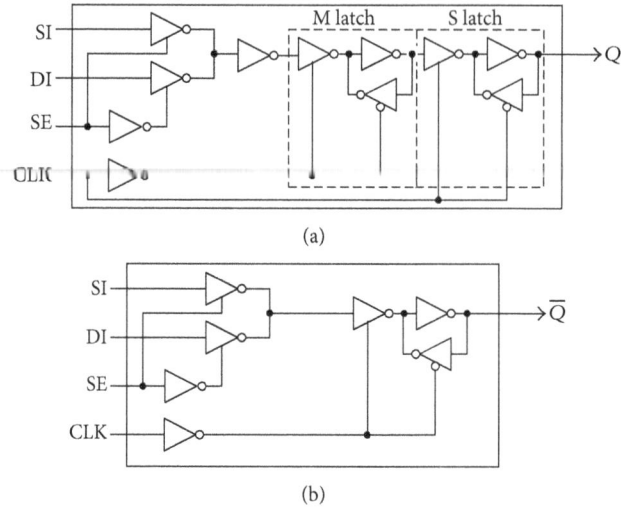

FIGURE 3: (a) Conventional scan architecture and (b) modified scan architecture [10].

severe area overhead. In addition, it is not able to cut off all the redundant switching activity in the combinational logic for the same reason we mentioned earlier about freezing the scan cell output to a constant logic during shifting. Another modified scan flip-flop for low power delay fault testing has been proposed in [9].

As it has been shown in Figure 2, it bypasses the slave latch with an alternative low cost dynamic latch in scan shifting path. Therefore, it can successfully eliminate all transitions to the combinational logic. For the application of stuck-at faults, slave latch is disabled by scan enable signal. However, no evaluation on benchmark circuits or comparison in terms of power consumption, performance, and area overhead to the existing gating methods has been presented.

Partial gating methods [18–22] have been proposed to reduce the full gating penalties in area overhead and performance degradation. By proper selection of scan cells on noncritical paths to be gated and their gating values, they try to maximize shift power reduction with acceptable performance degradation. Most recently reported partial gating methods such as those in [23–26] gate a subset of scan cells not only during shift mode but also during capture mode in order to reduce peak power besides shift power reduction. However, in large industrial designs, scan cells have large fan out cones. Thus, ungated scan cells in partial gating method can still cause a great amount of switching activities in the combinational logic. Moreover, in both existing full gating and partial gating techniques, significant power is consumed in the gating elements themselves, which causes the peak power to increase from 5% to 60% in all the benchmark circuits when the gating overhead was considered [9]. This is due to the large switching activity occurring in the gating logic when scan mode changes to capture mode or vice versa [17]. The missing part of both existing full gating and partial gating methods is that they have limited their approaches only to power reduction in combinational circuits during shift cycles without highlighting much that shift power consumption is also related to total amount of power consumed in scan chain besides combinational logic.

While aforementioned gating methods mainly have tried to reduce the level of switching activity in the combinational part, a few schemes can be found in the literature that have concentrated on reducing the level of switching activity in scan cell. A modified scan element has been presented in [10] to reduce scan cell switching activity. Figures 3(a) and 3(b) demonstrate the conventional and modified scan

architectures, respectively. As observed from Figure 3(b), master output $\overline{Q}$ of modified scan cell can be utilized instead of Q to form the scan chain which results in three less inverters in the scan output propagation path. In addition, an inverter in the internal multiplexer of scan cell has been removed, which leads to less propagation delay in the input path of the modified scan cell. However, the main drawback of this scheme is that the structure represents a latch instead of flip-flop which makes it an unsuitable element to be employed as a scan cell.

Finally, many previous works have successfully reduced power consumption in both combinational logic and scan chain by using separate methods for each section. In [27], FLS has been applied to the scan partitioning method which activates a part of scan cell at a time to reduce power consumption in scan chain. To overcome the limitation of the existing gating methods, we propose a gating scan architecture with the following contributions.

(i) The main novelty of the proposed gating scan architecture is an integrated solution targeting power reduction in combinational part as well as inside the scan cell itself which, to the best of our knowledge, have not been reported at the same time in any previous works.

(ii) Most of the previous gating methods did not consider the negative impact of the power consumed in the gating logics that result in peak power violation during capture mode. The proposed method can reduce the maximum level of peak power compared to existing gating methods as it will be explained in the next section.

(iii) The proposed gating scan architecture introduces a new short shift path that improves both shift and capture propagation delays as well as power consumption in scan chain during shift mode. This makes shifting at higher frequency possible in those cases

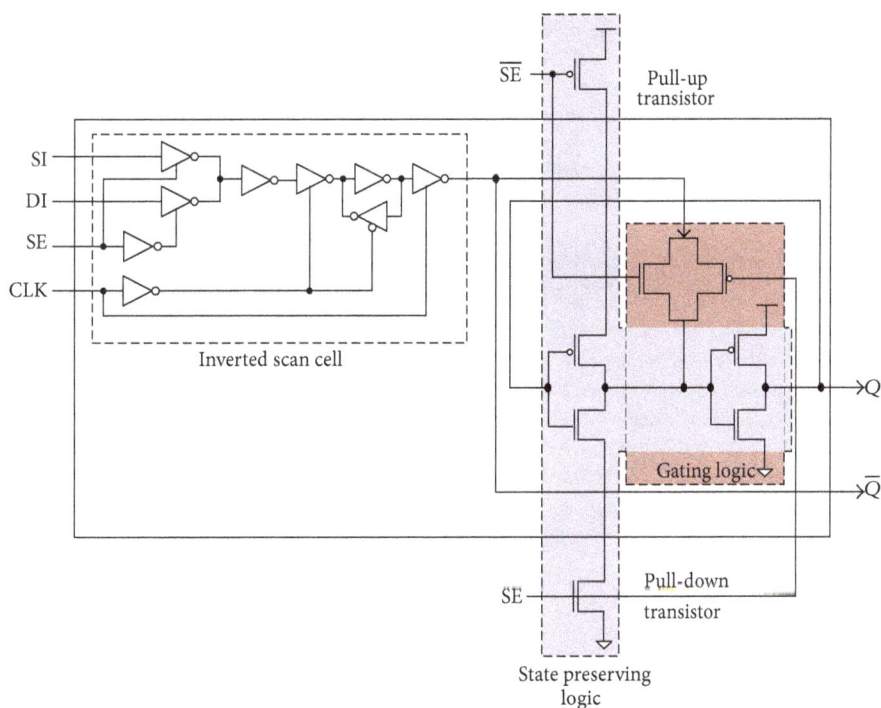

FIGURE 4: The proposed new gating scan cell structure.

that maximum shift frequency has been bounded by maximum allowable power consumption. Therefore, the proposed gating design improves test application time over existing gating solutions.

(iv) The proposed gating scan cell has minimum impact on the performance degradation due to smaller launch and capture propagation delay compared to existing gating methods. Therefore, in the application of partial gating more scan cells can be gated using the proposed gating scan cell without violating critical path timing and thus higher power reduction is achievable.

3. The Proposed Gating Scan Architecture

We propose a novel gating scan structure for shift power reduction considering both scan chain and combinational part with the following features:

(A) gating redundant transitions from scan cell to combinational logic during shift mode,

(B) reducing switching activity inside the scan cell during shift mode.

In order to achieve the above-mentioned features within an integrated structure, we have modified the slave latch in the conventional master/slave scan cell with state preserving and gating ability. In addition, to achieve second feature, a new shift path has been set up which at the same time speeds up shift and capture process over existing gating methods. Since state preserving and gating logics have been embedded as part of slave latch in the proposed gating scan cell, the

area overhead is as low as two transistors that are sharable by several scan cells. The proposed structure contributes to average power reduction in the scan architecture (combinational logic and scan chain) during shift mode while not causing high peak power during capture mode. In gating methods the main source of excessive peak power during capture mode is the switching activities in gating elements when the mode changes from shift to capture mode or vice versa. Excessive peak power can be avoided by reducing the level of switching activity during test [3]. Therefore, the proposed structure is able to control peak power violations by reducing switching activities in other parts of scan cells. A conceptual discussion about our architecture will be presented in the next section in order to preliminarily show the effectiveness of our proposed scan architecture with respect to total average power, propagation delay, and area overhead.

We have improved the modified scan element [10] and named it gating scan cell. According to Figure 4, each gating scan cell consists of three main substructures:

(i) gating logic,

(ii) state preserving logic,

(iii) inverted scan cell.

3.1. Gating Logic. During shift cycle, the rippling transitions cause great switching activities in the scan chain. The propagation of this switching activity into the combinational part contributes to large redundant transitions in the circuit lines. In order to suppress the scan chain transitions from propagating during shift cycles, we have proposed a scan structure which is augmented by a gating logic. For constructing the

Test vector: $v_1\ v_2\ v_3\ v_4\ v_5$
              1   0   1   0   1
Adaptive test vector: $\overline{v_1}\ v_2\ \overline{v_3}\ v_4\ \overline{v_5}$
                       0   0   0   0   0

```
                        ┌ 0 0 0 0 0 ┐
                   →    │ 0 1 0 0 0 │
                   →    │ 0 1 0 0 0 │ ←── Shift-in
                   →    │ 0 1 0 1 0 │      process
                   →    └ 0 1 0 1 0 ┘ ←── Test vector at
                                           inverted output Q̄
```

Test response: $r_1\ r_2\ r_3\ r_4\ r_5$
                1   0   1   0   1
Test response after shift-out: $\overline{r_1}\ r_2\ \overline{r_3}\ r_4\ \overline{r_5}$ →   ┌ 0 0 0 0 0 ┐

```
                          ┌ 1 0 1 0 1 ┐
                          │ 0 0 1 0 1 │
Shift-out    →            │ 0 0 1 0 1 │
 process                  │ 0 0 0 0 1 │
                          └ 0 0 0 0 1 ┘
```

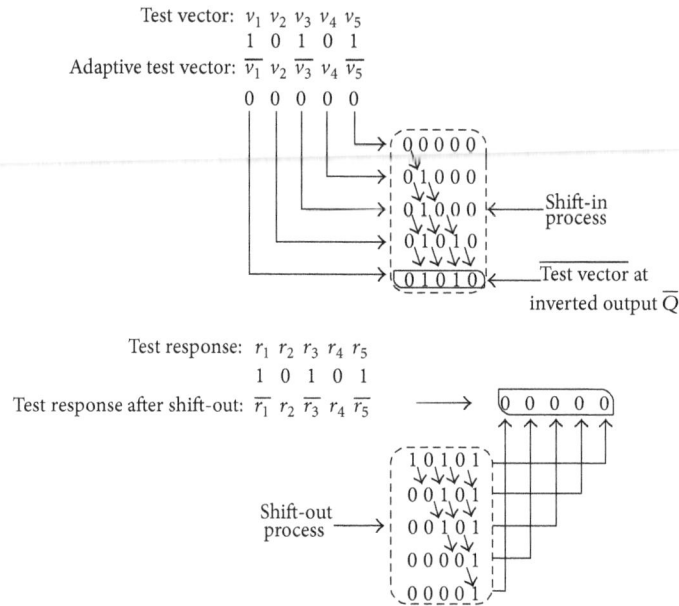

FIGURE 5: Adaptive scan process.

gating logic, we have utilized a transmission gate and an inverter to gate the scan output to the combinational logic. It uses the transmission gate to cut off the connection between the inverted scan cell output $\overline{Q}$ (used for stitching the scan cells) and the output Q of the scan cells during shift mode. As a result, the switching activities on the $\overline{Q}$ during shift mode does not affect the scan cell output Q which is used for driving the combinational logic. High resistance offered by an inactive transmission gate reduces the leakage current in the transmission gate during shift mode and response capture cycle since the transmission gate is idling in these intervals. In addition, transmission gate is a strong driver to feed the gating logic inverter and pseudo primary inputs during normal/capture mode.

3.2. State Preserving Logic. In order to totally prevent the unnecessary transitions to the combinational logic during shift mode, a state preserving logic has been proposed. It is a feedback structure that refreshes the scan output Q with the previous logic state. The two pull-up and pull-down sleep transistors are active during shift mode which makes the state preserving logic fixes the scan output logic to the same previous logic. However, unlike gating logic, this section is transparent in normal/capture mode of operation since the sleep transistors are inactive. During this mode, the state preserving logic consumes low leakage power since the two sleep transistors cut off the power rail. The two pull-up and pull-down transistors also contribute to active leakage reduction due to stacking effect [28, 29]. These can alleviate the effect of state preserving on peak power during normal/capture mode. The transmission gate and pull-up and pull-down sleep transistors are driven by shift enable signal SE so no extra control signal is required. Sharing the pull-up

and pull-down transistors of the state preserving logic among all the scan cells can alleviate the scan chain area overhead.

3.3. Inverted Scan Cell. We have removed part of slave latch in the conventional scan cell and exploited it as the inverted scan cell. The elimination of two by passed inverters and a transmission gate at the scan new shift path contributes to less switching activities inside the scan cell and consequently in scan chain. On the other hand, this reduces the number of switching activities needed for transferring the data to the scan cell's shift output. This results in average power reduction during shift mode while moderating the effect of gating and state preserving on peak power during capture mode. Although the inverted scan cell has master latch only, the overall scan architecture can still work as flip-flop because the inserted state preserving logic together with gating logic functions as slave latch in addition to their power reduction roles. Due to the shift path with less complexity, the scan chain speed has been accelerated during shifting. The reduced area and propagation delay due to the removal of two inverters and a transmission gate in the scan structure can moderate the area and delay overhead imposed by the augmented gating logic and also state preserving logic.

Eliminating the inverters in the scan cell structure does not affect the correctness of test patterns shifted into scan chain since adaptive test patterns can be used instead of original ones. The correct function of scan chain is achievable by special care of ATPG that generates the adaptive test patterns. The adaptive scan process has been summarized in Figure 5.

Let v_i represent a test pattern to be shifted to a scan cell. The adaptive test patterns are generated such that $\overline{v_i}$ is shifted to the destination of inverted scan cell after shift-in process. $\overline{v_i}$ is inverted to v_i after going through gating logic during test

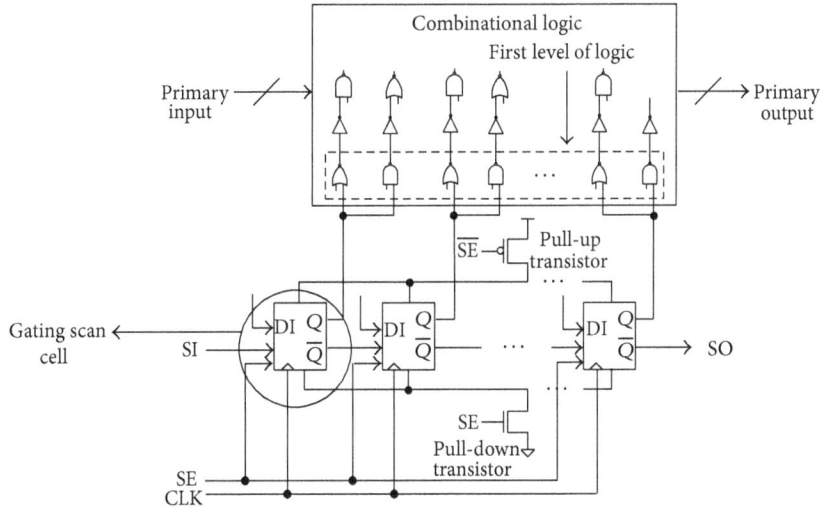

FIGURE 6: The proposed gating scan chain architecture.

application. Let v_i' be an adaptive test pattern and $1 \leq i \leq N$, where N is the number of scan cells in a scan chain. $v_i' = v_i$ for even index i and $v_i' = \overline{v_i}$ for odd index i whether N is an odd or even number. Adaptive test pattern formulation is as follows:

$$v_i' = v_i \quad \text{even index } i, \qquad (1)$$
$$v_i' = \overline{v_i} \quad \text{odd index } i.$$

Modifications are necessary for test responses after shift-out process since the responses are captured by the inverted scan cells. Let r_i represent the test response of a pseudo output. After being captured by the inverted scan cell, it becomes $\overline{r_i}$. Shifting process of $\overline{r_i}$ is taking place along the inverted scan cells. Let r_i' be the test response that reaches scan-out after shifting where $1 \leq i \leq N$ and N is the number of scan cells. $r_i' = \overline{r_i}$ for odd index i and $r_i' = r_i$ for even index i when N is an odd number while $r_i' = r_i$ for odd index i and $r_i' = \overline{r_i}$ for even index i when N is an even number. Modifying test pattern formulation is as follows.

For N number of scan cells when N is an odd number, consider the following:

$$r_i' = r_i \quad \text{even index } i, \qquad (2)$$
$$r_i' = \overline{r_i} \quad \text{odd index } i.$$

For N number of scan cells when N is an even number, consider the following:

$$r_i' = r_i \quad \text{odd index } i, \qquad (3)$$
$$r_i' = \overline{r_i} \quad \text{even index } i.$$

Therefore, inversion needs to be performed on r_i' when $r_i' = \overline{r_i}$ to retrieve the correct test response for response comparison.

Figure 6 illustrates the proposed scan chain architecture in full gating DFT method. In the application of partial gating,

$\overline{Q}$ output of either proposed gating scan cells or conventional scan cells should be used to form the scan chain for correct test pattern shifting operation. The area overhead imposed by proposed gating structure is only two extra transistors per several scan cells.

Similar to [9], our proposed method could eliminate all the transitions to the combinational logic. However, the difference between the proposed method and [9] is that we have replaced a part of slave latch with state preserving and gating logic and control it by scan enable (SE) signal which can result in two advantages compared to this method:

(1) less switching activity inside the proposed scan cell due to

(A) state preserving logic that is controlled by SE in the proposed gating scan cell, which causes less switching activity compared to the one controlled by clock,

(B) four less numbers of transistors in the scan cell structure;

(2) two pull-up and pull-down transistors have been used to control the state preserving, which can be shared among several scan cells. Thus, area overhead is reduced.

4. Experimental Results and Comparisons

To verify the effectiveness of the new gating scan cell architecture in terms of power and propagation delay over existing gating methods, we have conducted experiments to compare the proposed scan cell with the conventional scan cell, partial MUX gating scan cell, NOR gating scan cell, FLS, and modified scan cell reported in [9] which is one of the latest low power scan cells for gating. The initial simulations were performed using the Berkeley Predictive Technology Model (PTM) 45 nm CMOS technology (BSIM4) [30] in

TABLE 1: Comparison of total power consumption during shift and normal/capture mode.

	Prop. scan cell	FLS scan cell		Partial MUX gating scan cell		NOR gating scan cell		Modified scan cell		Conventional scan cell	
	Power con. (W)	Power con. (W)	% imp. over	Power con. (W)	% imp. over	Power con. (W)	% imp. over	Power con. (W)	% imp. over	Power con. (W)	% imp. over
Ave. total power	$1.5352E-06$	$2.6251E-06$	41.51	$2.5276E-06$	39.26	$2.7807E-06$	44.79	$1.8385E-06$	16.49	$2.1370E-05$	28.17

TABLE 2: Comparison of average power consumption during shift mode.

		Prop. scan cell	% imp.	FLS scan cell	% imp.	Partial MUX gating scan cell	% imp.	NOR gating scan cell	% imp.	Modified scan cell	% imp.
Avg. power (W)	Shift cycle #1	$1.7915E-06$	37.16	$3.4209E-06$	-19.98	$2.8857E-06$	-1.21	$3.5642E-06$	-25.01	$1.9972E-06$	29.94
	Shift cycle #2	$1.7880E-06$	37.42	$3.4172E-06$	-19.59	$2.8920E-06$	-1.21	$3.5699E-06$	-24.93	$1.9930E-06$	30.25
	Shift cycle #3	$1.7953E-06$	37.14	$3.4261E-06$	-19.94	$3.5013E-06$	-22.58	$3.5683E-06$	-24.92	$1.9920E-06$	30.25
	Shift cycle #4	$1.7947E-06$	37.15	$3.4282E-06$	-20.04	$3.5004E-06$	-22.57	$3.5680E-06$	-24.94	$1.9918E-06$	30.25

HSPICE by applying random patterns with the supply voltage of 1.0 Volt at room temperature. Since the main purpose of these primary experiments is to evaluate the proposed gating scan cell structure as the basic unit in the proposed gating scan chain in terms of power and propagation delay, all of the comparison parameters have been measured for the scan cell structure by applying random patterns and without considering combinational part. Due to the compact architecture of the gating scan cell with state preserving and gating logic, scan cells in FLS, partial MUX gating, and NOR gating schemes have been considered as a consistent entity with their gating structures. Table 1 contains comparisons of the mentioned scan cells with the proposed gating scan cell in terms of total average power during shift and normal/capture mode. In FLS scan cell, an inverter gate has been considered as the first level gate that is the most optimal condition for FLS. The percentage of improvements for proposed gating scan cell over other scan cells is displayed under each related scan cell column.

As observed from Table 1, the proposed gating scan cell exhibits 28.17% improvement in terms of average total power over the conventional scan cell. It means the power consumption in the proposed gating scan cell is even less than the original scan cell without any gating policy. This is expected because the proposed gating scan cell eliminates switching activity on the stimulus path of scan cell during shift mode. Also using shorter shift path that consists of less number of transistors compared to conventional scan cell intensifies power reduction impact of proposed scan cell. The proposed gating scan cell can save up to 44.79% average total power over existing gating scan cells such as NOR gating scan cell that has been used widely in most partial gating methods. It is noteworthy that, except proposed gating scan cell and modified scan cell, all of the compared scan cells experience increase of the total power consumption up to 30.12% compared to conventional scan cell. The proposed scan cell can save total power by 16.49% over modified scan cell since the gating and state preserving logic has been implemented with less number of switching activities.

In Table 2, the amount of power consumption for proposed gating scan cell and other existing gating scan cells during four-shift cycle has been shown. Each column contains the shift power consumption for related scan cell and the percentage of improvement over conventional scan cell. As it is observable from Table 2, the proposed gating scan cell outperforms existing gating scan cells in terms of shift power reduction inside the scan cell structure by 32.95% (on average) over conventional scan cell. Other gating schemes such as FLS, partial MUX gating, and NOR gating show increase in power consumption since they force the scan cell's driving path to a constant logic during shifting that can cause extra transient at this line (refer to Figure 7). Partial MUX gating consumes less shift power in the first two shift cycles compared to the other shift cycles. The reason is that in these shift cycles, the scan cell's driving path has the same value as the gating value when the mode changes to shift mode; thus, no redundant transition occurs.

Based on Table 2, unlike FLS scan cell, partial MUX gating scan cell, and NOR gating scan cell, the proposed gating scan cell can reduce power consumption in scan cell structure during shift mode. It is noteworthy that modified scan cell is able to reduce power only on the scan cell stimulus path by holding the logic during shift mode. However, besides power reduction on stimulus path via holding value on this line, the proposed gating scan cell can also reduce the switching activity inside the scan cell architecture. Hence, we achieve more power reduction compared to modified scan cell.

As mentioned before, one of the main concerns about gating designs is elevated peak power beyond the chip power

TABLE 3: Comparison of peak power consumption during normal/capture mode.

		Prop. scan cell	% incr.	FLS scan cell	% incr.	Partial MUX gating scan cell	% incr.	NOR gating scan cell	% incr.	Modified scan cell	% incr.
Peak power (W)	Capture cycle #1	5.3813E−05	−0.07	5.6058E−05	**4.09**	5.4240E−05	0.72	5.6006E−05	3.99	5.3778E−05	−0.13
	Capture cycle #2	5.3953E−05	−0.07	5.5041E−05	1.94	5.3995E−05	0.003	7.1611E−05	32.63	5.3959E−05	−0.07
	Capture cycle #3	5.3969E−05	0.003	5.4132E−05	0.30	5.3970E−05	0.005	7.1591E−05	32.65	5.3954E−05	−0.02
	Capture cycle #4	5.4454E−05	**0.94**	5.4746E−05	1.49	5.4242E−05	0.55	7.1578E−05	**32.69**	5.4100E−05	**0.29**
	Capture cycle #5	5.4589E−05	0.93	5.42010E−05	0.21	5.7592E−05	**6.48**	7.1567E−05	32.32	5.4089E−05	0.007

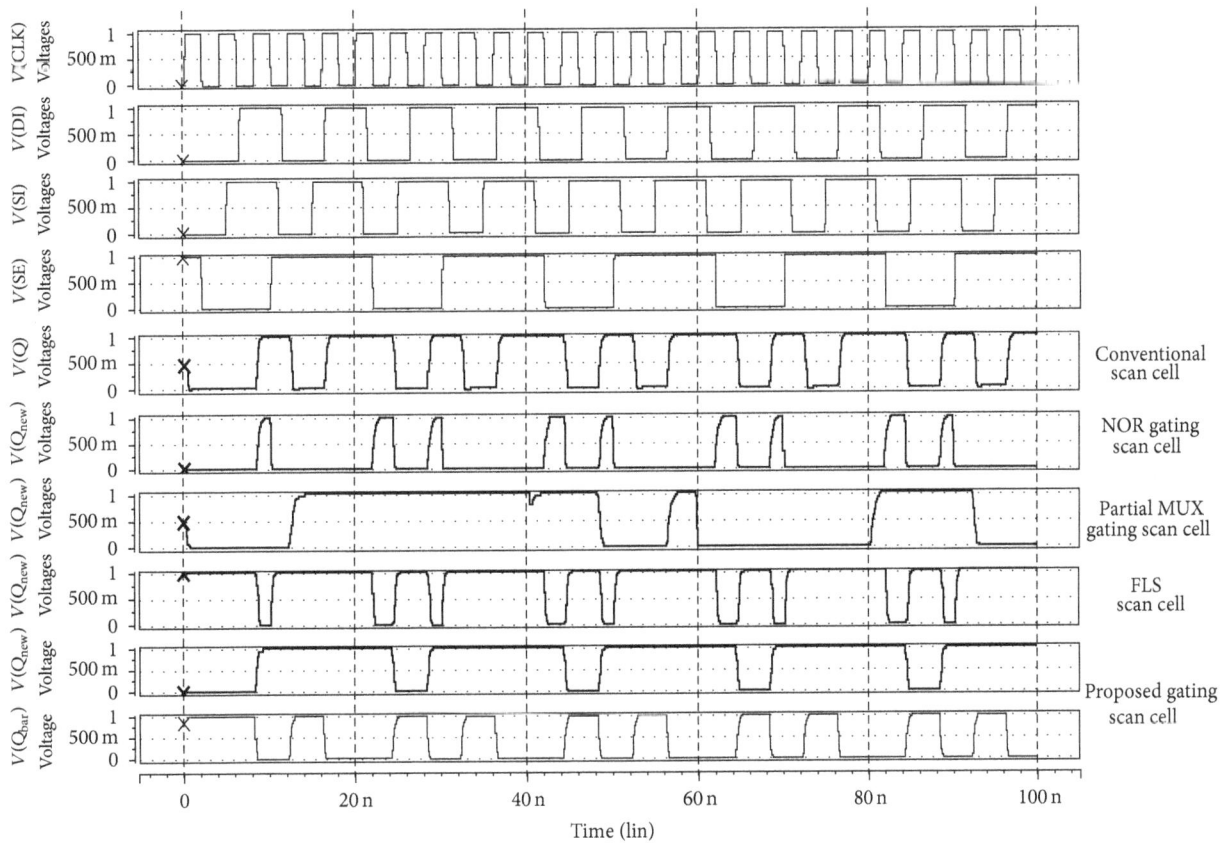

FIGURE 7: Driving path timing waveforms for existing gating architectures and the proposed gating scan cell.

budget caused by gating logic toggling between gating (shift) mode and transparent (normal/capture) mode. In order to show that the proposed gating scan cell has low impact on peak power increase, we have compared the percentage of peak power increase over conventional scan cell in the proposed gating scan cell and existing gating methods through five normal/capture cycles (each including one clock period).

The proposed gating scan cell and modified scan cell are able to reduce the highest peak power compared with existing gating scan cells. According to Table 3, the proposed scan cell elevates peak power up to 0.94% (less than 1%)

while the highest peak power increase in other gating scan cells is up to 32.69% (NOR gating scan cell). Unlike other compared gating scan cells, in case of NOR gating scan cell, peak power increase remains high during four successive capture cycles; thus, there is a strong possibility to elevate the temperature beyond the chip thermal budget. Figure 7 illustrates the impact of existing gating methods such as FLS, partial MUX gating along with proposed gating scan cell, on the stimulus path transients to combinational logic. As it is observable from the waveforms, in contrast with other gating methods and conventional scan cell, the state preserving and

TABLE 4: Comparison of propagation delay overhead.

	Proposed scan cell		FLS scan cell		Partial MUX gating scan cell		NOR gating scan cell		Modified scan cell	
	Ave. propa. delay (S)	% imp.	Ave. propa. delay (S)	% imp,	Ave. propa. delay (S)	% imp	Ave. propa. Delay (S)	% imp.	Ave. propa. delay (S)	% imp.
SI to shift output (shift delay)	$1.437E-09$	9.33	$1.587E-09$	-0.12	$1.587E-09$	-0.12	$1.586E-09$	-0.06	$1.448E-09$	8.64
DI to shift output (capture delay)	$9.458E-10$	12.82	$1.089E-09$	-0.36	$1.157E-09$	-6.63	$1.086E-09$	-0.09	$9.545E-10$	12.02
DI to Q_{new} (launch delay)	$1.107E-09$	-2.02	$1.247E-09$	-14.13	$1.271E-09$	-17.14	$1.249E-09$	-15.11	$1.135E-09$	-4.60
CLK to shift output	$1.373E-10$	51.75	$2.866E-10$	-0.70	$2.865E-10$	-0.66	$2.864E-10$	0.0	$1.477E-10$	48.1
CLK to Q_{new} (launch) output	$3.073E-10$	-7.97	$4.470E-10$	-57.06	$4.713E-10$	-65.60	$4.490E-10$	-17.84	$3.354E-10$	-17.84

gating logic is able to completely suppress all transients to the combinational logic. Q_{new} is the Q output of the gating scan cell that drives the combinational part and Q_{bar} represents the new shift output in proposed gating scan cell. As it is shown, the waveform of Q_{new} in the proposed scheme shows almost 50% reduction in transients for the same input patterns compared to FLS, partial MUX gating, NOR gating, and conventional scan cell, leading to less switching activity inside the combinational logic.

From the results it can be concluded that the proposed gating scan cell is able to improve power consumption compared to existing gating methods. This is not only because of power reduction inside the scan cell, but also according to Figure 7 due to power reduction in combinational logic by holding the logic on the scan cell's stimulus path during shifting. The proposed scan architecture also has contributed to accelerating shift because the inverted scan cell has less number of transistors and gating logic and state preserving logic do not involve the scan shift propagation path. Also, due to shorter shift output the captured data will be available faster on the shift output. Therefore, improvements in shift and normal/capture propagation delay over FLS, partial MUX gating scheme, and NOR gating have been observed. A comparative impact of the existing scan gating schemes including the proposed architecture on the scan cell delay parameters has been summarized in Table 4. For all compared scan cells the amount of average propagation delay is followed by the percentage of improvement over conventional scan cell.

According to Table 4, propagation delay improvements in shift and capture have been achieved over FLS scan cell, partial MUX gating scan cell, NOR gating scan cell, and modified scan cell. The clock to shift output delay has been improved in the proposed gating scan cell by 51.75% over conventional scan cell. Moreover, the proposed scan cell improves data shift and capture propagation delays by 9.33% and 12.82%, respectively, compared to conventional scan cell while the delay penalty on data launch and clock to launch output is 2.02% and 7.97%, respectively. However, it still shows significant reduction in launch delay penalties compared to

the scan cells in other gating schemes. Thus, we can expect that the proposed gating scan chain will be able to keep performance degradation caused by gating under acceptable threshold.

Since the layout rules for feature size of 45 nm gate length are not available, the measure that has been used for area overhead calculations is the total transistor active area ($W * L$ for a transistor). In order to evaluate the proposed gating scan cell in terms of area overhead, we have compared it with FLS and NOR gating that have the least area penalty in turn among existing gating techniques. The routing overhead has not been considered in area overhead. However, the routing overhead associated with the proposed gating structure should not be high since no additional control signal is required. In the proposed gating architecture, similar to NOR gating, area overhead increases as the number of scan cells in the design increases. The area overhead in FLS does not depend on the number of scan cells directly but the numbers of first level gates are driven by scan cells. This is because gating transistor and pull up transistor in FLS have been introduced on each gate in the first level of combinational gates. Therefore, we consider ISCAS'89 benchmark circuits to study the impact of proposed gating architecture and existing gating techniques on area overhead. Referring to [22], for random input patterns, approximately half of the first level gates are switching at the same time and the gating transistor for idle gates is not actually used. Therefore, the channel width for the shared gating transistor in FLS has been considered half of the sizes of unshared gating transistors and is given by

$$W_{gating} = 0.5 * FO * (10 * W_{min}), \qquad (4)$$

where FO is the number of first level gates. In the proposed gating scan cell the pull-up and pull-down transistors in state preserving logic have been shared among all scan cells. Like FLS, we have assumed that for random input patterns nearly half of the scan cells do not switch and the pull-up and pull-down transistors are practiced for almost half of the scan cells

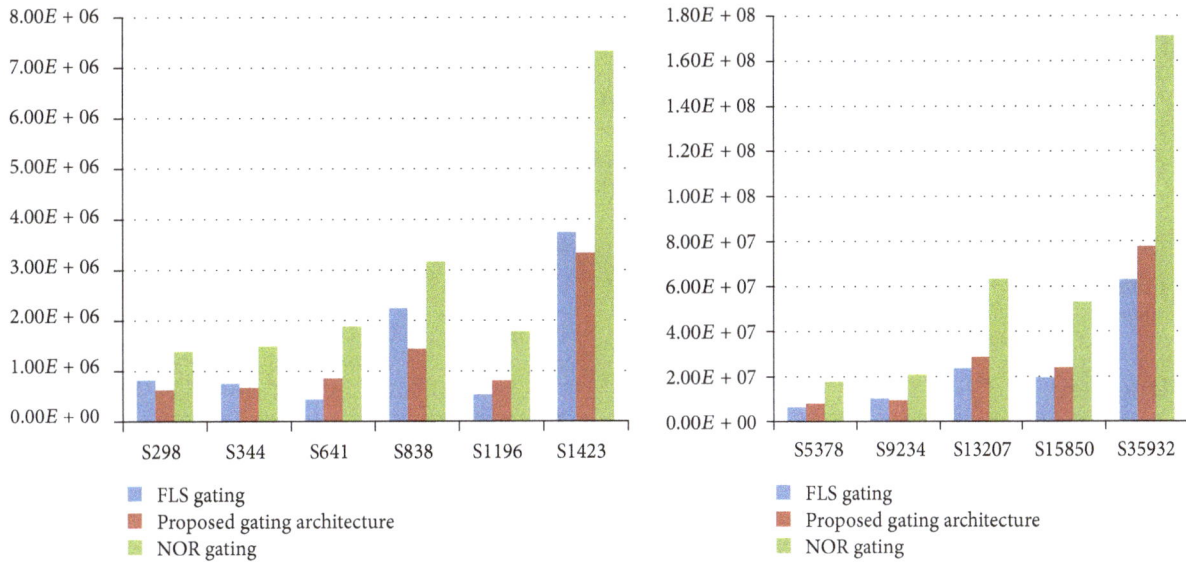

FIGURE 8: Area overhead comparisons.

at the same time. Thereby, the channel width for shared pull-up and pull-down transistors is chosen to be as follows:

$$W_{\text{pull up}} = 0.5 * FF * \left(5 * W_{\text{min(Pmos)}}\right),$$

$$W_{\text{pull down}} = 0.5 * FF * \left(5 * W_{\text{min(Nmos)}}\right), \tag{5}$$

where FF is the number of scan cells. The comparative impact of the proposed gating architecture and existing gating techniques on area increase in nm^2 scale for different benchmark circuits has been shown in Figure 8.

According to Figure 8, the proposed gating architecture exhibits 21.42% improvement on average in terms of area overhead compared to NOR gating for all benchmark circuits. As mentioned earlier, area overhead in FLS is highly affected by the number of unique fanouts (nonoverlapped fanout cones) of scan cells in which gating transistor and pull-up transistor are inserted. Therefore, as this ratio increases above two in benchmark circuits, the proposed gating architecture saves more area up to 11.29% (s838) compared to FLS. However, the proposed gating architecture loses its efficiency in area improvement over FLS in benchmark circuits with the ratio of the unique fanouts to number of scan cells less than two. The highest area penalty for the proposed gating architecture over FLS is 12.24% in s641 which has equal number of unique fanouts and scan cells (ratio = 1). However, in large industrial designs the number of fanouts of scan cell is usually more than one. Thus, the proposed gating architecture area overhead over FLS is not usually high. The area overhead degradation compared to FLS for all benchmark circuits is on average 1.18% for an average 1.8 unique fanouts per scan cell.

Power and delay improvements for scan cell can be extended to the whole scan chain in our future works of experiments. In order to verify that the proposed architecture keeps its efficiency in terms of power and performance for the scan chains with large number of scan cells, postlayout simulations will be conducted on the benchmark circuits.

Further simulations will be established on the various benchmark circuits by applying adaptive test patterns to clarify the exact amount of dynamic power improvement in the combinational logic and results will be presented in near future.

5. Conclusion

Scan gating approach offers simple yet effective solution to reduce shift power significantly independent of test set and is less intrusive to the design. However, gating logics have been exploited as the existing gating methods add significant delay to signal propagation paths. Moreover, they suffer from large overhead in terms of area and switching activity inside the gating logics in normal/capture mode that has high negative impact on peak power. Likewise, shift power reduction in most existing gating approaches has mostly been concentrated on combinational logic while scan part has received less attention. We proposed a novel gating scan architecture as an integrated solution for shift power reduction in both scan cells and combinational logic under area constraints. The proposed gating scan cells in this structure are modified scan cells augmented by gating and state preserving logics to gate and hold the scan cells stimulus path with low impact on peak power. Critical shift timing has been improved by using a less complex shift path in new gating scan cell. Compared to the lowest cost gating techniques the proposed gating scan cell has less DFT overhead with respect to average power, shift, and capture delay. Similar to other gating methods such as modified scan cell, the proposed gating architecture can be used for delay fault testing. The proposed gating scan cell can be applied effectively to both full gating and partial gating methods. The fault coverage is not degraded and it does not face routing problems since no additional control signal has been employed. The proposed scheme can efficiently be utilized in

BIST architecture. It can also be applied together with other scan-based power optimization methods such as scan chain partitioning and reordering techniques. Low power ATPG-based methods can be applied to the proposed structure efficiently. Therefore, these make the proposed approach a potential candidate for scan-based DFT architectures.

Conflict of Interests

The authors declare that there is no conflict of interests regarding the publication of this paper.

Acknowledgment

This research is partly sponsored by UTM Research University Grant vote no. 07H05.

References

[1] P. Girard, "Survey of low-power testing of VLSI circuits," *IEEE Design & Test of Computers*, vol. 19, no. 3, pp. 82–92, 2002.

[2] J. Altet and A. Rubio, *Thermal Testing of Integrated Circuits*, Springer, Boston, Mass, USA, 2002.

[3] A. Bosio, L. Dilillo, P. Girard, A. Todri, and A. Virazel, "Why and how controlling power consumption during test: a survey," in *Proceedings of the IEEE 21st Asian Test Symposium (ATS '12)*, pp. 221–226, November 2012.

[4] P. Girard, X. Wen, and N. A. Touba, "Low Power testing," in *System-on-Chip Test Architectures: Nanometer Design for Testability*, chapter 7, pp. 207–350, Morgan Kaufmann, 2006-2007.

[5] P. Narayanan, R. Mittal, S. Poddutur, V. Singhal, and P. Sabbarwal, "Modified flip-flop architecture to reduce hold buffers and peak power during scan shift operation," in *Proceedings of the 29th IEEE VLSI Test Symposium (VTS '11)*, pp. 154–159, May 2011.

[6] P. Rosinger, B. M. Al-Hashimi, and N. Nicolici, "Scan architecture with mutually exclusive scan segment activation for shift- and capture-power reduction," *IEEE Transactions on Computer-Aided Design of Integrated Circuits and Systems*, vol. 23, no. 7, pp. 1142–1153, 2004.

[7] S. Gerstendörfer and H.-J. Wunderlich, "Minimized power consumption for scan-based BIST," in *Proceedings of the IEEE International Test Conference (ITC '99)*, pp. 77–84, IEEE, Atlantic City, NJ, USA, September 1999.

[8] S. Bhunia, H. Mahmoodi, D. Ghosh, S. Mukhopadhyay, and K. Roy, "Low-power scan design using first-level supply gating," *IEEE Transactions on Very Large Scale Integration (VLSI) Systems*, vol. 13, no. 3, pp. 384–395, 2005.

[9] A. Mishra, N. Sinha, V. Satdev, S. Chakravarty, and A. D. Singh, "Modified scan flip-flop for low power testing," in *Proceedings of the 19th IEEE Asian Test Symposium (ATS '10)*, pp. 367–370, Shanghai, China, December 2010.

[10] K. Paramasivam, K. Gunavathi, and A. Nirmalkumar, "Modified scan architecture for an effective scan testing," in *Proceedings of the IEEE Region 10 Conference (TENCON '08)*, pp. 1–6, IEEE, Hyderabad, India, November 2008.

[11] M. L. Bushnell and V. D. Agrawal, *Essentials of Electronic Testing for Digital Memory, and Mixed-Signal VLSI Circuits*, Kluwer Academic Publishers, Boston, Mass, USA, 2000.

[12] S. P. Khatri and S. K. Ganeshan, "A modified scan-D flip flop to reduce test power," in *Proceedings of the 15th IEEE International Test Synthesis Workshop (ITSW '08)*, 2008.

[13] X. Zhang and K. Roy, "Power reduction in test-per-scan BIST," in *Proceedings of the 6th IEEE International On-Line Testing Workshop*, pp. 133–138, Palma de Mallorca, Spain, 2000.

[14] N. Parimi and X. Sun, "Design of a low power D flip-flop for test-per-scan circuits," in *Proceedings of the IEEE Canadian Conference on Electrical and Computer Engineering*, vol. 2, pp. 777–780, May 2004.

[15] S. Bhunia, H. Mahmoodi, A. Raychowdhury, and K. Roy, "First level hold: a novel low-overhead delay fault testing technique," in *Proceedings of the 19th IEEE International Symposium on Defect and Fault Tolerance in VLSI Systems*, pp. 314–315, October 2004.

[16] S. Bhunia, H. Mahmoodi, D. Ghosh, S. Mukhopadhyay, and K. Roy, "Arbitrary two-pattern delay testing using a low-overhead supply gating technique," *Journal of Electronic Testing*, vol. 24, no. 6, pp. 577–590, 2008.

[17] A. K. Suhag, S. Ahlawat, V. Shrivastava, and N. Singh, "Elimination of output gating performance overhead for critical paths in scan test," *International Journal of Circuits and Architecture Design*, vol. 1, no. 1, pp. 62–73, 2013.

[18] S. Sharifi, J. Jaffari, M. Hosseinabady, A. Afzali-Kusha, and Z. Navabi, "Simultaneous reduction of dynamic and static power in scan structures," in *Proceedings of the Design, Automation and Test in Europe (DATE '05)*, vol. 2, pp. 846–851, March 2005.

[19] M. Elshoukry, M. Tehranipoor, and C. P. Ravikumar, "A critical-path-aware partial gating approach for test power reduction," *ACM Transactions on Design Automation of Electronic Systems*, vol. 12, no. 2, Article ID 1230809, 2007.

[20] R. Sankaralingam and N. Touba, "Inserting test points to control peak power during scan testing," in *Proceedings of the 17th IEEE International Symposium on Defect and Fault Tolerance in VLSI Systems (DFT '02)*, pp. 138–146, IEEE, 2002.

[21] X. Kavousianos, D. Bakalis, and D. Nikolos, "Efficient partial scan cell gating for low-power scan-based testing," *ACM Transactions on Design Automation of Electronic Systems*, vol. 14, no. 2, article 28, 2009.

[22] D. Jayaraman, R. Sethuram, and S. Tragoudas, "Gating internal nodes to reduce power during scan shift," in *Proceedings of the 20th Great Lakes Symposium on VLSI (GLSVLSI '10)*, pp. 79–84, May 2010.

[23] X. Lin and Y. Huang, "Scan shift power reduction by freezing power sensitive scan cells," *Journal of Electronic Testing*, vol. 24, no. 4, pp. 327–334, 2008.

[24] X. Lin and J. Rajski, "Test power reduction by blocking scan cell outputs," in *Proceedings of the 17th Asian Test Symposium (ATS '08)*, pp. 329–336, November 2008.

[25] W. Zhao, M. Tehranipoor, and S. Chakravarty, "Power-safe test application using an effective gating approach considering current limits," in *Proceedings of the 29th IEEE VLSI Test Symposium (VTS '11)*, pp. 160–165, IEEE, Dana Point, Calif, USA, May 2011.

[26] Y.-T. Lin, J.-L. Huang, and X. Wen, "A transition isolation scan cell design for low shift and capture power," in *Proceedings of the IEEE 21st Asian Test Symposium (ATS '12)*, pp. 107–112, November 2012.

[27] S. Bhunia, H. Mahmoodi, D. Ghosh, and K. Roy, "Power reduction in test-per-scan BIST with supply gating and efficient scan partitioning," in *Proceedings of the 6th International Symposium*

on Quality Electronic Design (ISQED '05), pp. 453–458, March 2005.

[28] B. H. Calhoun, F. A. Honore, and A. Chandrakasan, "Design methodology for fine-grained leakage control in MTCMOS," in *Proceedings of the International Symposium on Low Power Electronics and Design (ISLPED '03)*, pp. 104–109, August 2003.

[29] K. Roy, S. Mukhopadhyay, and H. Mahmoodi-Meimand, "Leakage current mechanisms and leakage reduction techniques in deep-submicrometer CMOS circuits," *Proceedings of the IEEE*, vol. 91, no. 2, pp. 305–327, 2003.

[30] Predictive Technology Model, 2013, http://ptm.asu.edu/.

## 17

Low-Area Wallace Multiplier

Shahzad Asif and Yinan Kong

Department of Engineering, Macquarie University, Sydney, NSW 2109, Australia

Correspondence should be addressed to Shahzad Asif; shahzad.asif@mq.edu.au

Academic Editor: Yu-Cheng Fan

Multiplication is one of the most commonly used operations in the arithmetic. Multipliers based on Wallace reduction tree provide an area-efficient strategy for high speed multiplication. A number of modifications are proposed in the literature to optimize the area of the Wallace multiplier. This paper proposed a reduced-area Wallace multiplier without compromising on the speed of the original Wallace multiplier. Designs are synthesized using Synopsys Design Compiler in 90 nm process technology. Synthesis results show that the proposed multiplier has the lowest area as compared to other tree-based multipliers. The speed of the proposed and reference multipliers is almost the same.

1. Introduction

Multiplication is one of the most widely used arithmetic operations. Due to this a wide range of multiplier architectures are reported in the literature providing flexible choices for various applications. Among them the simplest is array multiplier [1] which is also the slowest. Some high performance multipliers are presented in [2–5]. The focus of this paper is Wallace multiplier [6]. Wallace multiplier uses full adders and half adders to reduce the partial product tree to two rows, and then a final adder is used to add these two rows of partial products. We call this design "TW (traditional Wallace) multiplier" in this text. TW multiplier performs its operation in three steps. (1) Generate all the partial products. (2) The partial product tree is reduced using full adders and half adders until it is reduced to two terms. (3) Finally, a fast adder is used to add these two terms.

Waters and Swartzlander [7] presented a reduced complexity Wallace multiplier by reducing the number of half adders in the reduction process. We call this design "RCW (reduced complexity Wallace) multiplier" from now on. The speed of the RCW multiplier is expected to be the same as of TW multiplier due to the equal number of reduction stages in both multipliers. The RCW uses a larger final adder as compared to the TW multiplier. A number of strategies

are reported in [8–10] to improve the speed of the RCW. However, the focus of their research is to reduce the delay by using a faster final adder while still using the same reduction tree as RCW. As a result, the final adder size for the multipliers in [8–10] is the same as that of RCW.

The focus of this paper is to optimize the reduction tree in a way that can reduce the size of the final adder. The reduced size of the final adder resulted in low area of the multiplier without incurring any extra delay. We call our design "PW (Proposed Wallace) multiplier." We also considered Dadda multiplier [11] for comparison due to its similarity with the Wallace multiplier.

This paper makes a contribution in the design of Wallace treed based multipliers by proposing a strategy to reduce the area of reduced complexity Wallace (RCW) multiplier. This innovative method allows for an effective utilization of half adders in such a way that the size of the final adder is reduced. It also provides a more regular structure of the reduction tree and the final adder.

The rest of the paper is organized as follows. Section 2 discusses some previous approaches for partial product tree reduction. In Section 3, the proposed Wallace multiplier is presented. In Section 4, the choice of final adder is discussed. Section 5 evaluates the results for all the designs synthesized in Synopsys. The work is concluded in Section 6.

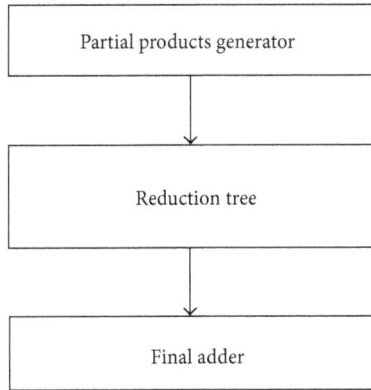

FIGURE 1: Block diagram of tree-based multipliers.

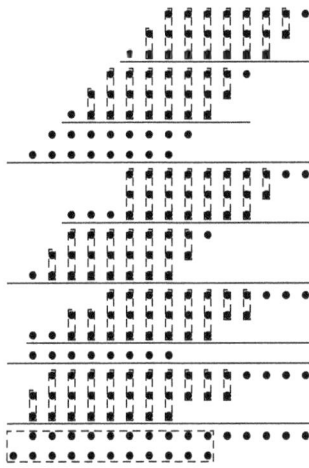

FIGURE 3: 8-bit reduced complexity Wallace reduction.

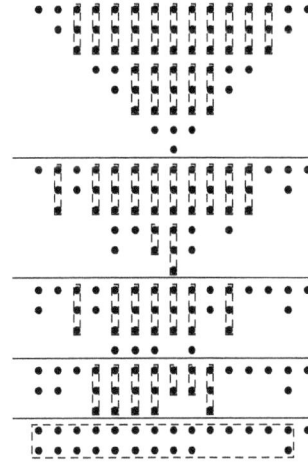

FIGURE 2: 8-bit traditional Wallace reduction.

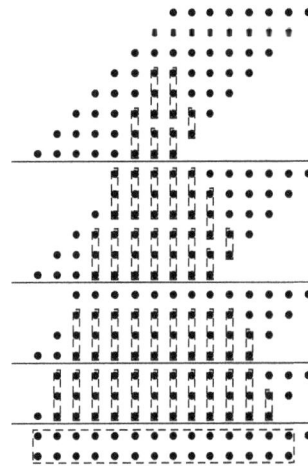

FIGURE 4: 8-bit Dadda reduction.

2. Previous Architectures

This section discusses some previous Wallace tree-based multiplier architectures. The general block diagram of tree-based multipliers is shown in Figure 1.

The dot notation [11] is used to represent the partial product tree in all the architectures discussed in this section as shown from Figures 2 to 5. The full adders and half adders are represented by boxes around the dot products. The box which encloses three dot products represents a full adder, whereas the box containing only two dot products is used to represent a half adder. The stages are separated by a thick horizontal line.

2.1. Traditional Wallace (TW) Multiplier.
In TW multiplier architecture, the partial product tree is divided into groups [6]. Each stage can have one or more groups as shown in the 8-bit TW reduction process in Figure 2.

The groups in a stage are separated by a thin horizontal line. Each group consists of three rows except the last group where the number of rows can be less than three. Equation (1) calculates the number of rows in the last group for each stage as

$$\text{Last\_Group}_i = r_i \bmod 3. \tag{1}$$

An N-bit multiplier has N rows in the first stage. The number of rows in remaining stages can be calculated by using

$$r_i = \left\lfloor \frac{2r_{i-1}}{3} \right\rfloor + r_{i-1} \bmod 3. \tag{2}$$

Reduction is performed using a full adder or a half adder depending on the number of elements in that particular column of the group. If a column has only one element then that is passed on to the next stage without any reduction. If the last group of a stage contains less than three rows then no reduction is performed on that group as shown in stage 1 of Figure 2.

The size of the final adder for an N-bit TW multiplier with S stages can be calculated by

$$\text{FinalAdder}_{\text{TW}} = (2N - 1) - S. \tag{3}$$

TABLE 1: Final adder size for different multipliers.

N	Final adder size				Logic levels
	TW	RCW	Dadda	PW	
8	11	14	14	10	4
16	25	30	30	24	5
24	40	46	46	39	6
32	55	62	62	54	6
64	117	126	126	116	7

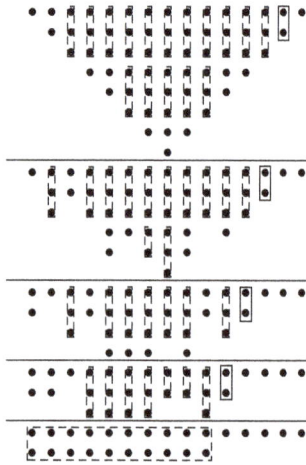

FIGURE 5: 8-bit proposed Wallace reduction.

3. Proposed Wallace (PW) Multiplier

In this section, we proposed a modification in the RCW multiplier to further reduce its area by reducing the size of the final adder. PW multiplier has the same number of stages and the same rule for maximum number of rows in a stage as in the other multipliers discussed in this paper.

An 8-bit PW reduction process is shown in Figure 5. PW uses an additional half adder in each stage in order to reduce the size of the final adder. The algorithm scans from the right side and starts the reduction by using a half adder when it finds the first column where the number of elements is greater than one. The additional half adders are shown in solid boxes at each stage of PW multiplier in Figure 5.

Compared with RCW in Figure 3, the introduction of half adders in Figure 5 makes the final adder in PW "less wide", namely, a smaller size. This is because, in each stage, the half adder that we introduced computes the final product bit for that particular column of the partial product tree. Therefore, the size of the required final adder is decreased by one in each stage. The least significant bit (LSB) of the product, P_0, is produced by the partial product generation block by computing $A_0 \times B_0$. In the first stage of the reduction process, product bit P_1 is computed by using the additional half adder. In the second stage, P_2 is computed. Similarly, stage 3 and stage 4 compute the product bits P_3 and P_4, respectively. Thus, when the partial product tree is reduced to two rows, five LSBs $(P_4 - P_0)$ of the product are already computed as shown in Figure 5. Therefore, the size of the final adder in 8-bit PW is reduced by four as compared to the final adder in RCW multiplier. The size of the final adder for an N-bit PW multiplier with S stages can be computed by

$$\text{FinalAdder}_{\text{PW}} = (2N - 2) - S. \qquad (5)$$

The comparison of (4) and (5) shows a reduction of S in the size of final adder from RCW to PW. This is achieved at the expense of an increased area for reduction process due to the insertion of additional half adders in the PW. However, the effect of additional half adders is very small as compared to the area saved by reducing the final adder size. Therefore, the overall area of the PW is less than that of the RCW.

The size of the final adder and their required logic levels for different multipliers are given in Table 1. The PW has the smallest final adder as compared to all other multipliers. All the multipliers need the same number of logic levels to implement the final adder, which means that all the multipliers will have almost the same delay. The architecture of the final adder is discussed in Section 4.

2.2. Reduced Complexity Wallace (RCW) Multiplier. Waters and Swartzlander [7] presented a modification in the TW multiplier to reduce the complexity of the reduction tree. An 8-bit RCW reduction process is shown in Figure 3.

The partial products are readjusted in a reverse pyramid style which makes it easy to analyse the tree for efficient reduction. The number of stages for RCW multiplier remains the same as that of TW multiplier. RCW tries to reduce the partial product tree using only full adders. Half adders are used only where they are necessary to satisfy the number of rows in a stage according to (2). This approach allows RCW multiplier to reduce the area of the reduction process. However, RCW multiplier uses a much larger final adder as compared to TW multiplier. The size of the final adder for an N-bit RCW multiplier can be computed by

$$\text{FinalAdder}_{\text{RCW}} = 2N - 2. \qquad (4)$$

2.3. Dadda Multiplier. Dadda multiplier [11] tries to reduce the number of full adders and half adders by performing the reduction only where it is essential to satisfy (2). An 8-bit Dadda reduction process is shown in Figure 4.

Dadda has the same number of stages as that of TW and RCW. We can see from Figure 4 that Dadda performed reduction only on four columns in stage 1. This is because the other columns already satisfy (2). The same approach is used in all stages to reduce the tree until we achieve a tree of only two rows. The size of the final adder is the same for Dadda and RCW multiplier as computed in (4).

Table 2: Synthesis parameters for Synopsys Design Vision.

Technology	90 nm CMOS
Supply voltage	1.2 V
Temperature	25°C
Process model	Typical
Interconnect model	Balanced tree

4. Final Adder Design

The third step of the Wallace tree-based multipliers is to add the remaining two rows using a fast adder. Some of the most widely used parallel-prefix adders used for high speed operations are Kogge-Stone [12], Sklansky [13], and Brent-Kung [14]. These adders use the same tree topology but differ in terms of logic levels, fanout, and interconnect wires. We used Kogge-Stone adder in all the multipliers discussed in this paper. The logic levels for implementation of an N-bit Kogge-Stone adder can be calculated by using

$$\text{LogicLevels} = \lceil \log_2(N) \rceil . \tag{6}$$

5. Results

In this section, we will discuss the verification of designs for correct operation, synthesis tool, and the results.

5.1. Functional Verification. The multipliers are implemented in VHDL with the test programs to verify the designs. All the possible input combinations are applied to thoroughly test the 8-bit multipliers. Since an exhaustive testing of bigger multipliers was not practical, they are tested with random inputs applied. The Galois-type linear feedback shift registers (LFSRs) are designed to generate pseudorandom binary sequence (PRBS) of maximum cycle for the multipliers under test [15]. All the designs are compiled and simulated using Synopsys VCS.

5.2. Synthesis Tool. All the multipliers are synthesized in Synopsys Design Compiler (DC) using 90 nm technology. The designs can be optimized for delay, power, and area by setting the appropriate options in the DC. The designer has the option of setting the various synthesis parameters such as fanout, wire load models, interconnect strategy, and PVT (process, voltage, and temperature).

The scripts are written to synthesize the TW, RCW, Dadda, and PW multipliers for optimized area. In order to have a fair comparison, the same synthesis parameters are specified for all the designs. Table 2 shows different parameters from SAED 90 nm library used for synthesis.

5.3. Synthesis Results. The detailed synthesis reports are generated by Design Compiler for area and timing. The area report includes number of cells, the area used by cells, and the interconnect area. The timing report shows the complete critical path along with the delay associated with each cell in the path. Table 3 shows the synthesis results for delay and area for different multipliers.

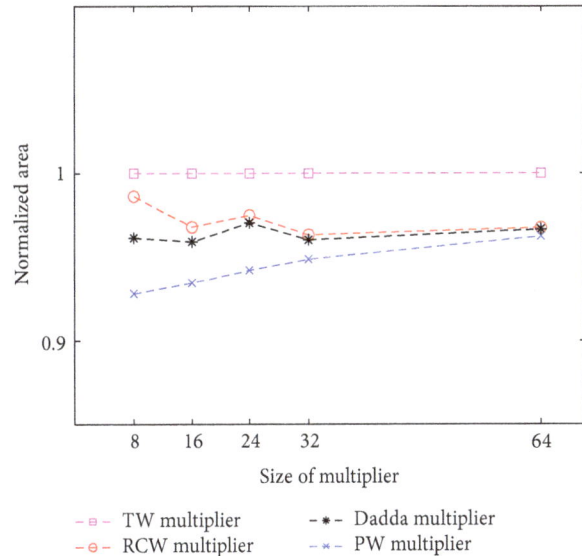

Figure 6: Area of different multipliers on Synopsys 90 nm technology.

Figure 6 shows the normalized area for each multiplier. The area of all multipliers is normalized with respect to TW multiplier by using

$$\text{Norm\_Value}_{\text{Mult\_XX}} = \frac{\text{Original\_Value}_{\text{Mult\_XX}}}{\text{Original\_Value}_{\text{TW\_Mult}}} . \tag{7}$$

It is clear from Figure 6 that the TW has the largest area as expected. Areas of RCW and Dadda are almost the same which also conforms to the results of [7]. PW has the lowest area for all the multiplier configurations. The reduction in area is more prominent when the size of the multiplier is small. As the size of the multiplier increases, the area of PW tends to asymptotically approach the area of RCW and Dadda multiplier. Therefore, the PW is particularly useful when the final adder has a significant area in the multiplier, while the area advantage might decrease in larger multipliers.

Figure 7 shows the normalized delay for each multiplier. The delays are normalized according to (7). All the multipliers use the same number of reduction stages and the same logic levels in the final adder. Therefore their delays are expected to be the same. However, the Design Compiler uses different cells to optimize the area in each design due to their different architectures and different final adder sizes. This can result in larger delays for the designs where the synthesizer can optimize the area by using relatively slower standard cells. It can be seen in Figure 7 that the PW has the least delay in 24-bit multiplier. In the rest of the multiplier sizes, PW has almost the same delay as of RCW which is less than the Dadda and TW. One exception to this is the 16-bit multiplier where PW has larger delay than RCW multiplier.

Figure 8 shows the normalized power consumption for each multiplier. The power consumption of all multipliers is normalized with respect to TW multiplier by using (7).

TABLE 3: Synthesis results from Synopsys DC on 90 nm technology.

Size	Delay (ns)				Area (μm^2)				Power (mW)			
	TW	RCW	Dadda	PW	TW	RCW	Dadda	PW	TW	RCW	Dadda	PW
8	2.81	2.64	2.64	2.66	3392	3346	3262	3148	1.92	2.04	1.94	1.96
16	4.13	3.65	3.80	3.75	14847	14372	14242	13876	11.09	11.41	11.22	11.20
24	4.64	4.58	4.62	4.44	34337	33479	33323	32352	27.69	28.24	28.32	28.04
32	14.83	14.62	14.82	14.62	61526	59271	59086	58375	51.85	52.74	53.11	52.48
64	22.88	21.57	21.98	21.62	246842	238843	238597	237553	216.50	219.17	222.64	218.92

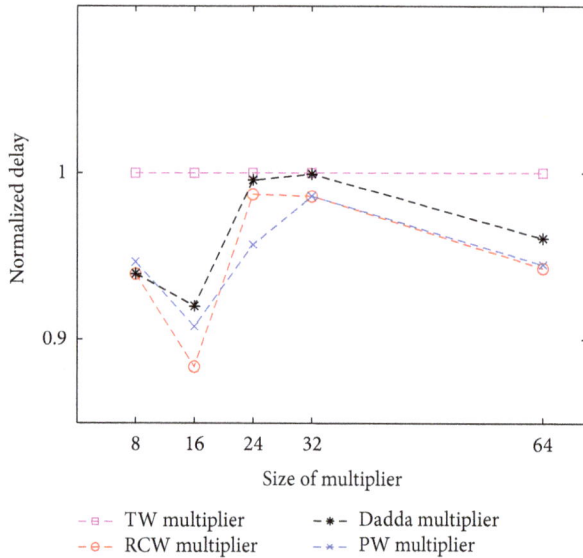

FIGURE 7: Delay of different multipliers on Synopsys 90 nm technology.

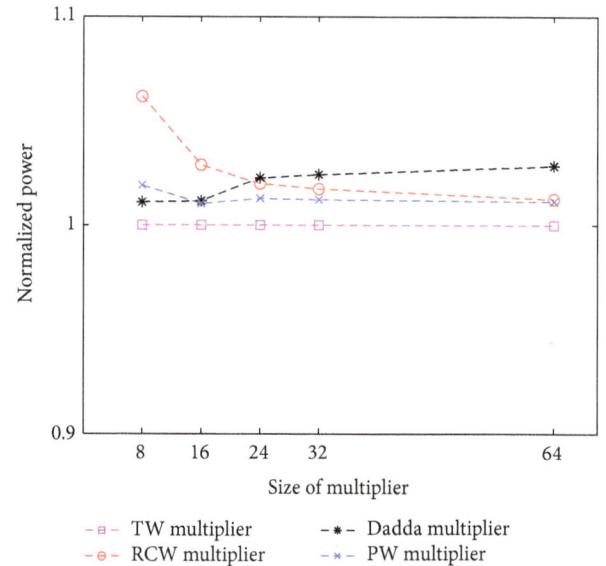

FIGURE 8: Power consumption of different multipliers on Synopsys 90 nm technology.

It is clear from Figure 8 that the TW multiplier has the lowest power consumption as compared to the other multipliers. One reason for this could be that the Design Compiler was able to find the low-power cells to synthesize the TW multiplier. The regular structure of TW multiplier could also be a reason of its low-power consumption. The power consumption of PW is less than that of the RCW multiplier due to the smaller final adder used in PW multiplier. It can be noted that the difference in power consumption of PW and RCW is very little for large multipliers, as expected, due to the small difference in their area.

6. Conclusion and Future Work

This paper presents a method to reduce the area of the Wallace multiplier. The proposed architecture, named as PW (proposed Wallace) multiplier, uses a smaller final adder to reduce the area of a multiplier. The designs are synthesized in Synopsys Design Compiler using 90 nm process technology. The synthesis results verify that the PW multiplier, as expected, has the smallest area as compared to the other Wallace based multipliers. The speed of the PW multiplier is almost the same as of other multipliers.

As our future work, we plan to implement the designs using Synopsys IC Compiler to analyze the postlayout results for area and delay. Synopsys Prime Time can be used to analyze the multipliers for their power consumption.

Conflict of Interests

The authors declare that there is no conflict of interests regarding the publication of this paper.

References

[1] N. H. E. Weste and D. M. Harris, *Integrated Circuit Design*, Pearson, 2010.

[2] J.-Y. Kang and J.-L. Gaudiot, "A fast and well-structured multiplier," in *Proceedings of the EUROMICRO Systems on Digital System Design (DSD '04)*, pp. 508–515, September 2004.

[3] C. R. Baugh and B. A. Wooley, "A twos complement parallel array multiplication algorithm," *IEEE Transactions on Computers*, vol. C-22, no. 12, pp. 1045–1047, 1973.

[4] S.-R. Kuang, J.-P. Wang, and C.-Y. Guo, "Modified booth multipliers with a regular partial product array," *IEEE Transactions on Circuits and Systems II: Express Briefs*, vol. 56, no. 5, pp. 404–408, 2009.

[5] B. C. Paul, S. Fujita, and M. Okajima, "ROM-based logic (RBL) Design: a low-power 16 bit multiplier," *IEEE Journal of Solid-State Circuits*, vol. 44, no. 11, pp. 2935–2942, 2009.

[6] C. S. Wallace, "A suggestion for a fast multiplier," *IEEE Transactions on Electronic Computers*, vol. EC-13, no. 1, pp. 14–17, 1964.

[7] R. S. Waters and E. E. Swartzlander, "A reduced complexity wallace multiplier reduction," *IEEE Transactions on Computers*, vol. 59, no. 8, pp. 1134–1137, 2010.

[8] S. Rajaram and K. Vanithamani, "Improvement of Wallace multipliers using parallel prefix adders," in *Proceedings of the International Conference on Signal Processing, Communication, Computing and Networking Technologies (ICSCCN '11)*, pp. 781–784, July 2011.

[9] P. Jagadeesh, S. Ravi, and K. H. Mallikarjun, "Design of high performance 64 bit mac unit," in *Proceedings of the International Conference on Circuits, Power and Computing Technologies (ICCPCT '13)*, pp. 782–786, March 2013.

[10] M. Kumaran and M. Kamarajan, "Multicore embedded system using parallel processing technique," *International Journal of Emerging Trands in Electrical and Electronics*, vol. 5, no. 3, 2013.

[11] L. Dadda, "Some schemes for parallel multipliers," *Alta Frequenza*, vol. 34, pp. 349–356, 1965.

[12] P. M. Kogge and H. S. Stone, "A parallel algorithm for the efficient solution of a general class of recurrence equations," *IEEE Transactions on Computers*, vol. C-22, no. 8, pp. 786–793, 1973.

[13] J. Sklansky, "Conditional-sum addition logic," *IRE Transactions on Electronic Computers*, vol. EC-9, pp. 226–231, 1960.

[14] R. P. Brent and H. T. Kung, "A regular layout for parallel adders," *IEEE Transactions on Computers*, vol. C-31, no. 3, pp. 260–264, 1982.

[15] R. Ward and T. Molteno, "Table of linear feedback shift registers," Datasheet, Department of Physics, University of Otago, 2007.

## 18

Optimization of Fractional-N-PLL Frequency Synthesizer for Power Effective Design

Sahar Arshad,[1] **Muhammad Ismail,**[1] **Usman Ahmad,**[2]
Anees ul Husnain,[3] **and Qaiser Ijaz**[3]

[1] Department of Electronic Engineering, University College of Engineering and Technology,
  The Islamia University of Bahawalpur, Bahawalpur 63100, Pakistan
[2] Scholar Teacher Research Alliance for Problem Solving (STRAPS), Bahawalpur 63100, Pakistan
[3] Department of Computer System Engineering, University College of Engineering and Technology,
  The Islamia University of Bahawalpur, Bahawalpur 63100, Pakistan

Correspondence should be addressed to Qaiser Ijaz; qaiser.ijaz@iub.edu.pk

Academic Editor: Yu-Cheng Fan

We are going to design and simulate low power fractional-N phase-locked loop (FNPLL) frequency synthesizer for industrial application, which is based on VLSI. The design of FNPLL has been optimized using different VLSI techniques to acquire significant performance in terms of speed with relatively less power consumption. One of the major contributions in optimization is contributed by the loop filter as it limits the switching time between cycles. Sigma-delta modulator attenuates the noise generated by the loop filter. This paper presents the implementation details and simulation results of all the blocks of optimized design.

1. Introduction

For many manufacturers and product developers, it is a good idea to reduce power consumption in electronic products. It is also an important idea to gain competitive advantage in an increasingly power hungry world. Low power consumption gives many benefits to designers and to users; for example, the main advantage is that it reduces stringent cooling requirements and it results in inexpensive and more compact products [1]. The rapid rise in power requirements has promoted governments and industry to increase energy efficiency and design low power components. The majority of frequency synthesis techniques fall into two categories: either direct frequency synthesis or indirect frequency synthesis [2]. To achieve fine frequency steps, the direct frequency synthesis technique is used because it is based on using digital techniques. To generate multiples (integer or noninteger) of a reference frequency, indirect frequency synthesis is used because it is based on a phase-locked loop (PLL). Here, the latter technique is used because we are going to implement PLL. It is used to generate a signal whose phase is related

to the phase of the input signal and this signal is called an output signal of the PLL. The input signal is called the "reference" signal. In a feedback loop, the oscillator is controlled by the output signal from the phase detector [3, 4]. The circuit compares the phase of a signal obtained from its output oscillator with the phase of the input signal to keep the phases matched by adjusting the frequency of its oscillator. A phase locked loop (PLL) architecture has two types, a Fractional-N PLL (FNPLL) and an integer-N PLL [5]. For a given frequency resolution, the latter has high reference frequency than the former, and, hence, the loop bandwidth which is limited to 10% of the reference frequency can be set larger in the FNPLL than in the integer-N-PLL. Therefore, the latter architecture is used for faster locking. This speed advantage of the FNPLL, however, comes at the price of increased design complexity [6]. This is because the fractional-N operation in steady state requires fractional spur reduction circuits whose quantization noise folds into the PLL spectrum via loop nonlinearities, demanding more significant design efforts to minimize the loop nonlinearities.

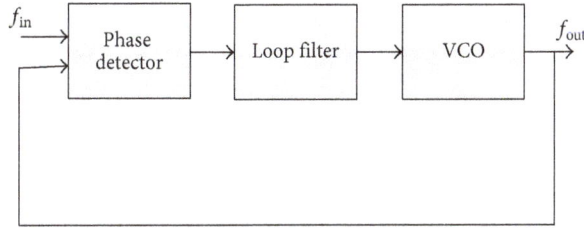

FIGURE 1: PLL system representation.

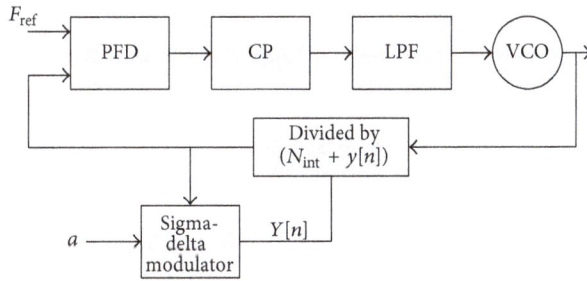

FIGURE 2: Sigma-delta FNPLL arrangement.

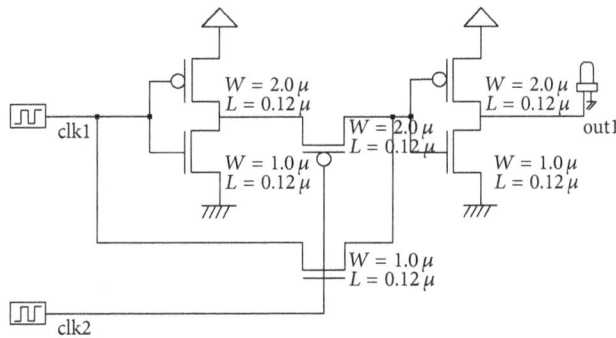

FIGURE 3: Schematic of phase detector (PD).

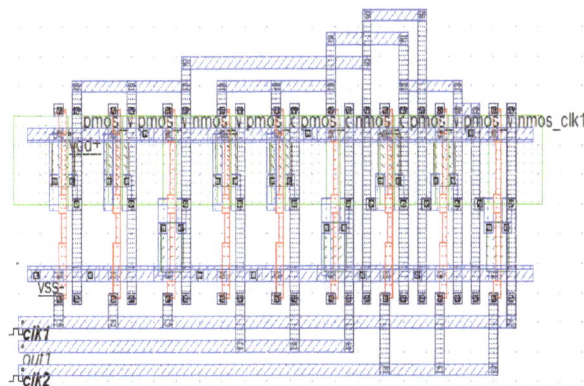

FIGURE 4: Layout of PD.

On the contrary, in the absence of fractional spurs, integer-N-PLLs involve less design complexity. Here, FNPLL is required. The expression of output frequency of the FNPLL is

$$\text{Freq}_{\text{FNPLL}} = (M \cdot n) * \text{Freq}_{\text{Ref}}. \qquad (1)$$

In this equation, M is an integer, and n is the fractional part. To obtain the desired fractional division ratio dual modulus is used [7]. Using the sigma-delta modulation technique, we can remove the fractional spurs. This technique generates a random integer number. The average of these random numbers will result in the desired ratio. A phase detector,

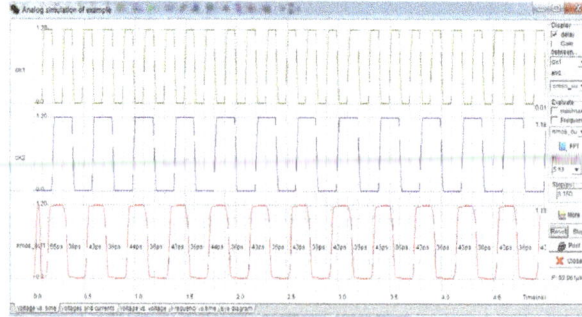

FIGURE 5: V versus T.

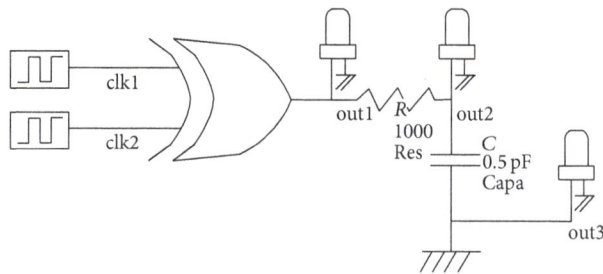

FIGURE 6: CMOS circuit of PD with LF.

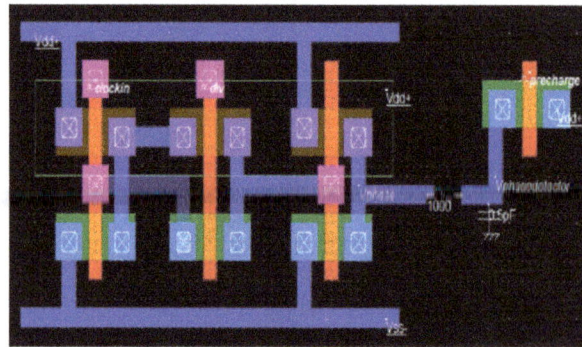

FIGURE 7: Layout of PD with LF.

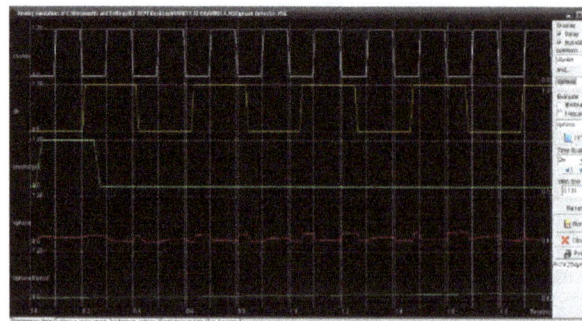

FIGURE 8: V versus T.

a loop filter, and a voltage controlled oscillator (VCO) are the main parts of phase-locked loop, as shown in Figure 1.

The important part of the phase-locked loop (PLL) is phase detector. It is also called a phase comparator, logic circuit, frequency mixer, or an analog multiplier that generates a voltage signal and this voltage signal shows the phase difference. Three units are coupled as a feedback system as shown in Figure 1. The periodic output signal is generated by the oscillator. The applications of PLL are versatile; for example, it can generate different stable frequencies or it can obtain a signal from noisy signals. A complete phase-locked loop block can be obtained from single integrated circuit.

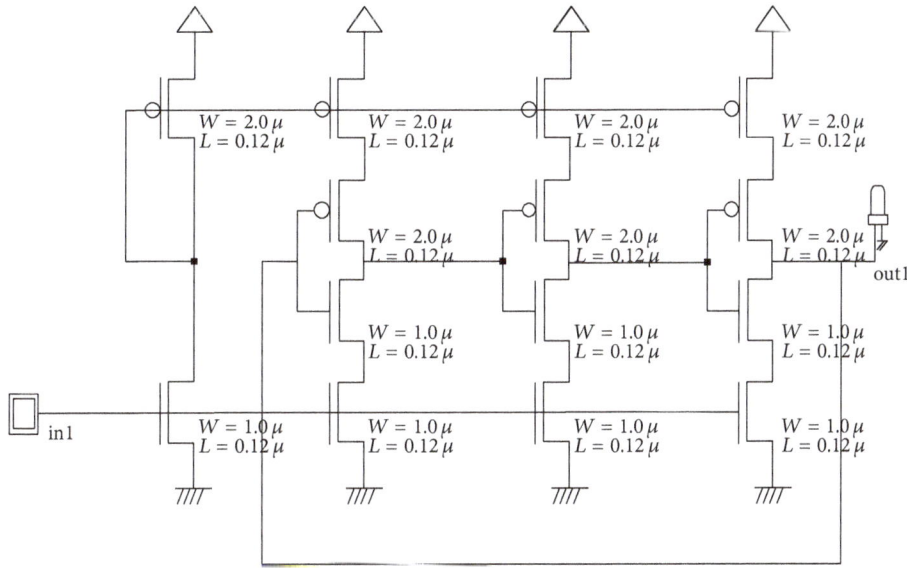

FIGURE 9: VCO schematic.

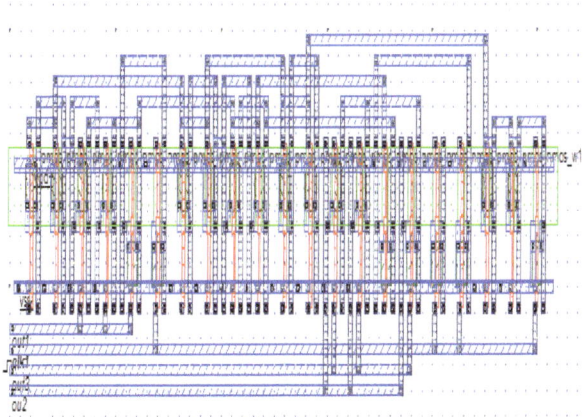

FIGURE 10: VCO layout.

This technique is used in advanced electronic products which have different output frequencies from some Hz to many Giga Hz [8]. To get low power consumption, high speed, and stability, we decide to design phase-locked loop of architecture fractional-n using 0.12 micrometer CMOS/VLSI design. As the demand of PLL is growing day by day in the field of communications, low leakage transistors will be used for maintaining low power but for this we have to make a little compromise on frequency.

The structure of FNPLL is depicted in Figure 2. We can control characteristics of PLL by using low pass filter, for example, transients response and bandwidth. The basic and essential functional unit of PLL is VCO. VCO is used for clock generation [9]. For synthesizing aspired frequencies, we use PLL with arbitrary frequency division (+N) method. This proposed technique has the ability to give fast settling time, reduce phase noise, and also reduce the effect of spurious frequencies when compared with existing FNPLL techniques.

2. PLL Design Using 0.12 Micrometer

2.1. Phase Detector. The first block has two inputs, the reference input and the feedback. It compares frequencies of input and produces an output using phase difference of inputs. To represent this block XOR gates are used. The gate produces a square wave when one-fourth of period shift of 90 degrees takes place at clock input, whereas output is different for all other angles. We apply output of the XOR gates to LPF which results in analog voltage, proportional to phase difference.

Figure 3 depicts a CMOS circuit of phase detector, Figure 4 describes layout, and Figure 5 represents the output waveform.

2.2. Loop Filter. To get pure DC voltage along with rectifiers filters, the electronic circuits are also used. The second block of PLL is loop filter and it has two distinct functions. First, maintains stability, that is defined by describing the

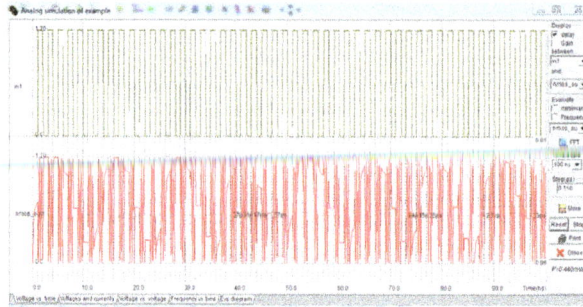

FIGURE 11: V versus T.

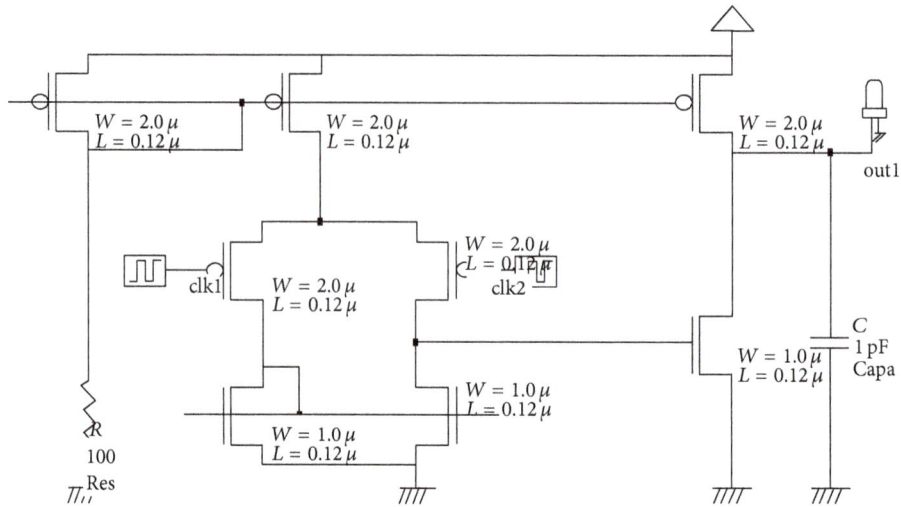

FIGURE 12. CMOS circuit of comparator.

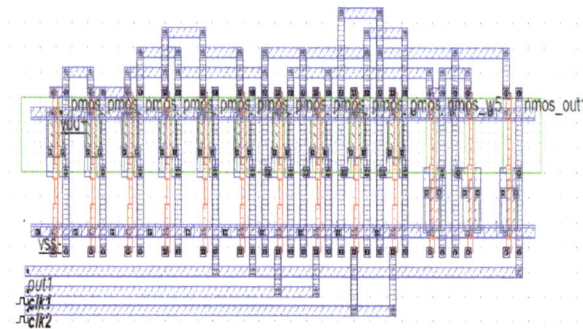

FIGURE 13: Layout of comparator.

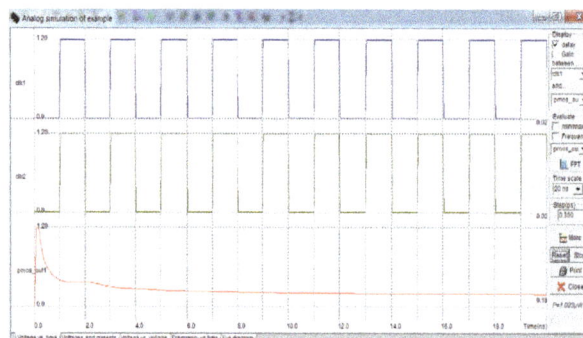

FIGURE 14: V versus T.

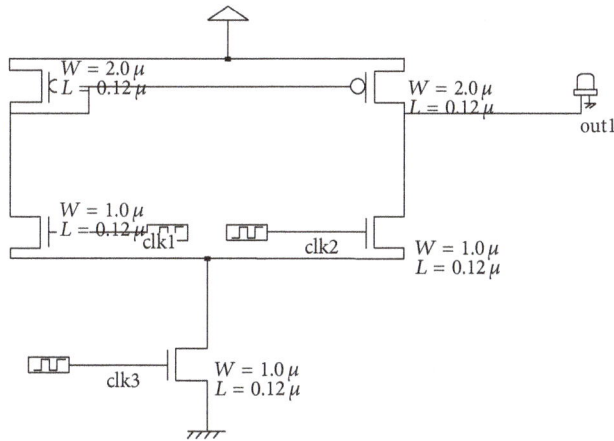

FIGURE 15: CMOS schematic of OTA.

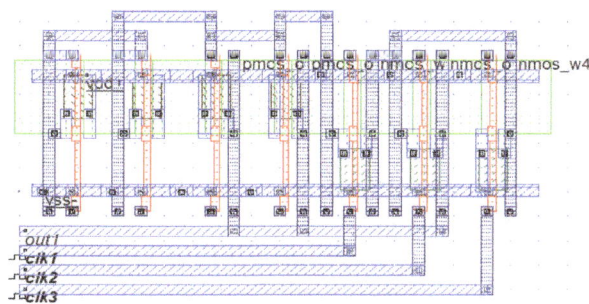

FIGURE 16: OTA layout.

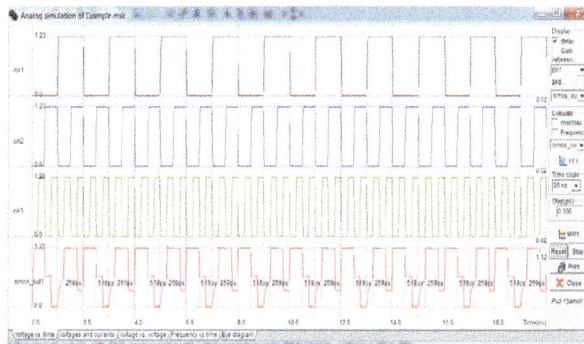

FIGURE 17: V versus T.

loop dynamics. This explains the response of the loop to uncertainties. The 2nd function is applied to the VCO control input which appears at the phase detector output. This frequency produces FM sidebands and modulates the VCO [10]. Other features of the PLL, for example, bandwidth, transient response, lock range, and capture range, can be controlled by LPF. The LPF is used to attenuate this energy, but it can also reject band. The low pass filter can be obtained by using a capacitor of large value and the capacitor is charged and discharged with the help of the switch resistance R_{on}. By the help of $R_{on}.C$ delay a low pass filter can be created. Figure 6 depicts a CMOS schematic of phase detector with loop filter, Figure 7 shows layout, and Figure 8 shows output waveform.

2.3. *Voltage Controlled Oscillator.* As VCO is a source of varying output signal so the frequency of the output signal is regulated over a DC voltage range. The output signal can be a square wave or a triangular wave form. The oscillation frequency is controlled by the value of input voltage [11]. Figure 9 shows a CMOS circuit of VCO, Figure 10 shows layout, and Figure 11 shows output waveform.

2.4. *Sigma-Delta Modulator.* Sigma-delta modulation technique is used to convert high definition signals to low definition signals in digital domain. We designed sigma-delta modulator using 0.12 micrometer feature size and then the layout was obtained. The input is the aspired fractional number (n) and the output is the sum of quantization noise and a DC part [12, 13]. By the use of integer divider quantization noise was generated. Figures 12 and 15 show the CMOS circuit; Figures 13 and 16 show the layout of comparator and operational transconductance amplifier. Figures 14 and 17 show the output waveforms.

3. Conclusion

Power usage and heat dissipation are one of the biggest challenges of VLSI industry today. In order to design the low power consuming component, without making significant change in performance, the design of FNPLL frequency synthesizer was implemented and simulated. The optimized design was implemented to 0.12 micrometer technology. Using CMOS logic, the schematics were designed and verified functionally and then prefabrication layout was sketched. The simulation curves of the layouts reflected reduction in power consumption, for the optimized design.

Conflict of Interests

The authors declare that there is no conflict of interests regarding the publication of this paper.

References

[1] R. Jacob Baker, *CMOS Circuit Design, Layout and Simulation*, IEEE Press, John Wiley & Sons, 3rd edition, 2010.

[2] A. Anil and R. K. Sharma, "A high efficiency charge pump for low voltage devices," *International Journal of VLSI Design & Communication Systems*, vol. 3, no. 3, 2012.

[3] U. L. Rohde, *Digital PLL Frequency Synthesis*, Prentice-Hall, Englewood Cliffs, NJ, USA, 1983.

[4] B. K. Mishra, S. Save, and S. Patil, "Design and analysis of second and third order PLL at 450 MHz," *International Journal of VLSI Design & Communication Systems*, vol. 2, no. 1, 2011.

[5] N. Weste and D. Harris, *CMOS VLSI Design—A Circuits and Systems Perspective*, Pearson Education, 3rd edition, 2005.

[6] U. A. Belorkar and S. A. Ladhake, "Design of low power phase lock loop using 45 nm VLSI technology," *International Journal of VLSI Design & Communication Systems*, vol. 1, no. 2, 2010.

[7] T. A. D. Riley, M. A. Copeland, and T. A. Kwasniewski, "Delta-Sigma modulation in fractional-n frequency synthesis," *IEEE Journal of Solid-State Circuits*, vol. 28, no. 5, pp. 553–559, 1993.

[8] M. H. Perrott, "Fractional-N Frequency Synthesizer Design Using The PLL Design Assistant and CppSim Programs," July 2008.

[9] S. Franssila, *Introduction to Microfabrication*, John Wiley & Sons, 2004.

[10] K. Woo, Y. Liu, E. Nam, and D. Ham, "Fast-lock hybrid PLL combining fractional-N and integer-N modes of differing bandwidths," *IEEE Journal of Solid-State Circuits*, vol. 43, no. 2, pp. 379–389, 2008.

[11] N. Fatahi and H. Nabovati, "Design of low noise fractional-N frequency synthesizer using sigma-delta modulation technique," in *Proceedings of the 27th International Conference on Microelectronics (MIEL '10)*, pp. 369–372, IEEE, May 2010.

[12] S. Borkar, "Obeying Moore's law beyond 0.18 micron," in *Proceedings of the 13th Annual IEEE International ASIC/SOC Conference*, pp. 26–31, September 2000.

[13] R. K. Krishnamurthy, A. Alvandpour, V. De, and S. Borkar, "High-performance and low-power challenges for sub-70 nm microprocessor circuits," in *Proceedings of the IEEE Custom Integrated Circuits Conference*, pp. 125–128, May 2002.

## 19

Gate-Level Circuit Reliability Analysis: A Survey

Ran Xiao and Chunhong Chen

Department of Electrical and Computer Engineering, University of Windsor, Windsor, ON, Canada N9B 3P4

Correspondence should be addressed to Chunhong Chen; cchen@uwindsor.ca

Academic Editor: Yu-Cheng Fan

Circuit reliability has become a growing concern in today's nanoelectronics, which motivates strong research interest over the years in reliability analysis and reliability-oriented circuit design. While quite a few approaches for circuit reliability analysis have been reported, there is a lack of comparative studies on their pros and cons in terms of both accuracy and efficiency. This paper provides an overview of some typical methods for reliability analysis with focus on gate-level circuits, large or small, with or without reconvergent fanouts. It is intended to help the readers gain an insight into the reliability issues, and their complexity as well as optional solutions. Understanding the reliability analysis is also a first step towards advanced circuit designs for improved reliability in the future research.

1. Introduction

As CMOS technology keeps scaling down to their fundamental physical limits, electronic circuits have become less reliable than ever before [1]. The reason is manifold. First of all, the higher integration density and lower voltage/current thresholds have increased the likelihood of soft errors [2, 3]. Secondly, process variations due to random dopant fluctuation or manufacturing defects have negative impacts on circuit performance and may cause circuits to malfunction [1]. These physical-level defects would statistically lead to probabilistic device characteristics. Also, some emerging nanoscale electronic components (such as single electron devices) have demonstrated their nondeterministic characteristics due to uncertainty inherent in their operation under high temperature and external random noise [4, 5]. This may further degrade the reliability of future nanoelectronic circuits. Thus, circuit reliability has been a growing concern in today's micro- and nanoelectronics, leading to the increasing research interest in reliability analysis and reliability-oriented circuit design.

For any reliability-aware architecture design, it is indispensable to estimate the reliability of application circuits both accurately and efficiently. However, analyzing the reliability (or the error propagation) for logic circuits could be computationally expensive in general (see Section 1.3 for details). Some approaches have been reported in literature, which tackle the problem either analytically or numerically (by simulation). The contribution of this paper is to provide an extensive overview and comparative study on typical reliability estimation methods with our simulation results and/or results reported in literature.

We first review the key concepts in reliability analysis and its role in circuit design and then describe and evaluate several existing mainstream approaches for reliability analysis by looking at their accuracy, efficiency, and flexibility. Examples and simulation results are also given in order to show their advantages and disadvantages. Finally, we provide some useful suggestions on how to choose an appropriate reliability analysis method under different circumstances, along with some remarks on possible future work.

1.1. Signal Probability and Reliability. The probability of a logic signal s is by default defined as the probability of the signal being logic "1" and is expressed as $P_s = \Pr\{s = "1"\}$. The reliability of the probabilistic signal s is defined as the probability that its value is correct (i.e., it is equal to its error-free value) and is expressed as $r_s = \Pr\{s = $ its error-free value$\}$. In gate-level design, the output signal of a gate may become unreliable due to its unreliable inputs and/or errors of gate itself. If we use the classical *von*

Neumann model [6] for gate errors, any gate can be associated independently with an error probability ε_i. In other words, the gate is modeled as a binary symmetric channel that generates a bit flip (from $0 \rightarrow 1$ or $1 \rightarrow 0$) by mistake at its output (known as *von Neumann* error [6]) symmetrically with the same probability. Thus, each gate i in the circuit has an independent gate reliability $r_i = 1 - \varepsilon_i$, which is assumed to be localized and statistically stable. Also, it is reasonable to assume that the error probability for any gate falls within $[0, 0.5]$ (or $r_i \in [0.5, 1]$).

The reliability for a combinational logic circuit (denoted by R_C) is defined as the probability of the correct functioning at its outputs (i.e., the joint signal reliability of all primary outputs). This reliability can be generally expressed as a function of gate reliabilities in the circuit (denoted by $\mathbf{r} = \{r_1, r_2, \ldots, r_{N_g}\}$, where N_g is the number of gates), as well as signal probabilities of all primary inputs(denoted by $\mathbf{P_{in}} = \{P_{in1}, P_{in2}, \ldots, P_{inN_{in}}\}$, where N_{in} is the number of primary inputs), that is,

$$R_C = \Pr\{\text{all outputs are correct}\} = f(\mathbf{r}, \mathbf{P_{in}}), \quad (1)$$

where the function f depends on the topology of the circuit under consideration. Note that the primary inputs are assumed to be fully reliable ($r_s = 1$ if s is a primary input). Under a particular case where all primary input probabilities are a constant (say 0.5), R_C turns out to be a function of $\mathbf{r}$ only.

It is worth noting that gate errors may come from either external noises (thermal noise, crosstalk, or radiation) [3] or inherent device stochastic behaviors [4]. In literature, the term "soft error" is used to emphasize the temporariness of the errors due to random external noises (e.g., glitches). In this paper, however, a more general term of *von Neumann* gate error model is used instead, as the probabilistic feature of gates is expected to exist widely and independently throughout the circuit. This differs from single-event upsets due to soft errors, where external noises are usually correlated temporally and spatially. In other words, our focus is the error propagation in combinational networks, where the gate-level logic masking is considered. For instance, some logic errors may not affect (or propagate to) final outputs if they occur in a nonsensitized portion of the circuit. Identifying these nonsensitized gates would be critical for reliability estimation and improvement.

1.2. Role of Reliability Analysis. In order to guide the IC design for reliable logic operations, it is required to develop tools that can accurately and efficiently evaluate circuit reliability, which is also a first step towards reliability improvement. However, reliability analysis is a nontrivial task due to the large size of IC circuits as well as the complexity of signal correlation and probability/reliability propagation within the circuit (as will become clear later in this paper). On the other hand, circuit reliability can be generally improved by increasing the gate reliabilities. This can be done by using redundant components. Classic redundancy techniques such as TMR [5] or NAND-multiplexing [7] achieve this by systematically replicating logic gates (other than sizing up the transistors) at the cost of increased area and power

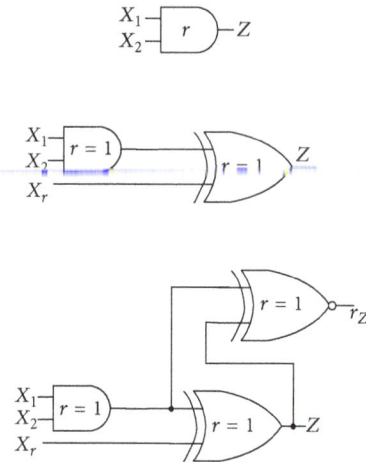

FIGURE 1: An AND gate and its equivalent circuit.

dissipation. One of the key issues in this context is to select the most critical (in terms of reliability and cost) components (or logic gates) in the circuit and improve the circuit reliability by increasing the robustness of only a few gates. In order to detect these critical gates, multiple cycles of reliability analysis are usually conducted for the whole circuit. In a more general term, accurate and efficient reliability analysis can provide a guideline for future reliability-oriented architecture design.

1.3. Complexity of Gate-Level Reliability Analysis. It is understood that the problem of determining whether the signal probability at a given node is nonzero is equivalent to the Boolean satisfiability (SAT) problem [8], a problem of determining whether there exists an interpretation that satisfies a given Boolean formula. A Boolean formula is called satisfiable if the variables of this given formula can be assigned in such a way as to make the formula evaluate to TRUE (3). The SAT has been proved to be an NP-complete problem (see [9]). The problem of computing all signal probabilities in a circuit can be formulated as a random satisfiability problem, which is to determine the probability that a random assignment of variables will satisfy a given Boolean formula [9]. The random satisfiability problem lies in a class of problems, called #P-complete, which is conjectured to be even harder than NP-complete. In the following, we show that the reliability evaluation problem is equivalent to the signal probability calculation problem and thus prove that it is also a #P-complete problem.

Let us consider a two-input AND gate ($Z = X_1X_2$) which has the gate reliability r, as shown in Figure 1. We first add an extra XOR gate at the output, as well as an extra input X_r, with an assumption that both the XOR gate and original AND gate are error-free. The signal probability of this extra input is equal to the original gate error rate ε (i.e., $P_{X_r} = \Pr\{X_r = \text{"1"}\} = 1 - r$). This ensures that the output Z of this extra XOR gate is equivalent to the original output of the AND gate.

For a combinational logic circuit, we first duplicate the whole circuit. In the original circuit, we make each gate error-free in order to compute the correct value at primary outputs. For the duplicated one, we extract the reliability of each gate using the aforementioned method (as a result, all gates are also error-free in the duplicated circuit and the gates' number is doubled). Then, we add 2-input XNOR gates for each pair of corresponding primary outputs in the original and duplicated circuits. Thus, the output reliability can be expressed as the signal probability at the output of the XNOR gates. By doing so (i.e., duplicating the circuit and extracting gate reliabilities), we see that the reliability estimation of original circuit is equivalent to the problem of computing the signal probabilities of the transformed circuit.

For a combinational logic circuit with N_{in} primary inputs, N_{out} primary outputs, and N_g logic gates, the problem of evaluating the signal reliability of all primary outputs and their joint reliability (i.e., the overall circuit reliability R_C) can be solved by exhaustively calculating all $2^{(N_{in}+N_g)}$ scenarios. In each scenario, the expected (correct) output and actual output values need to be calculated with the complexity of $O(N_g)$. The total complexity is then $O(N_g \cdot 2^{N_{in}+N_g})$. As circuits become very large, it would be difficult or even impossible to perform the exact analysis of the reliability due to the exponential complexity. Usually, some tradeoff has to be made between the accuracy and efficiency for reliability analysis.

In order to tackle this issue, a number of different approaches have been reported in literature, including probabilistic transfer matrix (PTM) method [10–12], Bayesian networks (BN) [13–15], Markov random field (MRF) [16–20], Monte Carlo (MC) simulation, testing-based method [3], stochastic computation model (SCM) [2, 21], probabilistic gate model (PGM) [22–25], observability-based analysis [26], Boolean difference-based error calculator (BDEC), and correlation coefficient method- (CCM-) based approaches [8, 26–28]. In the following, we overview some of these approaches and analyze their pros and cons in terms of accuracy, efficiency, and flexibility with simulation results.

2. Probabilistic Transfer Matrix (PTM) Method

An accurate analytical model for reliability analysis problem is based on the probabilistic transfer matrices (PTMs), which compute the circuit output reliability for all input patterns [10, 11]. This computational framework begins with the definition of a probability matrix which is used to represent the probability of a logic gate's output for each input pattern. For instance, the probability matrix representation for a two-input NAND logic gate is shown in Figure 2, where each column of the matrix $\mathbf{M}_g$ represents the probability of the gate output Z being "0" or "1" for all different input patterns (i.e., $X_1 X_2$ = "00," "01," "10," and "11"). For example, the element $\mathbf{M}_{11} = \Pr\{Z = 0 \mid X_1 X_2 = 00\} = 1 - r$, where r is the gate reliability. In general, the probability matrix for an n-input 1-output gate is a $2^n \times 2$ matrix.

For a circuit, all gate probability matrices shall be combined together to construct the PTM of the whole circuit. More specifically, the serial and parallel connections of gates correspond to a matrix product and tensor product [10],

respectively. The fanout behavior is represented by explicit fanout gates, where a 1-input m-output fanout gate is simply mimicked by a 1-input m-output buffer gate. A fault-free circuit has an ideal transfer matrix (ITM), where the correct value of the output occurs with the probability of 1. This means that, in each row of the PTM, there is single "1" for the correct output value and there are "0"s for other output combinations. The circuit reliability (i.e., the probability of outputs being correct) is evaluated by comparing its PTM and ITM.

The process of combining gate probability matrices implicitly takes into account the signal dependency between gates by considering the underlying joint and conditional probabilities within the circuit. As a result, the calculation of the circuit PTM is exact. However, the limited scalability is often a price that has to be paid for this computational framework to capture complex circuit behaviors. Consider a combinational logic circuit with N_{in} primary inputs, N_{out} primary outputs, and N_g logic gates. The circuit PTM is a matrix with $2^{N_{in}}$ rows and $2^{N_{out}}$ columns (i.e., $2^{N_{in}} \times 2^{N_{out}}$), which contains the transition probability from all input combinations toward all output combinations. In other words, its space complexity is $O(2^{N_{in}+N_{out}})$. This exponential space requirement is the main bottleneck of PTM approach. Particularly, for a computer with 2 GB memory, the maximum size of the circuit that can be handled is limited to 16 input/output signals. By utilizing some advanced computation methods (such as algebraic decision diagrams (ADDs) and encoding [10, 11]), the signal width may be extended up to ~50, where the signal width is defined as the largest number of signals at any level in the circuit. Unfortunately, this limit is still computationally unacceptable in the real world for large-scale benchmark circuits (e.g., C2670 which has 157 inputs and 64 outputs). Nonetheless, for small circuits, the PTM is a very good analytical method, as it provides exact results within a reasonable runtime and shows the probabilistic behavior of unreliable logic gates.

Also, this approach can serve as the foundation of many other heuristic approaches by providing other important information such as signal probabilities and observability, with the capability of analyzing the effect of electrical masking on error mitigation as well. For instance, in [10], the observability of a gate g is defined as the ratio of the error probability of the whole circuit and the error probability ε_i of this gate, that is, $(1 - R_C(\varepsilon_i))/\varepsilon$, where $R_C(\varepsilon_i)$ is the circuit reliability when the only unreliable gate is ith gate (with all other gates being error-free). Clearly, the gate with highest observability can be regarded as the most susceptible, meaning that it will impact (or decrease) the circuit reliability the most. It should be noted that this only represents the simplest case where only single gate failure is considered. In most real cases, however, the gate observabilities may not be independent, and thus the joint observabilities usually need to be considered instead.

The detailed algorithm with the PTM is summarized as follows.

Step 1. Levelize the circuit; compute PTMs of each logic component in each level denoted by $\mathbf{M}_{Lv_i}^j$.

$$\mathbf{M}_g = \begin{array}{c} \\ 00 \\ 01 \\ 10 \\ 11 \end{array} \begin{bmatrix} \begin{array}{cc} 0 & 1 \end{array} \\ 1-r & r \\ 1-r & r \\ 1-r & r \\ r & 1-r \end{bmatrix}$$

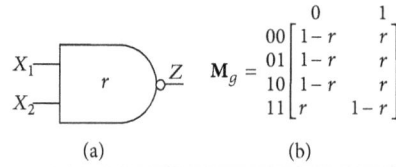

FIGURE 2: (a) A 2-input NAND gate and (b) its probability matrix $\mathbf{M}_g$ (according to [10]).

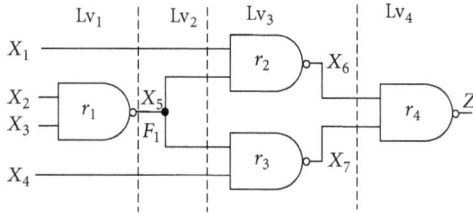

FIGURE 3: The example circuit schematic (a portion of C17 benchmark circuit).

Step 2. Within one level, the PTMs of each logic components (gates, wires, and fanout nodes) are tensored together to form the PTM of the current level; that is, $\mathbf{M}_{\mathrm{Lv}_i} = \mathbf{M}_{\mathrm{Lv}_i}^1 \otimes \mathbf{M}_{\mathrm{Lv}_i}^2 \cdots$;

Step 3. The PTMs of all levels are then multiplied together to get the circuit PTM; that is, $\mathbf{M} = \prod_i \mathbf{M}_{\mathrm{Lv}_i}$.

Step 4. Calculate the ideal transfer matrix $\mathbf{J}$ using the truth table of the logic function (error-free signal probabilities $p(i)$ for input patterns are evaluated with the computation complexity of $O(N_g \cdot 2^{N_{\mathrm{in}}})$).

Step 5. The circuit reliability is given by [11]:

$$R_C = \sum_{i,j}^{J(i,j)=1} M(i,j) \, p(i). \tag{2}$$

We take a simple circuit as an example to illustrate the analysis process of PTM approach. The circuit schematic is shown in Figure 3, where the circuit has 4 levels, and the fanout F_1 reconverges at gate number 4, generating the dependency between signal X_6 and X_7. Since there are four inputs (X_1, X_2, X_3, and X_4) and one single output Z, the circuit PTM would be a 16×2 matrix $\mathbf{M}$ which stores the probability of occurrence of all input-output vector pairs. The $\mathbf{M}$ is constructed by combining PTMs of all levels (using matrix product due to serial connection in this case), while the PTM of each level is calculated by combining PTMs of each logic components within the current level (using tensor

product due to their parallel connection). More specifically, we have (based on [10])

$$\mathbf{M} = (I \otimes \mathrm{NAND}_1 \otimes I)_{16\times8} \cdot (I \otimes F \otimes I)_{8\times16}$$
$$\cdot (\mathrm{NAND}_2 \otimes \mathrm{NAND}_3)_{16\times4} \cdot (\mathrm{NAND}_4)_{4\times2}$$

$$I = \begin{bmatrix} 1 & 0 \\ 0 & 1 \end{bmatrix}, \qquad F = \begin{bmatrix} 1 & 0 & 0 & 0 \\ 0 & 0 & 0 & 1 \end{bmatrix}, \tag{3}$$

$$\mathrm{NAND}_i = \begin{bmatrix} 1-r_i & r_i \\ 1-r_i & r_i \\ 1-r_i & r_i \\ r_i & 1-r_i \end{bmatrix},$$

where the matrix I refers to a 2×2 identity PTM, and each parenthesized term in (3) corresponds to a specific circuit level. Assuming the gate reliabilities are $r_1 = r_2 = r_3 = r_4 = 0.95$ and the probability of all input signals is equally 0.5, the circuit PTM and ideal transfer matrix are found using the above algorithm as follows:

$$\mathbf{M} = \begin{bmatrix} 0.8622 & 0.1378 \\ 0.1312 & 0.8688 \\ 0.8622 & 0.1378 \\ 0.1312 & 0.8688 \\ 0.8622 & 0.1378 \\ 0.1312 & 0.8688 \\ 0.8622 & 0.1378 \\ 0.8238 & 0.1762 \\ 0.1312 & 0.8688 \\ 0.0928 & 0.9073 \\ 0.1312 & 0.8688 \\ 0.0928 & 0.9073 \\ 0.1312 & 0.8688 \\ 0.0928 & 0.9073 \\ 0.8238 & 0.1762 \\ 0.8217 & 0.1783 \end{bmatrix}, \quad \mathbf{J} = \begin{bmatrix} 1 & 0 \\ 0 & 1 \\ 1 & 0 \\ 0 & 1 \\ 1 & 0 \\ 0 & 1 \\ 1 & 0 \\ 1 & 0 \\ 0 & 1 \\ 0 & 1 \\ 0 & 1 \\ 0 & 1 \\ 0 & 1 \\ 0 & 1 \\ 1 & 0 \\ 1 & 0 \end{bmatrix}. \tag{4}$$

It can be seen from $\mathbf{M}$ that the output reliability depends on input patterns. The lowest and highest values for the output reliability are 0.8217 and 0.9073, which occur when the input vector $(X_1 X_2 X_3 X_4) = (1111)$ and (1001, 1011, and 1101), respectively. The circuit reliability is found to be $R_C = 0.8658$ with the runtime of 0.2798 s.

The PTM algorithm has been implemented on some small circuits. The simulation results show that its performance is fairly good for circuits with less than 20 gates. If the circuit size increases to ~40, both runtime and memory cost will grow dramatically, making the PTM method computationally expensive. In order to handle large-scale circuits, a variant PTM method was proposed in [11], where the input vector sampling is used. The simulation results show that this does improve efficiency with reduced memory cost, while the accuracy remains to be seen.

In summary, the PTM method has two major limitations. First, the signal width of the circuit that can be analyzed is very limited. This is due to the fact that its space complexity grows exponentially with the number of inputs and outputs, leading to prohibitively massive matrix storage and manipulation overhead for large-scale circuits. Secondly, the circuit structure needs to be preprocessed (such as circuit levelization and identification of the fanout nodes and wire pairs) prior to the algorithm implementation. Also, the PTM assumes all signals are correlated, which makes the method less efficient for circuits with no or a few reconvergent fanouts.

3. Monte Carlo (MC) Simulation

MC is a widely known simulation-based approach, where experimental data are collected to characterize the behavior of a circuit by randomly sampling its activity [2]. It is usually used when an analytical approach is unavailable or difficult to implement. The obvious drawbacks of this approach lie in the fact that numerous pseudorandom numbers need to be generated, and a large number of simulation runs must be executed to reach a stable result. This makes the reliability analysis for large circuits a very time-consuming process. As a stochastic computation framework, the MC method makes the result gradually converge to its exact value as more simulation runs are performed. In the process of achieving relatively stable results, certain statistical parameters (such as standard deviation σ and/or coefficient of variance (CV) which is defined as the ratio of the standard deviation and the mean, i.e., σ/μ) are usually used as the stopping criteria. In [2], CV = 0.001 is used to represent an acceptable level of accuracy, and the number of simulation runs required is given by

$$N_{\text{MC}} = \frac{1 - R_C}{R_C} \cdot \frac{1}{\text{CV}^2} \approx 10^6 \cdot \left(\frac{1}{R_C} - 1 \right), \quad (5)$$

where R_C is again the circuit reliability. Since the circuit reliability usually decreases with the circuit size (N_g), the N_{MC} will increase with the circuit size for a given accuracy (measured by CV). Assuming that the R_C ranges from 0.1 to 0.9, the number of MC runs will vary around $10^5 \sim 10^7$. It should be mentioned that (5) only gives an approximated range of N_{MC}, and its actual value is usually determined experimentally for real circuits. Let us take the circuit of Figure 3 again as an example. From (5), the required N_{MC} is ~1.55 × 10^5 if $R_C = 0.8658$. Figure 4 shows the relative error at R_C against N_{MC}. It can be seen from the figure that after ~10^4 runs, the result becomes relatively stable around its final value. However, a small random fluctuation is inevitable. Even after ~10^5 simulation runs, the relative error of

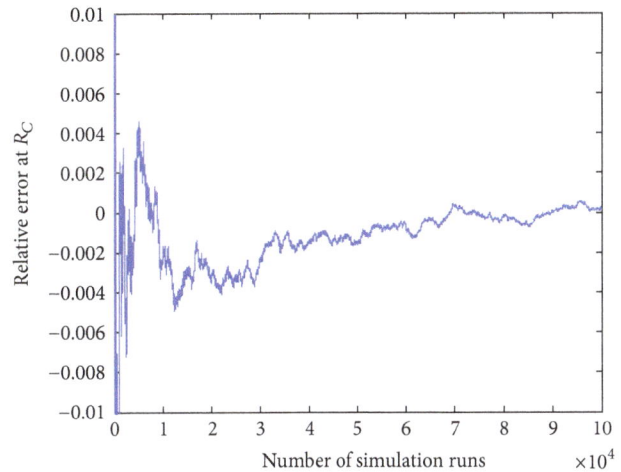

FIGURE 4: The relative error of circuit reliability R_C of Figure 3 versus the number of MC simulation runs N_{MC}.

the MC result is a nonzero value (5.636e − 04), indicating a low convergent rate with the MC. This is a common feature for stochastic computations.

4. Stochastic Computation Model (SCM)

Unlike the MC method which uses Bernoullisequences for simulation, the SCM approach takes non-Bernoulli sequences [2, 21]. In a non-Bernoulli sequence, for a given probability p and a sequence length N, the number of "1"s to be generated is fixed and given by $N \cdot p$, and only the positions of the "1"s are determined by a random permutation of binary bits. Therefore, in SCM approach, less pseudorandom numbers are generated for the same length of simulation, compared to MC simulation where pseudorandom numbers are independently generated for each gate or input to mimic the behavior of probabilistic circuits [2].

Consider a circuit with N_{in}, N_{out}, N_g, $\mathbf{P_{in}}$, and ε (refer to the previous sections for definitions of these variables). If we use a sequence length of N, the total required number of random numbers is given by $(N_{\text{in}} + N_g) \cdot N$ in MC simulation. In contrast, for the SCM approach with the same sequence length, only $N\varepsilon$ pseudorandom numbers need to be generated (for the positions of "1"s) for a gate with error rate ε. Therefore, the total number of random numbers is reduced to $(N_{\text{in}} \cdot p_{\text{in}} + N_g \cdot \varepsilon) \cdot N$. Since the gate error rate ε is usually a small value which can be viewed as a scale factor, the total required random number is significantly reduced. In other words, for a specific level of accuracy, the non-Bernoulli sequence requires a smaller sequence length than the Bernoulli sequence does. However, how to efficiently determine the required minimum sequence length for the SCM is still an open question. In [2], an empirical function (rather than an analytical expression) was used for this purpose.

Again, we took the example circuit of Figure 3 and used the same sequence length with MC (i.e., $N_{\text{SCM}} = N_{\text{MC}} = 10^5$) with gate error rate $\varepsilon = 0.001$. The SCM and MC

TABLE 1: Runtime comparison of MC and SCM on benchmark circuits ($\varepsilon = 0.01$).

Circuit	Size	MC (10^6 runs)	SCM (10^6 runs)	
		Runtime (s)	Runtime (s)	
			$\varepsilon = 0.01$	$\varepsilon = 0.1$
c432	160	183	31	38
c499	202	203	37	45
c880	383	373	63	77
c1355	546	472	92	111
c1908	880	842	183	215
c2670	1193	1151	265	311
c3540	1669	1616	409	505
c5315	2406	2548	786	961
c7552	3512	3732	1325	1495

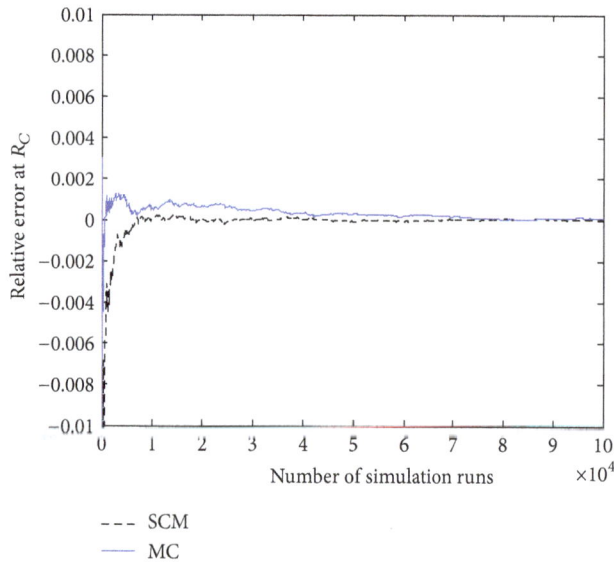

FIGURE 5: The relative error of circuit reliability R_C versus the number of simulation runs N_{MC}/N_{SCM}.

simulation results are compared in Figure 5, where both have a similar convergence rate. However, the runtimes with SCM and MC are $T_{SCM} = 0.0745$ s and $T_{MC} = 1.2528$ s, respectively, indicating that the SCM method is more efficient than the MC. This efficiency improvement is mainly due to less random numbers that are generated in the SCM simulation.

We also implemented both SCM and MC approaches in *Matlab* with the same sequence length of 10^6 (gate error rate $\varepsilon = 0.01$) and tested their performance on ISCAS'85 benchmark circuits. The results are shown in Table 1, where the runtime with the SCM is around $1/6 \sim 1/3$ of that with the MC. One of the disadvantages of SCM is the difficulty in determining its simulation sequence length N_{SCM}. Also, its runtime is proportional to gate error rate ε as well as input probabilities. If ε is relatively large (say 0.2), the runtime improvement of SCM over MC would be marginal (only scaled by a constant).

5. Probabilistic Gate Model (PGM)

The PGM is another reliability analysis method which is based on the probabilistic models of unreliable logic gates [22–25]. In the simple version of PGM, the input signals of each gate in the circuit are assumed to be independent. Under this assumption, the output probability of each gate can be easily calculated using the information of input signal probabilities and gate error rate. For instance, consider a 2-input NAND gate with input probabilities of X_1 and X_2 and gate error rate of ε. Its output signal probability can be expressed as (after [24])

$$Z = \Pr\left(\text{``1''} \mid \text{gate faulty}\right) \cdot \Pr\left(\text{gate faulty}\right)$$
$$+ \Pr\left(\text{``1''} \mid \text{gate not faulty}\right) \cdot \Pr\left(\text{gate not faulty}\right) \quad (6)$$
$$= (1 - \varepsilon) + (2\varepsilon - 1) X_1 X_2.$$

This output probability Z can be used recursively as the input information at next level of gates. One of the main features with PGM is that the circuit reliability is analyzed by exhaustively evaluating each input combination and output. For any given input combination, the error-free output value Z_{ef} is calculated, and then the output signal probability P_O is evaluated using the PGM of all gates in the circuit. Depending on the error-free output value, the output reliability R for this specific input combination is given by [24]

$$R = \begin{cases} P_O, & Z_{ef} = 1, \\ 1 - P_O, & Z_{ef} = 0. \end{cases} \quad (7)$$

Finally, the overall output reliability is the weighted sum of all conditional output reliabilities over all possible input combinations, where the weight is the probability of a specific input combination.

Intuitively, the operation process of PGM is similar to PTM in the sense that both of them consider all input combinations in a forward topological order. An obvious disadvantage with the PGM approach is that it is almost impossible to exhaustively enumerate all input combinations when the number of inputs increases (say to 30 and above). Therefore, a certain sampling technique is often necessary for large circuits. The input patterns sampling becomes another source of errors, in addition to the inaccuracy caused by signal independence assumption in constructing gate PGMs (it should be pointed out that while signal correlations due to fanouts originating from the primary inputs are eliminated by assigning the deterministic values (either "0" or "1") to all primary inputs, those caused by other reconvergent fanouts nodes are not).

In order to eliminate all signal correlations, an accurate PGM algorithm was proposed in [24] where deterministic values are assigned explicitly to all reconvergent fanout nodes within the circuit. More specifically, for each fanout, the original circuit is transformed to two auxiliary circuits [24], one with the fanout node being set to logic value "0" and the other to "1." In each of these two circuits, the output probability is computed by using conditional probabilities for the given value at the fanout. This procedure is executed iteratively until

TABLE 2: Simulation results for simple PGM in comparison with MC.

Circuit	Size	Simple PGM approach (10^3 samples)			Monte Carlo (10^6 runs)
		Average error (%)	Max error (%)	Runtime (s)	Runtime (s)
C432	160	0.54	1.59	4.66	183
C499	202	0.1	0.31	4.92	203
C880	383	0.61	2.83	9.17	373
C1355	546	1.26	1.66	12.21	472
C1908	880	0.39	0.85	18.69	842
C2670	1193	2.43	16.61	25.72	1151
C3540	1669	0.077	2.27	39.46	1616
C5315	2307	10.88	43.16	61.42	2548
C7552	3512	2.68	13.68	75.19	3732
Average	—	**2.18**	**9.22**	—	—

TABLE 3: Comparison of simple PGM and accurate PGM.

Circuit	Size	Simple PGM approach (10^3 samples)		Accurate PGM approach (10^3 samples) [24]
		Average error (%)	Runtime (s)	Runtime (s)
Cu	43	1.37	0.0277	0.10
z4ml	45	0.94	0.0039	0.05
x2	38	0.52	0.0275	0.22
Mux	50	0.52	0.282	0.10

all fanouts have been processed. If all input combinations are simulated, this procedure will lead to exact results for any circuits. However, for a circuit with N_f reconvergent fanouts, a total of 2^{N_f} auxiliary circuits are required and analyzed. Therefore, the computation complexity becomes $O(N_g \cdot 2^{N_{in}+N_f})$ [24]. However, in many real circuits, the number of reconvergent fanouts N_f is comparable to the number of gates (N_g). Thus, the complexity of the above accurate PGM algorithm is still an exponential function of the circuit size, making it infeasible in general for large circuits.

In an effort to improve the efficiency of the accurate PGM method, a modular PGM approach was also introduced in [24]. It is based on the observation that many large circuits contain a limited number of simple logic components that are used repeatedly. With this in mind, circuits can be decomposed into several modules whose reliabilities are calculated using the accurate PGM method. The circuit output reliability is then evaluated by combining these modules along the path from primary inputs. Unfortunately, the input sampling is still needed in this case for large-scale circuits.

For the example circuit of Figure 3 with 4 input signals, a total of 16 input combinations need to be considered. We plot the conditional output reliability for each input combination in Figure 6, which shows that the output reliability varies within a relatively small range (no more than ±10%) for different input combinations. In other words, the input vector sampling can be implemented effectively with small errors. The overall output reliability is given by a weighted sum over all input combinations and is found to be $R_C = 0.8701$ (with the runtime of $T_{PGM} = 0.0093$ s), compared to the accurate value of 0.8658 given by PTM (i.e., the relative error is as low as ~0.5%).

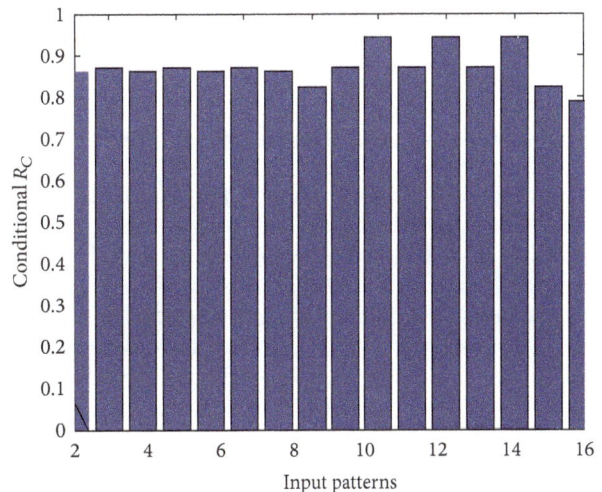

FIGURE 6: The conditional output reliability of Figure 3 for different input combinations (x-axis labels 1~16 indicate 16 input patterns from 0000~1111).

In order to see the performance of different PGM algorithms on large circuits, we implemented the simple PGM algorithm in *Matlab* and tested it on ISCAS'85 benchmarks. The results are shown in Table 2. We also compare the simple PGM with both accurate and modular PGM methods in Tables 3 and 4, where the simulation results for both accurate and modular PGM methods are taken from [24].

It can be seen from these tables that the simple PGM algorithm can provide highly accurate results if the circuits (such as C432 and C1355) have no or few reconvergent fanouts and/or if the fanouts originate from the primary inputs.

TABLE 4: Simulation results for modular PGM in comparison with MC.

Circuit	Size	Modular PGM approach (10^3 samples) [24]		Monte-Carlo (10^6 runs)
		Average error (%)	Runtime (s)	Runtime (s)
C432	160	9.21	0.25	183
C499	202	0.11	2.16	203
C1355	546	4.2	2.98	472
C2670	1193	0.43	4.26	1151

For those circuits with significant fanouts (such as C2670 and C5315), the average (or maximum) errors for the simple PGM can increase significantly (in particular, the maximum error is up to 43% for C5315, as shown in Table 2). From Table 3, the accurate PGM need longer runtimes than the simple PGM for small circuits. Results in Table 4 confirm that the modular PGM is very efficient while the accuracy may not always be good enough for some circuits (with an average error of 9% for C432).

In summary, for all the above three different versions of PGM, the input sampling is inevitable for improved efficiency if the number of primary inputs N_{in} is large ($\sim$30). This is mainly where the analysis errors come in. Thus, it can be concluded that they represent a good model only for circuits with a small number of primary inputs, where no input sampling is required. For the circuits without reconvergent fanouts, the input sampling in the PGM approach is unnecessary, because both signal probability and output reliability in this case can be computed within $O(N_g)$ time (see [29] for details).

6. Observability-Based Reliability Analysis

Another reliability analysis method was presented in [26], which is based on the observation that an error at the output of any gate is the cumulative effect of a local error component attributed to the error probability of the gate, and a propagated error component was attributed to the failure of gates in its transitive fan-in cone. In [26], the observability of a gate (or its output signal) is the conditional circuit error probability given the single error at current gate. The value of this observability can be simply defined as $o_i = (1 - R_C(\varepsilon_i = 1))$, where $R_C(\varepsilon_i = 1)$ is the circuit reliability given a single error with the current gate, and can be calculated using Boolean differences [29], symbolic techniques (such as BDDs), or simulation method. It can be expected that the gate observabilities are highly related to the input probabilities.

For a single-fault case (i.e., only one gate in the circuit is erroneous), the circuit reliability (assuming a single primary output) can be simply calculated by considering each fault case individually. Assume that the error rate and observability of the ith gate are ε_i and o_i, respectively. If gate i is erroneous while the other gates are fault-free, the output reliability simply is equal to o_i. Thus, the overall reliability can be easily calculated by

$$R_C = \sum_{\text{all } i} \left(\varepsilon_i \cdot o_i \cdot \prod_{j \neq i} (1 - \varepsilon_j) \right), \qquad (8)$$

which is exact for the single-fault case.

If a multiple-error case is considered, the complexity of computing the reliability will grow exponentially with N_g. In order to improve the efficiency in this case, the following two assumptions are used in [26]: (a) the impacts of gate failures on the primary output are decoupled, which implies that the output is erroneous if an odd number of gates are simultaneously observable and (b) the observabilities of all gates are independent. As a result, the simultaneous observability of multiple gates is simply the product of their individual observabilities.

We took the example circuit of Figure 3 for illustration. First, let us assume all four gates ($G_1 \sim G_4$) in the circuit are erroneous with the probabilities of ε_1, ε_2, ε_3, and ε_4, respectively (other cases can be analyzed similarly). Based on the above assumption (a), we only need to consider the cases where an odd number (1 or 3) of gates is simultaneously observable. This means that when an even number (0, 2, or 4) of gates is observable, the output signal Z will has correct value as gate errors are logically masked by one another. Secondly, under the assumption (b), the probability of only one gate being observable is given by $\sum_i (o_i \cdot \prod_{j \neq i} (1 - o_j))$ (the probability of three gates being simultaneously observable can be calculated similarly). Based on these assumptions, a closed-form expression for the circuit reliability of the circuit (assuming a single primary output) can be written generally as a function of error probabilities and observabilities of all gates [26]; that is,

$$R_C = \frac{1}{2} \left(1 + \prod_i (1 - 2\varepsilon_i o_i) \right), \qquad (9)$$

which can be computed efficiently if all gate observabilities are known (however, this analysis is only suitable for small circuits or large ones with small values of gate error probabilities, which will be clear later). The gate observability can be determined using the PTM method. For instance, the observability of gate G_1 in Figure 3 is calculated as the output reliability by setting $r_1 = 0$ and $r_2 = r_3 = r_4 = 1$. The results are $[o_1, o_2, o_3, o_4] = [0.25, 0.375, 0.375, 0]$. We calculate the circuit reliability R_C using the above expression and plot the results against the accurate values given by the PTM in Figure 7(a) for different values of gate reliability. The relative error is shown in Figure 7(b). It can be seen clearly from these figures that the observability-based analysis is only accurate for small gate error rates, in which case the probability for single gate failure is significantly higher than that for multiple gate failures.

To reduce the computational complexity of the above observability-based reliability analysis, [26] also proposed

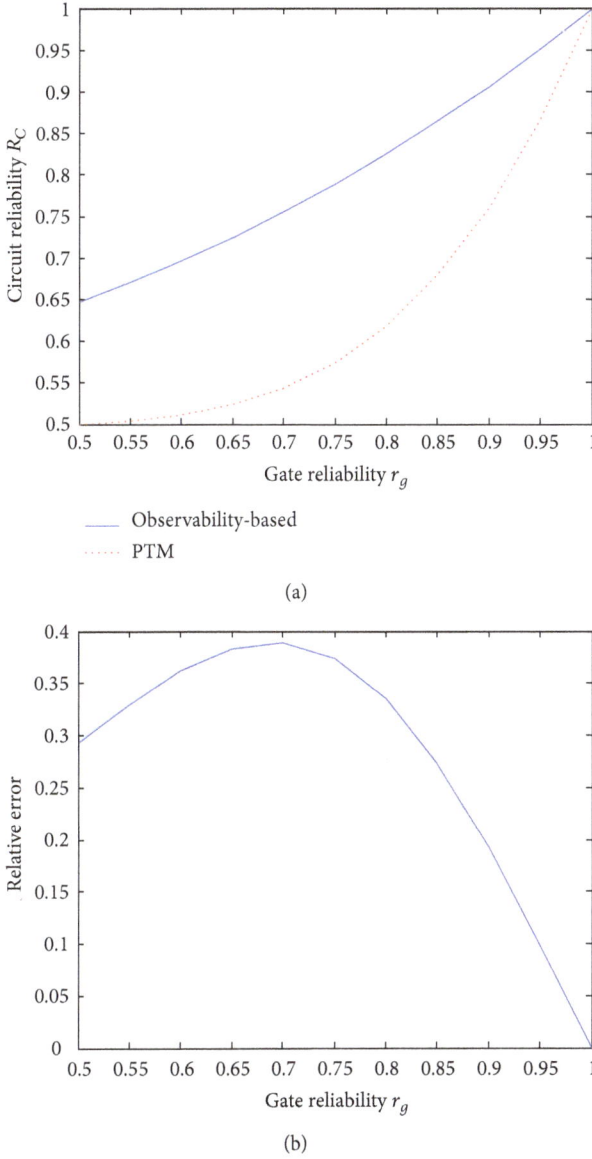

FIGURE 7: (a) Circuit reliability R_C versus gate reliability and (b) relative error versus gate reliability for the example circuit of Figure 3.

a sampling algorithm by considering the constraint that only a maximum of k gates can fail simultaneously. This algorithm first generates a set of samples for failed gates and guarantees that the total number of gates with error is no more than k. Then, a single-pass reliability analysis algorithm [26] was used to evaluate the error probability at the primary outputs, leading to the computational complexity of $O(N_g \cdot k^2)$, where N_g is the number of gates with error. For a specific sample, the reliabilities of gates in the sampling are set to be 0 and the rest are set to be 1. Finally, the overall circuit reliability is estimated by averaging the reliabilities over all samples. Therefore, this maximum-k gate failure model can be viewed as a hybrid method that makes a trade-off between the accuracy of simulation-based method and the efficiency of analytical approach. It provides more accurate results than

the single-pass algorithm [26] and takes the shorter runtime than MC or SCM.

7. Correlation Coefficient Method (CCM)

CCM is a widely used approach that evaluates the signal probabilities for (fault-free) combination circuits [30]. As mentioned before, the reliability analysis can be transformed to signal probability computation. Therefore, the CCM can be used to evaluate the reliability estimation [8, 26–28, 30]. The main idea of CCM is briefly described below.

In order to compute the signal probability, the correlation coefficient between two probabilistic signals (denoted by i and j) is defined as [8]

$$C_{i,j} = C_{j,i} = \frac{P(ij)}{P(i)P(j)} = \frac{P(j \mid i)}{P(j)}, \quad (10)$$

which is equal to 1 for signals i and j are independent. It should be noted that, here, only the first order correlation coefficients are considered, and the correlation of two signals with a third one (denoted by h) is approximated as $C_{ij,h} = C_{i,h} \cdot C_{j,h}$. For reliability computation, four correlation coefficients for a pair of signals are needed. Each coefficient corresponds to a combination of events (i.e., $0 \rightarrow 1$ or $1 \rightarrow 0$ error) on the signal pair. In other words, the signal error (or reliability) correlation coefficient between signals i and j is defined as [26]

$$C_{ij} = \frac{P(i_{0 \rightarrow 1} j_{0 \rightarrow 1})}{P(i_{0 \rightarrow 1}) P(j_{0 \rightarrow 1})},$$

$$C_{i\bar{j}} = \frac{P(i_{0 \rightarrow 1} j_{1 \rightarrow 0})}{P(i_{0 \rightarrow 1}) P(j_{1 \rightarrow 0})},$$

$$C_{\bar{i}j} = \frac{P(i_{1 \rightarrow 0} j_{0 \rightarrow 1})}{P(i_{1 \rightarrow 0}) P(j_{0 \rightarrow 1})}, \quad (11)$$

$$C_{\bar{i}\bar{j}} = \frac{P(i_{1 \rightarrow 0} j_{1 \rightarrow 0})}{P(i_{1 \rightarrow 0}) P(j_{1 \rightarrow 0})},$$

where the $P(i_{0 \rightarrow 1})$ is the probability that the value of signal i flips to 1 from its correct value 0, that is, the error probability of i given that its error-free value is 0. Once the error correlations and error-free signal probabilities are generated, the single-pass analysis is conducted using the forward topological order with the computational complexity of $O(N_g)$. Since the computation complexity of CCM is linear with the number of levels (L) and pseudoquadratic with the number of gates per level (N_L), the overall complexity of CCM-based reliability analysis turns out to be $O(N_g^{1.5})$ if a square circuit is assumed (i.e., $N_L = L = N_g^{0.5}$). This complexity is an upper bound as not all signals are correlated in real circuits.

In [26] which uses the CCM, an average relative error of up to ~13% over all outputs was reported for circuits with significant fanout (e.g., C499 and C1355) when the gate error rates range within [0, 0.5] (for other benchmark circuits, the error was around 2~6%). Also, the relative errors may not be mitigated significantly by using more correlation coefficients. For instance, by using 0, 4, and 16 correlation coefficients, the relative errors for C499 are only improved to 13.1%, 11.2%, and 11.11%, respectively [26], where the zero-coefficient case means that all signals are treated as independent with the computation complexity of $O(N_g)$. It is shown in [26] that the runtime of using 4 coefficients is several orders of magnitude longer than the zero-coefficient case (~100 s versus ~1 s, for circuit with ~1000 gates). Therefore, it may not be worthwhile to calculate more correlation coefficients for slightly improved accuracy. In [30], the relative error for large circuits (with hundreds of gates) was reported at ~7% on average with the runtime of ~10 s, which is comparable to those from [26].

8. Comparison and Future Work

In summary, the ultimate goal of existing approaches for reliability analysis is to achieve more accurate results with as low computational cost as possible. Both accuracy and efficiency depend on specific circuit structures and their size, and, in most cases, the tradeoff between them needs to be made. The main features of each approach are described as follows.

(a) If circuits have no reconvergent fanouts (e.g., a circuit with tree structure), both signal probability and reliability can be calculated exactly with linear time (i.e., $O(N_g)$). The readers are referred to [29] for further details.

(b) For those circuits with reconvergent fanouts, the PTM method and accurate PGM model can promise exact results, while their computation costs are exponentially high. The PTM approach requires the space complexity of $O(2^{N_{in}+N_{out}})$, and the accurate PGM has the computation complexity of $O(N_g \cdot 2^{N_{in}+N_f})$. Thus, some sampling techniques are usually needed to handle large-scale circuits in these computation frameworks, leading to less accurate results.

(c) Simulation-based methods (such as MC or SCM) can provide the results with high level of accuracy, as long as enough simulation sequences are applied. To achieve a required level of accuracy, the number of simulation runs need to be determined statistically or empirically. The time complexity can be estimated by $O(N_g \cdot N_{MC})$ or $O(N_g \cdot N_{SCM})$, where N_g is again the circuit size and N_{MC} (or N_{MC}) represents the number of simulation runs. The SCM is more efficient than MC especially for small gate error rates, as the runtime of the former is approximately scaled by a constant factor.

(d) The observability-based approach has some theoretical implications, since it gives reasonable results only for circuits with extremely-low gate error rates. The maximum-k method can be viewed as the combination of CCM-based and simulation-based methods.

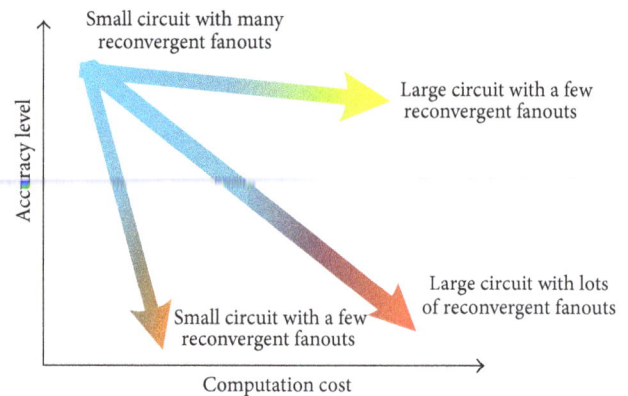

FIGURE 8: A general sketch of solution space for different circuit categories.

It shows better performance than the observability-based approach in terms of both accuracy and efficiency for lower gate error rates.

(e) If all reconvergent fanouts within circuits originate from primary inputs, the simple PGM method gives exact results with the computational complexity of $O(N_g \cdot 2^{N_{in}})$. For circuits consisting of a few logic modules that are repetitively used, the modular PGM method is a good option that can provide good accuracy with short runtime.

From the above discussions, it can be concluded that errors in reliability analysis are mainly due to the reconvergent fanouts (or signal correlation) inherent in many circuits under consideration in the sense that the accurate results can be obtained efficiently for circuits with no or a few reconvergent fanouts. On the other hand, the circuit size (i.e., a large number of primary inputs or a large number of gates or both) is the main contributor to high computational costs for reliability analysis. Therefore, the most challenging problem is to analyze the reliability for large-scale circuits with a lot of reconvergent fanouts. Figure 8 illustrates the expected solution space in general in terms of accuracy and computational cost for different circuit categories. Any existing approach for the reliability analysis corresponds to a specific point in this space, which represents a tradeoff between accuracy and efficiency. For instance, the results from PTM, PGM, or MC fall into the right-upper corner of this figure with expensive computation and high level of accuracy. An ideal approach should be able to provide results somewhere near the left-upper corner where both accuracy and efficiency can be ensured.

While gate-level reliability analysis methods are well documented, there are some other important issues that remain to be tackled. First of all, most existing methods only deal with the reliability of each individual output and/or the averaged reliability over all outputs. However, the joint reliability for multiple outputs (i.e., the probability that all outputs are error-free simultaneously) is what really matters. This joint reliability could be totally different from any individual output reliability or the averaged output reliability, depending on the possible correlation among individual output reliabilities. For an extreme case where all individual output reliabilities are independent, the joint reliability will simply be the product of all these reliabilities, which leads to a minimum value. As the correlation of output reliabilities becomes strong, the joint reliability tends to rise. In general, the complexity of computing the joint reliability would be an exponential function of the number of primary outputs. It is still an open question how to estimate the joint reliability for multiple-output circuits in an efficient way. Secondly, most of the current reliability analysis frameworks assume that the reliability for an error-free output being "0" (denoted by r_0) is the same as that for an error-free output being "1" (denoted by r_1). This is the so-called symmetric reliability model. However, this assumption does not always hold true in the real world. Thus, an asymmetric reliability model (where $r_0 \neq r_1$) would make more sense for better estimation of reliability. This requires further research work that can take the asymmetric model into consideration. Finally, there is also plenty of room for gate-level reliability improvement using reliability-critical gates as well as considering other performance metrics (such as circuit area and delay and power consumption). Unfortunately, to the best of authors' knowledge, little or limited study has been done so far in this regard.

9. Conclusion

We have reviewed the state-of-the-art methods for reliability analysis and shown their advantages and disadvantages. Some of these methods have been implemented on benchmark circuit examples to compare their performance in terms of accuracy and efficiency. While these methods seem to be effective for some specific cases/circuits, no single one of them stands out as an all-time winner due to the nature and complexity of the reliability analysis problem. Further work has also been suggested for the future research in this area.

Conflict of Interests

The authors declare that there is no conflict of interests regarding the publication of this paper.

References

[1] S. Borkar, "Designing reliable systems from unreliable components: The challenges of transistor variability and degradation," *IEEE Micro*, vol. 25, no. 6, pp. 10–16, 2005.

[2] J. Han, H. Chen, J. Liang, P. Zhu, Z. Yang, and F. Lombardi, "A stochastic computational approach for accurate and efficient reliability evaluation," *IEEE Transactions on Computers*, vol. 63, no. 6, pp. 1336–1350, 2014.

[3] S. Krishnaswamy, S. M. Plaza, I. L. Markov, and J. P. Hayes, "Signature-based SER analysis and design of logic circuits," *IEEE Transactions on Computer-Aided Design of Integrated Circuits and Systems*, vol. 28, no. 1, pp. 74–86, 2009.

[4] C. Chen and Y. Mao, "A statistical reliability model for single-electron threshold logic," *IEEE Transactions on Electron Devices*, vol. 55, no. 6, pp. 1547–1553, 2008.

[5] C. Chen, "Reliability-driven gate replication for nanometer-scale digital logic," *IEEE Transactions on Nanotechnology*, vol. 6, no. 3, pp. 303–308, 2007.

[6] J. von Neumann, "Probabilistic logics and the synthesis of reliable organisms from unreliable components," in *Automata Studies*, C. E. Shannon and J. McCarthy, Eds., pp. 43–98, Princeton University Press, Princeton, NJ, USA, 1956.

[7] J. Han and P. Jonker, "A system architecture solution for unreliable nanoelectronic devices," *IEEE Transactions on Nanotechnology*, vol. 1, no. 4, pp. 201–208, 2002.

[8] S. Ercolani, M. Favalli, M. Damiani, P. Olivo, and B. Ricco, "Estimate of signal probability in combinational logic networks," in *Proceedings of the 1st European Test Conference*, pp. 132–138, Paris, France, April 1989.

[9] M. R. Garey and D. S. Johnson, *Computers and Intractability: A Guide to the Theory of NP-Completeness*, W. H. Freeman, San Francisco, Calif, USA, 1979.

[10] S. Krishnaswamy, G. F. Viamontes, I. L. Markov, and J. P. Hayes, "Accurate reliability evaluation and enhancement via probabilistic transfer matrices," in *Proceedings of the Design, Automation and Test in Europe*, vol. 1, pp. 282–287, March 2005.

[11] S. Krishnaswamy, G. F. Viamontes, I. L. Markov, and J. P. Hayes, "Probabilistic transfer matrices in symbolic reliability analysis of logic circuits," *ACM Transactions on Design Automation of Electronic Systems*, vol. 13, no. 1, article 8, 2008.

[12] W. Ibrahim, V. Beiu, and M. H. Sulieman, "On the reliability of majority gates full adders," *IEEE Transactions on Nanotechnology*, vol. 7, no. 1, pp. 56–67, 2008.

[13] T. Rejimon, K. Lingasubramanian, and S. Bhanja, "Probabilistic error modeling for nano-domain logic circuits," *IEEE Transactions on Very Large Scale Integration (VLSI) Systems*, vol. 17, no. 1, pp. 55–65, 2009.

[14] T. Rejimon and S. Bhanja, "Scalable probabilistic computing models using Bayesian networks," in *Proceedings of the IEEE International 48th Midwest Symposium on Circuits and Systems (MWSCAS '05)*, pp. 712–715, August 2005.

[15] J. T. Flaquer, J. M. Daveau, L. Naviner, and P. Roche, "Fast reliability analysis of combinatorial logic circuits using conditional probabilities," *Microelectronics Reliability*, vol. 50, no. 9–11, pp. 1215–1218, 2010.

[16] R. I. Bahar, J. Chen, and J. Mundy, "A probabilistic-based design for nanoscale computation," in *Nano, Quantum and Molecular Computing: Implications to High Level Design and Validation*, S. Shukla and R. I. Bahar, Eds., chapter 5, Kluwer Academic, Norwell, Mass, USA, 2004.

[17] R. I. Bahar, J. Mundy, and J. Chen, "A probability-based design methodology for nanoscale computation," in *Proceedings of the International Conference on Computer-Aided Design*, pp. 480–486, November 2003.

[18] A. R. Kermany, N. H. Hamid, and Z. A. Burhanudin, "A study of MRF-based circuit implementation," in *Proceedings of the International Conference on Electronic Design (ICED '08)*, pp. 1–4, December 2008.

[19] D. Bhaduri and S. Shukla, "NANOLAB—a tool for evaluating reliability of defect-tolerant nanoarchitectures," *IEEE Transactions on Nanotechnology*, vol. 4, no. 4, pp. 381–394, 2005.

[20] X. Lu, J. Li, and W. Zhang, "On the probabilistic characterization of nano-based circuits," *IEEE Transactions on Nanotechnology*, vol. 8, no. 2, pp. 258–259, 2009.

[21] H. Chen and J. Han, "Stochastic computational models for accurate reliability evaluation of logic circuits," in *Proceedings*

of the 20th Great Lakes Symposium on VLSI (GLSVLSI '10), pp. 61–66, May 2010.

[22] J. B. Gao, Y. Qi, and J. A. B. Fortes, "Bifurcations and fundamental error bounds for fault-tolerant computations," *IEEE Transactions on Nanotechnology*, vol. 4, no. 4, pp. 395–402, 2005.

[23] J. Han, E. Taylor, J. Gao, and J. Fortes, "Faults, error bounds and reliability of nanoelectronic circuits," in *Proceedings of the IEEE 16th International Conference on Application-Specific Systems, Architectures, and Processors (ASAP '05)*, pp. 247–253, July 2005.

[24] J. Han, H. Chen, E. Boykin, and J. Fortes, "Reliability evaluation of logic circuits using probabilistic gate models," *Microelectronics Reliability*, vol. 51, no. 2, pp. 468–476, 2011.

[25] J. Han, E. R. Boykin, H. Chen, J. H. Liang, and J. A. B. Fortes, "On the reliability of computational structures using majority logic," *IEEE Transactions on Nanotechnology*, vol. 10, no. 5, pp. 1009–1022, 2011.

[26] M. R. Choudhury and K. Mohanram, "Reliability analysis of logic circuits," *IEEE Transactions on Computer-Aided Design of Integrated Circuits and Systems*, vol. 28, no. 3, pp. 392–405, 2009.

[27] L. Chen and M. B. Tahoori, "An efficient probability framework for error propagation and correlation estimation," in *Proceedings of the IEEE 18th International On-Line Testing Symposium (IOLTS '12)*, pp. 170–175, Sitges, Spain, June 2012.

[28] S. Ercolani, M. Favalli, M. Damiani, P. Olivo, and B. Ricco, "Testability measures in pseudorandom testing," *IEEE Transactions on Computer-Aided Design of Integrated Circuits and Systems*, vol. 11, no. 6, pp. 794–800, 1992.

[29] N. Mohyuddin, E. Pakbaznia, and M. Pedram, "Probabilistic error propagation in logic circuits using the boolean difference calculus," in *Proceedings of the 26th IEEE International Conference on Computer Design (ICCD '08)*, pp. 7–13, October 2008.

[30] S. Sivaswamy, K. Bazargan, and M. Riedel, "Estimation and optimization of reliability of noisy digital circuits," in *Proceedings of the 10th International Symposium on Quality Electronic Design (ISQED '09)*, pp. 213–219, March 2009.

## 20

Performance Analysis of Modified Drain Gating Techniques for Low Power and High Speed Arithmetic Circuits

Shikha Panwar, Mayuresh Piske, and Aatreya Vivek Madgula

School of Electronics Engineering (SENSE), VIT University, Vandalur-Kelambakkam Road, Chennai 600127, India

Correspondence should be addressed to Shikha Panwar; shikha.panwar24@gmail.com

Academic Editor: Yu-Cheng Fan

This paper presents several high performance and low power techniques for CMOS circuits. In these design methodologies, drain gating technique and its variations are modified by adding an additional NMOS sleep transistor at the output node which helps in faster discharge and thereby providing higher speed. In order to achieve high performance, the proposed design techniques trade power for performance in the delay critical sections of the circuit. Intensive simulations are performed using Cadence Virtuoso in a 45 nm standard CMOS technology at room temperature with supply voltage of 1.2 V. Comparative analysis of the present circuits with standard CMOS circuits shows smaller propagation delay and lesser power consumption.

1. Introduction

As we move on to finer MOSFET technologies, transistor delay has decreased remarkably which helped in achieving higher performance in CMOS VLSI processors. With technology scaling, it is required to reduce the threshold and power supply voltages. As square of power supply voltage is directly proportional to dynamic power dissipation, to achieve less consumption of power, supply voltage has to be reduced. Static power and dynamic power are two main components of total power dissipation. Static power consumption is calculated in the form of leakage current through each device. Substantial increase has been observed in subthreshold leakage current with scaling of threshold voltage [1]. Subthreshold current I_{ST} is given by [1]

$$I_{ST} = \mu_0 \text{Cox} \left(\frac{W}{L}\right)(m-1)(V_T)^2 \times e^{(V_g - V_{th})/mV_T} \tag{1}$$
$$\times \left(1 - e^{-V_{DS}/V_T}\right),$$

where

$$m = 1 + \frac{\text{Cdm}}{\text{Cox}}, \tag{2}$$

where thermal voltage, $V_T = KT/q$, μ_0 is the mobility, V_g is the gate voltage, V_{th} is the threshold voltage, V_{DS} is

termed as drain to source voltage, and m is the body effect coefficient. Cdm and Cox are the depletion layer and gate oxide capacitances, respectively.

To counteract the excessive leakage in CMOS circuit, many architectural techniques have been proposed over the years. Power gating [2] and stacking effect [3] are two well-known techniques for reducing leakage power dissipation. Power gating normally makes use of sleep transistors that are connected either between the power supply and the pull-up network (PUN) or between the pull-down network (PDN) and ground. Sleep transistors are switched on when the circuit is evaluating and they are switched off in standby mode to conserve the leakage power in the logic circuit. Multi-threshold-CMOS (MTCMOS) [4] technique is also an effective way to achieve considerable decline in leakage power consumption. In MTCMOS technique, high V_{th} sleep transistors are added in the circuit whereas PUN and PDN use low V_{th} devices. In dual threshold circuits [5], low V_{th} devices are used in the delay critical sections and high V_{th} devices are used to reduce the leakage current in the circuitry.

Stacking of transistor in series reduces the subthreshold leakage current when one transistor is in the off state. Stacking effect is used in sleepy stack technique [6] and force stack technique [7]. Sleepy stack technique provides better results than forced stack technique. In forced stack, an extra sleep

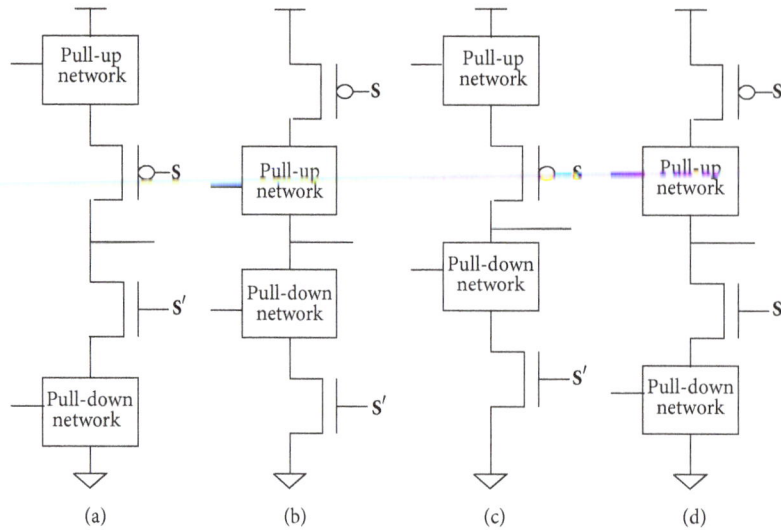

FIGURE 1: (a) Drain gating, (b) power gating, (c) drain-header and power-footer gating (DHPF), and (d) drain-footer and power-header gating (DFPH).

transistor is inserted for each input of the gate both in PUN and in PDN resulting in higher delay and area. In sleepy stack, an additional sleep transistor is connected in parallel with the transistor stack. This reduces the leakage current but at the same time delay in the circuit is increased.

LECTOR [8] and GALEOR [9] are also two leakage tolerant techniques. LECTOR makes use of two leakage control transistors (LCTs) that are connected between the PUN and PDN. In the same time GALEOR technique makes use of gated leakage transistors (GLTs). Both LCTs and GLTs reduce leakage by increasing the resistance between supply voltage and ground.

Another efficient technique to counter the leakage current problem is drain gating and its variation [10], explained in detail in Section 2. The modified circuits are proposed in Section 3. Simulation results taking NAND gate, 1-bit full adder, and 8-bit RCA (Ripple carry adder) as test bench circuits are enumerated in Section 4 and Section 5 provides the final conclusion.

2. Drain Gating Technique and Its Variant Circuits

In drain gating technique [10] shown in Figure 1(a), two sleep transistors are added between the PUN and PDN. PMOS transistor with sleep input (S) is connected between PUN and output node, whereas NMOS transistor with sleep input (S') is inserted between the output node and PDN. When the circuit is in evaluation mode, the NMOS and PMOS sleep transistors are turned on resulting in low resistance conducting path. When the circuit is in standby, both transistors are switched off to reduce the standby power. Other variant circuits of drain gating are, namely, power gating, drain-header and power-footer gating (DHPF), and drain-footer and power-header gating (DFPH). In power gating technique, PMOS sleep transistor with input (S) is

added between the power supply and the PUN, whereas NMOS sleep transistor with input (S') is added between the PDN and ground as shown in Figure 1(b). The two mixed techniques DHPF and DFPH are shown in Figures 1(c) and 1(d), respectively. As the name suggests, in DHPF, a PMOS sleep switch is inserted between PUN and output node and an NMOS sleep switch is inserted between the PDN and ground rail. DFPH consists of an NMOS sleep switch between output node and PDN and a PMOS sleep switch between the power supply and the PUN. Comparative results in Section 4 indicate that power gating technique is the best leakage tolerant technique whereas drain gating technique has the least delay among the previously proposed circuits.

3. The Proposed High Speed Circuit Techniques

The proposed circuits are aimed at reducing the propagation delay incurred by drain gating technique and its variations. Four different circuit techniques, namely *high speed drain gating (HS-drain gating)*, *HS-power gating*, *HS-DHPF*, and *HS-DFPH* as shown in Figures 2(a), 2(b), 2(c), and 2(d) respectively, are proposed in this section. In *HS-drain gating* technique an additional sleep transistor with sleep input (S) is connected at the output node parallel to the NMOS sleep transistor (S') and PDN. During the active mode, when the logic circuit evaluates the circuits output, the added NMOS sleep transistor (S) provides an additional discharging path in the circuit. This added transistor helps in speedy evaluation, hence providing higher speed. In a similar fashion, an additional NMOS sleep transistor with sleep input (S) is added to power gating, DHPF, and DFPH circuits.

The proposed cicuits have been verified by taking NAND gate, 1-bit full adder, and 8-bit RCA as test bench circuits. Experimental results in Section 4 prove that the modified HS-drain gating technique has the the least delay among

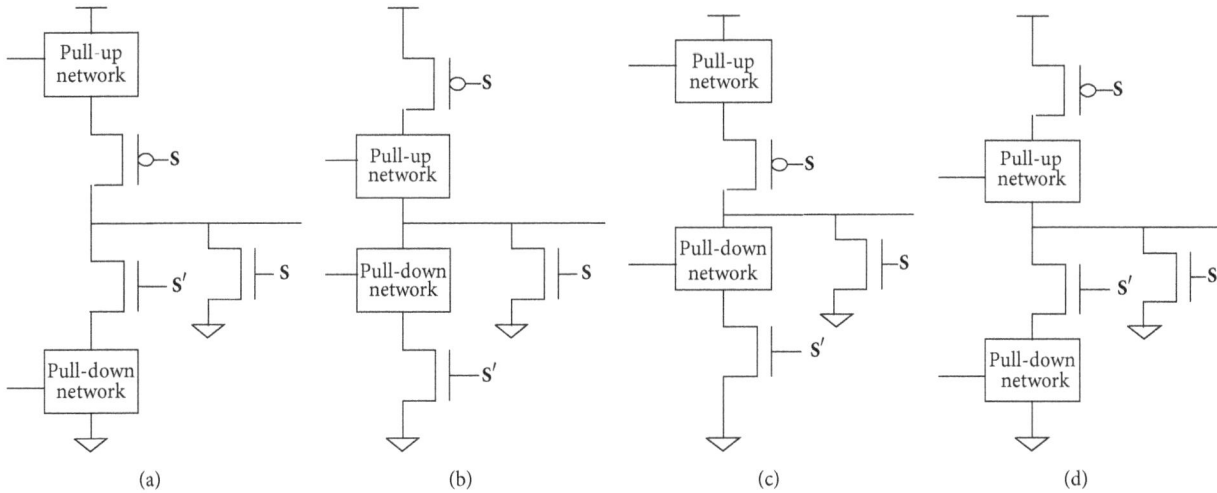

FIGURE 2: (a) HS-drain gating, (b) HS-power gating, (c) HS-DHPF, and (d) HS-DFPH.

TABLE 1: Power and delay values of NAND gate, FA, and 8-bit RCA using various techniques.

Circuit techniques	NAND gate		FA		8-bit RCA	
	Power (nW)	Delay (ps)	Power (nW)	Delay (ps)	Power (uW)	Delay (ps)
Standard CMOS	22.32	45e3	2.1e3	30e3	52.2	23.7e3
Drain gating	12.63	25e3	393	15e3	7.53	8.85e3
Power gating	8.73	205e3	238	150e3	2.39	20.5e3
DHPF	11.08	80e3	340	25e3	3.02	12e3
DFPH	8.71	175e3	245	150e3	4.02	15.5e3
HS-drain gating	18.42	2.22	250	15.08	6.97	10.3
HS-power gating	16.57	11.76	246	49.9	2.13	21.4
HS-DHPF	16.70	6.3	248	35.97	4.15	11.9
HS-DFPH	16.64	11.71	247	44.87	2.92	17

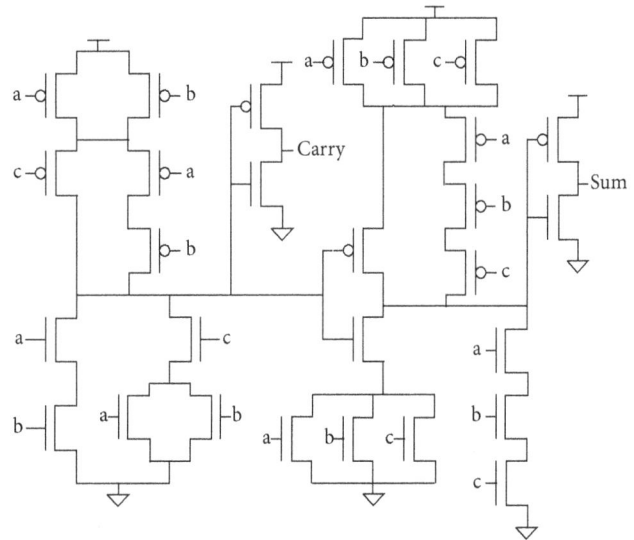

FIGURE 3: 1-bit CMOS full adder.

the existing and proposed architectural techniques as shown in Figure 5. Also HS-power gating technique has the lowest power as compared to standard CMOS circuit and the newly proposed circuits as shown in Table 1. The ratio of PMOS to NMOS size is set to be equal to 2.

Two-input NAND gate using HS-drain gating operates in two modes, namely, *sleep or standby mode* and *active mode*. When the circuit is in active mode, sleep input (**S**) is in low state, and output node gets charged to power supply voltage. Both NMOS and PMOS sleep transistors connected between PUN and PDN are turned on and output is evaluated. For example, if we provide input to the PUN as 0(XX) where XX stands for input vectors (00, 01, 10, 11), output will be high for the first three cases and low for the fourth case for the NAND gate. Sleep signal should be provided in the form of alternate high and low signals. When sleep signal (**S** = 1), both PMOS

and NMOS sleep transistors between PUN and PDN network turn off and additional NMOS sleep transistor is turned on, discharging the output node to ground thereby resulting in higher performance. A trade-off is achieved between power and delay so as to maintain high speed in the proposed circuits.

4. Simulations and Results

Two-input NAND gate, 1-bit full adder, and 8-bit RCA are implemented using the proposed high speed architectural techniques. The circuit diagrams for 1-bit full adder and 8-bit RCA are shown in Figures 3 and 4, respectively. Each stage in 8-bit RCA consists of a 1-bit full adder (FA). Each FA circuit consists of 28 transistors. In RCA, carry is propagated from one stage to another and final carry is obtained as C_8 shown in Figure 4.

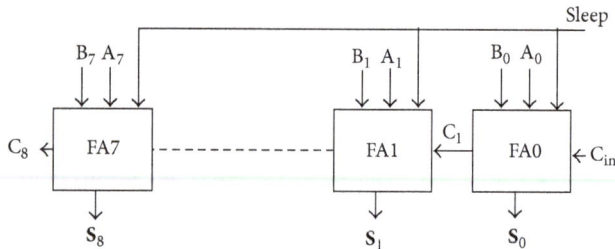

FIGURE 4: 8-bit RCA.

The total power consumption and propagation delay of various existing and proposed techniques for NAND gate, FA, and 8-bit RCA are compared in Table 1. HS-drain gating technique has the least delay. HS-power gating, HS-DFPH, and HS-DHPF suffer from 50%, 39%, and 13% propagation delay with respect to HS-drain gating technique. Standard drain gating and its variants circuit techniques suffer from 99% propagation delay in comparison with HS-drain gating technique. Circuits employing HS-power gating technique have very low power consumption. Power savings of nearly 85% are achieved in arithmetic architectures employing HS-power gating technique. HS-drain gating technique has the least power saving among the proposed circuits. HS-DHPF and HS-DFPH techniques optimize the power and delay in CMOS arithmetic circuits.

The corner analysis for the drain gating design and its variants is plotted along with that of the modified high speed counterparts. Figure 5 shows the temperature versus the propagation delay graph for 8-bit RCA using the existing techniques and the proposed techniques.

Similarly Figure 6 shows the plot of process corners versus the propagation delay of 8-bit RCA using the existing techniques and the proposed techniques.

On observing the comparative graph shown in Figures 5 and 6, we can infer that the designs made using the modified high speed drain gating technique and its corresponding variants have substantial reduction in the propagation delay when compared to the designs made using the CMOS, drain gating technique, and its variants.

5. Conclusions

In this paper, we have tabulated the total power consumption and the propagation delay for certain circuits using the existing low power and performance enhancing techniques and the newly proposed ones. Also we have made a comparative study of these techniques for the parameters like temperature, process corners, and propagation delay. Simulation results show that the proposed circuits work effectively even at extreme temperature and at different transistor configurations.

From the above mentioned experimental data, we can observe that, by implementing the high speed modified designs for the drain gating technique and its variants, we are able to enhance the performance of the design at lower power

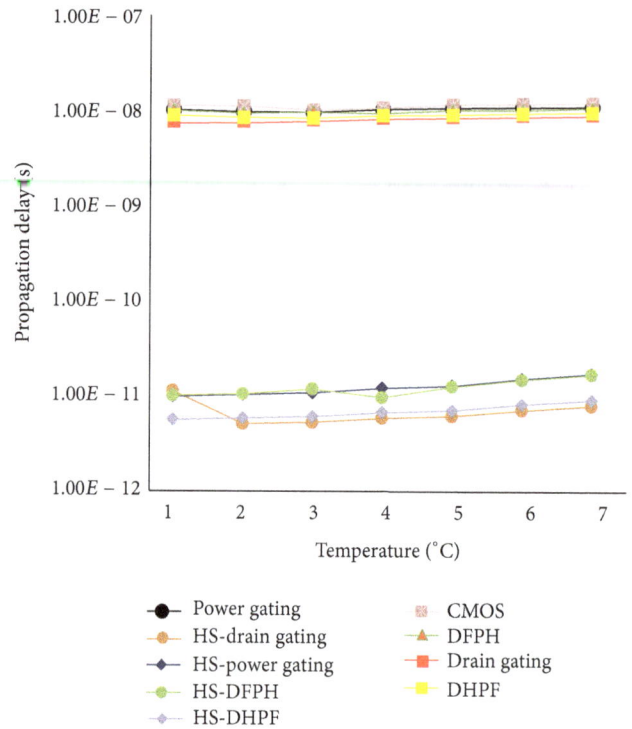

FIGURE 5: Temperature versus the propagation delay for the existing and the proposed techniques.

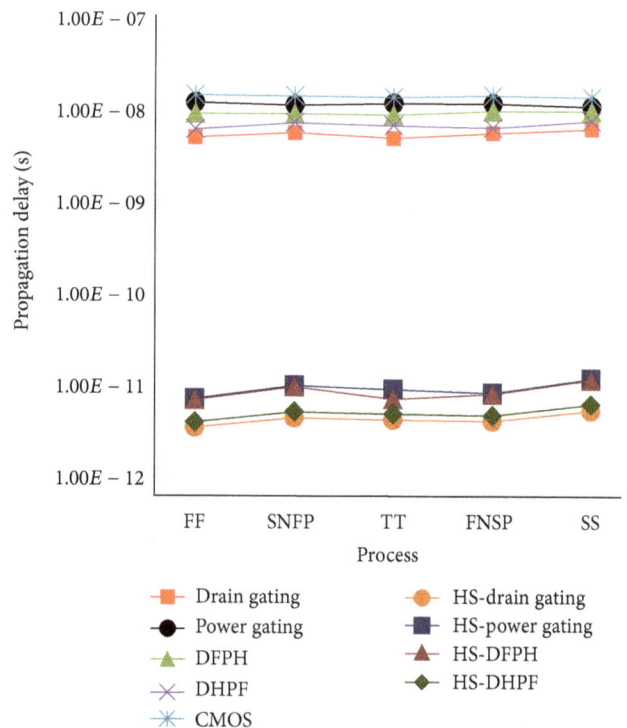

FIGURE 6: Process versus the propagation delay for 8-bit RCA using the existing and the proposed techniques.

consumption. Power consumption savings as observed in 8-bit RCA and 1-bit full adder are 95% and 88%, respectively, whereas propagation delay has been reduced by almost 99% in both RCA and full adder circuit.

Conflict of Interests

The authors declare that there is no conflict of interests regarding the publication of this paper.

References

[1] K. Roy, S. Mukhopadhyay, and H. Mahmoodi-Meimand, "Leakage current mechanisms and leakage reduction techniques in deep-submicrometer CMOS circuits," *Proceedings of the IEEE*, vol. 91, no. 2, pp. 305–327, 2003.

[2] M. Powell, S. Yang, B. Falsafi, K. Roy, and T. N. Vijaykumar, "Gated-Vdd: a circuit technique to reduce leakage in deep-submicron cache memories," in *Proceedings of the IEEE Symposium on Low Power Electronics and Design (ISLPED '00)*, pp. 90–95, July 2000.

[3] M. Johnson, D. Somasekhar, L. Y. Chiou, and K. Roy, "Leakage control with efficient use of transistor stacks in single threshold CMOS," *IEEE Transactions on VLSI Systems*, vol. 10, no. 1, pp. 1–5, 2002.

[4] S. Mutoh, T. Douseki, Y. Matsuya, T. Aoki, S. Shigematsu, and J. Yamada, "1-V power supply high-speed digital circuit technology with multithreshold-voltage CMOS," *IEEE Journal of Solid-State Circuits*, vol. 30, no. 8, pp. 847–854, 1995.

[5] L. Wei, Z. Chen, M. C. Johnson, K. Roy, Y. Ye, and V. K. De, "Design and optimization of dual-threshold circuits for low-voltage low-power applications," *IEEE Transactions on Very Large Scale Integration (VLSI) Systems*, vol. 7, no. 1, pp. 16–24, 1999.

[6] J. C. Park and V. J. Mooney III, "Sleepy stack leakage reduction," *IEEE Transactions on Very Large Scale Integration (VLSI) Systems*, vol. 14, no. 11, pp. 1250–1263, 2006.

[7] S. Narendra, V. De, D. Antoniadis, A. Chandrakasan, and S. Borkar, "Scaling of stack effect and its application for leakage reduction," in *Proceedings of the International Symposium on Low Electronics and Design (ISLPED '01)*, pp. 195–200, Huntington Beach, Calif, USA, August 2001.

[8] N. Hanchate and N. Ranganathan, "LECTOR: a technique for leakage reduction in CMOS dircuits," *IEEE Transactions on Very Large Scale Integration (VLSI) Systems*, vol. 12, no. 2, pp. 196–205, 2004.

[9] S. Katrue and D. Kudithipudi, "GALEOR: leakage reduction for CMOS circuits," in *Proceedings of the 15th IEEE International Conference on Electronics, Circuits and Systems (ICECS '08)*, pp. 574–577, September 2008.

[10] J. W. Chun and C. Y. R. Chen, "A novel leakage power reduction technique for CMOS circuit design," in *Proceedings of the International SoC Design Conference (ISOCC '10)*, pp. 119–122, November 2010.

## 21

Radix-$2^{\alpha}/4^{\beta}$ Building Blocks for Efficient VLSI's Higher Radices Butterflies Implementation

Marwan A. Jaber and Daniel Massicotte

Laboratory of Signals and Systems Integrations, Electrical and Computer Engineering Department, Université du Québec à Trois-Rivières, QC, Canada G9A 5H7

Correspondence should be addressed to Daniel Massicotte; daniel.massicotte@uqtr.ca

Academic Editor: Dionysios Reisis

This paper describes an embedded FFT processor where the higher radices butterflies maintain one complex multiplier in its critical path. Based on the concept of a radix-r fast Fourier factorization and based on the FFT parallel processing, we introduce a new concept of a radix-r Fast Fourier Transform in which the concept of the radix-r butterfly computation has been formulated as the combination of radix-$2^{\alpha}/4^{\beta}$ butterflies implemented in parallel. By doing so, the VLSI butterfly implementation for higher radices would be feasible since it maintains approximately the same complexity of the radix-2/4 butterfly which is obtained by block building of the radix-2/4 modules. The block building process is achieved by duplicating the block circuit diagram of the radix-2/4 module that is materialized by means of a feed-back network which will reuse the block circuit diagram of the radix-2/4 module.

1. Introduction

For the past decades, the main concern of the researchers was to develop a fast Fourier transform (FFT) algorithm in which the number of operations required is minimized. Since Cooley and Tukey presented their approach showing that the number of multiplications required to compute the discrete Fourier transform (DFT) of a sequence may be considerably reduced by using one of the fast Fourier transform (FFT) algorithms [1], interest has arisen both in finding applications for this powerful transform and for considering various FFT software and hardware implementations.

The DFT computational complexity increases according to the square of the transform length and thus becomes expensive for large N. Some algorithms used for efficient DFT computation, known as fast DFT computation algorithms, are based on the divide-and-conquer approach. The principle of this method is that a large problem is divided into smaller subproblems that are easier to solve. In the FFT case, dividing the work into subproblems means that the input data $x_{[n]}$ can be divided into subsets from which the DFT is computed, and then the DFT of the initial data is reconstructed from these intermediate results. Some of these methods are known

as the Cooley-Tukey algorithm [1], split-radix algorithm [2], Winograd Fourier transform algorithm (WFTA) [3], and others, such as the common factor algorithms [4].

The problem with the computation of an FFT with an increasing N is associated with the straightforward computational structure, the coefficient multiplier memories' accesses, and the number of multiplications that should be performed. The overall arithmetic operations deployed in the computation of an N-point FFT decreases with increasing r as a result; the butterfly complexity increases in terms of complex arithmetic computation, parallel inputs, connectivity, and number of phases in the butterfly's critical path delay. The higher radix butterfly involves a nontrivial VLSI implementation problem (i.e., increasing butterfly critical path delay), which explains why the majority of FFT VLSI implementations are based on radix 2 or 4, due to their low butterfly complexity. The advantage of using a higher radix is that the number of multiplications and the number of stages to execute an FFT decrease [4–6].

The most recent attempts to reduce the complexity of the higher radices butterfly's critical path was achieved by the concept of a radix-r fast Fourier transform (FFT) [8, 9], in which the concept of the radix-r butterfly computation has been formulated as composed engines with identical structures and a systematic means of accessing the corresponding multiplier coefficients. This concept enables the design of butterfly processing element (BPE) with the lowest rate of complex multipliers and adders, which utilizes r or $r - 1$ complex multipliers in parallel to implement each of the butterfly computations. Another strategy was based on targeting hardware oriented radix 2^α or 4^β which is an alternative way of representing higher radices by means of less complicated and simple butterflies in which they used the symmetry and periodicity of the root unity to further lower down the coefficient multiplier memories' accesses [10–20].

Based on the higher radices butterfly and the parallel FFT concepts [21, 22], we will introduce the structure of higher multiplexed 2^α or 4^β butterflies that will reduce the resources in terms of complex multiplier and adder by maintaining the same throughput and the same speed in comparison to the other proposed butterflies structures in [13–20].

This paper is organized as follows. Section 2 describes the higher radices butterfly computation and Section 3 details the FFT parallel processing. Section 4 elaborates the proposed higher radices butterflies; meanwhile Section 5 draws the performance evaluation of the proposed method and Section 6 is devoted to the conclusion.

2. Higher Radices' Butterfly Computation

The basic operation of a radix-r PE is the so-called butterfly computation in which r inputs are combined to give the r outputs via the following operation:

$$\mathbf{X} = \mathbf{B}_r \mathbf{x}_{\text{in}},$$

$$\mathbf{x}_{\text{in}} = \left[x_{(0)}, x_{(1)}, \ldots, x_{(r-1)} \right]^T, \tag{1}$$

$$\mathbf{X} = \left[X_{(0)}, X_{(1)}, \ldots, X_{(r-1)} \right]^T,$$

where $\mathbf{x}_{\text{in}}$ and $\mathbf{X}$ are, respectively, the butterfly's input and output vectors. $\mathbf{B}_r$ is the butterfly matrix ($\dim(\mathbf{B}_r) = r \times r$) which can be expressed as

$$\mathbf{B}_r = \mathbf{W}_N \mathbf{T}_r, \tag{2}$$

for decimation in frequency (DIF) process, and

$$\mathbf{B}_r = \mathbf{T}_r \mathbf{W}_N, \tag{3}$$

for decimation in time (DIT) process. In both cases the twiddle factor matrix, $\mathbf{W}_N$, is a diagonal matrix which is defined by $\mathbf{W}_N = \text{diag}(1, w_N^p, w_N^{2p}, \ldots, w_N^{(r-1)p})$ with $p = 0, 1, \ldots, N/r^s - 1$ and $s = 0, 1, \ldots, \log_r N - 1$ and $\mathbf{T}_r$ is the *adder tree* matrix within the butterfly structure expressed as [4]

$$\mathbf{T}_r = \begin{bmatrix} w_N^0 & w_N^0 & w_N^0 & \cdots & \cdots & w_N^0 \\ w_N^0 & w_N^{N/r} & w_N^{2N/r} & \cdots & \cdots & w_N^{(r-1)N/r} \\ w_N^0 & w_N^{2N/r} & w_N^{4N/r} & \vdots & \vdots & w_N^{2(r-1)N/r} \\ \vdots & \vdots & \vdots & \vdots & \vdots & \vdots \\ \vdots & \vdots & \vdots & \vdots & \vdots & \vdots \\ w_N^0 & w_N^{(r-1)N/r} & w_N^{2(r-1)N/r} & \cdots & \cdots & w_N^{(r-1)^2 N/r} \end{bmatrix}. \tag{4}$$

As seen from (2) and (3), the adder tree, $\mathbf{T}_r$, is almost identical for the two algorithms, with the only difference being the order in which the twiddle factor and the adder tree multiplication are computed. A straightforward implementation of the adder tree is not effective for higher radices butterflies due to the added complex multipliers in the higher radices butterflies' critical path that will complicate its implementation in VLSI.

By defining the element of the lth line and the mth column in the matrix $\mathbf{T}_r$ as $[\mathbf{T}_r]_{l,m}$,

$$[\mathbf{T}_r]_{l,m} = w_N^{[\![(lmN/r)]\!]_N}, \tag{5}$$

where $l = 0, 1, \ldots, r - 1$, $m = 0, 1, \ldots, r - 1$, and $[\![x]\!]_N$ represents the operation x modulo N. By defining $\mathbf{W}_{N(m,v,s)}$ the set of the twiddle factor matrix as

$$[\mathbf{W}_N]_{l,m(v,s)} = \text{diag}\left(w_{N(0,v,s)}, w_{N(1,v,s)}, \ldots, w_{N(r-1,v,s)} \right), \tag{6}$$

where the index r is the FFT's radix, $v = 0, 1, \ldots, V - 1$ represents the number of words of size r ($V = N/r$), and $s = 0, 1, \ldots, S$ is the number of stages (or iterations $S = \log_r N - 1$). Finally, the twiddle factor matrix in (2) and (3) can be expressed for the different stages of an FFT process as [7, 8]

$$[\mathbf{W}_N]_{l,m(v,s)} = \begin{cases} w_N^{[\![\lfloor v/r^s \rfloor lr^s]\!]_N} & \text{for } l = m \\ 0 & \text{elsewhere,} \end{cases} \tag{7}$$

for the DIF process and (3) would be expressed as

$$[\mathbf{W}_N]_{l,m(v,s)} = \begin{cases} w_N^{[\![\lfloor v/r^{(S-s)} \rfloor lr^{(S-s)}]\!]_N} & \text{for } l = m \\ 0 & \text{elsewhere,} \end{cases} \tag{8}$$

for the DIT process, where $l = 0, 1, \ldots, r - 1$ is the lth butterfly's output, $m = 0, 1, \ldots, r - 1$ is the mth butterfly's input, and $\lfloor x \rfloor$ represents the integer part operator of x.

As a result, the lth transform output during each stage can be illustrated as

$$X_{(v,s)}[l] = \sum_{m=0}^{r-1} x_{(v,s)}[m] \, w_N^{[\![lmN/r + \lfloor v/r^s \rfloor lr^s]\!]_N}, \tag{9}$$

for the modified DIF process, and

$$X_{(v,s)}[l] = \sum_{m=0}^{r-1} x_{(v,s)}[m] \, w_N^{[\![lmN/r + \lfloor v/r^{(S-s)} \rfloor mr^{(S-s)}]\!]_N}, \tag{10}$$

for the modified DIT process.

The conceptual key to the modified radix-r FFT butterfly is the formulation of the radix-r as composed engines with identical structures and a systematic means of accessing the corresponding multiplier coefficients [8, 9]. This enables the design of an engine with the lowest rate of complex multipliers and adders, which utilizes r or $r - 1$ complex multipliers in parallel to implement each of the butterfly computations. There is a simple mapping from the three indices m, v, and s (FFT stage, butterfly, and element) to the addresses of the multiplier coefficients needed by using the proposed FFT address generator in [24]. For a single processor environment, this type of butterfly with r parallel multipliers would result in decreasing the time delay for the complete FFT by a factor of $O(r)$. A second aspect of the modified radix-r FFT butterfly is that they are also useful in parallel multiprocessing environments. In essence, the precedence relations between the engines in the radix-r FFT are such that the execution of r engines in parallel is feasible during each FFT stage. If each engine is executed on the modified processing element (PE), it means that each of the r parallel processors would always be executing the same instruction simultaneously, which is very desirable for SIMD implementation on some of the latest DSP cards.

Based on this concept, Kim and Sunwoo proposed a proper multiplexing scheme that reduces the usage of complex multiplier for the radix-8 butterfly from 11 to 5 [25].

3. Parallel FFT Processing

For the past decades, there were several attempts to parallelize the FFT algorithm which was mostly based on parallelizing each stage (iteration) of the FFT process [26–28]. The most successful FFT parallelization was accomplished by parallelizing the loops during each stage or iteration in the FFT process [29, 30] or by focusing on memory hierarchy utilization that is achieved by the combination of production and consumption of butterflies' results, data reuse, and FFT parallelism [31].

The definition of the DFT is represented by

$$X_{(k)} = \sum_{n=0}^{N-1} x_{(k)} w_N^{nk}, \quad k \in [0, N-1], \tag{11}$$

where $x_{(n)}$ is the input sequence, $X_{(k)}$ is the output sequence, N is the transform length, and w_N is the Nth root of unity: $w_N = e^{-j2\pi/N}$. Both $x_{(n)}$ and $X_{(k)}$ are complex valued sequences.

Let $x_{(n)}$ be the input sequence of size N and let p_r denote the degree of parallelism which is multiple of N; therefore, we can rewrite (11) by considering $k_1 = 0, 1, \ldots, V-1$, $p = 0, 1, \ldots, p_r - 1$, $q = 0, 1, \ldots, p_r - 1$, $V = N/p_r$, and $k = k_1 + qV$ as [9]

$$X_{(k_1 + qN/p_r)}$$

$$= \left[w_N^0 \sum_{n=0}^{N/p_r - 1} x_{(p_r n)} w_{N/p_r}^{n(k_1 + q_r N/p_r)} \right.$$

$$+ w_N^{(k_1 + qN/p_r)} \sum_{n=0}^{N/p_r - 1} x_{(p_r n+1)} w_{N/p_r}^{n(k_1 + qN/p_r)}$$

$$\left. + \cdots + w_N^{(p_r-1)(k_1 + qN/p_r)} \sum_{n=0}^{N/p_r - 1} x_{(p_r n+(p_r-1))} w_{N/p_r}^{n(k_1 + qN/p_r)} \right]. \tag{12}$$

If $X_{(k)}$ is the Nth order Fourier transform $\sum_{n=0}^{N-1} x_{(n)} w_N^{nk}$, then, $X_{(0)_{(k_1)}}$, $X_{(1)_{(k_1)}}, \ldots,$ and $X_{(p_r-1)_{(k_1)}}$ will be the Nth/p_r

order Fourier transforms given, respectively, by the following expressions: $\sum_{n=0}^{V-1} x_{(p_r n)} w_V^{nv}$, $\sum_{n=0}^{V-1} x_{(p_r n+1)} w_V^{nv}, \ldots,$ and $\sum_{n=0}^{V-1} x_{(p_r n+(p_r-1))} w_V^{nv}$.

4. The Proposed Higher Radices Butterflies

Most of the FFTs' computation transforms are done within the butterfly loops. Any algorithm that reduces the number of additions and multiplications in these loops will reduce the overall computation speed. The reduction in computation is achieved by targeting trivial multiplications which have a limited speedup or by parallelizing the FFTs that have a significant speedup on the execution time of the FFT. In this section we will be limited in the elaboration of the proposed butterfly's radix-$2^\alpha/4^\beta$ (the radix-2/4 families) for the DIT FFT process. By rewriting (3) as

$$\mathbf{X} = \mathbf{W}_N \sum_{m=0}^{r-1} x_{(m)} w_N^{lmN/r} = \mathbf{W}_N \sum_{m=0}^{r-1} x_{(m)} w_r^{lm} \tag{13}$$

and by applying the concept of the parallel FFT (introduced in Section 3) on the kernel $\mathbf{B}_r$, therefore, (13) will be expressed as

$$\mathbf{X} = \mathbf{W}_N \sum_{m=0}^{r-1} x_{(m)} w_r^{lm}$$

$$= \mathbf{W}_N \left[\sum_{m=0}^{r/\alpha-1} x_{(\alpha m)} w_r^{lm\alpha} + \cdots \right.$$

$$\left. + \sum_{m=0}^{r/\alpha-1} x_{(\alpha m+(\alpha-1))} w_r^{l(\alpha m+(\alpha-1))} \right] \tag{14}$$

$$= \mathbf{W}_N \left[X_{(0)} + w_r^l X_{(1)} + \cdots + w_r^{l(\alpha-1)} X_{(\alpha-1)} \right]$$

$$\text{for } l = 0, \ldots, \frac{r}{\alpha} - 1.$$

It is to be noted that the notation w_x in all figures of this paper represents the set of twiddle factor associated with the butterfly input defined by $[w_0, \ldots, w_{(r-2)}] = \text{diag}(w_N^p, w_N^{2p}, \ldots, w_N^{(r-1)p})$.

For the radix-4 butterfly ($r = 2$ and $\alpha = 2$), we can express (13) as

$$\mathbf{X} = \mathbf{W}_N \left[\sum_{m=0}^{1} x_{(2m)} w_2^{lm} + w_4^l \sum_{m=0}^{1} x_{(2m+1)} w_2^{lm} \right] \tag{15}$$

$$= \mathbf{W}_N \left[X_{(0)} + w_4^l X_{(1)} \right],$$

and the conventional radix-2^2 (MDC-R2^2) BPE in terms of radix-2 butterfly is illustrated in Figure 1.

The use of resources could also be reduced by a feedback network and a multiplexing network where the feedback network is for feeding the ith output of the jth radix-2 adder network to the jth input of the ith butterfly and the multiplexers selectively pass the input data or the feedback, alternately, to the corresponding radix-2 adder network as

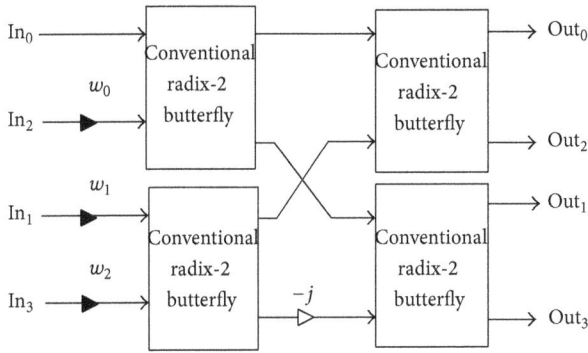

FIGURE 1: Conventional radix-2^2 (MDC-R2^2) BPE (butterfly processing element).

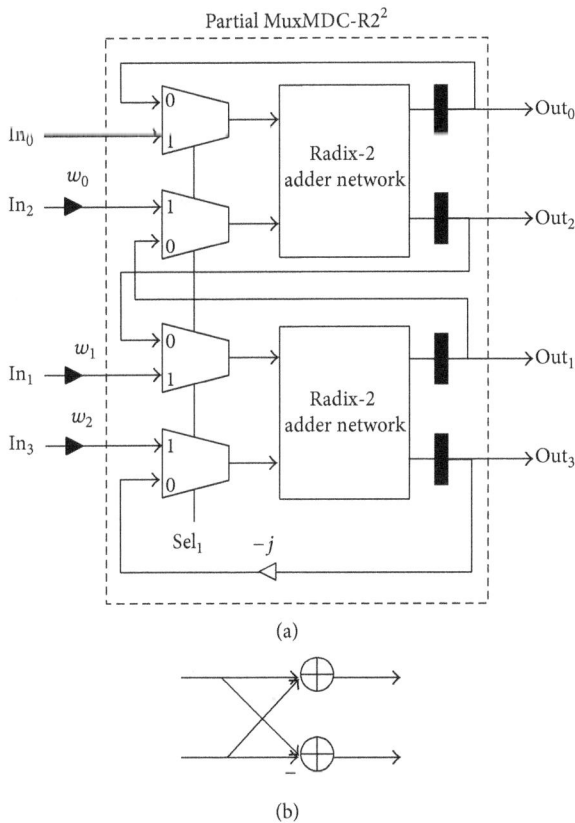

FIGURE 2: (a) Proposed multiplexed radix-2^2 (MuxMDC-R2^2) BPE and (b) block circuit diagram of the radix-2 adder network [7].

illustrated in Figure 2(a) [23]. The circuit block diagram of the radix-2 adder network is illustrated in Figure 2(b) that consists of two complex adders only.

With the rising edge of the clock cycle the inputs data are fed to the butterfly's input of the system presented in Figure 1. In order to complete the butterfly's operations within one clock cycle, the following conditions should be satisfied:

$$T_{CLK} > T_{CM} + 2T_{CA},$$

$$\text{Throughput} = \frac{4}{T_{CLK}},$$ (16)

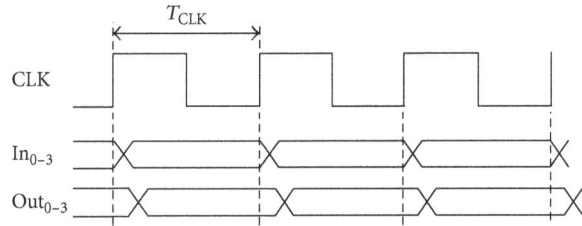

FIGURE 3: Timing block diagram of Figure 1.

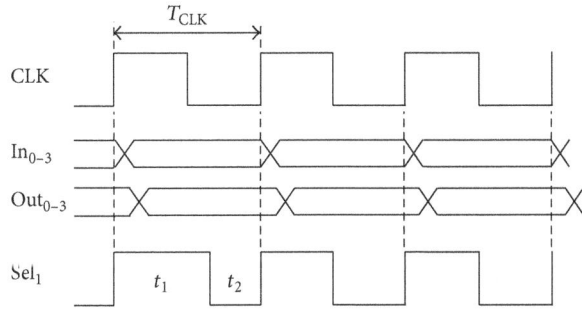

FIGURE 4: Timing block diagram of Figure 2(a).

where T_{CM}/T_{CA} is the time required to perform one complex multiplication/addition and the timing block diagram of Figure 1 is sketched in Figure 3.

With the rising edge of the clock cycle the inputs data are fed to the butterfly's input of the system presented in Figure 2(a) and with the falling edge of the clock cycle the feedback data are fed to the butterfly's input. In order to complete the butterfly's operations within one clock cycle, the following conditions should be satisfied:

$$t_1 > T_{CM} + T_{CA},$$

$$t_2 > T_{CA},$$

$$T_{CLK} > (t_1 + t_2) > T_{CM} + 2T_{CA},$$ (17)

$$\text{Throughput} = \frac{4}{T_{CLK}},$$

and the timing block diagram of Figure 2(a) is illustrated in Figure 4.

Further block building of these modules could be achieved by duplicating the block circuit diagram of Figure 2(a) and combining them in order to obtain the radix-8 MDC-R2^3 BPE; therefore, for this case ($r = 4$ and $\alpha = 2$), (4) could be expressed as

$$\mathbf{X}_{(l)} = \mathbf{W}_N \left[\sum_{m=0}^{3} x_{(2m)} w_4^{lm} + w_8^l \sum_{m=0}^{3} x_{(2m+1)} w_4^{lm} \right]$$

$$= \mathbf{W}_N \left[X_{(0)} + w_8^l X_{(1)} \right],$$ (18)

and the signal flow graph (SFG) of the DIT conventional MDC-R2^3 BPE butterfly is illustrated in Figure 5. The resources in the conventional MDC-R2^3 BPE could also be

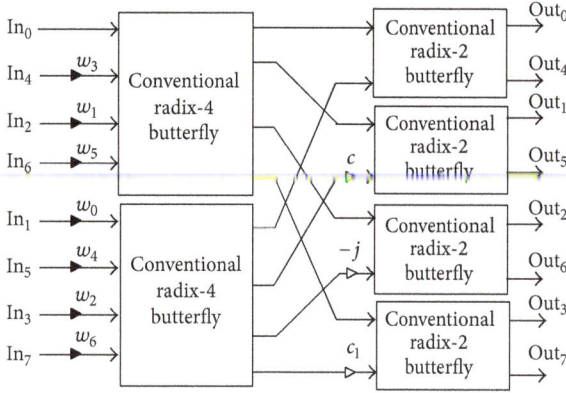

FIGURE 5: Conventional MDC-R2³ BPE.

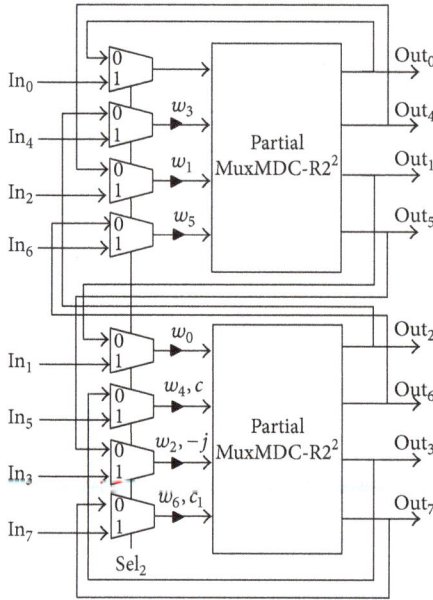

FIGURE 6: Proposed MuxMDC-R2³ BPE based on the partial MuxMDC-R2².

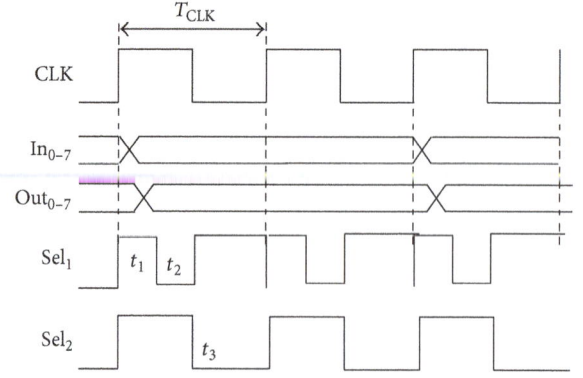

FIGURE 7: Timing block diagram of Figure 6.

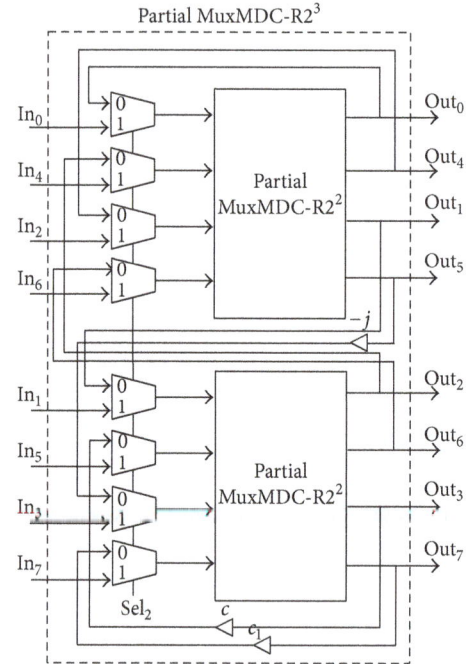

FIGURE 8: Proposed Partial MuxMDC-R2³.

reduced by means of the partial multiplexed radix 2^2 and a feedback network yielding to the proposed MuxMDC-R2³ BPE structure in Figure 6.

The clock timing of Figure 5 is computed as

$$T_{\text{CLK}} > T_{\text{CM}} + t_{\text{pm}} + 3T_{\text{CA}},$$
$$\text{Throughput} = \frac{8}{T_{\text{CLK}}}, \tag{19}$$

where t_{pm} is the time required to execute one complex multiplication on a constant multiplier and the clock timing of the proposed MuxMDC-R2³ is estimated as

$$t_1 > T_{\text{CM}} + T_{\text{CA}},$$
$$t_2 > T_{\text{CA}},$$
$$t_3 = t_1,$$
$$T_{\text{CLK}} > (t_1 + t_2 + t_3) > 2T_{\text{CM}} + 3T_{\text{CA}}, \tag{20}$$
$$\text{Throughput} = \frac{8}{T_{\text{CLK}}}.$$

The overall timing block diagram of the proposed MuxMDC-R2³ is sketched in Figure 7. In Figure 6, the inputs are multiplied by the twiddle factors w_i when $S_2 = 1$ and by the constant factors $-j$, c, c_1 or 1 for $S_2 = 0$.

Further block building of these modules could be achieved by combining two radix-8 butterflies with eight radix-2 butterflies in order to obtain the conventional MDC-R2⁴ BPE; therefore, for this case ($r = 8$ and $\alpha = 2$), (4) could be expressed as

$$\mathbf{X}_{(l)} = \mathbf{W}_N \left[\sum_{m=0}^{7} x_{(2m)} w_8^{lm} + w_8^l \sum_{m=0}^{7} x_{(2m+1)} w_8^{lm} \right] \tag{21}$$
$$= \mathbf{W}_N \left[X_{(0)} + w_{16}^l X_{(1)} \right],$$

and the signal flow graph (SFG) of the proposed DIT radix-2⁴ MuxMDC-R2⁴ based on the partial MuxMDC-R2³ (Figure 8) is illustrated in Figure 9.

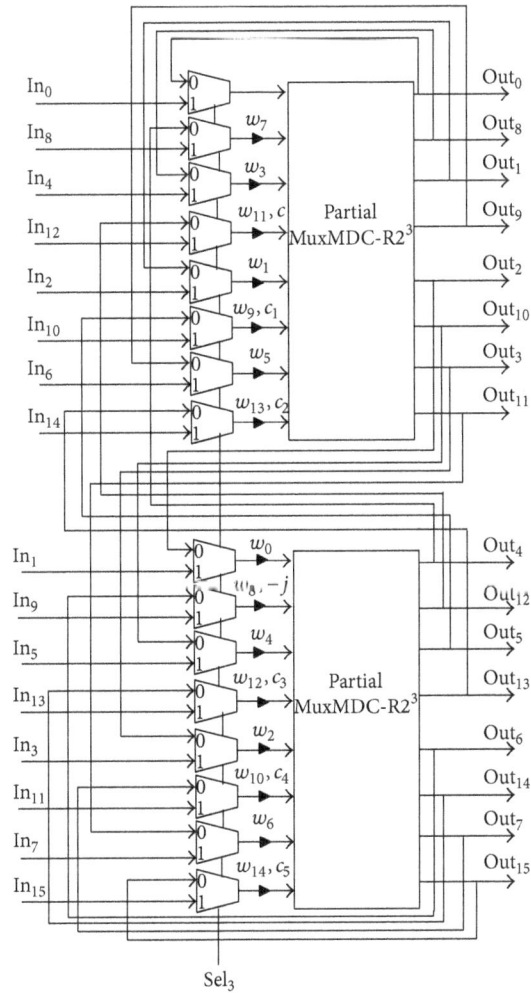

FIGURE 9: Proposed MuxMDC-R2^4 BPE based on the Partial MuxMDC-R2^3.

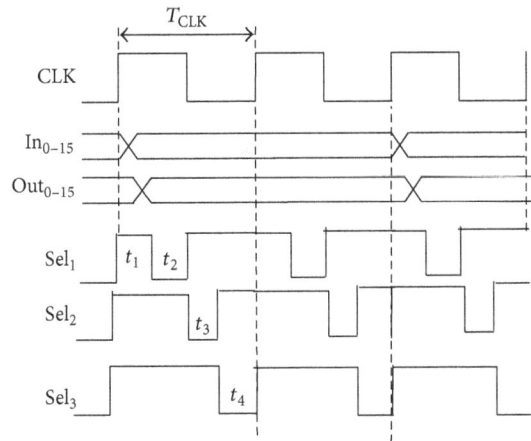

FIGURE 10: Timing block diagram of Figure 6.

The clock timing of the conventional MDC-R2^4 BPE is computed as

$$T_{CLK} > T_{CM} + 2t_{pm} + 4T_{CA},$$

$$\text{Throughput} = \frac{16}{T_{CLK}},$$

(22)

and the clock timing of the proposed MuxMDC-R2^4 is estimated as

$$t_1 > T_{CM} + T_{CA},$$

$$t_2 > T_{CA},$$

$$t_3 = t_{pm} + T_{CA},$$

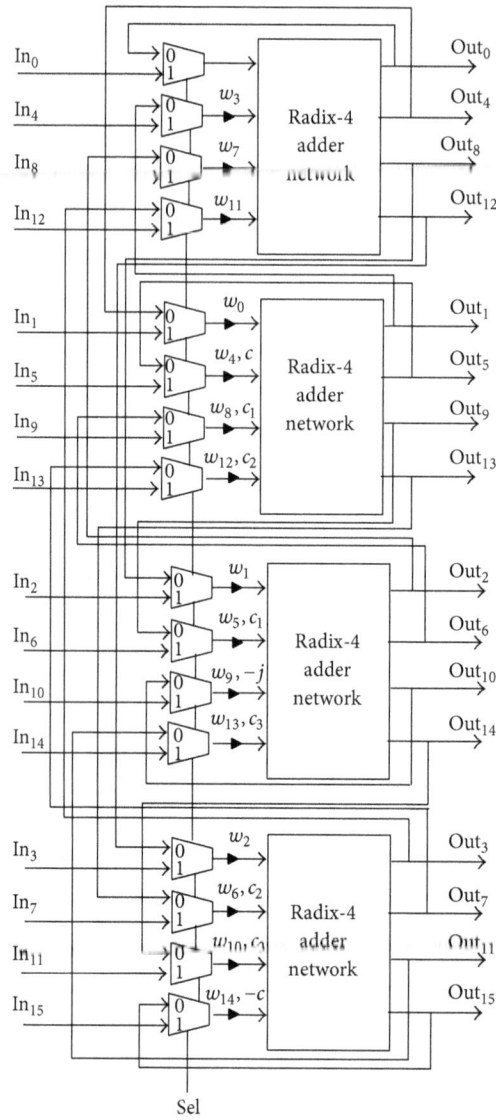

FIGURE 11: The proposed DIT MuxMDC-R4^2.

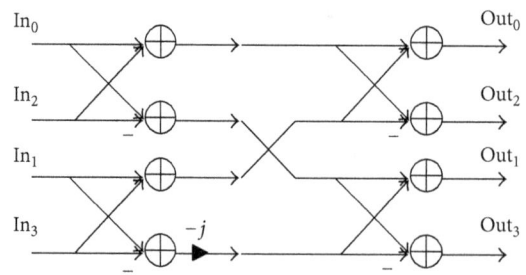

FIGURE 12: Block circuit diagram of the radix-4 adder network.

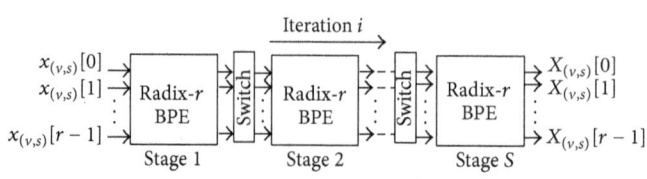

FIGURE 13: S stages radix-r pipelined FFT.

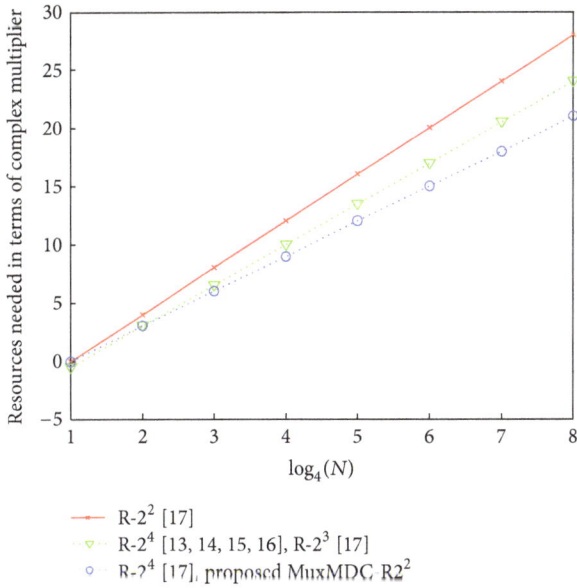

FIGURE 14: Comparison between the different butterflies' structures in terms of complex multiplier needed to compute the 4 parallel BPE pipelined FFTs of size N.

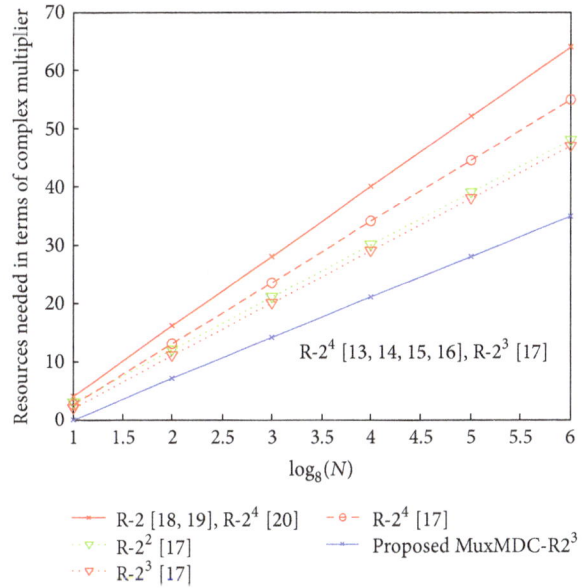

FIGURE 16: Comparison between the different butterflies' structures in terms of complex multiplier needed to compute the 8 parallel BPE pipelined FFTs of size N.

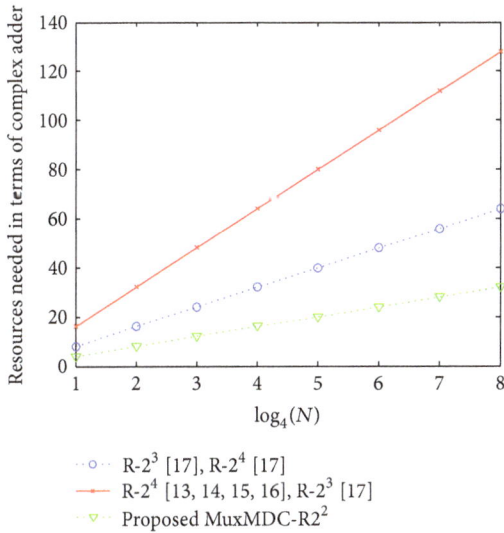

FIGURE 15: Comparison between the different butterflies' structures in terms of complex adder needed to compute the 4 parallel BPE pipelined FFTs of size N.

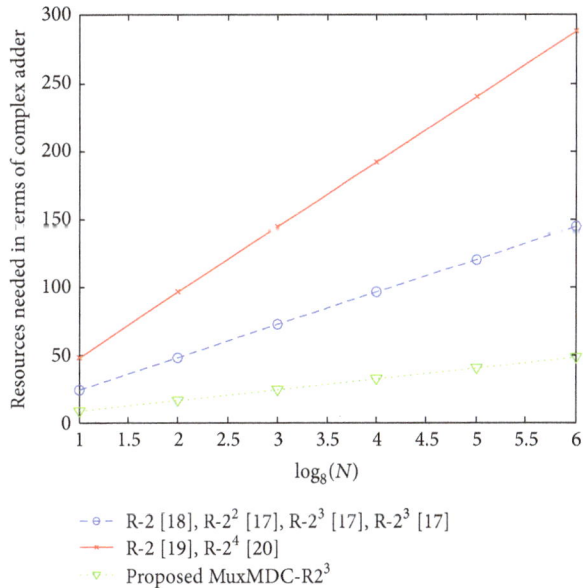

FIGURE 17: Comparison between the different butterflies' structures in terms of complex adder needed to compute the 8 parallel BPE pipelined FFTs of size N.

$$t_4 = t_1,$$

$$T_{\text{CLK}} > \left(t_1 + t_2 + t_3 + t_4\right) > 2T_{CM} + t_{\text{pm}} + 4T_{CA},$$

$$\text{Throughput} = \frac{16}{T_{\text{CLK}}}. \tag{23}$$

The overall timing block diagram of the proposed MuxMDC-R2^4 is sketched in Figure 10.

With the same reasoning as above, we will be limited in the elaboration of the proposed butterfly's radix-4$^\alpha$ family to the DIT FFT process.

For the radix-16 butterfly ($r = 4$ and $\alpha = 4$), we can express (4) as

$$
\mathbf{X} = \mathbf{W}_N \left[\sum_{m=0}^{3} x_{(4m)} w_4^{lm} + w_{16}^{l} \sum_{m=0}^{3} x_{(4m+1)} w_4^{lm} \right.
$$
$$
\left. + w_{16}^{2l} \sum_{m=0}^{3} x_{(4m+2)} w_4^{lm} + w_{16}^{3l} \sum_{m=0}^{3} x_{(2m+1)} w_4^{lm} \right]
$$
$$
= \mathbf{W}_N \left[X_{(0)} + w_{16}^{l} X_{(1)} + w_{16}^{2l} X_{(2)} + w_{16}^{3l} X_{(3)} \right],
$$

$$\tag{24}$$

TABLE 1: Resources needed to compute an FFT of size N.

Butterfly structure	Complex multiplier	Complex adder	Latency (cycles)	T (Spc)
4 parallel BPE architectures				
R-2^4 [13], [14]	$4(\log_4 N - 1)$	$16\log_4 N$	$N/4$	4
R-2^4 [15]	$4(\log_4 N - 1)$	$16\log_4 N$	$N/4$	4
R-2^4 [16]	$4(\log_4 N - 1)$	$16\log_4 N$	$N/4$	4
R-2^2 [17]	$3(\log_4 N - 1)$	$16\log_4 N$	$N/4$	4
R-2^3 [17]	$4(\log_4 N - 1)$	$8\log_4 N$	$N/4$	4
R-2^4 [17]	$3.5(\log_4 N - 4)$	$8\log_4 N$	$N/4$	4
Proposed MuxMDC-R2^2	$3(\log_4 N - 1)$	$4\log_4 N$	$N/4$	4
8 parallel BPE architectures				
R-2 [18]	$8(\log_4 N - 1)$	$16\log_4 N$	$N/8$	8
R-2 [19]	$8(\log_4 N - 1)$	$32\log_4 N$	$N/8$	8
R-2^4 [20]	$8(\log_4 N - 1)$	$32\log_4 N$	$N/8$	8
R-2^2 [17]	$6(\log_4 N - 1)$	$16\log_4 N$	$N/8$	8
R-2^3 [17]	$6\log_4 N - 7$	$16\log_4 N$	$N/8$	8
R-2^4 [17]	$7\log_4 N - 8$	$16\log_4 N$	$N/8$	8
Proposed MuxMDC-R2^3	$7(\log_8 N - 1)$	$8\log_8 N$	$N/8$	8
16 parallel BPE architectures				
Proposed MuxMDC-R2^4	$17(\log_{16} N - 1)$	$16\log_{16} N$	$N/16$	16
Proposed MuxMDC-R4^2	$15(\log_{16} N - 1)$	$32\log_{16} N$	$N/16$	16

TABLE 2: Resources needed in terms of FA to compute an FFT of size N.

Butterfly structure	FA
4 parallel BPE architectures	
R-2^4 [13], [14]	$12n^2\left(\log_4 N - 1\right) + 20\left(p\log_4 N - 1\right) + 32p\log_4 N$
R-2^4 [15]	$12n^2\left(\log_4 N - 1\right) + 20p\left(\log_4 N - 1\right) + 32p\log_4 N$
R-2^4 [16]	$12n^2\log_4\left(N - 1\right) + 20p\log_4\left(N - 1\right) + 32p\log_4 N$
R-2^2 [17]	$9n^2\left(\log_4 N - 1\right) + 15p\left(\log_4 N - 1\right) + 32p\log_4 N$
R-2^3 [17]	$12n^2\left(\log_4 N - 1\right) + 20p\left(\log_4 N - 1\right) + 16p\log_4 N$
R-2^4 [17]	$10.5n^2\left(\log_4 N - 1\right) + 17.5p\left(\log_4 N - 1\right) + 16p\log_4 N$
Proposed MuxMDC-R2^2	$9n^2\left(\log_4 N - 1\right) + 15p\left(\log_4 N - 1\right) + 8p\log_4 N$
8 parallel BPE architectures	
R-2 [18]	$24n^2\left(\log_4 N - 1\right) + 40p\left(\log_4 N - 1\right) + 32p\log_4 N$
R-2 [19]	$24n^2\left(\log_4 N - 1\right) + 40p\left(\log_4 N - 1\right) + 64p\log_4 N$
R-2^4 [20]	$24n^2\left(\log_4 N - 1\right) + 40p\left(\log_4 N - 1\right) + 64p\log_4 N$
R-2^2 [17]	$18n^2\left(\log_4 N - 1\right) + 30p\left(\log_4 N - 1\right) + 32p\log_4 N$
R-2^3 [17]	$18n^2\left(\log_4 N - 1\right) + 30p\left(\log_4 N - 1\right) + 32p\log_4 N - 21n^2 - 35p$
R-2^4 [17]	$21n^2\left(\log_4 N - 1\right) + 35p\left(\log_4 N - 1\right) + 32p\log_4 N - 24n^2 - 40p$
Proposed MuxMDC-R2^3	$21n^2\left(\log_8 N - 1\right) + 35p\left(\log_8 N - 1\right) + 16p\log_8 N$
16 parallel BPE architectures	
Proposed MuxMDC-R2^4	$51n^2\left(\log_{16} N - 1\right) + 85p\left(\log_{16} N - 1\right) + 32p\log_{16} N$
Proposed MuxMDC-R4^2	$45n^2\left(\log_{16} N - 1\right) + 75p\left(\log_{16} N - 1\right) + 64p\log_{16} N$

and the proposed MDC-R4^2 in terms of radix-4 network is illustrated in Figure 11 where the feedback network is for feeding the ith output of the jth radix-4 network to the jth input of the ith butterfly and the switches selectively pass the input data or the feedback, alternately, to the corresponding radix-4 butterfly. The circuit block diagram of the radix-4 network is illustrated in Figure 12.

5. Performance Evaluation

FFTs are the most powerful algorithms that are used in communication systems such as OFDM. Their implementation is very attractive in fixed point due to the reduction in cost compared to the floating point implementation. One of the most powerful FFT implementations is the pipelined FFT

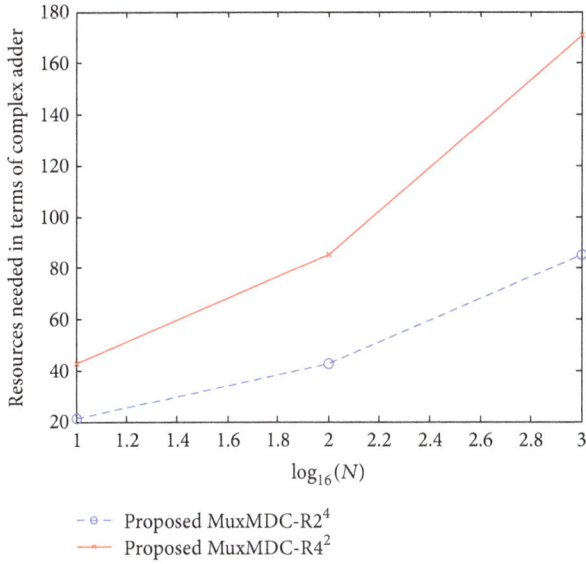

FIGURE 18: Comparison between the different butterflies' structures in terms of complex adder needed to compute the 16 parallel BPE pipelined FFTs of size N.

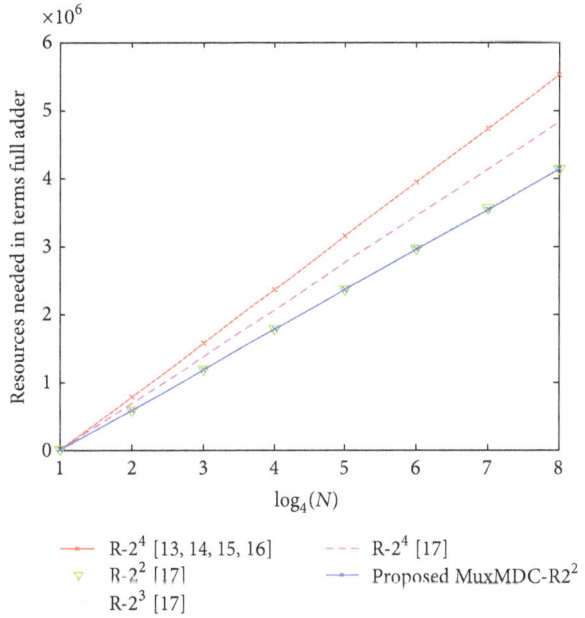

FIGURE 19: Comparison between the different butterflies' structures in terms of complex multiplier needed to compute the 16 parallel BPE pipelined FFTs of size N.

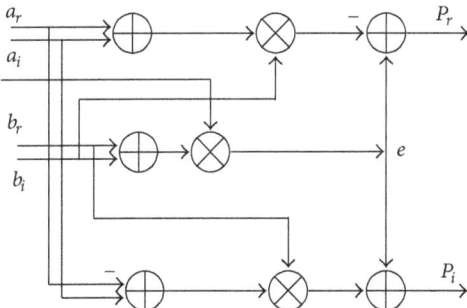

FIGURE 20: Complex multiplier using three real multipliers and five real adders.

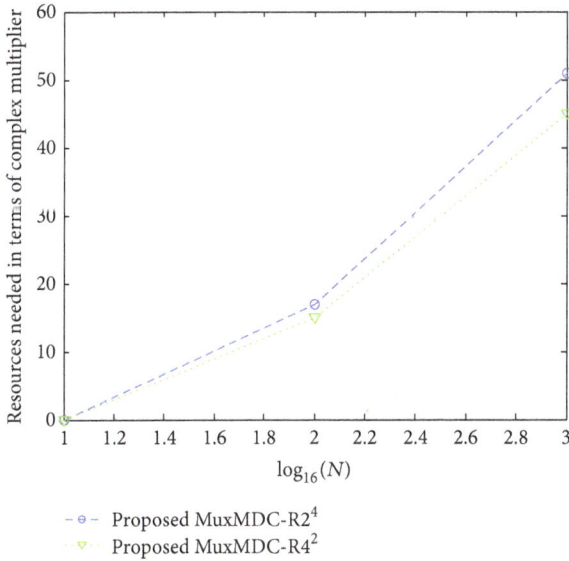

FIGURE 21: Comparison between the different butterflies' structures in terms of full adder needed to compute the 4 parallel pipelined FFTs of size N (multiplier on 16 bits and adder on 32 bits).

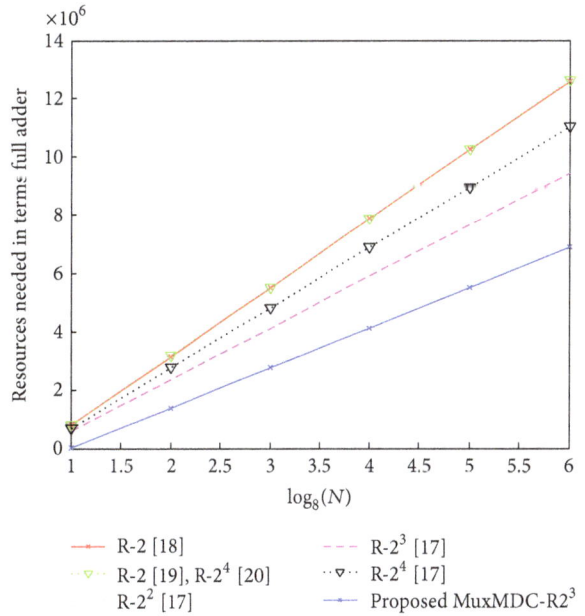

FIGURE 22: Comparison between the different butterflies' structures in terms of full adder needed to compute the 8 parallel pipelined FFTs of size N (multiplier on 16 bits and adder on 32 bits).

which is highly implemented in the communication systems; see Figure 13.

Since the objective of this paper is mainly concentrated on the higher radices butterflies structures, in our performance study we will be limited to the impact of the butterfly structure. Once the pipeline is filled, the butterflies will produce r output each clock cycle (throughput T in samples per cycle (Spc)). Therefore, Table 1 will draw the comparison between

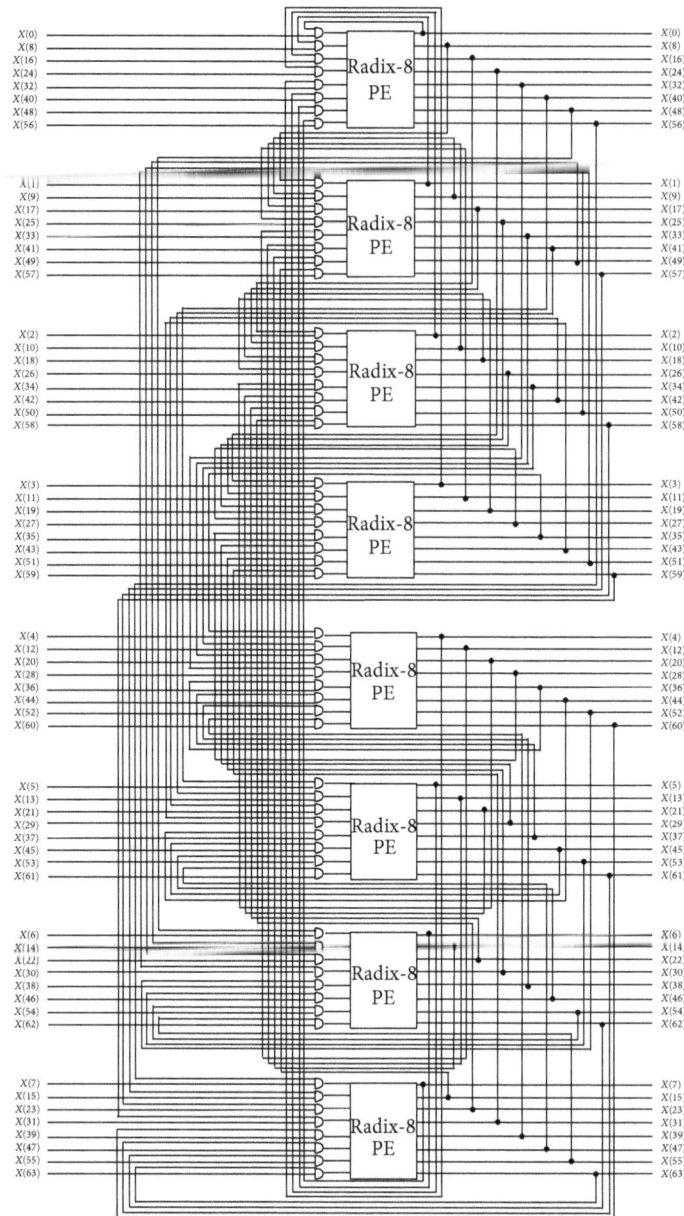

FIGURE 23: Two stage pipelined FFT (or array structure) with a feedback network [23].

the different butterflies' structures in terms of resources needed to compute an FFT of size N.

As shown in Figure 14, we could clearly see that the proposed MuxMDC-R2^2 for the four parallel pipelined FFTs of size N will have the same amount of complex multiplier compared to the radix 2^4 cited in [30]. Furthermore, our proposed MuxMDC-R2^2 achieves a reduction in the usage of complex multiplier by a factor that ranges between 1.1 and 1.4 compared to the other cited butterflies.

For the 4 parallel pipelined FFTs of size N, the reduction in the usage of complex adder for our proposed method MuxMDC-R2^2 ranges between 1.9 and 3.9 compared to the cited butterflies as shown in Figure 15.

For the 8 parallel pipelined FFTs of size N, the reduction factor in the usage of complex multiplier for our proposed

MuxMDC-R2^3 could range from 1.3 to 2.1 compared to the cited butterflies as illustrated in Figure 16.

For the same structure, the reduction factor in the usage of complex adder for our proposed method MuxMDC-R2^3 could range from 3.0 to 5.4 compared to the cited butterflies (Figure 17).

It seems that the proposed MuxMDC-R2^4 uses less complex adders than the proposed MuxMDC-R4^2 as shown in Figure 18 where the proposed MuxMDC-R2^4 achieves a reduction in the usage of complex adder by a factor of 2 but the proposed MuxMDC-R4^2 achieves a reduction in the usage of complex multiplier by a factor of 1.1 as shown in Figure 19.

Since one complex multiplication is counted as 3 real multiplications and 5 real additions as shown in Figure 20,

Table 2 will illustrate the required resources in terms of full adder (FA) that will be computed as (a) n^2 for two n-digit real multiplier and (b) p for two p-digit real adder.

For the four parallel pipelined FFTs of size N, it seems that the R-2^2 butterfly cited in [30] will have approximately the same amount of FA as the proposed MuxMDC-R2^2 according to Figure 21. Our proposed MuxMDC-R2^2 will achieve a reduction in the usage of FA by a factor that ranges between 1.17 and 1.34 (Figure 21).

With regard to the eight parallel pipelined FFTs of size N, it seems that the proposed MuxMDC-R2^3 will achieve a reduction in the usage of FA by a factor that ranges between 1.4 and 1.9 in comparison to the other cited butterflies as shown in Figure 22.

Since the implementation of higher radices by means of the radix-$2^\alpha/4^\beta$ butterfly is feasible, the optimal pipelined FFT is achieved by the two stage FFT as shown in Figure 23 where the use of complex memories between the different stages is completely eliminated and the delay required to fill up the pipeline is totally absent.

6. Conclusion

It has been shown that the higher radix FFT algorithms are advantageous for the hardware implementation, due to the reduced quantity of complex multiplications and memory access rate requirements. This paper has presented an efficient way of implementing the higher radices butterflies by means of the radix-$2^\alpha/4^\beta$ kernel where serial parallel models have been represented. The proposed optimized different structures with a scheduling scheme of complex multiplications are suitable for embedded FFT processors. Furthermore, it has been proven that the higher radices butterflies could be obtained by reusing the block circuit diagram of the radix-$2^\alpha/4^\beta$ butterfly. Based on this concept, the hardware resources needed could be reduced which is highly desirable for low power consumption FFT processors. The proposed method is suitable for large pipelined FFTs implementation where the performance gain will increase with an increasing FFTs' radix size. This structure is also appropriate for SIMD implementation on some of the latest DSP cards.

Conflict of Interests

The authors declare that there is no conflict of interests regarding the publication of this paper.

Acknowledgments

The authors would like to thank the financial support from the Natural Sciences and Engineering Research Council of Canada and from JABERTECH's Shareholders Trevor Hill from Alberta and Bassam Kabbara from Kuwait.

References

[1] W. Cooley and J. W. Tukey, "An algorithm for the machine calculation of complex Fourier series," *Mathematics of Computation*, vol. 19, pp. 297–301, 1965.

[2] P. Duhamel and H. Hollmann, "Split radix FFT algorithm," *Electronics Letters*, vol. 20, no. 1, pp. 14–16, 1984.

[3] S. Winograd, "On computing the discrete Fourier transform," *Proceedings of the National Academy of Sciences of the United States of America*, vol. 73, no. 4, pp. 1005–1006, 1976.

[4] T. Widhe, *Efficient Implementation of FFT Processing Elements*, Linkoping Studies in Science and Technology no. 619, Linkoping University, Linköping, Sweden, 1997.

[5] T. Widhe, J. Melander, and L. Wanhammar, "Design of efficient radix-8 butterfly PEs for VLSI," in *Proceedings of the IEEE International Symposium on Circuits and Systems (ISCAS '97)*, pp. 2084–2087, June 1997.

[6] J. Melander, T. Widhe, K. Palmkvist, M. Vesterbacka, and L. Wanhammar, "An FFT processor based on the SIC architecture with asynchronous PE," in *Proceedings of the IEEE 39th Midwest Symposium on Circuits and Systems*, vol. 3, pp. 1313–1316, Ames, Iowa, USA, August 1996.

[7] M. Jaber and D. Massicotte, "The self-sorting JMFFT algorithm eliminating trivial multiplication and suitable for embedded DSP processor," in *Proceedings of the 10th IEEE International NEWAS Conference*, Montreal, Canada, June 2012.

[8] M. Jaber, "Butterfly processing element for efficient fast Fourier transform method and apparatus," US Patent No. 6, 751, 643, 2004.

[9] M. A. Jaber and D. Massicotte, "A new FFT concept for efficient VLSI implementation: part I—butterfly processing element," in *16th International Conference on Digital Signal Processing (DSP '09)*, pp. 1–6, Santorini, Greece, July 2009.

[10] Y. Wang, Y. Tang, Y. Jiang, J. Chung, S. Song, and M. Lim, "Novel memory reference reduction methods for FFT implementations on DSP processors," *IEEE Transactions on Signal Processing*, vol. 55, no. 5, pp. 2338–2349, 2007.

[11] S. He and M. Torkelson, "Design and implementation of a 1024-point pipeline FFT processor," in *Proceedings of the IEEE Custom Integrated Circuits Conference*, pp. 131–134, May 1998.

[12] S. He and M. Torkelson, "New approach to pipeline FFT processor," in *Proceedings of the 10th International Parallel Processing Symposium (IPPS '96)*, pp. 766–770, April 1996.

[13] E. E. Swartzlander, W. K. W. Young, and S. J. Joseph, "A radix 4 delay commutator for fast Fourier transform processor implementation," *IEEE Journal of Solid-State Circuits*, vol. 19, no. 5, pp. 702–709, 1984.

[14] J. H. McClellan and R. J. Purdy, *Applications of Digital Signal Processing*, Applications of Digital Signal Processing to Radar, chapter 5, Prentice Hall, New York, NY, USA, 1978.

[15] H. Liu and H. Lee, "A high performance four-parallel 128/64-point radix-2^4 FFT/IFFT processor for MIMO-OFDM systems," in *Proceedings of the IEEE Asia Pacific Conference on Circuits and Systems (APCCAS '08)*, pp. 834–837, Macao, China, December 2008.

[16] S.-I. Cho, K.-M. Kong, and S.-S. Choi, "Implementation of 128-point fast fourier transform processor for UWB systems," in *Proceedings of the International Wireless Communications and Mobile Computing Conference (IWCMC '08)*, pp. 210–213, Crete Island, Greece, August 2008.

[17] M. Garrido, J. Grajal, M. A. Sanchez, and O. Gustafsson, "Pipelined radix-2k feedforward FFT architectures," *IEEE Transactions on Very Large Scale Integration (VLSI) Systems*, vol. 21, no. 1, pp. 23–32, 2013.

[18] J. A. Johnston, "Parallel pipeline fast Fourier transformer," *IEE Proceedings F: Communications, Radar and Signal Processing*, vol. 130, no. 6, pp. 564–572, 1983.

[19] E. H. Wold and A. M. Despain, "Pipeline and parallel-pipeline FFT processors for VLSI implementations," *IEEE Transactions on Computers*, vol. 33, no. 5, pp. 414–426, 1984.

[20] S.-N. Tang, J.-W. Tsai, and T.-Y. Chang, "A 2.4-GS/s FFT processor for OFDM-based WPAN applications," *IEEE Transactions on Circuits and Systems II: Express Briefs*, vol. 57, no. 6, pp. 451–455, 2010.

[21] M. Jaber, "Parallel multiprocessing for the fast Fourier transform with pipeline architecture," US Patent No. 6, 792, 441.

[22] M. A. Jaber and D. Massicotte, "A new FFT concept for efficient VLSI implementation: part II—parallel pipelined processing," in *Proceedings of the 16th International Conference on Digital Signal Processing (DSP 2009)*, pp. 1–5, Santorini, Greece, July 2009.

[23] M. Jaber, "Fourier transform processor," US Patent No. 7, 761, 495.

[24] M. Jaber, "Address generator for the fast Fourier transform processor," US-6, 993, 547 82 and European Patent Application Serial no: PCT/USOI /07602.

[25] E. J. Kim and M. H. Sunwoo, "High speed eight-parallel mixed-radix FFT processor for OFDM systems," in *Proceedings of the IEEE International Symposium of Circuits and Systems (ISCAS '11)*, pp. 1684–1687, Rio de Janeiro, Brazil, May 2011.

[26] P. Li and W. Dong, "Computation oriented parallel FFT algorithms on distributed computer," in *Proceedings of the 3rd International Symposium on Parallel Architectures, Algorithms and Programming (PAAP '10)*, pp. 369–373, Dalian, China, December 2010.

[27] D. Takahashi, A. Uno, and M. Yokokawa, "An implementation of parallel 1-D FFT on the K computer," in *Proceedings of the IEEE International Conference on High Performance Computing and Communication*, pp. 344–350, Liverpool, UK, June 2012.

[28] R. M. Piedra, "Parallel 1-D FFT Implementation with TMS320C4x DSPs," Texas Instruments SPRA108, Digital Signal Processing Semiconductor Group, 1994.

[29] http://www.fftw.org/.

[30] V. Petrov, "MKL FFT performance—comparison of local and distributed-memory implementations," Intel Report, 2012, http://software.intel.com/en-us/node/165305?wapkw=fft.

[31] V. I. Kelefouras, G. S. Athanasiou, N. Alachiotis, H. E. Michail, A. S. Kritikakou, and C. E. Goutis, "A methodology for speeding up fast fourier transform focusing on memory architecture utilization," *IEEE Transactions on Signal Processing*, vol. 59, no. 12, pp. 6217–6226, 2011.

## 22

A Discrete Event System Approach to Online Testing of Speed Independent Circuits

P. K. Biswal, K. Mishra, S. Biswas, and H. K. Kapoor

Department of Computer Science and Engineering, Indian Institute of Technology, Guwahati 781 039, India

Correspondence should be addressed to S. Biswas; santoshbiswas402@yahoo.com

Academic Editor: Marcelo Lubaszewski

With the increase in soft failures in deep submicron ICs, online testing is becoming an integral part of design for testability. Some techniques for online testing of asynchronous circuits are proposed in the literature, which involves development of a checker that verifies the correctness of the protocol. This checker involves Mutex blocks making its area overhead quite high. In this paper, we have adapted the Theory of Fault Detection and Diagnosis available in the literature on Discrete Event Systems to online testing of speed independent asynchronous circuits. The scheme involves development of a state based model of the circuit, under normal and various stuck-at fault conditions, and finally designing state estimators termed as detectors. The detectors monitor the circuit online and determine whether it is functioning in normal/failure mode. The main advantages are nonintrusiveness and low area overheads compared to similar schemes reported in the literature.

1. Introduction

With the advancement of VLSI technology for circuit design, there is a need to monitor the operations of circuits for detecting faults [1]. These needs have increased dramatically in recent times because, with the widespread use of deep submicron technology, there is a rise in the probability of development of faults during operation. Performing tests before operation of a circuit and assuming continued fault-free behaviour may decrease the reliability of operation. In other words, there is a need for *online testing (OLT) of VLSI circuits, whereupon they are designed to verify, during normal operation, whether their output response conforms to the correct behaviour.*

Most of the circuits used in VLSI designs are synchronous. Compared to synchronous circuits, asynchronous designs offer great advantages such as no clock skew problem, low power consumption, and average case performances rather than the worst case performances. Testing asynchronous circuits as compared to synchronous circuits is considered difficult due to the absence of the global clock [2].

OLT has been studied for the last two decades and can be broadly classified into the following main categories:

 (i) Self-checking design.

 (ii) Signature monitoring in FSMs.

 (iii) Duplication.

 (iv) Online BIST.

The approach of self-checking design consists of encoding the circuit outputs using some error detecting code and then checking some code invariant property (e.g., Parity) [3–5]. Some examples are Parity codes [6], m-out-of-n codes [7], and so forth. The area overhead for making circuits self-checkable is usually not high. These techniques, termed as "intrusive OLT methodologies," require some special properties in the circuit structure to limit the scope of fault propagation. These properties can be achieved by resynthesis and redesign of the original circuit, which may affect the critical paths in the circuit.

Signature monitoring techniques for OLT [8, 9] work by studying the state sequences of the circuit FSM model during its operation. These schemes detect faults that lead to illegal paths in the control flow graph, that is, paths having transitions which do not exist in the specified FSM. To make the runtime signature of the fault-free circuit FSM different from the one with fault, a *signature invariant property* is forced during FSM synthesis, making the technique intrusive. Further, the state explosion problem in FSM models makes the application of this scheme difficult for practical circuits.

Duplication based OLT technique works by simply replicating the original circuit and comparing the output responses [10]; a fault is detected if the outputs do not match. The major advantage of duplication based scheme is nonintrusivity; however, area overhead is more than double. To address this issue, partial duplication technique is applied [11, 12]. This scheme first generates a complete set of test vectors for all the faults possible, using Automatic Test Pattern Generation (ATPG) algorithms. After that, a subset of faults are selected (based on required coverage) and a subset of test vectors (based on tolerable latency) for the selected faults are taken and synthesized into a circuit which is used for OLT. It may be noted that ATPG algorithms are optimized to generate the minimum number of test vectors that detect all faults. As the scheme applies ATPG algorithms in a reverse philosophy, it becomes prohibitively complex for large circuits.

The technique of designing circuits with additional on-chip logic, which can be used to test proper operation of the circuit before it starts, is called off-line BIST. Off-line BIST resources are now being used for online testing [13–16]. This technique utilizes the idle time of various parts of the circuit to perform online BIST. Idle times are reducing in circuits because of pipelining and parallelism techniques and so online BIST scheme is of limited utility.

Motivation of the Work. From the above discussion we may state that an efficient approach for OLT should have the following metrics:

(i) The OLT scheme should be nonintrusive. This is the most important constraint as designers meet requirements, like frequency, area, power, and so forth, of the circuit to produce an efficient design and do not want the test engineers to change them.

(ii) The OLT technique should support well accepted fault models.

(iii) The scheme should be computationally efficient so that it can handle reasonably large circuits.

Most of the papers cited in the above discussion are for OLT of synchronous circuits and only a few of them [4, 5, 10] are applicable to asynchronous circuits. Now, we elaborate on these three works on OLT of asynchronous circuits and derive motivation of the present work.

Traditionally, double redundancy methods were used for OLT of asynchronous designs [10]. In this scheme, two copies of the same circuit work in parallel and the online tester checks whether they generate the same output. This scheme results in more than 100% area and power overheads. Further, both being the same circuit, they are susceptible to similar nature of failures. The schemes reported in [4, 5] basically

work by checking whether the output of the asynchronous circuit maintains a predefined protocol (i.e., there is no premature or late occurrence of transitions). The checker circuit is implemented using David cells (DCs), Mutual Exclusion (Mutex) elements, C-elements, and logic gates. The checker circuit has two modes: operation-normal mode and self-test mode. In normal mode, the checker is used to detect whether there is any violation in the protocol being executed by the CUT. On the other hand, in self-test mode, the checker is used to detect faults that may occur within the checker itself. Mutex elements (component of asynchronous arbiter) were used to grant exclusive access to the shared DCs between different modes of operation. The area overhead of the Mutex blocks is high, even compared to the original circuit. So, area overhead of the online tester in this case would be much higher than that of the original circuit and even the redundancy based methods. Further, this tester only checks the protocol and so fault coverage or detection latency cannot be guaranteed.

Discrete event system (DES) model based methods are used for failure detection for a wide range of applications because of the simplicity of the model as well as the associated algorithms [17]. A DES is characterized by a discrete state space and some event driven dynamics. Finite state machine (FSM) based model is a simple example of a DES. In the state based DES model, the model is partitioned according to the failure or normal condition. The failure detection problem consists in determining, in finite time after occurrence of the failure, whether the system is traversing through normal or failure subsystem. A fault is detectable by virtue of certain transitions (in the failure states) which are called fault detecting transitions (FD-transitions). FD-transition is a transition of the faulty subsystem, for which there is no corresponding equivalent transition in the normal subsystem. Using the FD-transitions, a DES fault detector is designed, which is a kind of state estimator of the system. For OLT of circuits, the detector is synthesized as a circuit which is executed concurrently with the circuit under test (CUT). Biswas et al. in [18, 19] have developed an OLT scheme for synchronous circuits using the FSM based DES theory, which satisfies most of the metrics mentioned above for an efficient online tester design. In this paper, we aim at using the theory of failure detection of DES models for OLT of SI circuits.

Just like synchronous circuits, the basic FSM framework is also used to model asynchronous circuits with slight modification. In case of synchronous circuits, state changes in the FSM occur only at the active edge of the register clock, irrespective of the time of change of the inputs. On the other hand, in asynchronous circuits, state changes can occur immediately after transition in the inputs. FSM used to model asynchronous circuit is called AFSM [20]. An alternative to AFSM is burst-mode (BM) state machines [20]. AFSM and BM state machine are similar from the modeling perspective; however, in case of BM state machine transitions are labeled with signal changes rather than their explicit values, which is the case in AFSMs. AFSMs and BM state machines assume that first inputs change followed by outputs and finally new state is reached. Due to the strict sequence of signal changes, all asynchronous protocols cannot be modeled using AFSMs or BM state machines. Extended BM state machines address this modeling issue by allowing some inputs in a burst to change monotonically along a sequence of bursts rather than in a particular burst. Petri net (PN) is widely accepted modeling framework for highly concurrent systems [21]. PN models a system using interface behaviors which are represented by allowed sequence of transitions or traces. The view of an asynchronous circuit as a concurrent system makes PN based models more appropriate than AFSMs and BM state machines for their analysis and synthesis. There are several variants of PNs among which signal transition graph (STG) is generally used to model asynchronous circuits. The major reason is that the STG interprets transitions as signal transitions and specifies circuit behavior by defining casual relations among these signal transitions [22].

In this paper, we aim at using the theory of failure detection of DES model for OLT of SI asynchronous circuits. Several modifications are made in the DES framework used for synchronous circuits [18, 19] when applied for SI circuits. The modifications are as follows:

(i) We first model SI circuits along with their faults as STGs and then translate them into state graphs. State graphs are similar to FSM based DES models from which FD-transitions can be determined.

(ii) In case of synchronous circuits, the fault detector is an FSM which detects the occurrence of FD-transitions. A synchronous circuit can be synthesized in a straightforward way from the FSM specification [18, 19] that performs online testing. Why the same design cannot be synthesized as an asynchronous circuit will be discussed in this paper. As the use of synchronous circuit for OLT of asynchronous modules is not desirable, we propose a new technique for detector design which can be synthesized as a SI circuit. The detector is designed as state graph model which is live and has complete state coding (CSC); these properties ensures its synthesizability as a SI circuit.

The paper is organized as follows. In Section 2, we present some definitions and formalisms of the DES framework. Section 3 illustrates DES modeling for a speed independent circuit under normal and stuck-at faults. In Section 4, the DES detector for the SI asynchronous circuit is designed. Synthesizing the DES detector as online tester circuit is also discussed in the same section. Section 5 presents experimental results regarding area overhead and fault coverage of the DES detector based online tester. Also, comparison of area overhead of the proposed approach with other similar schemes is reported. Finally, we conclude in Section 6.

2. DES Modeling Framework: Definitions and Formalisms

A *discrete event system (DES) model G* is defined as $G = \langle V, X, \mathfrak{I}, X0 \rangle$ where $V = \{v_1, v_2, \ldots, v_n\}$ (In case of modeling SI circuits as DES, the state variables are values of the I/O

signals. So, in this work, we will interchangeably use the terms signal and variable.) is a finite set of discrete variables assuming values from the set $\{0, 1\}$, called the domains of the variables, X is a finite set of states, $\mathfrak{I}$ is a finite set of transitions, and $X0 \subseteq X$ is the set of initial states. A state x is a mapping of each variable to one of the elements of the domain of the variable. A *transition* $\tau \in \mathfrak{I}$ from a state x to another state x^+ is an ordered pair $\langle x, x^+ \rangle$, where x is denoted by *initial*(τ) and x^+ is denoted as *final*(τ).

2.1. Failure Modeling. The failure of the system is modeled by dividing the DES model into submodels and each submodel is used to model the system under normal or failure conditions. To differentiate between the submodels, each state x is assigned a failure label by a status variable S with its domain being equal to $\{N \cup F_1 \cup F_2 \cup \cdots \cup F_d\}$, where N is normal status, F_i, $1 \le i \le d$, is failure status, and d is the number of possible faults.

Definition 1 (normal G-state). A G-state x is normal if $x(S) = \{N\}$. The set of all normal states is denoted by X_N.

Definition 2 (F_i-G-state). A G-state x is failure state or synonymously an F_i-state, if $F_i \in x(S)$. The set of all F_i-states is denoted by X_{F_i}.

Definition 3 (normal G-transition). A G-transition $\langle x, x^+ \rangle$ is called a normal G-transition if $x, x^+ \in X_N$.

Definition 4 (F_i-G-transition). A G-transition $\langle x, x^+ \rangle$ is called an F_i-G-transition if $x, x^+ \in X_{F_i}$.

Definition 5 (equivalent states). Two states x and y are said to be equivalent, denoted by xEy, if $x|_V = y|_V$ and $x(S) \ne y(S)$.

In other words, two states are said to be equivalent if they have the same values for state variables and different value for status variable.

A transition $\langle x, x^+ \rangle$, where $x(S) \ne x^+(S)$, is called a *failure* transition indicating the first occurrence of some failure in the system. Since failures are assumed to be *permanent*, there is no transition from any state in X_{F_i} to any state in X_N or from any state in X_{F_i} to any state in X_{F_j}.

Definition 6 (equivalent transitions). Two transitions $\tau_1 = \langle x_1, x_1^+ \rangle$ and $\tau_2 = \langle x_2, x_2^+ \rangle$ are equivalent, denoted by $\tau_1 E \tau_2$, if $x_1 E x_2$, $x_1^+ E x_2^+$ and they must associate with the same signal change.

Suppose that there is a transition in failure DES model for which there is no corresponding equivalent transition in normal DES model, then that transition is called failure detecting transition (FD-transition). The failure is detected when the system traverses through the FD-transition. Thus, we can define FD-transition as follows.

Definition 7 (FD-transition). A F_i-G-transition of faulty DES model $\tau' = \langle x', x'^+ \rangle$ is an FD-transition, if there is no G-transition $\tau = \langle x, x^+ \rangle$ in the normal DES model such that $\tau' E \tau$.

The motivation of failure detection using DES model is to find out such FD-transitions and design DES detector using these transitions. In the next section, we discuss how to model SI circuits using DES.

3. DES Model of a Speed Independent Circuit: Normal and Faulty

As already discussed, the first step to design a DES based online tester is to obtain the normal and faulty state based model of the CUT. However, the traditional state based DES paradigm cannot be directly used for modeling SI circuits. So in this case we will start with signal transition graph (STG), which is a type of Petri net based DES, to specify fault-free and faulty conditions. The STGs will be converted into state graphs (similar to FSMs) using the concept presented in [22].

We first discuss fault modelling at the STG level using an example and concepts from [22]. In addition to the models (i.e., faults in transistors of the C-elements) given in [22], we have also modeled stuck-at faults on all wires (i.e., input/output of gates).

3.1. Fault Modeling. The SI asynchronous CUT example being considered to illustrate the proposed scheme is shown in Figure 1 (taken from [22]). Traditionally, synchronous circuits consist of blocks of combinational logic connected with clocked latches or registers, while, in case of SI circuit designs, we basically have logic gates as building blocks with C-elements, which act as storage elements. Transistor level diagram of C-element is shown in Figure 2; logic function of the C-element can be described by the Boolean equation $C = AB + AC' + BC'$, where C is the next state and C' is the old state value [22, 23]. The output of C-element becomes logically high (low) when both the inputs are logically high (low); otherwise it keeps its previous logic value. There are two types of C-elements used in SI circuits: static C-element and dynamic C-element. The static version of C-element promises that the information inside it can be stored for unbounded periods. However, dynamic versions of C-element provide gains in terms of area, power, and delay [23–26]. Since the circuits having high operating speed, low area, and power consumption are preferred in modern days, we have chosen SI circuits with dynamic C-elements instead of static ones.

Figure 3 shows the STG for the CUT being considered. Rising (falling) transitions on signals, indicated by + (−), are shown in the STG. The dark circles along the arcs are called tokens. The token indicates one of possibly a set of signals that enable transition to fire. If all input arcs for a signal transition have tokens then that signal transition is said to be enabled. For example, when signal R_{in} goes high (denoted by $R_{in}+$) and R_{out} goes high (denoted by $R_{out}+$), only then $A_{out}+$ transition can take place. Upon firing $A_{out}+$, a token is placed on each of its outgoing arcs, thus enabling $R_{in}-$. Note that $R_{out}-$ is enabled after $A_{out}+$ and $A_{in}+$.

In this paper, we have considered SI circuits that contain C-elements (we assumed dynamic version) and logic gates. For the logic gates, the most popular fault model is the stuck-at fault model, which is at the gate level. However,

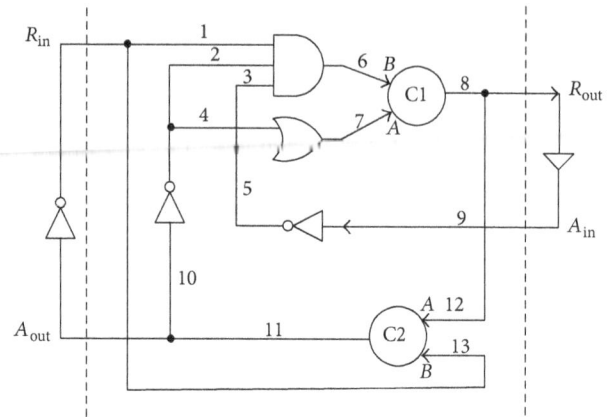

FIGURE 1: Example of speed independent asynchronous CUT.

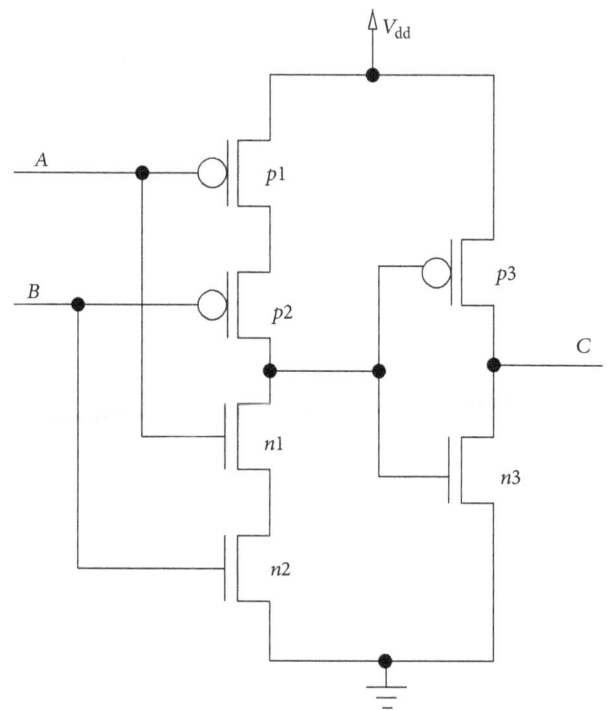

FIGURE 2: Transistor diagram of dynamic C-element.

for the C-elements stuck-on and stuck-off faults for each transistor are an accepted fault model [22]. So we have chosen a mixed gate/transistor level description for modeling the faults. To illustrate fault modeling at both C-elements and basic gates, we consider the circuit example from [22] which is shown in Figure 1.

3.1.1. STG Based Modeling. In this work, we model single stuck-at faults in the gates and transistors (for the C-elements) and map them to STGs of the circuit. For the analysis, the signals attached to the inputs A and B of the C-elements are also indicated in the gate level circuit diagram of Figure 1. Now, we consider some of these faults (one at a time), analyze their effects, and finally modify the STG to model the faults.

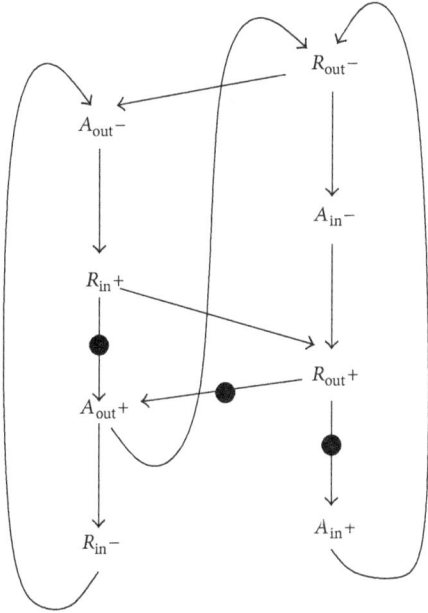

FIGURE 3: Signal transition graph of sample circuit.

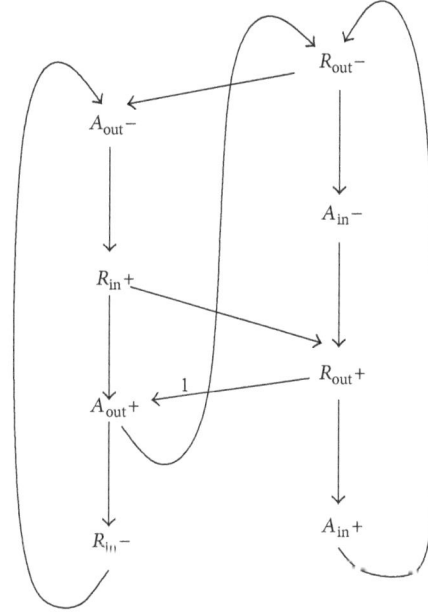

FIGURE 4: STG level model of stuck-on fault $n1$ of C2 [22].

Consider the C-element C2 of Figure 1 and refer to transistor level circuit of Figure 2. The C-element C2 has R_{out} and R_{in} as inputs and A_{out} as output. If the transistor $n1$ has a stuck-on fault, this leads to error in the circuit that it needs to wait for only $n2$ to be enabled to generate output. When $n2$ turns on, then a path to ground via $n1$ and $n2$ gets established, which makes $p3$ on and $n3$ off, making C high. So C2 has to wait only for $n2$ (i.e., $R_{in}+$ which corresponds to input B of C2 to become 1) to turn on and change the output. In other words, it has to wait for only $R_{in}+$ (and not also for $R_{out}+$, which is the requirement under normal condition) before it can generate $A_{out}+$. Thus, the fault in $n1$ leads to premature firing of the $A_{out}+$ transition. We represent this by including a token on the arc connecting $R_{out}+$ to $A_{out}+$. Availability of this token will enable $A_{out}+$ to fire as soon as $R_{in}+$ arrives, without waiting for $R_{out}+$. This token is denoted by a "1" shown on the arc in Figure 4.

Now, consider C-element C1 producing output R_{out}, with transistor $p1$ having stuck-on fault. As $p1$ is on, the gate has to wait for $p2$ to turn on before it can change the output. When $p2$ turns on (by virtue of $B = 0$) then there is no path to ground as $n2$ is off, which makes $p3$ off and $n3$ on, making C low. Here, C1 has to wait for the input $B = 0$ to generate $R_{out}-$. Referring to Figure 1, for B to become 0, we need either A_{out} to become 1 (the same as R_{in} becoming 0) or A_{in} to become 1. Thus, as soon as we have either $A_{out}+$ or $A_{in}+$, $R_{out}-$ would fire. It may be noted that, under normal condition, for $R_{out}-$ to fire, we also require $A = 0$, which mandates both $A_{out}+$ and $A_{in}+$. This failure condition is indicated in the STG by adding a "1" to the input arcs of $R_{out}-$, which is shown in Figure 5. To elaborate, Figure 5(a) (Figure 5(b)) shows that $R_{out}-$ can be fired as soon as $A_{in}+$ ($A_{out}+$) fires and does not wait for $A_{out}+$ ($A_{in}+$).

As the third fault, let C1 have stuck-on fault at $n2$. The stuck-on fault at $n2$ enforces the circuit to wait only for $n1$ to

be enabled for generating output $R_{out}+$. As $n1$ is connected to input A, which is logical ORing of $A_{out}-$ and $A_{in}-$, transition $R_{out}+$ can fire after $A_{out}-$ or $A_{in}-$ (without requiring to wait for $R_{in}+$). This premature firing of transition $R_{out}+$ is indicated in the STG by adding a "1" to $R_{in}+$, which is shown in Figure 6.

For the gates, stuck-at-0 and stuck-at-1 faults are considered at their inputs and output nets. Let Line 2 of the AND gate from Figure 1 have a stuck-at-0 fault. This makes Line 6 stuck-at-0 fault. As Line 6 is connected to the B input of the C-element C1, we have transistor $p2$ on and transistor $n2$ off. Note that as $n2$ is always off, there is no path to the ground. So the output R_{out} can never become 1 because $p3$ can never turn on. In other words, we will never have the $R_{out}+$ transition. This is indicated by adding a "0" on the output arcs of $R_{out}+$ in Figure 7.

If Line 9 gets stuck at 1, this will lead to Lines 3 and 5 being stuck at 0, further leading to Line 6 being stuck at 0. As Line 6 is connected to the B input of the C-element, we will have the fault manifestation similar to the case of Line 2 stuck at 0. Now, we consider a stuck-at-0 fault at Line 13. As this line is connected to the B input of the C-element C2, it will lead to output A_{out} never becoming 1. The effect is shown by adding a "0" to the output arcs of $A_{out}+$ in Figure 8.

Now, we consider an example of a redundant fault; that is, no logical difference is observed in the operation of the circuit after fault. An instance of such a fault is $n1$ stuck-on fault in C1. This fault enforces the circuit to wait only for $n2$ to be enabled (i.e., B to be 1) for generating output $R_{out}+$. As $n2$ is connected to input B, which is logical ANDing of $R_{in}+$, $A_{out}-$, and $A_{in}-$, $R_{out}+$ can fire only after three transitions, namely, $R_{in}+$ and $A_{out}-$ and $A_{in}-$ fire. It may be noted that $A_{out}-$ and $A_{in}-$ also imply that input A (connected to $n1$) of C1 is 1, which in turn implies condition for $n1$. As fault and normal condition both imply $n1$ to be on, stuck-on fault at $n1$

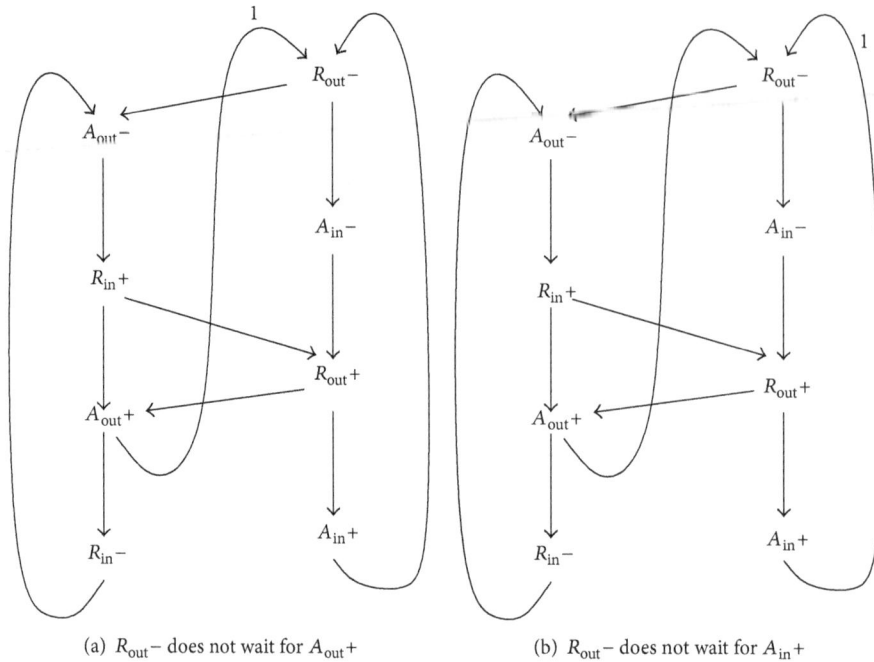

(a) $R_{out}-$ does not wait for $A_{out}+$

(b) $R_{out}-$ does not wait for $A_{in}+$

FIGURE 5: STG level model of stuck-on fault in $p1$ of C1 [22].

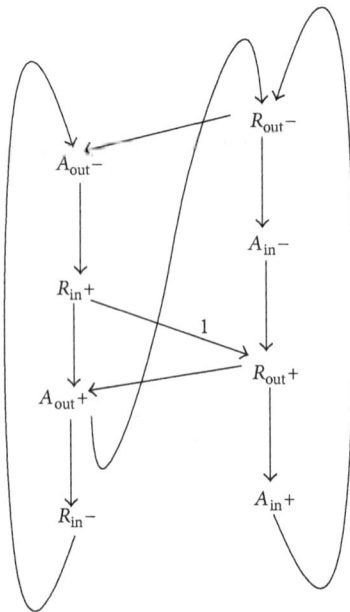

FIGURE 6: STG level model of stuck-on fault $n2$ of C1 [22].

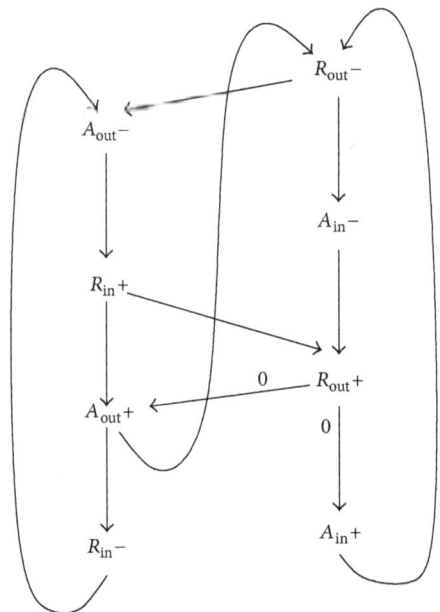

FIGURE 7: STG level model of stuck-at-0 fault in Line 2.

of C1 does not generate any behavioral difference. Obviously, such faults cannot be detected and under the single stuck-at fault assumption do not cause significant reliability issues.

For the fault model considered, the total number of faults in a SI circuit having dynamic C-elements is equal to 12 times the number of C-elements (each C-element consists of 6 transistors and each transistor can have stuck-on and stuck-off faults) plus twice the number of I/O lines of the gates (each line has either stuck-at-0 or stuck-at-1 fault).

So the number of faults in case of the circuit considered in Figure 1 is not too small and listing them all would make a tabular representation long. So a partial list of faults and their effects on the STG is given in Table 1.

3.1.2. State Graph Based Fault Modeling and FD-Transitions. As already discussed, the first step of DES based OLT design is to generate the normal and faulty models. For SI circuits, first the STGs under normal and faulty conditions are

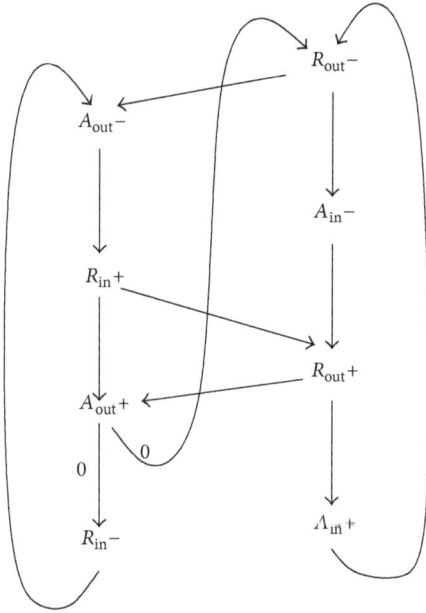

FIGURE 8: STG level model of stuck-at-0 fault in Line 13.

TABLE 1: A partial list of faults and their effects on STG.

Fault type	Effect of fault on the transitions
Transistor $n1$ stuck-on fault for C1	$R_{out}+$ can only be fired after $R_{in}+$, $A_{out}-$, and $A_{in}-$ (the same as normal condition)
Transistor $n2$ stuck-on fault for C1	$R_{out}+$ can be fired after either $A_{out}-$ or $A_{in}-$
Transistor $p1$ stuck-on fault for C1	$R_{out}-$ can be fired after either $A_{out}+$ or $A_{in}+$
Transistor $p2$ stuck-on fault for C1	$R_{out}-$ can only be fired after $A_{out}+$ and $A_{in}+$
Transistor $n1$ stuck-on fault for C2	$A_{out}+$ can be fired after $R_{in}+$
Transistor $n2$ stuck-on fault for C2	$A_{out}+$ can be fired after $R_{out}+$
Transistor $p1$ stuck-on fault for C2	$A_{out}-$ can be fired after $R_{in}-$
Transistor $p2$ stuck-on fault for C2	$A_{out}-$ can be fired after $R_{out}-$
Transistor $n1$ stuck-off fault for C1	All arcs from $R_{out}+$ are always 0
Transistor $n2$ stuck-off fault for C1	All arcs from $R_{out}+$ are always 0
Transistor $p1$ stuck-off fault for C1	All arcs from $R_{out}-$ are always 0
Transistor $p2$ stuck-off fault for C1	All arcs from $R_{out}-$ are always 0
Transistor $n1$ stuck-off fault for C2	All arcs from $A_{out}+$ are always 0
Transistor $n2$ stuck-off fault for C2	All arcs from $A_{out}+$ are always 0
Transistor $p1$ stuck-off fault for C2	All arcs from $A_{out}-$ are always 0
Transistor $p2$ stuck-off fault for C2	All arcs from $A_{out}-$ are always 0
Transistor $n3$ stuck-on fault for C1 Transistor $p3$ stuck-off fault for C1	All arcs from $R_{out}+$ are always 0
Transistor $p3$ stuck-on fault for C1 Transistor $n3$ stuck-off fault for C1	All arcs from $R_{out}-$ are always 0
Transistor $p3$ stuck-on fault for C2 Transistor $n3$ stuck-off fault for C2	All arcs from $A_{out}-$ are always 0
Transistor $n3$ stuck-on fault for C2 Transistor $p3$ stuck-off fault for C2	All arcs from $A_{out}+$ are always 0
Stuck-at-0 fault at Line 1 or Line 2 or Line 3 Stuck-at-0 fault at Line 6 or Line 7 Stuck-at-1 fault at Line 9 or Line 10	All arcs from $R_{out}+$ are always 0

obtained and then converted into state based models. In this subsection, we explain the concept using the example circuit of Figure 1 under normal condition and two faults, namely, stuck-on fault in $n1$ of C2 and stuck-on fault in $p1$ of C1.

The state based DES model for the normal circuit is shown in Figure 9. It may be noted that in the circuit there are 4 I/O signals, namely, R_{in}, R_{out}, A_{in}, and A_{out}. In the DES model corresponding to each signal, there is a discrete variable: $\langle v_1 = R_{in}, v_2 = R_{out}, v_3 = A_{in}, v_4 = A_{out}\rangle$ which assumes values from the set $\{0, 1\}$. The set of states are $X0$ to $X13$ and $X0$ is the initial state. State mappings and transitions are shown in Figure 9; for example, state $X0$ maps variables $\langle R_{in}, R_{out}, A_{in}, A_{out}\rangle$ to $\langle 1, 1, 0, 0\rangle$. In states $X0$ and $X1$ mappings are $\langle 1, 1, 0, 0\rangle$ and $\langle 1, 1, 0, 1\rangle$, respectively. So transition from $X0$ to $X1$ changes A_{out} from 0 to 1; this is indicated by transition $A_{out}+$. Now, if we look at the STG for the normal circuit in Figure 3 we note that $A_{out}+$ can fire if $R_{in}+$ and $R_{out}+$ have a token (i.e., $R_{in} = 1$ and $R_{out} = 1$). In state $X0$ as $R_{in} = 1$ and $R_{out} = 1$, $A_{out}+$ can fire. Similarly, the whole DES can be constructed.

The state based DES model for the circuit under $n1$ stuck-on fault in C2 is shown in Figure 10. The set of states are $X'0$ to $X'13$ and $X'0$ is the initial state. The transitions and state mapping are shown in the figure. As discussed in the previous subsection, $n1$ stuck-on fault in C2 results in premature firing of $A_{out}+$ (i.e., it need not wait for $R_{out}+$ and can fire only if $R_{in}+$ holds). If we observe the failure DES model in Figure 10, we note that there are two dotted transitions, which correspond to the failure condition, that is, premature firing of $A_{out}+$. One dotted transition is between $X'11$ and $X'6$. It may be noted that in $X'11$ we have $R_{in} = 1$, $R_{out} = 0$, $A_{in} = 1$, and $A_{out} = 0$, where even though $R_{out}+$ is not enabled, because $R_{in}+$ is enabled, $A_{out}+$ fires. A similar premature firing of $A_{out}+$ occurs between $X'13$ and

TABLE 1: Continued.

Fault type	Effect of fault on the transitions
Stuck-at-0 fault at Line 4 and Line 5	All arcs from $R_{out}+$ are always 0
Stuck-at-1 fault at Line 4 or Line 5	
Stuck-at-1 fault at Line 6 or Line 7	All arcs from $R_{out}-$ are always 0
Stuck-at-0 fault at Line 9 or Line 10	
Stuck-at-0 fault at Line 8	All arcs from $R_{out}+$ are always 0 and all arcs from $A_{out}+$ are always 0
Stuck-at-1 fault at Line 8	All arcs from $R_{out}-$ are always 0 and all arcs from $A_{out}-$ are always 0
Stuck-at-0 at Line 11	All arcs from $R_{out}-$ are always 0 and all arcs from $A_{out}+$ are always 0
Stuck-at-1 fault at Line 11	All arcs from $R_{out}+$ are always 0 and all arcs from $A_{out}-$ are always 0
Stuck-at-0 fault at Line 12 or Line 13	All arcs from $A_{out}+$ are always 0
Stuck-at-1 fault at Line 12 or Line 13	All arcs from $A_{out}-$ are always 0

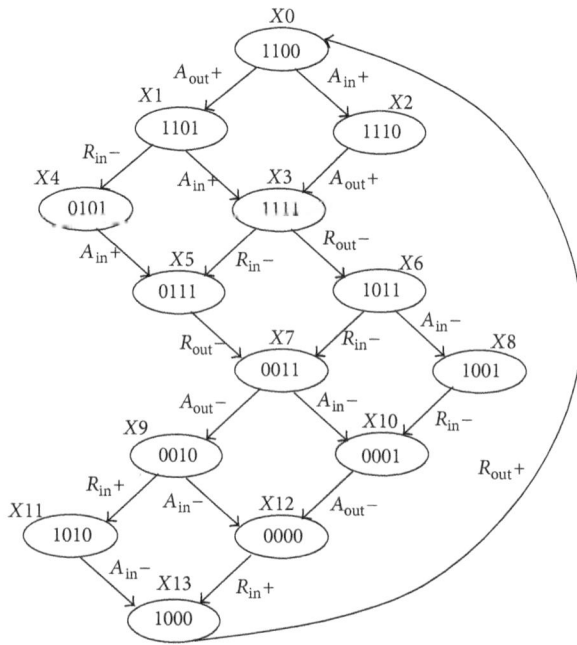

FIGURE 10: DES model for circuit with $n1$ stuck-on fault in C2 (STG of Figure 4).

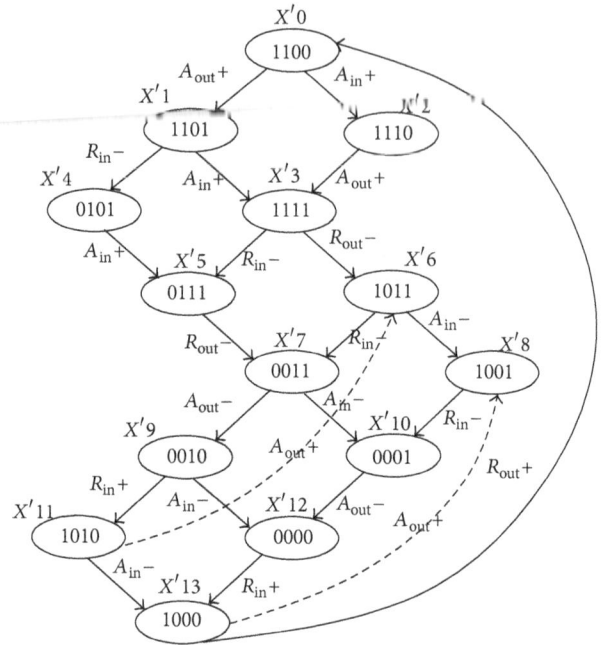

FIGURE 9: DES model for normal circuit (STG of Figure 3).

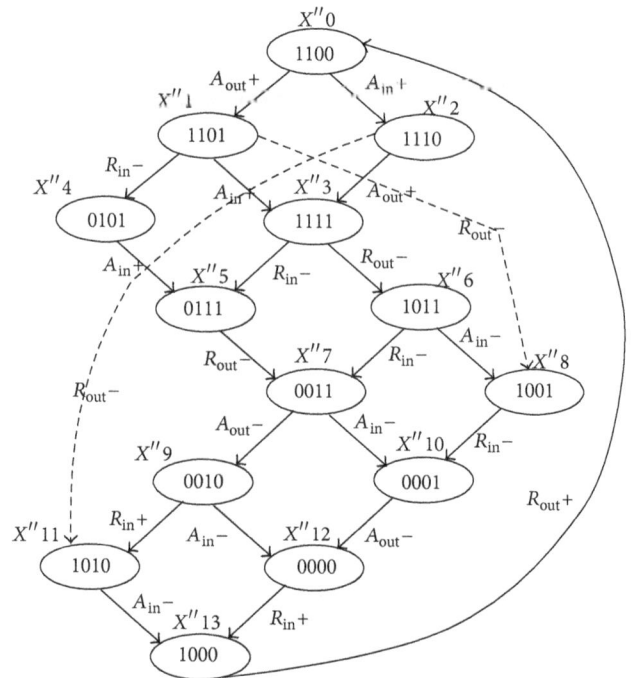

FIGURE 11: DES model for circuit with $p1$ stuck-on fault in C1 (STGs of Figure 5).

$X'8$. So, for the fault $n1$ stuck-on fault in C2, there are two FD-transitions, namely, $\langle X'11, X'6 \rangle$ and $\langle X'13, X'8 \rangle$.

The state based DES model for the circuit under $p1$ stuck-on fault in C1 is shown in Figure 11. The set of states are $X''0$ to $X''13$ and $X''0$ is the initial state. As discussed in the previous subsection, $p1$ stuck-on fault in C1 results in premature firing of $R_{out}-$ triggered by either $A_{in}+$ or $A_{out}+$. This is captured by the dotted transitions $\langle X''2, X''11 \rangle$ and $\langle X''1, X''8 \rangle$ in failure DES model in Figure 11. The transition $\langle X''2, X''11 \rangle$ represents firing of $R_{out}-$ by $A_{in}+$ in spite of $A_{out}+$ not being

enabled and the transition $\langle X''1, X''8 \rangle$ represents firing of $R_{out}-$ by $A_{out}+$ in spite of $A_{in}+$ not being enabled. Thus, for this fault there are two FD-transitions, namely, $\langle X''1, X''11 \rangle$ and $\langle X''1, X''8 \rangle$.

The set of FD-transitions is shown in Table 2.

TABLE 2: FD-transitions.

FD-transitions	Start state	Final state	Effect
1	0001	0101	Premature fire of $R_{out}+$ before $R_{in}+$ (by $A_{in}-$)
2	0000	0100	Premature fire of $R_{out}+$ before $R_{in}+$ (by $A_{in}-$)
3	1110	1010	Premature fire of $R_{out}-$ before $A_{out}+$ (by $A_{in}+$)
4	1010	1011	Premature fire of $A_{out}+$ before $R_{out}+$ (by $R_{in}+$)
5	1000	1001	Premature fire of $A_{out}+$ before $R_{out}+$ (by $R_{in}+$)
6	0111	0110	Premature fire of $A_{out}-$ before $R_{out}-$ (by $R_{in}-$)
7	0101	0100	Premature fire of $A_{out}-$ before $R_{out}-$ (by $R_{in}-$)
8	1011	1010	Premature fire of $A_{out}-$ before $R_{in}-$ (by $R_{out}-$)
9	1001	1000	Premature fire of $A_{out}-$ before $R_{in}-$ (by $R_{out}-$)
10	1101	1001	Premature fire of $R_{out}-$ before $A_{in}+$ (by $A_{out}+$)

In the next section, we will discuss the procedure for design of the DES detector from FD-transitions and its synthesis as a SI circuit.

4. DES Detector Based Online Tester

A DES detector is basically a state estimator which predicts whether the Circuit Under Test (CUT) traverses through normal or faulty states/transitions. Broadly speaking, the detector is constructed using transitions which can manifest the fault effects. In other words, such a transition is a faulty transition for which there is no corresponding equivalent normal transition. As already mentioned, we call such transitions failure detecting transitions (i.e., FD-transitions). In the circuit under consideration, comparing the normal (Figure 9) and $n1$ stuck-on fault at C2 DES models (Figure 10), we may note that there are two transitions (dotted) $\langle X'11, X'6\rangle$ and $\langle X'13, X'8\rangle$ which manifest the fault effect. Corresponding to these transitions, there are no equivalent transitions in the normal model. These two transitions are FD-transitions for the fault and are used in DES detector construction.

If the CUT is a synchronous circuit then obviously the online tester is also a synchronous circuit that can be designed from the FD-transitions using straightforward FSM synthesis philosophy [18, 19]. The detector FSM has three classes of states, namely, initial, intermediate, and final. The detector measures the I/O signals of the CUT (i.e., variables) to determine whether the following happens.

On startup the detector is in its initial state and it checks whether the CUT is in the initial state of any FD-transition. For example, if we consider only two faults in the circuit under consideration, stuck-on fault in $n1$ of C2

and stuck-on fault in $p1$ of C1, then the FD-transition set is $\{\langle X'11, X'6\rangle, \langle X'13, X'8\rangle, \langle X''2, X''11\rangle\}$. So in the initial state the detector checks whether the signals R_{in}, R_{out}, A_{in}, and A_{out} are 1, 0, 1, and 0 or 1, 0, 0, and 0 or 1, 1, 1, and 0. If so, the detector moves to an intermediate state (in the next clock edge) corresponding to the value matched. For each of the FD-transitions, there is a corresponding intermediate state in the detector. For example, if R_{in}, R_{out}, A_{in}, and A_{out} are measured to be 1, 0, 1, and 0 in the initial state of the detector, the detector moves to the intermediate state corresponding to FD-transition $\langle X'11, X'6\rangle$. However, if the signals do not match initial state of any FD-transition the detector loops in the initial state. In the intermediate state whether the values of the signals of the CUT match with the final state of the corresponding FD-transition is checked; if so, the fault is detected and the detector moves to the final state and is deadlocked there. Otherwise, it moves to the initial state. Continuing with the example, if the values of R_{in}, R_{out}, A_{in}, and A_{out} are 1, 0, 1, and 1 from the intermediate state, FD-transition $\langle X'11, X'6\rangle$ is detected (i.e., stuck-on fault in $n1$ of C2) and detector moves to final state in the next clock edge. Otherwise, if R_{in}, R_{out}, A_{in}, and A_{out} are 1, 0, 0, and 0 then CUT is normal and the detector moves to the initial state.

The above mentioned philosophy of constructing the detector and then synthesizing it into a synchronous system is widely used in the DES theory [17] and has been applied for OLT of synchronous circuits [18, 19]. Obviously, if the CUT is an asynchronous circuit and so must be the detector circuit. However, it may be noted that the same philosophy cannot be directly used in the case of SI circuits. The reason is that the FSM of the detector designed above has liveness issue in the final state and has complete state coding (CSC) violations in the intermediate states.

Now we propose a new technique for detector design which can be synthesized as a SI circuit. The detector is designed as state graph model which is live and has complete state encoding, that ensures its synthesizability as a SI circuit. Before formalizing the algorithm for the design of the state graph of the detector, we first introduce the basic philosophy of its working using the examples from the previous section.

An FD-transition in SI circuit design paradigm can be stated as, "under failure, a signal s can change in the presence of signals $y1, y2, \ldots, yn$, $(1 \leq n)$ which is not possible under normal condition." For example, in the case of $n1$ of C2 stuck-on fault, $\langle X'11, X'6\rangle$ is an FD-transition which changes signal A_{out} from 0 to 1 (i.e., $A_{out}+$) and the other signals are $R_{in} = 1$, $R_{out} = 0$, and $A_{in} = 1$ (Figure 10). It may be noted that in normal condition for changing A_{out} from 0 to 1 we need $R_{in} = 1$, $R_{out} = 1$, and $A_{in} = 0$ (Figure 9). Comparing with the faulty condition we may state that, "under $n1$ of C2 stuck-on fault, signal A_{out} can change from 0 to 1 in presence of signals $R_{in} = 1$, $R_{out} = 0$, and $A_{in} = 1$ which is not possible under normal condition." So, to detect whether FD-transition $\langle X'11, X'6\rangle$ has occurred, the detector needs to tap lines R_{out}, A_{in}, and A_{out} (R_{in} is not required to be monitored as its value is same under normal and faulty case) of the CUT and determine whether $A_{out}+$ has fired and at that time whether $R_{out} = 0$ or $A_{in} = 1$ or both; if so, a status output line is made 1. For optimization of the detector in terms of number of states, tap lines, and so forth, without loss of fault detection capability, we may consider checking $A_{out}+$ in presence of either $R_{out} = 0$ or $A_{in} = 1$ but not both. If we consider the other FD-transition $\langle X'13, X'8 \rangle$ for the fault, it can be detected by checking whether $A_{out}+$ has fired and at that time whether $R_{out} = 0$; R_{in} and A_{in} are not required to be monitored as their values are the same under normal and faulty case. So, it may be stated that to detect the fault by FD-transitions $\langle X'11, X'6 \rangle$ and $\langle X'13, X'8 \rangle$ we need to check whether $A_{out}+$ has fired and at that time whether $R_{out} = 0$. The design and flow of the detector for these two FD-transitions are as follows:

(1) The state encoding tuple is $\langle R_{out}, A_{out}, Status \rangle$. The initial state of the detector d_0 is encoded as 100. The first two bits represent the complement of $R_{out} = 0$ and $A_{out} = 1$, that is, complement of value of R_{out} in state $X'11$ and complement of change of A_{out} by the FD-transition. The third bit represents $Status$ output of the detector which is 0 until FD-transition is detected.

(2) The detector waits for signal R_{out} to become 0 and if so it moves to state d_1 say, which is encoded as $\langle 000 \rangle$. However, from state d_0, if A_{out} becomes 1, FD-transition cannot be detected because this is normal situation (state $X0$ in Figure 9) where $A_{out}+$ fires when $R_{out} = 1$; detector moves to state d_5 having encoding $\langle 110 \rangle$. When A_{out} becomes 0 in state d_5, the detector moves back to d_0 from where it again waits to detect whether the FD-transition occurs.

(3) From state d_1 the following may happen:

(a) If R_{out} becomes 1, then FD-transition cannot be detected and so the detector moves back to d_0.

(b) If A_{out} becomes 1, then FD-transition and hence fault are detected. The detector moves to state d_2 having encoding $\langle 010 \rangle$. Following that detector makes $Status$ output high and moves to state d_3 with encoding $\langle 011 \rangle$.

Once $status$ line is 1, that is, fault is detected online, the system should switch to an alternative backup circuit, as under the single stuck-at fault model faults are assumed to be permanent [22]. By that logic, the detector should stop or loop in d_3 indefinitely; however, it would lead to deadlock and is nonsynthesizable as a SI circuit. To avoid this deadlock, a simple modification is made in the detector state graph without affecting the fault detection performance. We wait at state d_3 for any signal to change (i.e., R_{out} from 0 to 1 or A_{out} from 1 to 0) and we move to state d_4; let us select R_{out} for this purpose. State encoding of d_4 is $\langle 111 \rangle$. From state d_4 we have a transition to state d_3 on change of R_{out} from 1 to 0.

Figure 12 illustrates the state graph for detecting $n1$ stuck-on fault at C2 by FD-transitions $\langle X'11, X'6 \rangle$ and $\langle X'13, X'8 \rangle$.

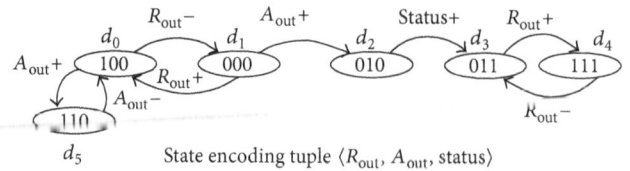

FIGURE 12: State graph for detecting FD-transitions $\langle X'11, X'6 \rangle$ and $\langle X'13, X'8 \rangle$.

In similar way we can design detectors for the other FD-transitions shown in Table 2. However, it may be noted that different circuits may be required for the other FD-transitions because merging all FD-transitions in a single detector state graph will lead to CSC problems. As shown in the example above, some FD-transitions can be merged into a single state graph maintaining CSC. Figures 13–16 illustrate the SGs for all the FD-transitions shown in Table 2. Also, the FD-transitions which could be merged are mentioned in the figures.

Before discussing the algorithm for generating the detector for a set of FD-transitions we introduce the notion of compatible FD-transitions.

Definition 8 (compatible FD-transitions). Two FD-transitions $\tau'1 = \langle X'1, X'1^+ \rangle$ and $\tau'2 = \langle X'2, X'2^+ \rangle$ are compatible if the following holds:

(i) If $s1$ is the signal change by $\tau'1$ and $s2$ is the signal change by $\tau'2$, then $s1$ is the same as $s2$. In other words, signal change by both FD-transitions is the same.

(ii) Let $V'1 \subseteq V$ (for FD-transition $\tau'1 = \langle X'1, X'1^+ \rangle$) be the set of variables whose values at $X'1$ are different compared to state(s) Xi, where Xi is any state under normal condition (normal DES) from which $s1$ is the signal change. Similarly, the set $V'2 \subseteq V$ is calculated for FD-transition $\tau'2 = \langle X'2, X'2^+ \rangle$. Then, $V'1 \cap V'2 \neq \emptyset$. In other words, there exists at least one signal (i.e., a variable) whose value is the same in initial state of both FD-transitions and that is different compared to the initial state(s) of the corresponding transition(s) under normal condition.

For example, consider two FD-transitions $\langle X'11, X'6 \rangle$ ($= \tau'1$, say) and $\langle X'13, X'8 \rangle$ ($= \tau'2$, say). We calculate $V'1$ for $\tau'1$ as follows. The value of variables at initial state $X'11$ is $\{R_{in} = 1, R_{out} = 0, A_{in} = 1, A_{out} = 0\}$. The signal change for $= \tau'1$ is $A_{out}+$. We get two states ($X0 = \{R_{in} = 1, R_{out} = 1, A_{in} = 0, A_{out} = 0\}$ and $X2 = \{R_{in} = 1, R_{out} = 1, A_{in} = 1, A_{out} = 0\}$) in normal condition from which the signal change $A_{out}+$ occurs. Thus, $V'1 = \{R_{out}(=0)\}$ because R_{out} is the only variable that is different in $X'11$ compared to normal states $X0$ and $X2$. Similarly, we calculate $V'2$ for $\tau'2$ as $V'2 = \{R_{out}(=0)\}$. Since $V'1 \cap V'2 \neq \emptyset = \{R_{out}(=0)\}$, thus, these two transitions are compatible and can be merged (as shown in state graph of Figure 12).

Algorithm 9. Algorithm for construction of detectors given the set of FD-transitions.

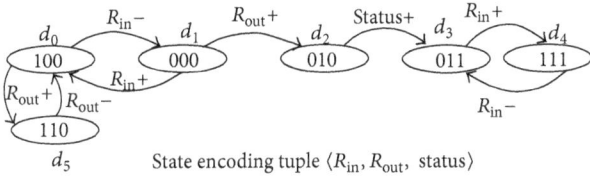

FIGURE 13: State graph for detecting FD-transitions–Sl. numbers 1 and 2 of Table 2.

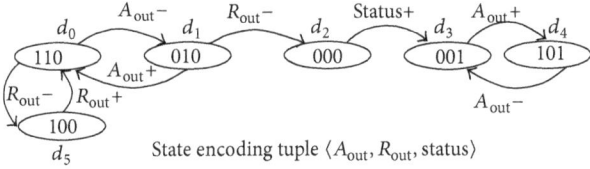

FIGURE 14: State graph for detecting FD-transition–Sl. numbers 5 of Table 2.

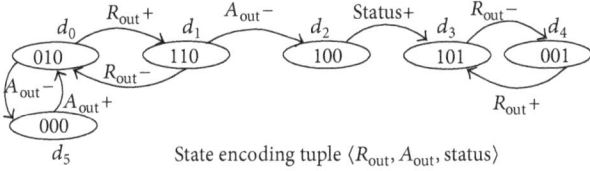

FIGURE 15: State graph for detecting FD-transitions–Sl. numbers 6 and 7 of Table 2.

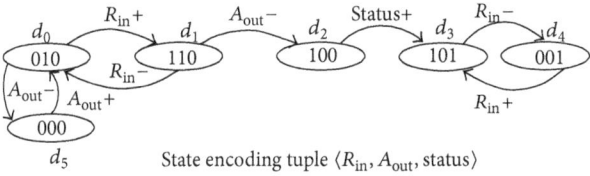

FIGURE 16: State graph for detecting FD-transitions–Sl. numbers 8 and 9 of Table 2.

Input. $\mathfrak{I}_{FD}$ is set of FD-transitions.

Output. Detectors for determining occurrence of FD-transitions are as follows:

(1) Partition $\mathfrak{I}_{FD}$ into equivalence classes. Let $\mathfrak{I}1_{FD}$, $\mathfrak{I}2_{FD}, \ldots, \mathfrak{I}l_{FD}$ be the sets generated.

(2) For each of these sets $(\mathfrak{I}i_{FD}, (1 \leq i \leq l))$, generate a detector state graph using Step (3) to Step (10).

(3) s is the signal changed by $\tau' \in \mathfrak{I}i_{FD}$ and y is any signal whose value is the same in initial states of all $\tau' \in \mathfrak{I}i_{FD}$. Further, signal y is different in the initial state of the corresponding normal transition which also makes the same change in s.

(4) Let state encoding of the detector be the tuple $\langle y, s, Status \rangle$.

(5) Create the initial state d_0. The values of the variables in d_0 are as follows: (i) y in the tuple for d_0 is complement of the value of the variable y in $initial(\tau')$, $\tau' \in \mathfrak{I}i_{FD}$, (ii) s in the tuple for d_0 is complement of the value of the variable s after its change by $\tau' \in \mathfrak{I}i_{FD}$, and (iii) *Status* in d_0 is 0.

(6) Create state d_1, with transition from d_0 to d_1 labeled as $y+$ $(y-)$ if value of y in d_0 is 0 (1). Also, create a transition from d_1 to d_0 labeled with inverse of the signal change as in transition from d_0 to d_1. Accordingly, encode state d_1.

(7) Create state d_2, with transition from d_1 to d_2 labeled as $s+$ $(s-)$ if value of s in d_1 is 0 (1). Accordingly, encode state d_2.

(8) Create state d_3, with transition from d_2 to d_3 labeled as *Status*+. Accordingly, encode state d_3.

(9) Create state d_4, with transition from d_3 to d_4 labeled as $y+$ $(y-)$ if transition from d_0 to d_1 is $y-$ $(y+)$. Add a transition from d_4 to d_3 with inverse of the signal change as in transition from d_3 to d_4. Accordingly, encode state d_4.

(10) Create state d_5, with transition from d_0 to d_5 labeled as $s+$ $(s-)$ if transition from d_1 to d_2 is $s+$ $(s-)$. Add a transition from d_5 to d_0 with inverse of the signal change as in transition from d_0 to d_5. Accordingly, encode state d_4.

4.1. Circuit Synthesis for DES Detector. It is clear from the construction of the state graphs of detectors that they have complete state coding [27, 28] and are live. So they can be synthesized as SI circuits using C-elements and logic gates by applying standard asynchronous circuit synthesis procedures [29, 30].

Figure 17 illustrates some snapshots regarding the steps of synthesizing the state graph of the detector shown in Figure 14 using CAD tool Petrify [31]; Figure 17(a) is the description of the state graph that is input to Petrify, Figure 17(b) illustrates the output of Petrify showing CSC and no liveness issues, and Figure 17(c) shows the equations obtained from Petrify. The circuit schematic (of the DES detector) that is synthesized for this state is shown in Figure 18.

Now, we explain some details of the Petrify equations. *INORDER* is a keyword of Petrify to represent all I/O signals of the corresponding state graph. *OUTORDER* is the keyword to represent the output signals of the state graph. Each subsequent line (denoted by $[0, 1] \ldots$) represents a gate of the circuit in terms of the function it implements. In case of the circuit of Figure 18, $INORDER = A_{out}, R_{out}, Status$ and $OUTORDER = Status$. The equations $[0] = A'_{out} R'_{out}$ and $[1] = 0$ represent the logic expressions of the internal Gate 4 and Gate 3, respectively. The equation $[Status] = [1]'([0] + Status) + Status[0]$ represents the output of the circuit.

In similar way, all state graphs for the FD-transitions have been synthesized into different circuits. Then, the final DES detector circuit for CUT is constructed by simply ORing

```
.model FN.g
.inputs  Aout Rout
.outputs Status
.state graph
d0_110 Rout- d5_100
d5_100 Rout+ d0_110
d0_110 Aout- d1_010
d1_010 Aout+ d0_110
d1_010 Rout- d2_000
d2_000 Status+ d3_001
d3_001 Aout+ d4_101
d4_101 Aout- d3_001
.marking {d0_110}
.end
```

(a) State graph

```
pradeep@pradeep-desktop:~/Desktop/asy work/examples$ ./petrify FN.sg -gc -eqn FN.
eqn
The STG has CSC.
# File generated by ./petrify 4.2 (compiled 15-Oct-03 at 3:06 PM)
# from <FN.sg> on 25-Apr-14 at 1:09 PM
# CPU time: 0.01 sec -- Host: pradeep-desktop
#   0.00(trav)+0.00(init)+0.00(min)+0.00(enc)+0.00(CSC)+0.00(map)+0.00(regs)+0.0
0(irred)
# The original TS had (before/after minimization) 6/6 states
# Original STG:     9 places,     8 transitions,   13 arcs ( 5 pt +  5 tp +  3 tt)
# Current STG:      5 places,     7 transitions,   20 arcs ( 10 pt +  9 tp +  1 tt)
# It is a Petri net with 2 self-loop places
.model FN
.inputs  Aout Rout
.outputs Status
.graph
Rout- p2
Rout+/1 p0 p3
Aout-/2 p3 p1
Aout+/3 p0 p3
Aout-/7 p4
Status+/5 p4
Aout+/6 Aout-/7
p0 Rout+/1 Aout-/2
p1 Aout+/3 Status+/5
p2 Rout+/1 Status+/5
p3 Rout- Aout-/2 Aout+/3
p4 Aout+/6
.marking { p0 p3 }
.end
pradeep@pradeep-desktop:~/Desktop/asy work/examples$ ▮
```

(b) Output of Petrify

```
# EQN file for model FN
# Generated by ./petrify 4.2 (compiled 15-Oct-03 at 3:06 PM)
# Outputs between brackets "[out]" indicate a feedback to input "out"
# Estimated area = 7.00

INORDER = Aout Rout Status;
OUTORDER = [Status];
[0] = Aout' Rout';
[1] = -0-;
[Status] = [1]' ([0] + Status) + Status [0];      # mappable onto gC
# The initial state is unstable. No reset information generated.
# Signal [1] enabled in the initial state.
```

(c) Output equations of Petrify

FIGURE 17: Screenshot showing the synthesis of DES detector from state graph using Petrify.

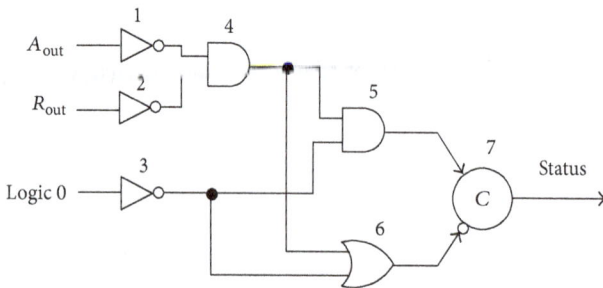

FIGURE 18: DES detector circuit for state graph shown in Figure 14.

the outputs of these circuits. The output of the detector circuit becomes high when output of at least one individual detector becomes high, thereby detecting the fault.

5. Experimental Evaluation

To validate the efficacy of the scheme, we analyze the area overhead ratio of the online tester circuit to that of the circuit under test. Further, we also compare the overhead with other techniques reported in the literature. In our experiments, we have considered some standard SI benchmark circuits [32]. Further, for comparison, we have also implemented our scheme on the circuits used in [5].

The algorithm discussed in previous section is used to design a CAD tool OLT-ASYN which generates detectors for

OLT given an asynchronous circuit specification. The design procedure includes the following steps.

(i) First, the behavior description of the SI circuit is represented using STG.

(ii) The STG representation of the SI circuit is converted into its corresponding state graph using Petrify.

(iii) Using Petrify, the state graphs are implemented as SI circuits using generalized C-element.

(iv) s-a-0 and s-a-1 faults are inserted in all the nets of the gates and stuck-on and stuck-off faults are inserted in the transistors of the C-elements of the circuit (one at a time) and the corresponding faulty state graphs are generated.

(v) FD-transitions are generated for all the faults using the DES theory.

(vi) Using Algorithm 9 state graphs of the detector(s) are generated.

(vii) The state graphs are implemented as SI circuits using generalized C-element implementation using Petrify.

Area overhead ratio of the online tester circuit to that of the circuit under test is computed using the formula

Area overhead ratio

$$= \frac{\text{Combined area of the Detector circuits}}{\text{Area of the CUT}}. \quad (1)$$

Area of gates and C-elements are considered in terms of number of transistors used in their CMOS implementation.

TABLE 3: Fault coverage, area overhead ratio, and execution time for the online detector designed using the proposed approach.

CUT	Number of gates	Number of faults	Fault coverage (%)	Area overhead	Execution (sec)*
Circuit 1	3	22	95	0.8	0.12
Circuit 2	5	32	96	1.5	0.23
Circuit 3	5	54	98	1.2	0.34
2 David cells [5]	6	44	94	1.6	0.22
4 David cells [5]	12	88	93	1	0.40
chu172	9	47	96	0.24	0.36
alloc-outbound	15	130	98	0.13	0.42
sbuf-read-ctl	19	152	95	0.12	0.51
sbuf-send-ctl	18	140	94	0.087	0.47
ram-read-sbuf	23	197	93	0.025	0.56

*Executed in AMD Phenom IIX3 710 Processor with 4 GB RAM in Linux OS.

For example, a two-input NAND gate has four transistors (two PMOS and two NMOS). Fault coverage is calculated as ratio of number of faults detected by the tester to the total number of faults.

Table 3 shows the number of gates, number of faults, fault coverage, area overhead ratio, and execution time of the proposed approach for the different SI circuits being considered.

The first three circuits in the table are simple examples whose gate level designs are shown in Figures 19, 20, and 21. The fourth and fifth circuits have been used in [5]. The others are standard asynchronous benchmarks [32] which are complex in terms of area, states, and signal compared to first five circuits in the table.

Broadly speaking, it can be observed from Table 3 that area overhead decreases with the increase of the size of the circuit. In [12], Drineas and Makris have identified that the area overhead ratio for partial replication based OLT for stuck-at faults is approximately $\alpha + 1/\beta$, where α is the fraction of test patterns incorporated in detector design ($\alpha = 1$ when all FD-transitions are incorporated in detector design) and β is the number of state bits required for circuit representation (i.e., proportion to circuit size). In this work, we have taken all possible FD-transitions and the obtained area overhead ratio acts (approximately) in accordance with the fact mentioned above.

From the discussion in the last section regarding design of the detector from the FD-transitions, it may appear that a large number of such detectors may be required for complex circuits. In the worst case, the number of detectors may be equal to the number of FD-transitions. Further, in case of large circuits as the number of nets and C-elements are high the number of FD-transitions may be proportionally large. However, interestingly, the experiments illustrated reverse trends. Large circuits have larger number of possible stuck-at faults; however, many of them are mapped to similar effects and hence the same FD-transitions; this can be observed from Table 1 for the running example. Further, using the principle of compatible of FD-transitions, it was found that multiple FD-transitions fall in the same clusters thereby resulting in the fact that a single detector suffices for more than one FD-transition. To conclude, it was observed that a few detectors

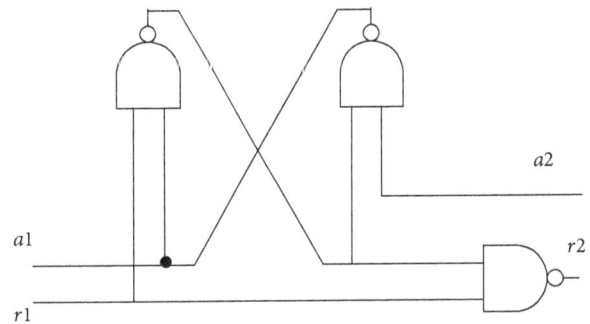

FIGURE 19: Circuit 1.

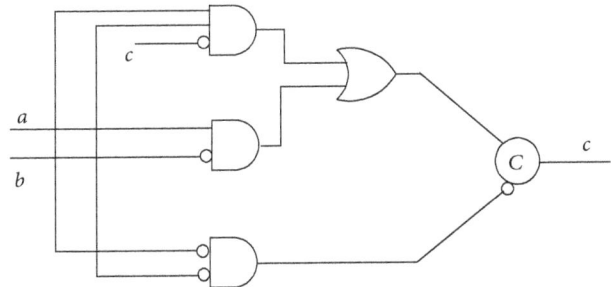

FIGURE 20: Circuit 2.

are actually required to cover all the faults. To the best of our knowledge, such facts regarding OLT of asynchronous circuits were not reported in the literature.

It may be noted that percentage of fault coverage is more than 95% in average. The number of faults that could not be detected was found to be redundant.

5.1. Mutex Approach to Testing. Now, we will discuss, in brief, the Mutex approach for online testing proposed in [5] and compare its area overhead ratio with our scheme. In [5], the scheme is demonstrated on the following specification of a handshaking protocol:

$$req+ \longrightarrow ack+ \longrightarrow req- \longrightarrow ack-. \qquad (2)$$

Online testing is performed using checkers which verify that sequencing of the signals as per the protocol is maintained;

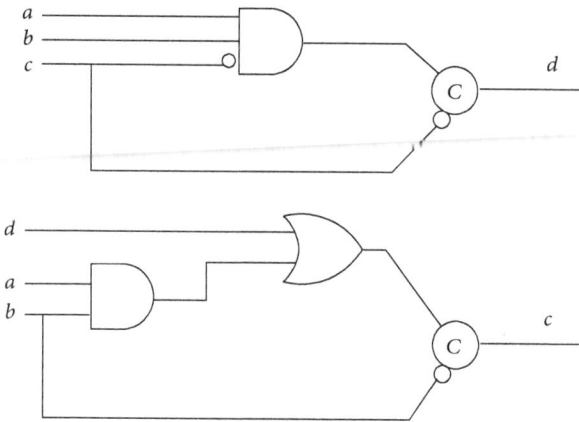

FIGURE 21: Circuit 3.

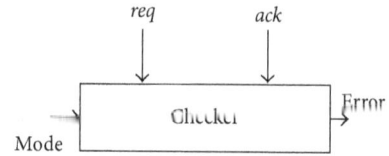

FIGURE 22: Block diagram of the checker.

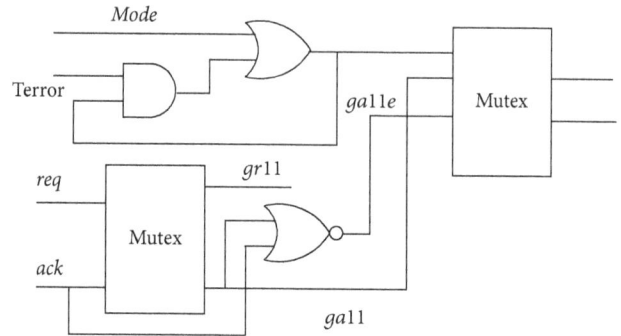

FIGURE 23: A part of the checker circuit [5].

that is, there is no premature or late firing of signals, no signal is stuck-at-0/stuck-at-1 fault, and so forth. All the protocol signals are used as the input for the checker. The checker has two functionalities, namely, self-checking and online testing of the handshaking protocol. The signal "mode" decides this selection. The block diagram of the checker is given in Figure 22. A part of the checker circuit is shown in Figure 23 which will be used to present the basic working of the checker. The full circuit diagram and functionality can be found in [5]. As shown in Figure 23, there are two Mutex components; one is used for the arbitration between $req-$ and $ack-$, while the other is used to arbitrate between $mode+$, $ga11+$, and $ga11e+$; the details of the signals $ga11+$ and $ga11e+$ are given below.

The initial state is 11 (both req and ack are high). This is guaranteed by the previous state. In the previous cycle of operation, $req+$ occurs before $ack+$, which sets $gr11 = 1$, thus providing the appropriate initial condition for the given cycle. The next signal change should be $req-$ (as per the handshaking protocol). So, once $req-$ occurs, $ga11$ signal is set to 1 (left Mutex), as ack signal is still high. As a result, the checker goes to next state 01 indicating no errors. If there is an error in the protocol under test, then $ack-$ precedes $req-$, and $ga11e$ is set to 1. This moves the checker along the fault branch.

When the mode signal is set to logic 1, the three-input arbiter (right Mutex in Figure 23) is used to arbitrate $mode+$, $ga11+$, and $ga11e+$. This Mutex is used for self-testing of the checker.

The asynchronous circuit under test (which realizes the above handshaking protocol) is implemented using David cells [5]. The partial checker circuit illustrated in Figure 23 basically tests the handshake protocol between a pair of David cells and performs self-checking. Along with the partial circuit shown in Figure 23, there are four David cells (not shown), which together test the handshaking protocol (between a pair of David cells) and self-testing of the checker. Among the two Mutex blocks and four David cells, half of them are used for handshaking protocol and the other half is used for self-testing. The logic gates are shared resources for both types of testing. In [5], both the circuit under test and

the checker circuity are implemented using David cells, using the flow of converting a Petri net model to asynchronous circuits based on David cells [33].

5.2. Comparison with the Mutex Approach. The proposed DES based approach for online testing does not involve self-testing of the detector. So, for comparison of area overhead of the proposed scheme with [5], we require only half of the resources used in [5]. Table 4 shows area overhead ratio of the checker (for online testing only) for two circuits implementing handshaking protocols involving two and four David cells, respectively. The table also reports the number of David cells, Mutex elements, and logic gates involved in the checker (for online testing). From Tables 4 and 3 (fourth and fifth circuit), we can deduce that the area overhead requirement for the Mutex method is higher compared to that of the purposed scheme. The advantages of the proposed method over the Mutex approach are as follows:

(1) The area overhead for the online tester circuit is less as compared to that of the Mutex approach by about 50%.

(2) There is flexibility to trade off area overhead, by reducing fault coverage depending upon the testability requirements. Such flexibility is not easy to be achieved by the Mutex approach. It may be noted that the proposed scheme verifies that there are no stuck-at faults while the Mutex approach verifies a protocol. In particular, it checks for the correct sequencing of the outputs. Fault coverage can be easily traded off with area overhead in our approach, while it is difficult to achieve something like "partial verification of protocol," avoiding checking certain incomplete output sequences.

TABLE 4: Area ratio for the Mutex approach.

CUT	Circuits for OLT	Area overhead ratio for Mutex method
2 David cells	2 David cells + 2 Mutex elements + gates	3.3
4 David cells	4 David cells + 4 Mutex elements + gates	4.9

(3) In the detector of the proposed scheme, there is no dependency on Mutex elements. The Mutex element itself can undergo a metastable state which needs to be handled by the metastability detector, adding to area overhead.

In this work, we could not compare fault coverage of our approach with the Mutex approach. The Mutex approach verifies online whether the output of the CUT follows the specified handshaking protocol. So, the Mutex approach basically follows functional testing. The scheme proposed in our paper works on structural testing and hence fault coverage can be given, while it is not possible for functional testing.

It may be noted that the circuits considered in the Mutex-based OLT scheme [5] were simple. For comparison with our scheme, we have done manual implementation of Mutex-based tester design on those circuits.

6. Conclusion

In the present paper, a method for the online testing of SI circuits has been proposed. We start by obtaining the STG of the SI circuit under test using the tool Petrify. After that, effects of stuck-at faults were modeled in the STGs. The normal and faulty STGs were transformed into state graphs and a DES detector was designed. The detector is capable of determining, online, whether any of the modeled stuck-at faults have occurred in the circuit. Finally, the detector is synthesized as a SI circuit with C-elements which is to be placed on chip. Several circuits were considered as case study and area overhead ratio of the detector was studied for these circuits. Results illustrated that area requirement of the detector of the proposed scheme is less than that of the Mutex approach [5] by about 50% on the average. Apart from this, there are several other advantages of the proposed approach, namely, independence of circuit functionality, nonintrusiveness, liveness, and CSC of the detector to ensure synthesizability and so forth.

The present scheme is applicable only for SI circuits with dynamic implementation of C-elements. As a further direction of research, this technique can be extended for other types of asynchronous circuits like Delay Insensitive (DI) circuits. It is required to verify whether this proposed DES based OLT scheme can be directly applied for DI circuits or some modifications would be needed.

Also, in case of SI circuits, the present scheme can be extended for static C-elements. OLT of static C-elements may be comparatively more complex than that of dynamic C-element because the dynamic C-element comprises less number of transistors than that of static C-element. Additionally, in the present work, it is found that transistor stuck-off or stuck-on faults lead to premature/nonoccurrence of a transition in the STG. Whether similar fault manifestation remains for static C-elements or there is a change needs to be verified. Clearly further research is required to solve these issues.

Conflict of Interests

The authors declare that there is no conflict of interests regarding the publication of this paper.

References

[1] M. Nicolaidis and Y. Zorian, "On-line testing for VLSI—a compendium of approaches," Journal of Electronic Testing: Theory and Applications, vol. 12, no. 1 2, pp. 7–20, 1998.

[2] H. Hulgaard, S. M. Burns, and G. Borriello, "Testing asynchronous circuits: a survey," Integration, the VLSI Journal, vol. 19, no. 3, pp. 111–131, 1995.

[3] M. Nicolaidis, "Design for soft-error robustness to rescue deep submicron scaling," in Proceedings of the IEEE International Test Conference, pp. 1140–1149, IEEE, Washington, DC, USA, October 1998.

[4] D. Shang, A. Yakovlev, F. P. Burnsand, F. Xia, and A. Bystrov, "Low-cost online testing of asynchronous handshakes," in Proceedings of the 11th IEEE European Test Symposium (ETS '06), pp. 225–232, IEEE, Southampton, UK, May 2006.

[5] D. Shang, A. Bystrov, A. Yakovlev, and D. Koppad, "On-line testing of globally asynchronous circuits," in Proceedings of the 11th IEEE International On-Line Testing Symposium (IOLTS '05), pp. 135–140, July 2005.

[6] K. De, C. Natarajan, D. Nair, and P. Banerjee, "RSYN: a system for automated synthesis of reliable multilevel circuits," IEEE Transactions on Very Large Scale Integration Systems, vol. 2, no. 2, pp. 186–195, 1994.

[7] W.-F. Chang and C.-W. Wu, "Low-cost modular totally self-checking checker design for m-out-of-n code," IEEE Transactions on Computers, vol. 48, no. 8, pp. 815–826, 1999.

[8] R. Leveugle and G. Saucier, "Concurrent checking in dedicated controllers," in Proceedings of the IEEE International Conference on Computer Design, pp. 124–127, October 1989.

[9] R. Leveugle and G. Saucier, "Optimized synthesis of concurrently checked controllers," IEEE Transactions on Computers, vol. 39, no. 4, pp. 419–425, 1990.

[10] T. Verdel and Y. Makris, "Duplication-based concurrent error detection in asynchronous circuits: shortcomings and remedies," in Proceedings of the 17th IEEE International Symposium on Defect and Fault Tolerance in VLSI Systems, pp. 345–353, IEEE, 2002.

[11] P. Drineas and Y. Makris, "Non-intrusive design of concurrently self-testable FSMs," in Proceedings of the 11th Asian Test Symposium (ATS '02)., pp. 33–38, November 2002.

[12] P. Drineas and Y. Makris, "Selective partial replication for concurrent fault-detection in FSMs," IEEE Transactions on Instrumentation and Measurement, vol. 52, no. 6, pp. 1729–1737, 2003.

[13] Y. Balasubrahamanyam, G. L. Chowdary, and T. J. V. S. Subrahmanyam, "A novel low power pattern generation technique

for concurrent BIST architecture," *International Journal of Computer Technology and Applications*, vol. 3, no. 2, pp. 561–565, 2012.

[14] P. Daniel and R. Chandel, "Dynamic self programming architecture for concurrent fault detection," *International Journal of Research in IT, Management and Engineering*, vol. 2, no. 12, pp. 67–81, 2012.

[15] I. Voyiatzis, A. Paschalis, D. Gizopoulos, C. Halatsis, F. S. Makri, and M. Hatzimihail, "An input vector monitoring concurrent BIST architecture based on a precomputed test set," *IEEE Transactions on Computers*, vol. 57, no. 8, pp. 1012–1022, 2008.

[16] V. Gherman, J. Massas, S. Evain, S. Chevobbe, and Y. Bonhomme, "Error prediction based on concurrent self-test and reduced slack time," in *Proceedings of the 14th Design, Automation and Test in Europe Conference and Exhibition (DATE '11)*, pp. 1626–1631, March 2011.

[17] M. Sampath, R. Sengupta, S. Lafortune, K. Sinnamohideen, and D. Teneketzis, "Diagnosability of discrete-event systems," *IEEE Transactions on Automatic Control*, vol. 40, no. 9, pp. 1555–1575, 1995.

[18] S. Biswas, S. Mukhopadhyay, and A. Patra, "A formal approach to on-line monitoring of digital VLSI circuits: theory, design and implementation," *Journal of Electronic Testing*, vol. 21, no. 5, pp. 503–537, 2005.

[19] S. Biswas, G. Paul, and S. Mukhopadhyay, "Methodology for low power design of on-line testers for digital circuits," *International Journal of Electronics*, vol. 95, no. 8, pp. 785–797, 2008.

[20] C. J. Myers, *Asynchronous Circuit Design*, John Wiley & Sons, New York, NY, USA, 2001.

[21] J. Cortadella, M. Kishinevsky, A. Kondratyev, L. Lavagno, and A. Yakovlev, "Hardware and petri nets: application to asynchronous circuit design," in *Application and Theory of Petri Nets 2000*, vol. 1825 of *Lecture Notes in Computer Science*, pp. 1–15, 2000.

[22] D. Lu and C. Q. Tong, "High level fault modeling of asynchronous circuits," in *Proceedings of the 13th IEEE VLSI Test Symposium*, pp. 190–195, May 1995.

[23] M. Shams, J. C. Ebergen, and M. I. Elmasry, "Modeling and comparing CMOS implementations of the C-element," *IEEE Transactions on Very Large Scale Integration (VLSI) Systems*, vol. 6, no. 4, pp. 563–567, 1998.

[24] M. Shams, J. C. Ebergen, and M. I. Elmasry, "A comparison of CMOS implementations of an asynchronous circuits primitive: the C-element," in *Proceedings of the International Symposium on Low Power Electronics and Design (ISLPED '96)*, pp. 93–96, Monterey, Calif, USA, August 1996.

[25] M. Moreira, B. Oliveira, F. Moraes, and N. Calazans, "Impact of C-elements in asynchronous circuits," in *Proceedings of the 13th International Symposium on Quality Electronic Design (ISQED '12)*, pp. 437–443, March 2012.

[26] M. T. Moreira, F. G. Moraes, and N. L. Calazans, "Beware the Dynamic C-Element," *IEEE Transactions on Very Large Scale Integration (VLSI) Systems*, vol. 22, no. 7, pp. 1644–1647, 2014.

[27] C. N. Liu, "A state variable assignment method for asynchronous sequential switching circuits," *Journal of the ACM*, vol. 10, no. 2, pp. 209–216, 1963.

[28] J. Cortadella, M. Kishinevsky, A. Kondratyev, L. Lavagno, and A. Yakovlev, "A region-based theory for state assignment in speed-independent circuits," *IEEE Transactions on Computer-Aided Design of Integrated Circuits and Systems*, vol. 16, no. 8, pp. 793–812, 1997.

[29] J. Gu and R. Puri, "Asynchronous circuit synthesis with Boolean satisfiability," *IEEE Transactions on Computer-Aided Design of Integrated Circuits and Systems*, vol. 14, no. 8, pp. 961–973, 1995.

[30] T. Chu, "Automatic synthesis and verification of hazard-free control circuits from asynchronous finite state machine specifications," in *Proceedings of the IEEE International Conference on Computer Design: VLSI in Computers and Processors*, pp. 407–413, Cambridge, Mass, USA, 1992.

[31] J. Cortadella, M. Kishinevsky, A. Kondratev, L. Lavagno, and A. Yakovlev, "Petrify: a tool for manipulating concurrent specifications and synthesis of asynchronous controllers," *IEICE Transactions on Information and Systems*, vol. E80-D, no. 3, pp. 315–325, 1997.

[32] Myers Research Group, *Atacs Online Demo*, Myers Research Group, 1999, http://www.async.ece.utah.edu/.

[33] D. Shang, F. Xia, and A. Yakovlev, "Asynchronous circuit synthesis via direct translation," in *Proceedings of the IEEE International Symposium on Circuits and Systems (ISCAS '02)*, vol. 3, pp. 369–372, Phoenix-Scottsdale, Ariz, USA, May 2002.

## 23

High-Efficient Circuits for Ternary Addition

Reza Faghih Mirzaee,[1] **Keivan Navi,**[2] **and Nader Bagherzadeh**[3]

[1] *Department of Computer Engineering, Islamic Azad University, Shahr-e-Qods Branch, Tehran 37541-374, Iran*
[2] *Faculty of Electrical and Computer Engineering, Shahid Beheshti University, G.C., Tehran 1983963113, Iran*
[3] *Department of Electrical and Computer Engineering, Center for Pervasive Communications and Computing (CPCC),*
  University of California, Irvine, CA 92697, USA

Correspondence should be addressed to Reza Faghih Mirzaee; r.faghih@shahryariau.ac.ir

Academic Editor: Mohamed Masmoudi

New ternary adders, which are fundamental components of ternary addition, are presented in this paper. They are on the basis of a logic style which mostly generates binary signals. Therefore, static power dissipation reaches its minimum extent. Extensive different analyses are carried out to examine how efficient the new designs are. For instance, the ternary ripple adder constructed by the proposed ternary half and full adders consumes 2.33 μW less power than the one implemented by the previous adder cells. It is almost twice faster as well. Due to their unique superior characteristics for ternary circuitry, carbon nanotube field-effect transistors are used to form the novel circuits, which are entirely suitable for practical applications.

1. Introduction

On-chip interconnections have become a serious challenge as more and more modules are packed into a chip. They dissipate lots of energy, increase response time, and cause coupling effects by adding more capacitance, resistance, and inductance to a circuit [1]. Multiple-valued logic (MVL) is an alternative solution to interconnect complexity and growing power dissipated by wires [2]. It reduces the amount of wires inside and outside a chip dramatically as more complex designs require a large number of wires for connecting circuit components. In addition, MVL has the high potential for increasing computational speed, reducing switching activity, and implementing many arithmetic and logic functions in a single chip [2, 3]. Among many MVL systems, ternary logic (also known as three-valued logic) has soared in popularity due to its simplicity and efficiency [4, 5].

In spite of potential superiorities of ternary logic, binary is still the dominant logic for circuit design in the industry. One of the main reasons is the intrinsic behaviour of transistors. The on-off characteristic of a transistor makes it an ideal device to implement Boolean algebra. However, dualism does not correspond to real-world applications effectively.

Another reason why ternary logic is not as popular as its binary counterpart is mainly because of the lack of sufficient practical, high-performance logic gates and computational components. To make ternary logic applicable in practice, efficient circuits must be developed before all else.

Voltage-mode MVL circuits are based on multithreshold designs [6, 7]. Therefore, traditional metal-oxide-semiconductor field-effect transistor (MOSFET) is not entirely suitable candidate for MVL implementation due to the fact that MOS devices are inherently single-threshold [8]. Since the introduction of new nanoscale devices such as quantum-dot cellular automata (QCA) and carbon nanotube field-effect transistor (CNTFET), many worthwhile endeavours have been made to present novel ternary circuits with high efficiency. The unique characteristic which makes CNTFET technology highly appropriate for ternary circuitry is the ability of adjusting threshold voltage by altering the diameter of CNTs under the gate terminal [9]. The tuneable threshold voltage brings essential flexibility which is a great necessity for ternary designs. Furthermore, CNTFETs operate far faster and even consume less power in comparison with traditional MOS devices [10, 11]. Although commercial CNTFET chips are not ready yet, many valuable achievements have been made so far. The implementation of CNTFET-based logic gates has been reported in [12, 13]. In addition, the first carbon nanotube computer has been recently developed [14].

A gate-level implementation for ternary half adder (THA) has been presented by Dhande and Ingole [15]. The main drawback is having a very large number of transistors. Lin et al. [16] have replaced some ternary gates with binary ones to reduce transistor count and decrease static power dissipation. Their design has 158 transistors (THA-158T). The final attempt is a NAND-based structure presented by Moaiyeri et al. [17]. In spite of a great reduction, it still needs 112 transistors (THA-112T). A new ternary full adder (TFA) has been presented by Ebrahimi et al. [18]. It is on the basis of two cascaded so-called THAs, in which the output carry is not produced. A carry generator subcircuit produces the final carry from the initial inputs and the output of the first pseudo-THA. The entire block requires 106 transistors (TFA-106T). Another TFA, which directly generates both outputs (Sum and C_{out}) from the input variables, has recently been presented by Keshavarzian and Sarikhani [19]. It needs 132 transistors to form the whole adder cell (TFA-132T).

In this paper, new ternary adders are presented on the basis of a logic style where a large portion is founded upon binary structures. As a result, static power dissipation reaches its minimum extent. The new adder cells operate very rapidly and have a reasonable number of transistors compared with the ones presented in the literature so far. Moreover, they benefit from full-swing operation, capability of working in high frequencies, and strong driving power.

Due to the inaccurate chip fabrication of CNTFET technology, diversity of using CNTs with different diameters decreases the manufacturability issue. Nevertheless, fabrication of multichirality CNTs is inevitable for ternary circuitry due to the fact that ternary circuits are based on multi-V_t designs [6, 7]. The entire novel ternary circuits are developed by CNTs with only three different diameters as it is very common in ternary logic circuitry [16–19]. The proposed designs show low sensitivity to undesired environmental and process variations.

The rest of the paper is organized as follows: Section 2 will express how we are motivated to design new circuits. The proposed ternary adders are presented in Section 3. Section 4 includes simulation results and comparisons. Eventually, Section 5 concludes the paper.

2. Motivation

Implementation of ternary logic is based on an additional voltage level in comparison with binary logic. The voltage level of $V_{dd}/2$ stands for the logic value "1" in the unbalanced ternary notation [20], whereas zero and V_{dd} voltages represent the logic values of "0" and "2," respectively. Voltage dividers such as resistors [21] or capacitors [22] are used to divide voltage. Current flows through the path established from the power supply to the ground each time voltage division occurs. A significant portion of the total power consumption in ternary circuits is static power. Although the usage of capacitors leads to less power dissipation, they provide weak current drivability.

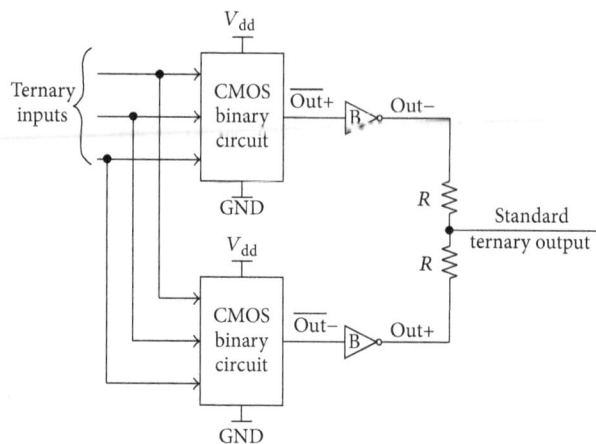

FIGURE 1: The utilized logic style for implementing new adder cells.

A great advantage of complementary-symmetry metal-oxide-semiconductor (COS-MOS, or CMOS) technology for implementing logic functions is the elimination of continuous static current in binary circuits, due to the fact that either the pull-down or the pull-up network is switched off. This is the reason why some recent works have replaced as many ternary gates as possible with binary ones [16]. A comparison between the average power consumption of THAs presented by Dhande and Ingole [15] and Lin et al. [16] demonstrates that it is more beneficial to use CMOS-based binary circuits as much as possible. The fewer times voltage division takes place, the less power dissipates. In this paper, one of the main targets is to use a logic style in which voltage division occurs as few times as possible so that the static power is reduced to its smallest amount.

The logic style directly influences delay, power consumption, and area characteristics, which are the most important parameters for performance evaluation. There are two definitions of a ternary function other than the standard one. The first (second) interpretation is negative (positive) ternary, denoted by − (+), in which the logic value "1" is replaced with "0" ("2") [23]. Therefore, they are in fact binary functions. Figure 1 illustrates the utilized logic style, which is based upon (1). Positive and negative complementary outputs ($\overline{Out+}$ and $\overline{Out-}$) are first generated. Then, binary inverters convert $\overline{Out+}$ and $\overline{Out-}$ to Out− and Out+, respectively. Finally, two transistors perform voltage division to generate STOut (1). Therefore, voltage division takes place only once for implementing a ternary function. This logic style is employed in this paper to design new ternary adders.

$$\text{STOut} = \frac{\left(\overline{\overline{Out+}}\right) + \left(\overline{\overline{Out-}}\right)}{2} = \frac{(Out-) + (Out+)}{2}. \quad (1)$$

3. New Single-Bit Ternary Adders

3.1. New Ternary Half Adder. Adder is a fundamental component for all arithmetic operations such as subtraction, multiplication, and division. Half adder is the simplest adder block which performs addition of two input signals. The

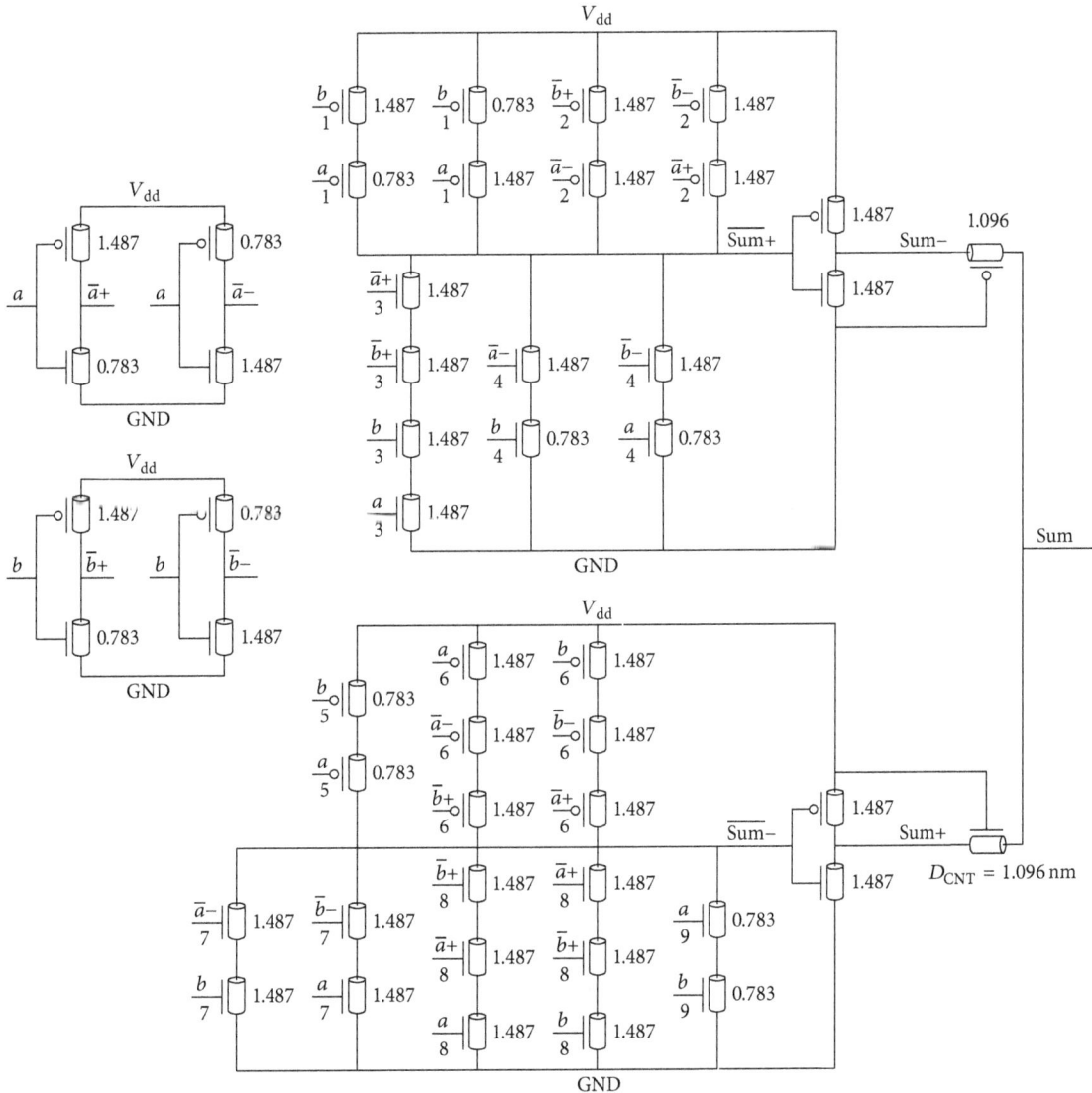

(a)

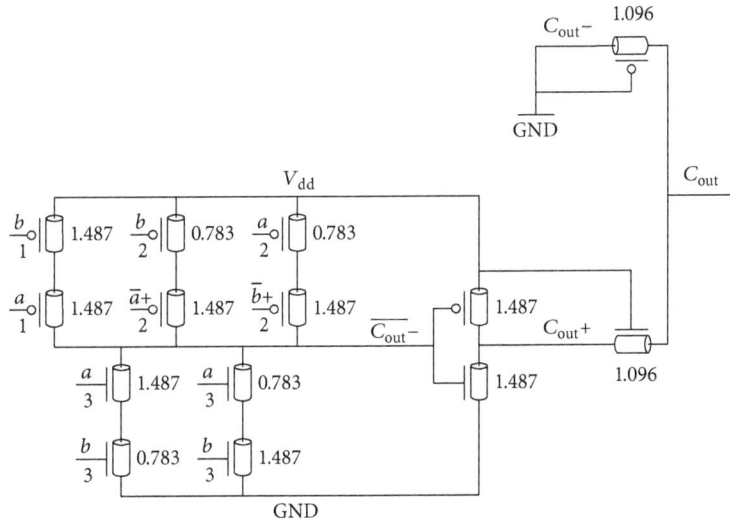

(b)

FIGURE 2: The proposed ternary half adder (#Tube = 3), (a) Sum generator subcircuit, (b) Carry generator subcircuit.

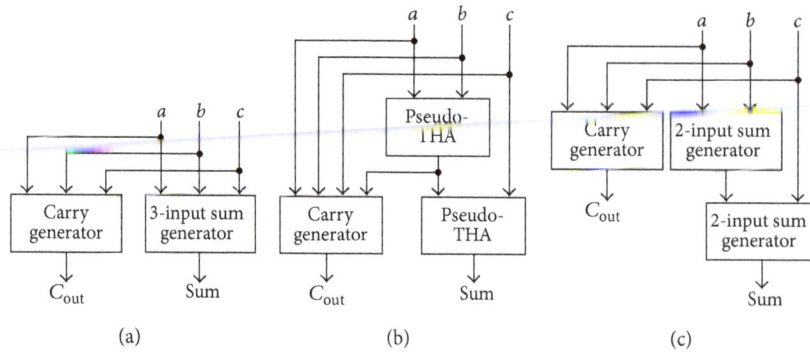

FIGURE 3: Strategies of building a ternary full adder block, (a) the block diagram of TFA presented by Keshavarzian and Sarikhani [19], (b) the block diagram of TFA presented by Ebrahimi et al. [18], and (c) the block diagram of the proposed TFA.

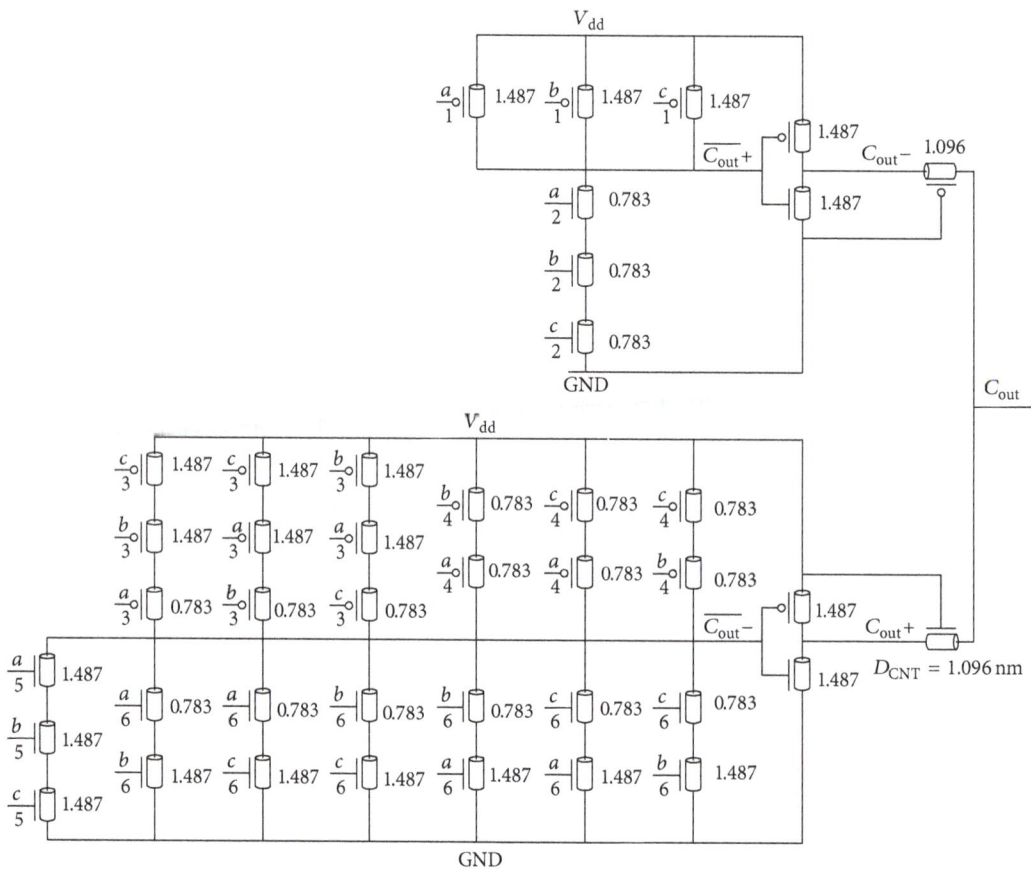

FIGURE 4: The proposed carry generator subcircuit for TFA (#Tube = 3).

proposed ternary half adder is illustrated in Figure 2, in which diameters of CNTs are indicated for each transistor. For CNTFETs with diameters of 1.489 nm, 1.096 nm, and 0.783 nm, the chirality numbers are (19, 0), (14, 0), and (10, 0), and subsequently the threshold voltages are 0.289 V, 0.392 V, and 0.549 V, respectively, ((2), (3)) [24]. The (n_1, n_2) indices are the chirality numbers which indicate wrapping vector along which a sheet of graphite is rolled up to form a carbon nanotube (CNT). These CNTs are used as the channel of the transistor. The diameter of an SWCNT can be as small as

0.4 nm [25]. The typical diameters change between 0.7 nm and 3 nm with mean diameter of 1.7 nm [26].

$$D_{CNT} = 0.0783 \times \sqrt{n_1^2 + n_2^2 + n_1 \times n_2}, \tag{2}$$

$$V_{th} \approx \frac{0.43}{D_{CNT}\,(nm)}. \tag{3}$$

The final outputs (Sum and C_{out}) are generated in a parallel manner. Two different subcircuits create $\overline{Sum+}$ and

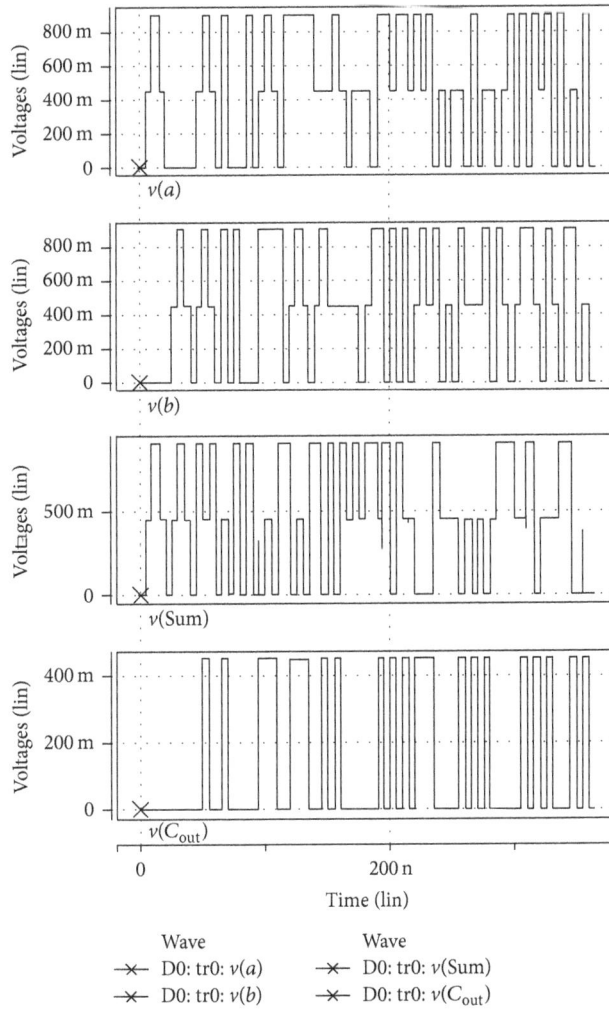

FIGURE 5: Transient response of the proposed THA.

TABLE 1: Midoutput values of the Sum generator subcircuit.

a	b	$\bar{a}+$	$\bar{a}-$	$\bar{b}+$	$\bar{b}-$	Sum	$\overline{\text{Sum}}+$	Path(s)	$\overline{\text{Sum}}-$	Path(s)
0	0	2	2	2	2	0	2	**1**	2	**5**
0	1	2	2	2	0	1	2	**1**	0	**7, 8**
0	2	2	2	0	0	2	0	**4**	0	**7**
1	0	2	0	2	2	1	2	**1**	0	**7, 8**
1	1	2	0	2	0	2	0	**3**	0	**8**
1	2	2	0	0	0	0	2	**2**	2	**6**
2	0	0	0	2	2	2	0	**4**	0	**7**
2	1	0	0	2	0	0	2	**2**	2	**6**
2	2	0	0	0	0	1	2	**2**	0	**9**

$\overline{\text{Sum}}-$ (Figure 2(a)) by taking input values shown in Table 1. The first set of transistors, which are marked with "1," connects the node $\overline{\text{Sum}}+$ to the power supply when either $(a, b) = (0, 0)$ or $(a, b) = (0, 1) \mid (1, 0)$, considering input permutations. Two other parallel paths, on which transistors are marked with "2," are supplemented in order to connect the output node to V_{dd} when either $(a, b) = (2, 2)$ or $(a, b) = (1, 2) \mid (2, 1)$. Within the pull-down network, the third path connects the node $\overline{\text{Sum}}+$ to the ground whenever $(a, b) = (1, 1)$. Finally, the fourth set of transistors is switched on when

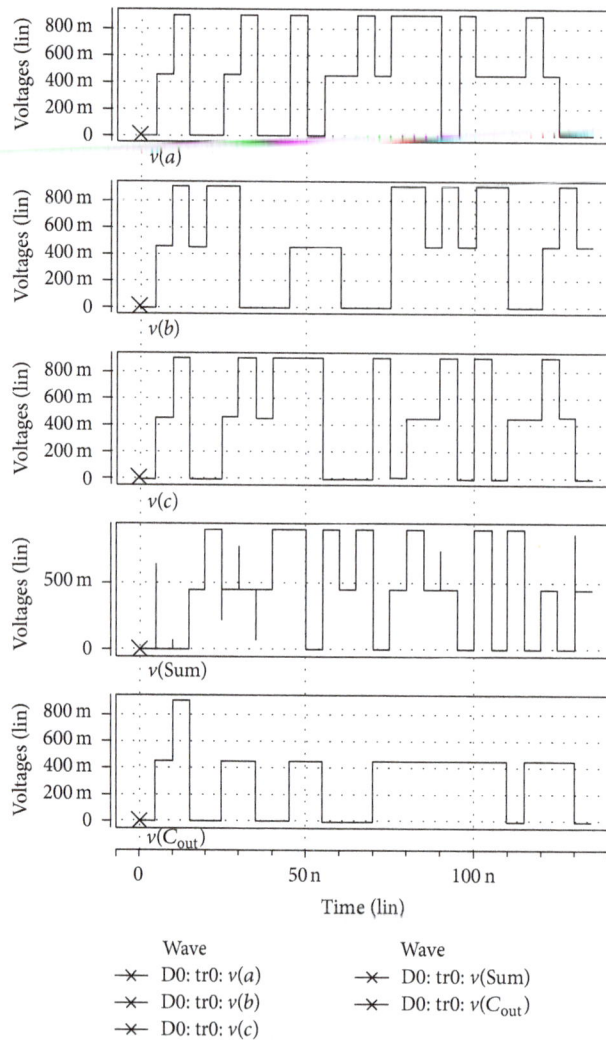

FIGURE 6: Transient response of the proposed TFA.

TABLE 2: Midoutput values of the carry generator subcircuit for THA.

a	b	$\bar{a}+$	$\bar{a}-$	$\bar{b}+$	$\bar{b}-$	C_{out}	$C_{out}-$	$\overline{C_{out}}-$	Path(s)
0	0	2	2	2	2	0	0	2	1
0	1	2	2	2	0	0	0	2	1
0	2	2	2	0	0	0	0	2	2
1	0	2	0	2	2	0	0	2	1
1	1	2	0	2	0	0	0	2	1
1	2	2	0	0	0	1	0	0	3
2	0	0	0	2	2	0	0	2	2
2	1	0	0	2	0	1	0	0	3
2	2	0	0	0	0	1	0	0	3

$(a, b) = (0, 2) \mid (2, 0)$. Table 1 summarizes the way paths connect the nodes $\overline{Sum}+$ and $\overline{Sum}-$ to the appropriate voltage source, in light of different input patterns.

The same concept leads us to the output carry generator subcircuit (Figure 2(b)). Table 2 shows which transistors set up the proper path to connect the midoutput $\overline{C_{out}}-$ to the

proper voltage source. $C_{out}-$ is always "0." Therefore, it is constantly connected to GND. PT and NT inverters (PTI and NTI) are also required to produce $\bar{a} + /\bar{b}+$ and $\bar{a} - /\bar{b}-$, respectively (Figure 2(a)). The entire block has 64 transistors, and it is mostly composed of binary parts, in which either the pull-down or pull-up network is switched off. Therefore,

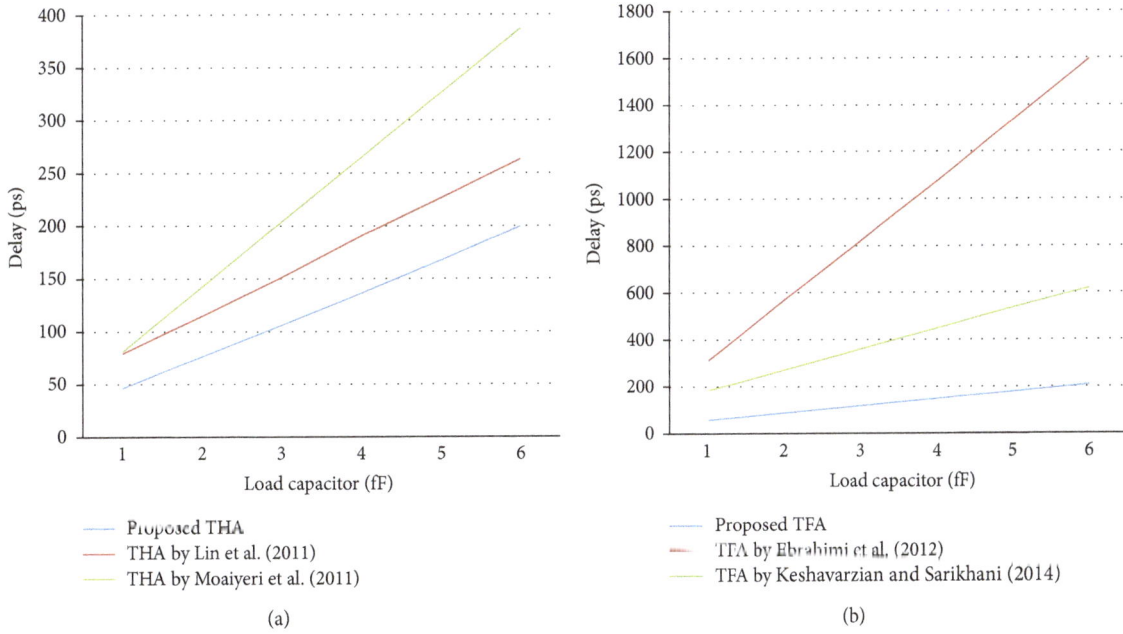

FIGURE 7: Delay versus load capacitors, (a) THAs, (b) TFAs.

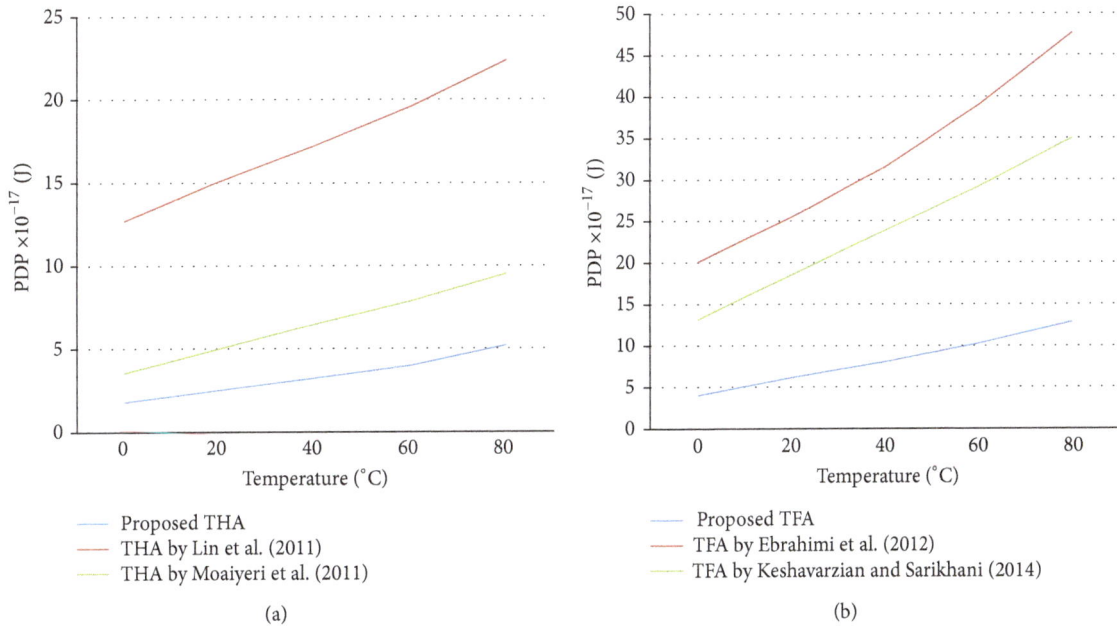

FIGURE 8: PDP versus temperature variations, (a) THAs, (b) TFAs.

static current does not flow within the subcircuits which generate midoutputs.

3.2. New Ternary Full Adder.
Full adder performs addition of three input signals. A ternary full adder, whose block diagram is depicted in Figure 3(a), has been presented by Keshavarzian and Sarikhani [19]. Both outputs (Sum and C_{out}) are directly generated from the input variables. Ebrahimi et al. [18] have proposed another TFA. Figure 3(b) reveals how the final outputs are generated within its block. The output carry is

considered as a function of the initial inputs as well as the output of the first pseudo-THA.

The output carry of the proposed TFA is produced directly from the initial inputs (Figure 3(c)). In this manner, output carry is generated far faster, which is a great advantage especially in a ripple adder structure. On the other hand, the proposed Sum generator subcircuit (Figure 2(a)) is cascaded twice to create the final output Sum. The direct approach of generating the output Sum requires multiple pass-transistors in series, which cause slow operation. This is the reason

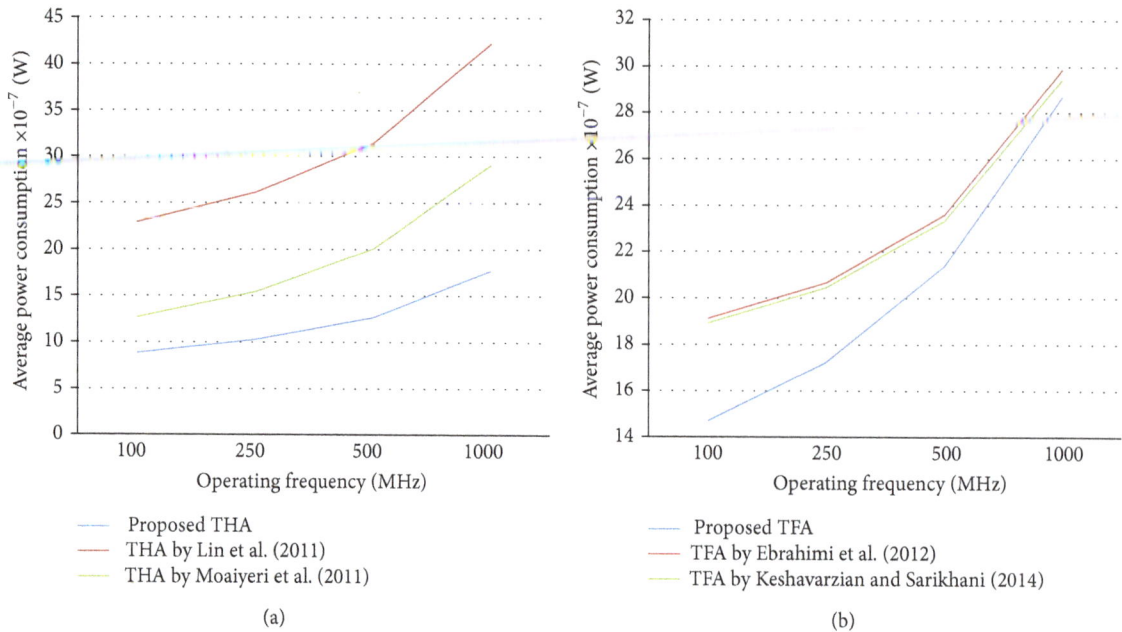

FIGURE 9: Average power consumption versus operating frequencies, (a) THAs, (b) TFAs.

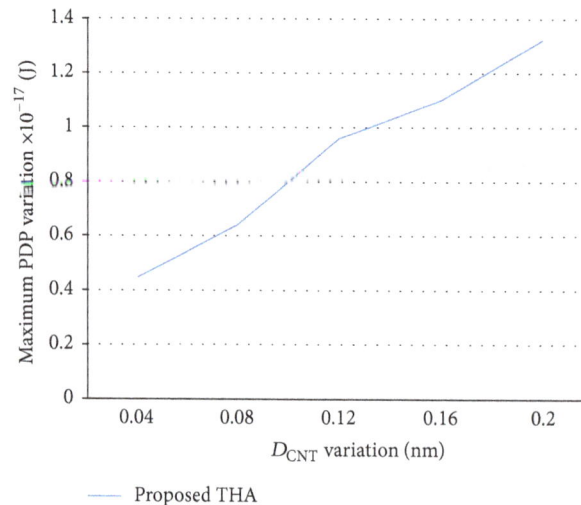

FIGURE 10: Maximum PDP variation versus D_{CNT} variation.

why Keshavarzian and Sarikhani [19] supplement a ternary buffer in order to rectify the drawback to some extent. Unlike the output Sum, C_{out} is a simple function, which is entirely appropriate to implemented directly from the initial inputs.

The proposed carry generator subcircuit for TFA is shown in Figure 4. Table 3 shows how transistors connect the midoutputs $\overline{C_{out}}+$ and $\overline{C_{out}}-$ to the proper voltage source depending on what input pattern is considered (Table 3, Second Column), regardless of its permutations. Two-input Sum generator subcircuits (Figure 2(a)) are cascaded in series to create the output Sum (Figure 3(c)). Although it is also possible to obtain the final output Sum directly from the initial inputs, the number of pass-transistors in series

increases inside the body of the subcircuit, and hence it leads to deficient overall performance. As a result, cascaded Sum generators are preferable. The entire full adder cell has 142 transistors.

4. Simulation Results

High-performance and state-of-the-art CNTFET-based designs are selected for comparison. Extensive simulation setups are taken into account to examine new adder cells in several aspects. All circuits are simulated with Synopsys HSPICE and 32 nm CNTFET technology [27, 28] in three power supply voltages (1 V, 0.9 V, and 0.8 V) at room temperature. This compact SPICE model includes all

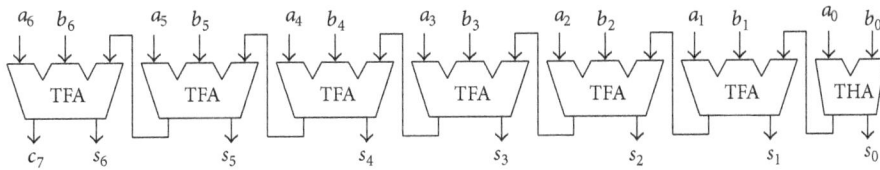

FIGURE 11: A 7-TIT ternary ripple adder constructed by a THA and six TFAs.

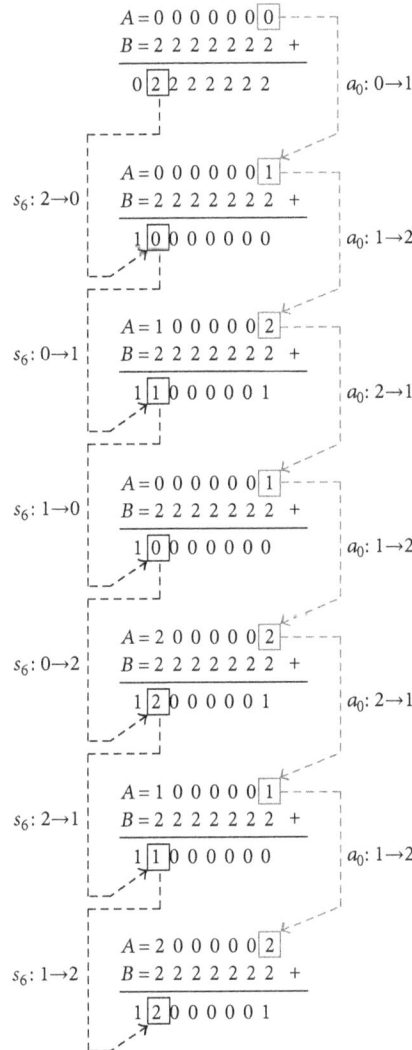

FIGURE 12: Input patterns which are fed to the ternary ripple adders to measure maximum delay.

nonidealities such as Schottky barrier effects, parasitic Drain/Source/Gate capacitances and resistances, and CNT charge Screening Effects. A brief description of the parameters of the CNTFET model, which has been designed for unipolar MOSFET-like devices with one or more CNTs, is shown in Table 4.

Fan-out of 4 ternary inverters (FO4) was employed as the output load in order to provide a realistic simulation setup. Transient responses of the proposed THA and TFA in 100 MHz operating frequency are plotted in Figures 5 and 6, respectively. Average power consumption during all transitions is also measured. Finally, power-delay product (PDP) is a balance between delay and power factors (4). Simulation results are shown in Table 5. The best results are shown in boldface for better clarification.

$$PDP = Max\left(Delay\right) \times Avg\left(Power\right). \qquad (4)$$

There are three nanotubes under the gate terminal of all transistors (#Tubes = 3). Channel length is 32 nm (L_g = 32 nm), and the distance between the centers of two adjacent CNTs under the gate of a transistor is set 20 nm (Pitch = 20 nm). Transistor width of a CNTFET can be approximately

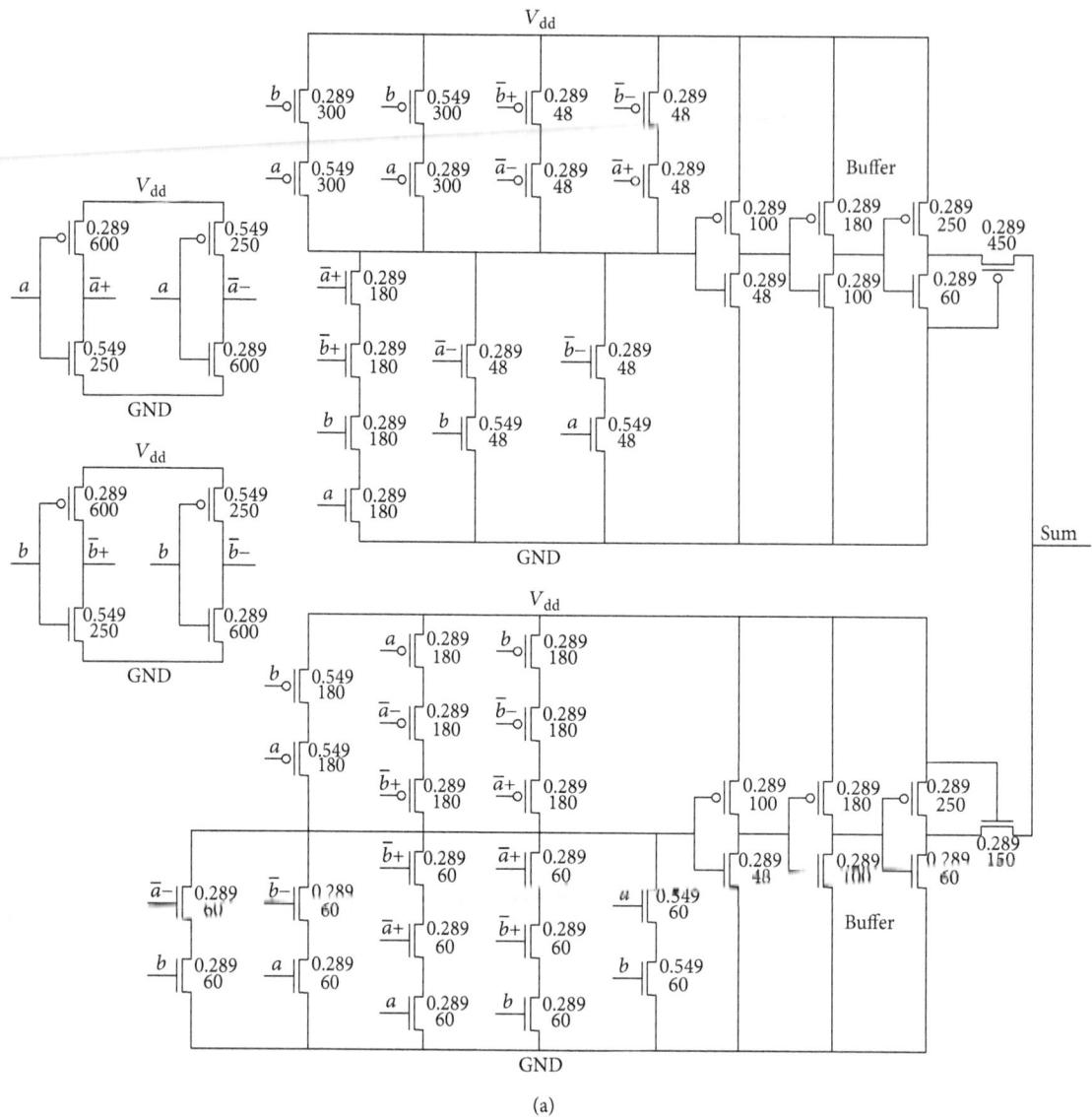

(a)

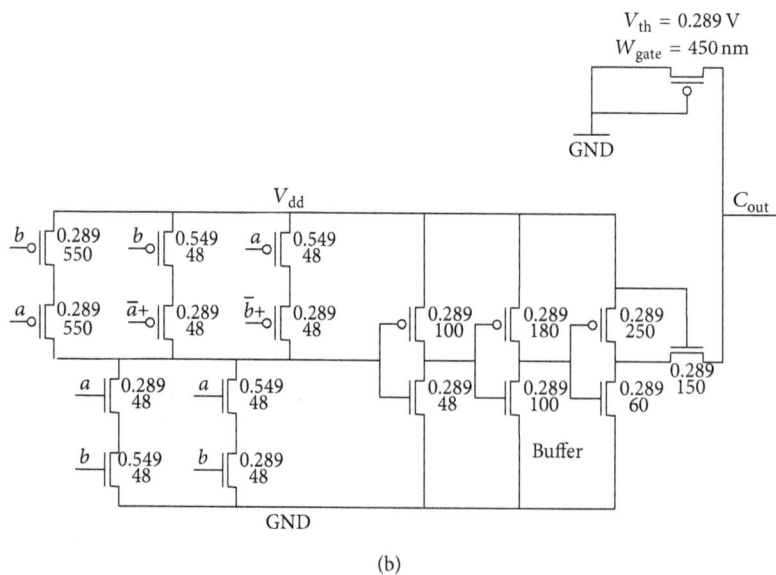

(b)

FIGURE 13: MOSFET implementation of the proposed ternary half adder.

TABLE 3: Midoutput values of the carry generator subcircuit for TFA.

$\sum$ (Inputs)	Input pattern	C_{out}	$\overline{C_{out}}+$	Path(s)	$\overline{C_{out}}-$	Path(s)
0	(0, 0, 0)	0	2	1	2	**3, 4**
1	(0, 0, 1)	0	2	1	2	**3, 4**
2	(0, 0, 2)	0	2	1	2	**4**
2	(0, 1, 1)	0	2	1	2	**3**
3	(0, 1, 2)	1	2	1	0	**6**
3	(1, 1, 1)	1	2	1	0	**5**
4	(0, 2, 2)	1	2	1	0	**6**
4	(1, 1, 2)	1	2	1	0	**5, 6**
5	(1, 2, 2)	1	2	1	0	**5, 6**
6	(2, 2, 2)	2	0	2	0	**5, 6**

TABLE 4: CNTFET model parameters.

Parameter	Description	Value
L_{ch}	Physical channel length	32 nm
L_{geff}	The mean free path in the intrinsic CNT channel	100 nm
L_{dd}/L_{ss}	The length of doped CNT drain/source-side extension region	32 nm
K_{gate}	The dielectric constant of high-k top gate	16
T_{ox}	The thickness of high-k top gate dielectric material	4 nm
C_{sub}	The coupling capacitance between the channel region and the substrate	20 pF/m
Efi	The Fermi level of the doped S/D tube	6 eV

calculated by (5) [29], where W_{min} is the minimum width of the gate. Therefore, by considering this equation, each transistor width is 60 nm. In addition to delay, power, and PDP, total number of transistors and total width of the adder cells (6) are also reported in Table 6 as reasonable criteria of area competence.

$$W_{gate} \approx Max\left(W_{min}, \#Tubes \times Pitch\right), \quad (5)$$

$$Total\ Cell\ Width = \sum_i Width\left(T_i\right). \quad (6)$$

Simulation results demonstrate the absolute superiority of the proposed circuits. The new designs consume the least power due to their unique structure which is mainly composed of binary parts. For example, the given THA consumes 1.411 μW and 0.384 μW less power in 0.9 V power supply than the designs presented by Lin et al. [16] and Moaiyeri et al. [17], respectively. Voltage division occurs only twice to create the final outputs (Sum and C_{out}), whereas it happens several times in each ternary component of previous THAs. In addition, the performance of the proposed THA is approximately twice higher than the design introduced in [17], while it has also 48 fewer transistors. Moreover, previous ternary half adders require a $V_{dd}/2$ voltage reference. It causes additional on-chip interconnection, which is in contrast with initial MVL targets. The logic style utilized in this paper eliminates the requirement of any extra voltage source.

It takes four successive pass-transistors to charge or discharge the output nodes of the design given by Ebrahimi et al. [18]. It causes poor driving power especially when the circuit faces long wires or high load capacitors. As a result, it operates very slowly. In the proposed structures,

the binary inverters situated in the middle isolate CMOS binary circuits from load capacitors, and they bring sufficient driving power for charging and discharging output loads. This is exactly what the ternary buffer in TFA-132T does. In spite of six pass-transistors in series, it operates faster than TFA presented in [18]. To examine driving capability more accurately, capacitors ranged from 1 fF up to 6 fF are applied as output loads. The results of this experiment are plotted in Figure 7. Delay parameter does not increase sharply as output capacitors enlarge, and the new structures operate efficiently despite the existence of large load capacitors.

In addition, the output value of the first pseudo-THA has to be generated first in TFA-106T to produce the output carry. The same output is only generated from the initial inputs in the new design. Therefore, it gets ready 76% faster than in the structure presented in [18]. The delay of the output carry for the given TFA in 0.9 V power supply is 22.192 psec, while the same parameter is 94.942 psec for the previous design.

To observe how temperature variation affects the performance of the proposed circuits, simulations are repeated with different ambient temperatures ranged from 0°C to 80°C. The result of this experiment is depicted in Figure 8. The performance of the proposed designs does not alter sharply in the presence of temperature fluctuations.

Since static power is a large fraction of average power consumption in ternary circuits, it is also measured separately for all designs. To measure static power dissipation, static (DC) input signals are applied to the circuits so that switching activity does not occur. This measurement must be repeated nine times for different input patterns of a 2-input ternary function (THA). It has to be repeated $3^3 = 27$ times if

TABLE 5: Simulation results of one-digit adders.

Design	$V_{dd} = 0.8$ V			$V_{dd} = 0.9$ V			$V_{dd} = 1$ V		
	Delay (psec)	Power (μW)	PDP (aJ)	Delay (psec)	Power (μW)	PDP (aJ)	Delay (psec)	Power (μW)	PDP (aJ)
	Ternary half adder								
Proposed THA	**38.74**	**0.282**	**10.94**	**28.42**	**0.880**	**25.01**	**25.43**	**2.778**	**70.66**
THA by Lin et al. [16]	91.40	0.771	70.43	66.64	2.291	152.7	54.03	5.817	314.3
THA by Moaiyeri et al. [17]	50.35	0.415	20.88	39.49	1.264	49.91	35.08	3.806	133.5
	Ternary full adder								
Proposed TFA	**58.94**	**0.456**	**26.89**	**43.95**	**1.472**	**64.68**	**35.66**	**4.595**	**163.9**
TFA by Ebrahimi et al. [18]	189.5	0.597	113.1	139.4	1.893	263.9	116.8	5.463	638.3
TFA by Keshavarzian and Sarikhani [19]	119.2	0.607	72.32	102.4	1.912	195.9	81.98	5.413	443.8

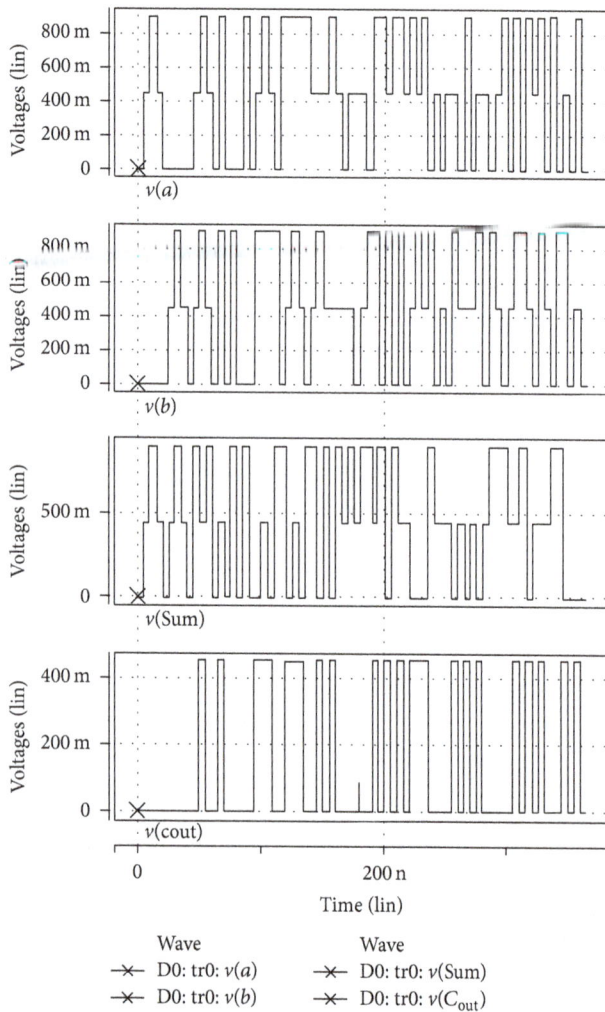

FIGURE 14: Transient response of the proposed THA with MOSFET technology.

TABLE 6: Area comparison.

Design	#Transistors	Total cell width (nm)
	Ternary half adder	
Proposed THA	**64**	**3840**
THA by Lin et al. [16]	158	9480
THA by Moaiyeri et al. [17]	112	6720
	Ternary full adder	
Proposed TFA	142	8520
TFA by Ebrahimi et al. [18]	**106**	**6360**
TFA by Keshavarzian and Sarikhani [19]	132	7920

a ternary function has three input variables (TFA). The average and the maximum static powers are reported in Table 7. Static power dissipation reaches its minimum extent in the proposed designs due to their unique structure which is mostly composed of binary parts. For instance, voltage division occurs only twice for the presented THA when all of the input variables equal "2," whereas it happens six times and four times in THA-158T and THA-112T, respectively.

Capability of working in high frequencies is put into practice for the proposed structures as well. Figure 9 shows how sharply power increases by the increase of operating frequency from 100 MHz to 1 GHz. The new circuits operate efficiently in high frequencies, and they consume less power than previous designs.

One of the challenges in CNTFET fabrication is that the diameter of carbon nanotubes cannot be set very precisely. D_{CNT} varies with a standard deviation from 0.04 nm to 0.2 nm for each mean diameter value [30]. To observe how tolerant the proposed designs are against process variation, Monte Carlo transient analysis is taken into consideration.

TABLE 7: Static power comparison.

Design	Average static power ×10⁻⁷ (W)	Maximum static power ×10⁻⁷ (W)
	Ternary half adder	
Proposed THA	**5.2661**	**11.058**
THA by Lin et al. [16]	16.540	49.030
THA by Moaiyeri et al. [17]	6.4400	18.767
	Ternary full adder	
Proposed TFA	**6.4724**	**13.829**
TFA by Ebrahimi et al. [18]	10.897	21.250
TFA by Keshavarzian and Sarikhani [19]	11.290	26.604

TABLE 8: Simulation results of 7-TIT ternary ripple adders.

Design	Delay (psec)	Power (μW)	PDP (fJ)	#Transistors	Total width (nm)
Ripple adder by the proposed THA and TFA	**187.84**	**7.5778**	**1.4234**	916	54960
Ripple adder by THA-112T and THF-132T	371.84	9.9079	3.6841	**904**	**54240**

This analysis is performed with a reasonable number of 30 iterations, in which the simulation is repeated 10 times and the largest deviation is reported. The statistical significance of 30 iterations is quite high. There is a 99% probability that over 80% of all possible component values operate properly if a circuit operates correctly for all of the 30 iterations [31]. Distribution of the diameter is assumed as Gaussian with 6-sigma distribution, which is a reasonable assumption for large number of fabricated CNTs [32]. The results of this experiment are shown in Figure 10 for the proposed ternary half adder, which is highly robust and tolerant against process variation.

Single-bit adder cells are used to form larger adders. In order to test the practicability of the new designs in a large circuit, a seven-Ternary digIT (7-TIT) ripple adder is constructed by combining a THA and six TFAs (Figure 11). THA-112T and TFA-132T are also put together to form another ternary ripple adder for comparison.

To measure worst-case delay, different input patterns are fed to the adder blocks (Figure 12) in a way that a transition propagates from the input of the first stage (a_0) to the outputs of the last stage (s_6). This input pattern is designed in a way it causes all possible transitions in the output Sum of the last stage (s_6). A long duration of iterative input pattern is also fed to the adder blocks to measure average power consumption. Simulation results are reported in Table 8. Although the proposed ripple adder has a few more transistors, it consumes less power and it operates far faster than the structure built by the previous adder cells.

Eventually, to provide a comparison between MOSFET and CNTFET technologies, the proposed ternary half adder is also implemented with 32 nm channel length bulk CMOS [33]. A brief description of the parameters of this model is shown in Table 9. The MOSFET implementation of the proposed THA is illustrated in Figure 13. Each transistor is

marked with a pair of numbers. The upper number indicates threshold voltage and the lower number is the width of the transistor (W_{gate}). Due to weaker on-current driving capability of bulk CMOS technology [34], buffers are supplemented, after the midoutputs are generated, to strengthen output signals. Driving power in MOSFETs is 3-4 times weaker than CNTFETs [35]. Therefore, transistor widths are also extended enough to overcome this deficiency, paying the price of enlarging area. Transient response of this adder cell is plotted in Figure 14. The whole cell is simulated under the same conditions as mentioned before for the CNTFET-based THAs. Table 10 shows the simulation results of both technologies. They demonstrate the fact that CNTFETs are absolutely more promising candidates for ternary circuitry. The CNTFET-based THA operates 44 times faster, consumes approximately 10 times less power, and occupies 3.3 times less area than its MOSFET counterpart.

5. Conclusions

New ternary adders have been proposed in this paper based on a logic style which is mostly composed of binary parts. Therefore, static power consumption reaches its minimum amount. Extensive different analyses have been carried out to examine efficiency in all aspects. The proposed designs benefit from low power consumption, high driving power, full-swing operation, and capability of working in low voltages and high frequencies. They can be used in larger circuits and practical environments.

A comparison between MOSFET and CNTFET has been also provided to conclude the superior technology for ternary circuitry. Due to more flexibility of adjusting the desired threshold voltage and high on-current driving capability, CNTFETs are definitely more promising devices for implementing ternary circuits in the future. Simulation

TABLE 9: Bulk CMOS model parameters.

Parameter	Description	Value
L_{ch}	Physical channel length	32 nm
L_{eff}	The effective gate channel length	12.6 nm
R_{dsw}	The source and drain resistance per unit channel width	150 Ω-μm
T_{ox}	The gate oxide thickness	1 nm
C_{gbo}	The gate-to-bulk overlap capacitance per unit channel length	25.6 pF/m
C_{gdl}/C_{gsl}	The overlap capacitance between gate and lightly doped drain/source region	265.3 pF/m

TABLE 10: Simulation results of the new THA with MOSFET and CNTFET technologies.

Design	Delay (psec)	Power (μW)	PDP (fJ)	#Transistors	Total width (nm)
The proposed THA with MOSFET technology	1248	8.257	10.30	76	12762
The proposed THA with CNTFET technology	**28.42**	**0.880**	**0.025**	**64**	**3840**

results confirm that a CNTFET-based ternary design surpasses MOSFET implementation in terms of speed, power consumption, and area.

Conflict of Interests

The authors declare that there is no conflict of interests regarding the publication of this paper.

References

[1] H. O. Ron, K. W. Mai, and A. Fellow, "The future of wires," *Proceedings of the IEEE*, vol. 89, no. 4, pp. 490–504, 2001.

[2] E. Özer, R. Sendag, and D. Gregg, "Multiple-valued logic buses for reducing bus energy in low-power systems," *IEE Proceedings: Computers and Digital Techniques*, vol. 153, no. 4, pp. 270–282, 2006.

[3] E. Dubrova, "Multiple-valued logic in VLSI: challenges and opportunities," in *Proceedings of the NORCHIP '99 Conference*, pp. 340–349, 1999.

[4] B. Hayes, "Third base," *American Scientist*, vol. 89, pp. 490–494, 2001.

[5] S. L. Hurst, "Multiple-valued logic—its status and its future," *IEEE Transactions on Computers*, vol. 33, no. 12, pp. 1160–1179, 1984.

[6] Y.-B. Kim, "Integrated circuit design based on carbon nanotube field effect transistor," *Transactions on Electrical and Electronic Materials*, vol. 12, no. 5, pp. 175–188, 2011.

[7] Y. Yasuda, Y. Tokuda, S. Zaima, K. Pak, T. Nakamura, and A. Yoshida, "Realization of quaternary logic circuits by n-channel MOS devices," *IEEE Journal of Solid-State Circuits*, vol. 21, no. 1, pp. 162–168, 1986.

[8] H. Inokawa, A. Fujiwara, and Y. Takahashi, "A multiple-valued logic with merged single-electron and MOS transistors," in *Proceedings of the IEEE International Electron Devices Meeting (IEDM '01)*, pp. 7.2.1–7.2.4, December 2001.

[9] S. Lin, Y.-B. Kim, and F. Lombardi, "A novel CNTFET-based ternary logic gate design," in *Proceedings of the 52nd IEEE International Midwest Symposium on Circuits and Systems (MWSCAS '09)*, pp. 435–438, Cancun, Mexico, August 2009.

[10] J. Appenzeller, "Carbon nanotubes for high-performance electronics: progress and prospect," *Proceedings of the IEEE*, vol. 96, no. 2, pp. 201–211, 2008.

[11] A. Rahman, J. Guo, S. Datta, and M. S. Lundstrom, "Theory of ballistic nanotransistors," *IEEE Transactions on Electron Devices*, vol. 50, no. 9, pp. 1853–1864, 2003.

[12] A. Bachtold, P. Hadley, T. Nakanishi, and C. Dekker, "Logic circuits with carbon nanotube transistors," *Science*, vol. 294, no. 5545, pp. 1317–1320, 2001.

[13] V. Derycke, R. Martel, J. Appenzeller, and P. Avouris, "Carbon nanotube inter- and intramolecular logic gates," *Nano Letters*, vol. 1, no. 9, pp. 453–456, 2001.

[14] M. M. Shulaker, G. Hills, N. Patil et al., "Carbon nanotube computer," *Nature*, vol. 501, pp. 526–530, 2013.

[15] A. P. Dhande and V. T. Ingole, "Design & implementation of 2-bit ternary ALU slice," in *Proceedings of the International Conference on IEEE Science of Electronics, Technology of Information and Telecommunication*, pp. 17–21, 2005.

[16] S. Lin, Y.-B. Kim, and F. Lombardi, "CNTFET-based design of ternary logic gates and arithmetic circuits," *IEEE Transactions on Nanotechnology*, vol. 10, no. 2, pp. 217–225, 2011.

[17] M. H. Moaiyeri, A. Doostaregan, and K. Navi, "Design of energy-efficient and robust ternary circuits for nanotechnology," *IET Circuits, Devices and Systems*, vol. 5, no. 4, pp. 285–296, 2011.

[18] S. A. Ebrahimi, P. Keshavarzian, S. Sorouri, and M. Shahsavari, "Low power CNTFET-based ternary full adder cell for nanoelectronics," *International Journal of Soft Computing and Engineering*, vol. 2, pp. 291–295, 2012.

[19] P. Keshavarzian and R. Sarikhani, "A novel CNTFET-based ternary full adder," *Circuits, Systems, and Signal Processing*, vol. 33, no. 3, pp. 665–679, 2014.

[20] J. A. Mol, J. van der Heijden, J. Verduijn, M. Klein, F. Remacle, and S. Rogge, "Balanced ternary addition using a gated silicon nanowire," *Applied Physics Letters*, vol. 99, no. 26, Article ID 263109, 2011.

[21] A. Raychowdhury and K. Roy, "Carbon-nanotube-based voltage-mode multiple-valued logic design," *IEEE Transactions on Nanotechnology*, vol. 4, no. 2, pp. 168–179, 2005.

[22] M. H. Moaiyeri, R. F. Mirzaee, A. Doostaregan, K. Navi, and O. Hashemipour, "A universal method for designing low-power

carbon nanotube FET-based multiple-valued logic circuits," *IET Computers and Digital Techniques*, vol. 7, no. 4, pp. 167–181, 2013.

[23] E. Trias, J. Navas, E. S. Ackley, S. Forrest, and M. Hermenegildo, *Negative Ternary Set-Sharing*, vol. 5366 of *Lecture Notes in Computer Science*, 2008.

[24] M. H. Moaiyeri, R. F. Mirzaee, K. Navi, and A. Momeni, "Design and analysis of a high-performance CNFET-based Full Adder," *International Journal of Electronics*, vol. 99, no. 1, pp. 113–130, 2012.

[25] Z. K. Tang, L. Y. Zhang, N. Wang et al., "One-dimensional superconductivity in 0.4 nm single-walled carbon nanotubes," *Proceedings of the Electrochemical Society*, pp. 587–595, 2002.

[26] M. S. Dresselhaus, G. Dresselhaus, and P. Avouris, *Carbon Nanotubes: Synthesis, Structure, Properties, and Applications*, Springer, 2001.

[27] J. Deng, *Device modeling and circuit performance evaluation for nanoscale devices: silicon technology beyond 45nm node and carbon nanotube field effect transistors [Ph.D. thesis]*, Stanford University, 2007.

[28] "Stanford University CNFET Model," http://nano.stanford.edu/models.phpwebsite.

[29] Y. B. Kim and Y.-B. Kim, "High speed and low power transceiver design with CNFET and CNT bundle interconnect," in *Proceeding of the 23rd IEEE International SOC Conference (SOCC '10)*, pp. 152–157, Las Vegas, Nev, USA, September 2010.

[30] H. Shahidipour, A. Ahmadi, and K. Maharatna, "Effect of variability in SWCNT-based logic gates," in *Proceedings of the 12th International Symposium on Integrated Circuits (ISIC '09)*, pp. 252–255, December 2009.

[31] S. Lin, Y.-B. Kim, and F. Lombardi, "Design and analysis of a 32 nm PVT tolerant CMOS SRAM cell for low leakage and high stability," *Integration, the VLSI Journal*, vol. 43, no. 2, pp. 176–187, 2010.

[32] K. El Shabrawy, K. Maharatna, D. Bagnall, and B. M. Al-Hashimi, "Modeling SWCNT bandgap and effective mass variation using a Monte Carlo approach," *IEEE Transactions on Nanotechnology*, vol. 9, no. 2, pp. 184–193, 2010.

[33] Predictive Technology Model, http://ptm.asu.edu.

[34] C. García and A. Rubio, "Manufacturing variability analysis in carbon nanotube technology: a comparison with bulk CMOS in 6T SRAM scenario," in *Proceedings of the 14th IEEE International Symposium on Design and Diagnostics of Electronic Circuits and Systems (DDECS '11)*, pp. 249–254, April 2011.

[35] F. Ali Usmani and M. Hasan, "Carbon nanotube field effect transistors for high performance analog applications: an optimum design approach," *Microelectronics Journal*, vol. 41, no. 7, pp. 395–402, 2010.

## 24

Design of Smart Power-Saving Architecture for Network on Chip

Trong-Yen Lee and Chi-Han Huang

Department of Electronic Engineering, National Taipei University of Technology, Taipei 10608, Taiwan

Correspondence should be addressed to Trong-Yen Lee; tylee@ntut.edu.tw

Academic Editor: Yu-Cheng Fan

In network-on-chip (NoC), the data transferring by virtual channels can avoid the issue of data loss and deadlock. Many virtual channels on one input or output port in router are included. However, the router includes five I/O ports, and then the power issue is very important in virtual channels. In this paper, a novel architecture, namely, Smart Power-Saving (SPS), for low power consumption and low area in virtual channels of NoC is proposed. The SPS architecture can accord different environmental factors to dynamically save power and optimization area in NoC. Comparison with related works, the new proposed method reduces 37.31%, 45.79%, and 19.26% on power consumption and reduces 49.4%, 25.5% and 14.4% on area, respectively.

1. Introduction

In recent years, the 3-dimensional IC and TSV (Through-Silicon Via) technology are proposed to solve area issues. The 3-dimensional IC of Intel Ivy Bridge processor and the 16-core multicore architecture can be implemented in 22 nm [1]. Therefore, the multicore and heterogeneous systems are popular research in SoC (system-on-chip). These architectures require high throughput and performance to transfer data in a multicore SoC. Therefore, the NoC (network-on-chip) can be proposed to solve this requirement, but it derived new problems such as power consumption and area [2, 3].

The NoC architecture [1] consists of processing element (PE), network interface (NI), router, and topology which is shown in Figure 1. The PEs transfer information to NI, the NI packages the information into flits then passes to routers. The routers have difference corner router (CR), edge router (ER), and router (R); the CR, ER and R has three, four, and five I/O ports to access information then each port includes *n* virtual channels. Router includes transmission channel, routing computation (RC), virtual channel arbiter (VA), switch arbiter (SA), and crossbar (XBAR). The flits includes header, body, and tail; the header flit has PE priority, source address, destination address, and so forth. The RC uses header flit and routing algorithms to find transmission path. VA uses two stages arbitration to select most high priority packet transmission and then will sign transmission channel. SA uses

two stages arbitration and will select most body flits into XBAR to transmit. The VA will be working when the packet is arrival. The SA operation when the flit is arrival. The tail flit represents last flit, and then the router will unregister transmission channel. The router topology includes mesh, star, and fat tree [4, 5].

Yoon et al. [6] analysis of virtual channels (VCs) can avoid routing and protocol deadlock and improve the routing performance when the packet traffic is congested. The VCs can solve packet switch hard issue but it leads the power and area and so forth issue in NoC.

Nicopoulos et al. [2] proposed IntelliBuffer architecture to solve PV (process variation) to reduce the power consumption in layer 1 [7]. It differs from the conventional architecture in two fundamental ways. First, these slots use clock-gating to reduce the power consumption when slots are empty. In order to avoid data loss transmission, one of slots clock keeps to access data in each I/O port. Second, the router creates a leakage classification register (LCR) table; then the write and read pointer always accesses the lowest power consumption slots from the LCR table.

Taassori et al. [3] proposed an adaptive data compression technology to reduce the number of packet bits in layer 3 [7]. It reduces of the number of transmissions. Therefore, it can improve power consumption of router. Palma et al. [8] use T-Bus-Invert technology to reduce the hamming distance transition activity rate to improve the power consumption.

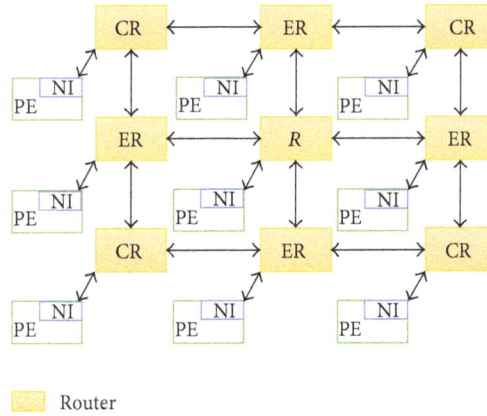

FIGURE 1: NoC architecture.

Jafarzadeh et al. [9] use end-to-end data coding technology to minimize switching activity rate and routing path to improve NI power consumption.

Lee et al. [10] proposed buffer clock-gating architecture and used clock-gating to reduce the transmit power consumption when slots are empty and full. Ezz-Eldin et al. [11] proposed an adaptive virtual channel with two sections in layer 1 [7]. First, the work used hierarchical multiplexing tree for Virtual Channels (VCs) to reduce area. Second, it uses clock-gating to reduce power consumption. Rosa et al. [12] proposed dynamic frequency scaling in PE for NoC. It considers the communication and loading rate to control the router frequency to reduce the power consumption.

Huaxi et al. [13] proposed fat tree-based optical NoC; this architecture includes topology, placement, layout, and protocol. This paper proposed low power and cost router optical turnaround router to improve the power consumption. Gu et al. [14] proposed Cygnus router to optimize the router algorithms to reduce the power consumption. Swaminathan et al. [15] create two FIFOs in NI. Use two FIFO dynamic configuration data access to improve throughput and power consumption.

In the next section we analyse the power consumption under the difference VCs access. Section 3 we introduce the topology and router packet architecture, we addition the SPS in router to save power. In Section 4 we present SPS with router design. Section 5 contains experimental results and Section 6 concludes this paper.

2. Power Issue with Virtual Channels

The multicore architecture and big data communication are more popular in next generation. Traditional communication technologies cannot meet a large amount of traffic on multicore and heterogeneous chip. The NoC can solve this issue. It uses network transmission method to make the difference core communication at same time. The NoC can solve the communication issue but the big data access enhances the power consumption.

The router composed of the arbitration and transmission unit [16] is illustrated in Figure 2. The arbitration unit selects the highest priority packet sent to next router. The arbitration unit includes routing computation (RC), VC arbiter (VA), and switch arbiter (SA). The RC is the calculation of routing paths and priorities. The VA contains a number of two-stage arbitrations to select packet and sign up VCs. First stage selects the local highest priority packet from input VCs to crossbar and signs up VCs. Second stage selects the global highest priority packet from input crossbar to output VCs and signs up VCs. The SA also contains a number of two-stage arbitrations to select flits for transmission. First stage selects the local highest priority flits from input VCs to crossbar. Second stage selects the global highest priority flits from input crossbar to output VCs. The VA executed prepacket and the SA executed preflits.

The router with transmission unit is illustrated in Figure 3. In this unit, it includes nVCs to access large packet from input physical channel to output physical channel. A power consumption calculation to VCs is shown in (1). The variable of n represents the number of access packets or flits in VCs. The variable of f represents access frequency in VCs. The variable of c represents capacitance and v represents voltage in VCs. Nicopoulos et al. [2] and Katabami et al. [17] proposed clock-gating to solve this issue.

In this paper, we proposed a dynamic control of each virtual channel clock in different transmission environments. Whether packet transfer is complete, the SPS can effectively reduce the power consumption and does not affect the transmission performance. Consider

$$P_{n\text{VCs}} = \sum_{n=1}^{\infty} 1^n \times (f \times c \times v)^2. \tag{1}$$

3. Router and Topology with SPS

3.1. Relation of Topology and Router. The relation of topology and router is illustrated in Figure 4. The router uses different transmission mode with topologies. For example, the mesh uses the X-Y routing to transmit. The X-Y routing flow chart for 2 × 2 meshes is illustrated in Figure 5, when the MSB of destination router address (R_{dm}) is equal to the MSB of current router address (R_{cm}) and if the LSB of router

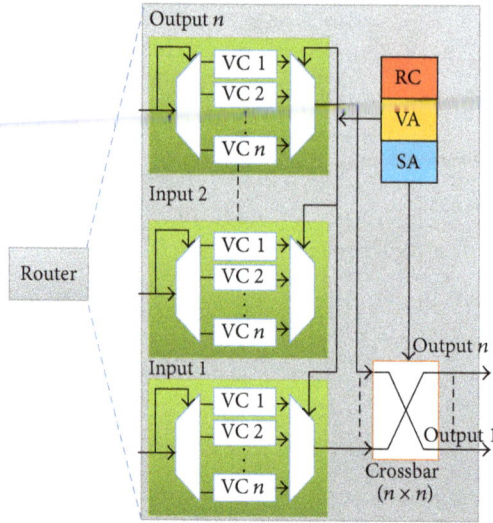

FIGURE 2: Router architecture with NoC.

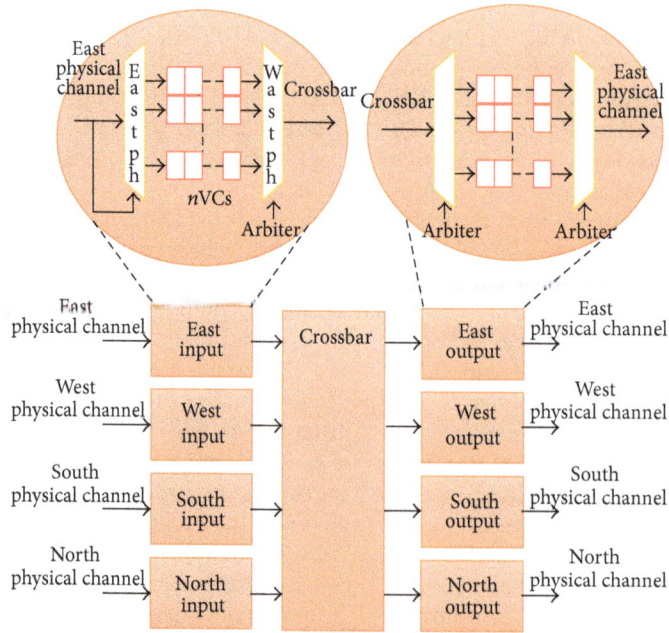

FIGURE 3: Transmission unit.

addresses (R_{dl} and R_{cl}) is equal then it means the flits arrival. Otherwise, the X-Y routing algorithm includes two-stage flows. In stage one, the flits are sent until that the R_{dm} equals of R_{cm} on the x-axis routers. In stage two, the flits are sent to the destination by y-axis routers. The virtual channel will be initialed under packet transmit on two routers, which procedure is shown on Algorithm 1.

The control method of arbiter architecture uses different transmission mode to design. The VC arbiter and switch bar are by the topology and priority to design the routing computation unit. Algorithm 2 constructs VC two stages arbitration of prepackets. Stage 1 decided high priority packet into crossbar from local VCs (input VCs) of each packet at lines 3 to 4 and lines 8 to 10. Stage 2 decided most important packet to transmission from global VCs (output VCs) of each packet at lines 5 to 6 and lines 11 to 13.

Sign up Algorithm
Input: R_{roth} and E_{mp}.
(1) while (flits arrival) do
(2)   if (R_{rothf2} is header and adx is free channel)
(3)   {sign up the channel and select the channel to output}
(4)   else if (R_{rothf2} is body and $adx = R_{roths2}$)
(5)   {select the channel to output}
(6)   else if (R_{rothf2} is tail and $adx = R_{roths2}$)
(7)   {clear the channel and select the channel to output;}
(8)   else
(9)   {read back flit to virtual channel}
(10) end while

ALGORITHM 1: Channel sign up algorithm.

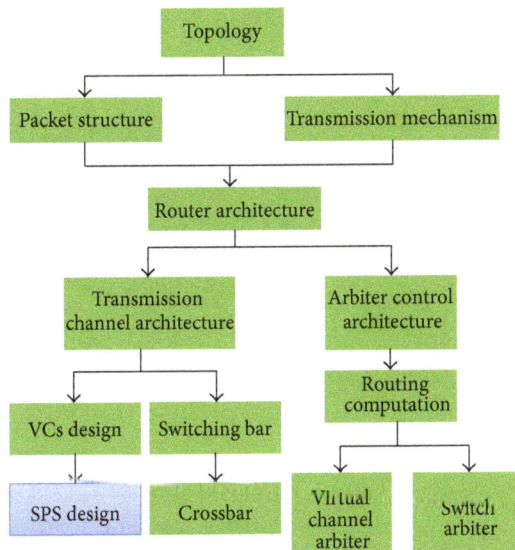

FIGURE 4: Topology and router relation with SPS.

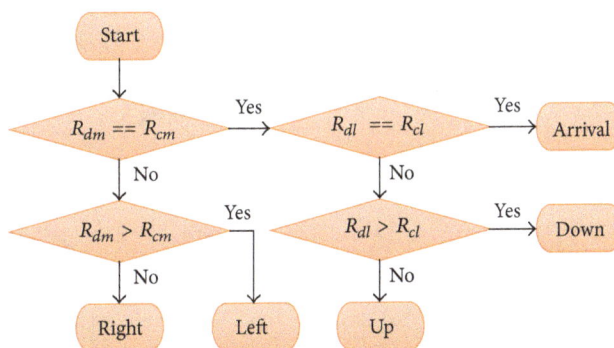

FIGURE 5: *X-Y* routing flow chart.

Virtual channel arbitration
Input: header flits
/∗**Control signal enable**∗/
(1) while (header flits) do
(2)   use lottery arbitration to select local and global highest priority flits
(3)   if (local)
(4)   {V_{ai} = local input virtual channel address}
(5)   if (global)
(6)   {V_{ao} = global input virtual channel address}
(7) end while
/∗**Channel switch**∗/
(8) Case V_{ai}
(9) {C_{ri1} = local packet of V_{ai}}
(10) end case
(11) Case V_{ao}
(12) {R_{it} = global packet of V_{ao}}
(13) end case

ALGORITHM 2: VC arbitration algorithm.

```
Switch arbitration
Input: body and tail flits
/∗Control signal enable∗/
(1)  while (body or tail flits) do
(2)  use channel sign up register to select local and global highest priority flits
(3)      if (local)
(4)      {S_ai = local input virtual channel address}
(5)      if (global)
(6)      {S_ao = global input virtual channel address}
(7)  end while
/∗Channel switch∗/
(8)  Case S_ai
(9)  {C_ri2 = local packet of S_ai}
(10) end case
(11) Case S_ao
(12) {R_ot = global packet of S_ao}
(13) end case
```

ALGORITHM 3: Switch arbitration algorithm.

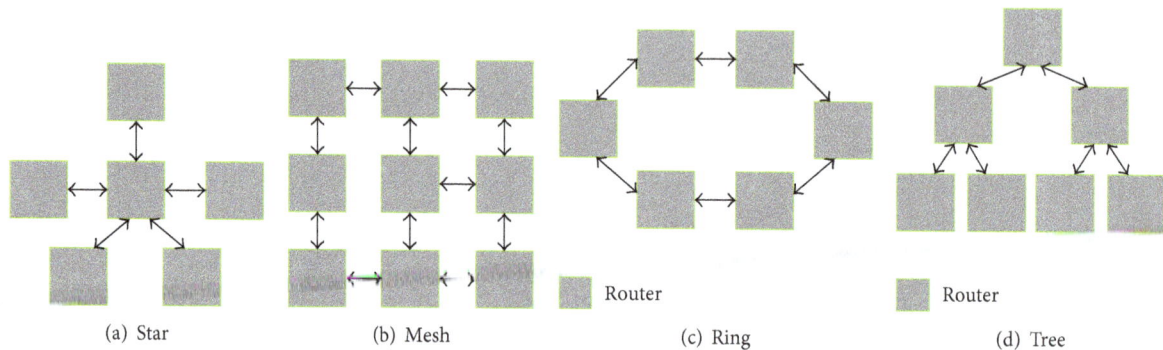

| (a) Star | (b) Mesh | (c) Ring | (d) Tree |

FIGURE 6: Router connection topology architecture.

Algorithm 3 constructs VC two stages arbitration of preflits. Stage 1 decided high priority flit into crossbar from local VCs (input VCs) of each flit at lines 3 to 4 and lines 8 to 10. Stage 2 decided most important flit to transmit from global VCs (output VCs) of each flits at lines 5 to 6 and lines 11 to 13.

The router includes four directions to connect other routers and one local physical channel to connect PE in transmission channel architecture. There have been nVCs of each physical channel without local physical channel. The switch bar support for transmission the most important packet to output channel. The SPS controls each VCs power consumption when the channel status changes. The SPS architecture is introduced in next section.

3.2. Topology Architecture.
The topology is definition of the packet transmission path between router and link. The router connection topology architecture is shown in Figure 6; they include star, mesh, ring, and tree topologies. The RC algorithms depend on topology architecture in arbitration unit. The VA and SA algorithms depend on packet priority in arbitration unit. In this paper, the topology is the 2×2 mesh,

the RC algorithm is X-Y routing, and the VA and SA algorithms are lottery [18].

The router that connects with PE is shown in Figure 7; so that the PE and router access information, use the network interface (NI). It handles the information between router and PE. The NI includes two level designs [19] as shown in Figure 8. It contains three modules to meet the specifications of the different layers. The shell module needs to meet IP specification. The kernel module needs to meet the NoC topology specification.

3.3. Flits with Router Architecture.
The flit specification with router is shown in Figure 9; the flit type of 2-bit 00 represents the one packet; this flit type does not sign up VCs. The 2-bit 01 represents the *header* flit which includes routing information and address; this flit type always is determined in sign up channel. The 2-bit 10 represents the *body* flit which includes transmission information; this flit payload records the segment packet. The 2-bit 11 represent the *tail* as last transmission information; this flit not only records the last segment packet but also cleans the VCs.

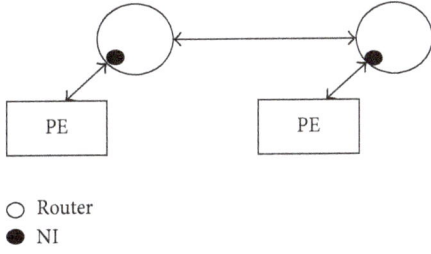

FIGURE 7: Router connection with PE.

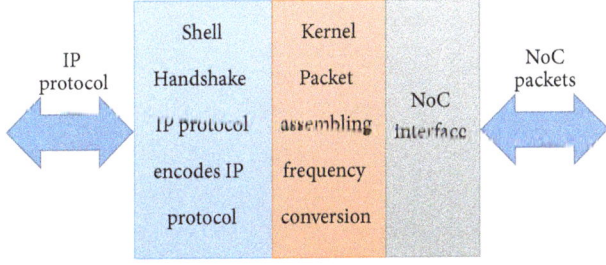

FIGURE 8: NI breakdown into Shell, Kernel, and interface.

FIGURE 9: Flits type of router.

4. SPS with Router Design

The VC that contains many slots to access data led to extra power consumption. In this paper, we propose SPS architecture to reduce the power consumption.

4.1. Router with SPS Architecture.

The proposed router with SPS architecture is illustrated in Figure 10. The physical channel (PC) is used to connect other routers and access information. The input VCs (IVC) is used to store information from PCs. It always is designed by FIFO or other sequential logic. The arbiter decides the flits priority to control input switch logic (ISL) and output switch logic (OSL) to transmit flits. It includes RC, VA, and SA. The crossbar (CR) connects IVC to OVC, the switch signal form arbiter. The output VCs (OVC) store information from CR. The proposed SPS uses the transmission channel status to dynamic control IVC and OVC clock in essential operating.

The VCs with SPS architecture are illustrated in Figure 11. It controls system clock into I/O VC to reduce power consumption. In this architecture, the VC contains 0 to $i - 1$ slots to access data.

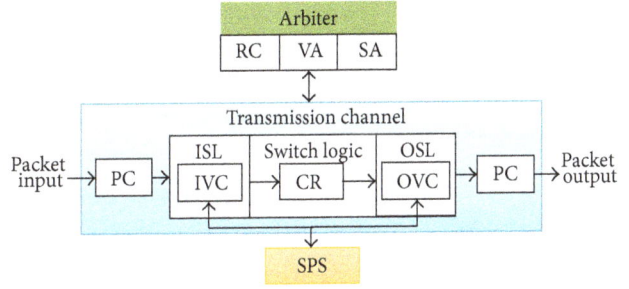

FIGURE 10: Router with SPS architecture.

4.2. Design of SPS Control Timing.

The VCs access timing diagrams of SPS architecture are illustrated in Figure 12. The Clock Block A indicates that the VCs have no information to transmit. The Clock Block B indicates that the VCs are writing information. The Clock Block C indicates that the data in VCs are waiting to transmit. Our analysis for unused clock-gating architecture is shown in (2). The slots access information of power consumption is denoted by P_a. The slot content full and empty of power consumption are denoted by P_f and P_e, respectively. The P_s is power consumption except for P_f, P_e, and P_a. The unused clock-gating architecture does not control clock for sequential logic in nVCs. Therefore, the logic will generate power consumption in high transmission structure.

The clocking gating consumes power in Clock Block B and Clock Block C. Our analysis for clock-gating architecture is shown in (3). The P_{g1} is power consumption of empty gating. The clock-gating architecture does not control clock when VCs is full stage. The VCs always store flits to wait for transmission.

The SPS consumes power in Clock Block B. Our analysis for SPS architecture is shown in (4). The P_{g2} is power consumption of SPS. It saves the power consumption of empty and full gating for nVCs. Consider

$$P_{r1} = P_a + P_f + P_e + P_s, \tag{2}$$

$$P_{r2} = P_a + P_f + P_s + P_{g1}, \tag{3}$$

$$P_{r3} = P_a + P_s + P_{g2}. \tag{4}$$

4.3. Design of SPS.

The proposed SPS uses the VCs status to dynamic control clock of each VC. The CFSM of SPS with VCs is illustrated in Figure 13; it contains two CFSM in this architecture.

The first CFSM includes initial, empty, full, and waiting status. *Initial status*: when the VC is reset, the structure is into the initial status until the flit arrive. *Empty status*: when the user resets the VCs or the flits transport to next storage unit, the structure is into this status. *Full status*: the store flit in VC is full. *Waiting status*: When the user reset the VCs or the store flit is complete.

The VCs with SPS algorithm is illustrated in Algorithm 4. In line 3, the VCs will initialize the VCs count and flags. The VCs will access flits to change VCs count when channel packet or arbiter signal arrive at line 4 to 9. When the VCs

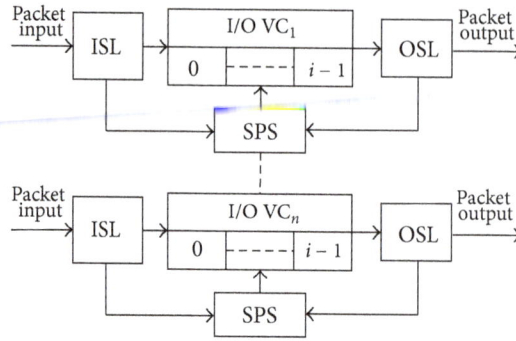

FIGURE 11: VCs with SPS architecture.

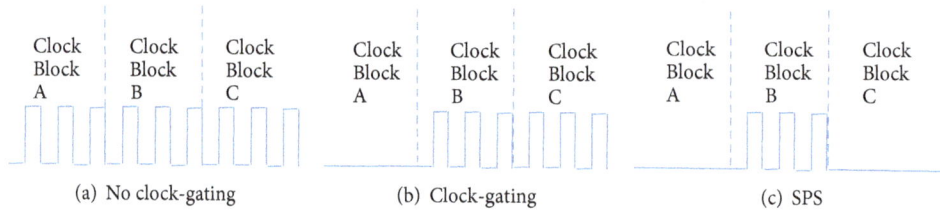

FIGURE 12: VCs power with clock diagram.

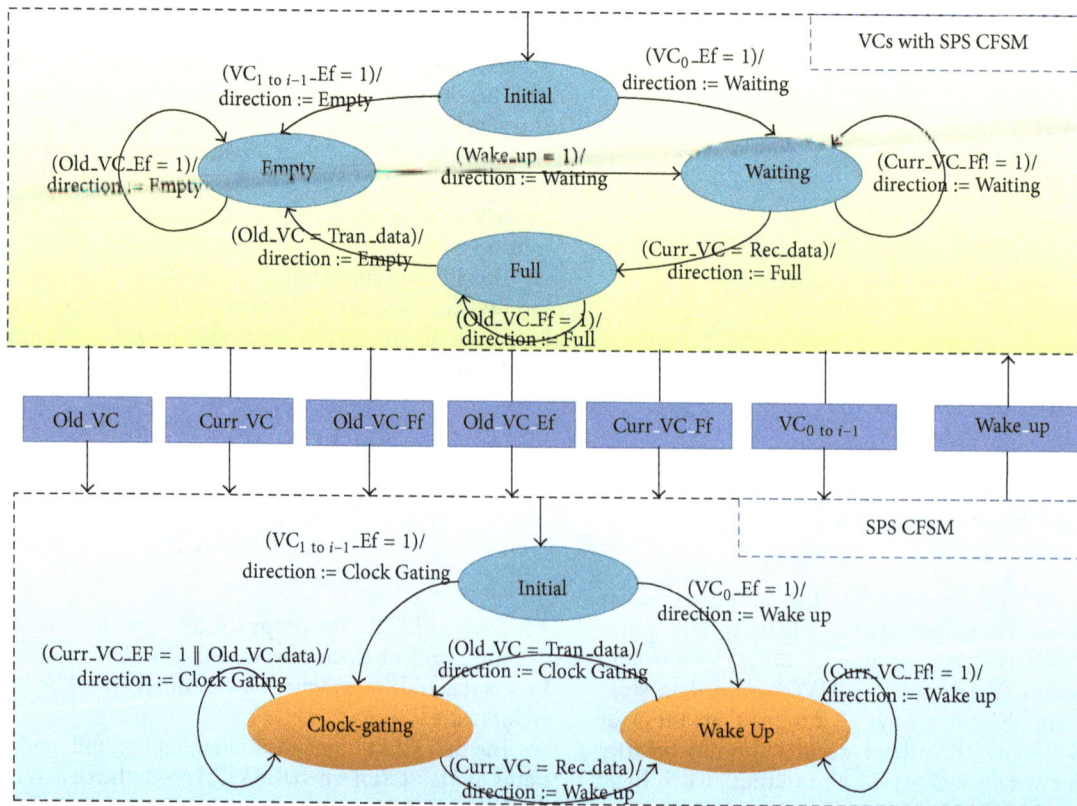

FIGURE 13: CFSM of SPS with VCs.

count can be changed, then the VCs flag will be changed at line 10 to 17.

The second CFSM includes initial, clock-gating, and wake up status. *Initial status*: this principle is the first CFSM of initial state. *Clock-gating*: when the VC changes to full or empty,

then SPS will disable this VC clock and change to this status. *Wake up*: when the VC want to store flit, one VC will wake up.

The SPS algorithm is illustrated in Algorithm 5. In line 3, the SPS will initialize VCs clock and access status from VCs

VCs with SPS Algorithm

Input: VCs clock, channel packet, arbiter signal and reset.

Output: channel packet, channel status

(1) VC_{count} is integer and range is $1 \leq VC_{count} \leq n$

(2) VC_{flag} includes full flag and empty flag

(3) initial VC_{count} and VC_{flag}

(4) while (channel packet or arbiter signal be arrival) do

(5)   if (channel packet be arrival and full flag $!= 1$)

(6) $\{VC_{count} = VC_{count} + 1$ and packet store in VCs$\}$

(7)   if (arbiter signal be arrival and empty flag $!= 1$)

(8) $\{VC_{count} = VC_{count} - 1$ and packet be read from VCs$\}$

(9) end while

(10) while (VC_{count} be change) do

(11)   if ($VC_{count} = n$)

(12) $\{$assign full flag to $1\}$

(13)   else if ($VC_{count} = 1$)

(14) $\{$assign empty flag to $1\}$

(15)   else

(16) $\{$assign full flag and empty flag to $0\}$

(17) end while

ALGORITHM 4: VCs with SPS algorithm.

SPS Algorithm

Input: system clock, channel packet, arbiter signal and reset.

Output: VCs clock

(1) VC_{group} is VCs group of 4 direction port

(2) VC_{flag} includes full flag and empty flag

(3) Initial VCs clock and access VCs count and stage flag

(4) follow LCR to arrangement all slots priority;

(5) VC_{clki} is VCs clock of each VC_{group} //where $1 \leq i \leq n$

(6) Example VC_{group} = East port

(7) initial $VC_{clki} = 0$; //where $1 \leq i \leq n$

(8) while (virtual channel be write) do

(9)   if (VC_{flag} = empty)

(10) $\{VC_{clki}$ = system clock$\}$

(11)   If (VC_{flag} = full flag)

(12) $\{VC_{clki} = 0$ and VC_{clki+1} = system clock$\}$

(13) end while

(14) while (virtual channel be read) do

(15)   if (empty flag = 1)

(16) $\{VC_{clki} = 0\}$

(17) end while

ALGORITHM 5: SPS algorithm.

with VC flags. The slots priority from LCR [2] and each VCs clock can be initialized at lines 4, 5, and 7. The SPS controls VCs clock to reduce the VCs power consumption when VCs is accessed and flags changed at lines 8 to 17.

5. Experimental Results

In this section, we proposed autotesting architect for router with SPS. This architect includes four modules of autotesting. The first module is test-vector generator (TVG); the FSM is illustrated in Figure 14; the Idle status is waiting for the

Router with SPS Algorithm

Input: system clock, start, Lottery Input.

Output: test-start, Implement-results

(1) If start testing

(2) $\{$test-start = 1; pass VD$\}$

(3) While (read test data from and start bit set-up to one) do

(4) Lottery Input = Test-vector

(5) Implement-results = Test-vector use Router with SPS to transmission;

(6) Test-vector address = Test-vector address + 1;

(7) If (test finish or start = 0)

(8) $\{$test-start = 0$\}$

(9) End while

ALGORITHM 6: Router with SPS testing algorithm.

requirement of start testing, when the requirement arrives, TVG then will change status from idle to generator. When the requirement is cancelled, the status be changed from generator to idle. The generator status will generate test-vector and compare-vector; this is illustrated in Figure 15; we use c language to generate lottery arbitration [18] in test-vector at control step 1. We use HDL to design the conventional router to generate the compare-vector and the input pattern from the test-vector at control step 2. When the compare-vector and test-vector functions are complete then the status will be changed from generator to vector output (VO) at control step 3. The VO status will transform test-vector and compare-vector to Xilinx memory IP files, through memory to control data output to test and compare only one clock.

The second module is vector database (VD); the control flow graph is illustrated in Figure 16; the module writes VO status vector in memory. The database includes two vectors to test and analyze the proposed circuit. The lottery database is provided test packet for router with SPS. The compare database is provided analysis for router with SPS.

The third module is router with SPS; we use VD to propose the test-vector to implement this module. The testing algorithm is illustrated in Algorithm 6, when the start signal set up to one from I/O, then the module starts to test and pass this signal to VD at lines 1 to 2. When testing is started, the input signal will be read from VD, shown at lines 3 and 4 in Algorithm 6. The read test-vector delay time is one clock from VD to router with SPS. The router with SPS uses VD test-vector to compute at line 6. When this pattern computation is finish, the next pattern will be read from VD at line 6. When the test pattern computation is finished or start signal is cancelled, test-start set up and stop testing at lines 7 and 8.

The final module, verification module, is illustrated in Figure 17; we verify the function in this module. The function verification is comparing of compare-vector and implement-results from VD and router with SPS. If the pattern is error, then verification result returns error signal.

The hardware experimental environment uses Xilinx FPGA xc5vlx50t-1ff1136 to verify SPS architecture. The software experimental environment uses Xilinx ISE 12.3 and the

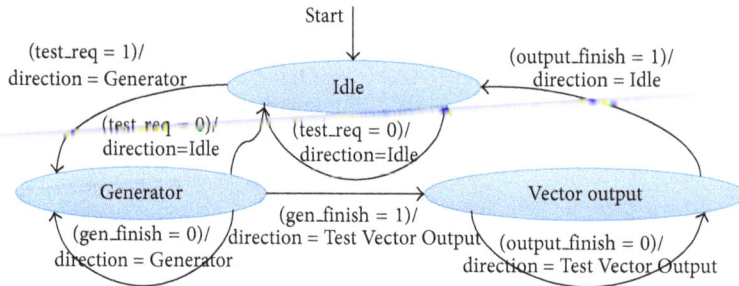

FIGURE 14: Test-vector generator (TVG) module FSM.

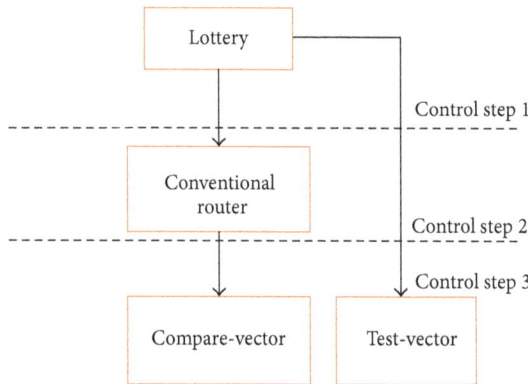

FIGURE 15: Generator status control and data flow graphs.

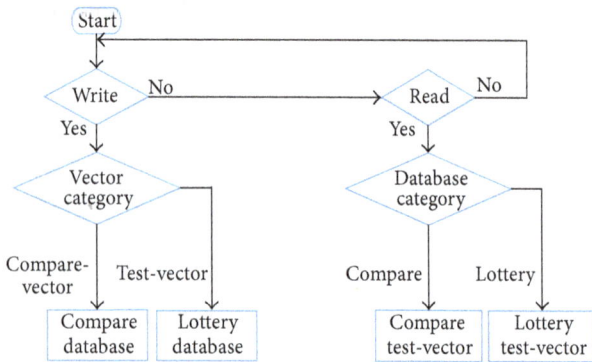

FIGURE 16: Vector database (VD) control flow graph.

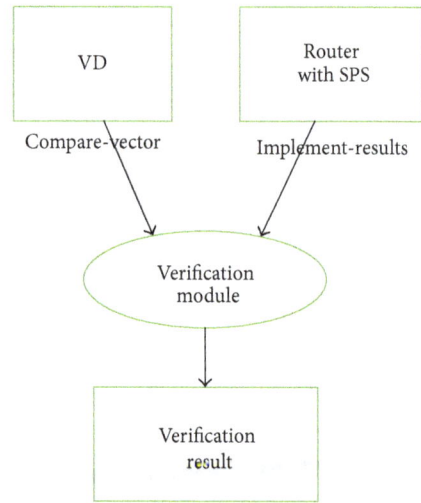

FIGURE 17: Verification module data flow graphs.

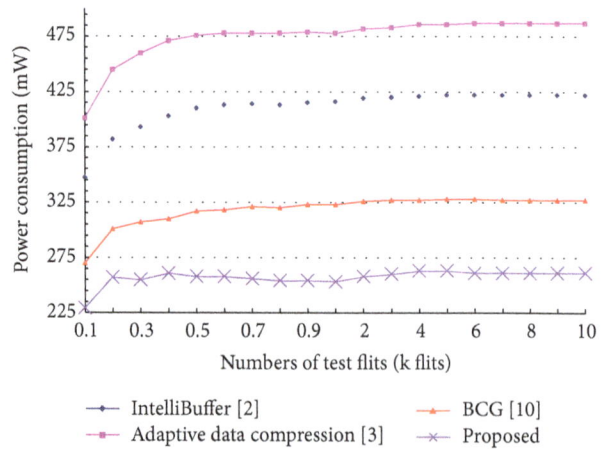

FIGURE 18: Power consumption distribution.

analysis tools use Modelsim 6.6, Xilinx Chipscope ILA, and Xpower 12.3, which are supported by Xilinx. The test experimental environment uses 2×2 mesh and X-Y routing; the PC have 4 VCs to access flits. The power consumption distribution is illustrated in Figure 18; the number of test packets is from 100 to 10000. The packet format is flit and packet length is 18 bits.

Comparing related works, as shown in Table 1, IntelliBuffer [2], adaptive data compression [3], and buffer clock-gating [10], the proposed method reduces 37.31%, 45.79%, and 19.26% on power consumption, respectively, and reduces 49.4%, 25.5% and 14.4% on area, respectively.

6. Conclusions

The *Smart Power-Saving (SPS)* architecture for network-on-chip was presented. A clock control circuit and SPS algorithm are demonstrated to reduce the power consumption on the NoC architecture. From experimental results, the proposed

TABLE 1: Comparison of power consumption and area.

Methods	Constraints			
	Power consumption (mW)	Area (number of slices)	Improved power	Improved area
IntelliBuffer [2]	410.42	1551	37.31%	49.4%
Adaptive data compression [3]	474.53	1054	45.79%	25.5%
Buffer clock-gating [10]	318.63	917	19.26%	14.4%
Newly proposed	257.05	785		

SPS architecture is more efficient to reduce the power consumption than IntelliBuffer [1], adaptive data compression [3], and buffer clock-gating [10] in the NoC architecture.

Conflict of Interests

The authors declare that there is no conflict of interests regarding the publication of this paper.

Acknowledgment

The authors would like to thank the Ministry of Science and Technology of the Republic of China, Taiwan, for partially supporting this research.

References

[1] D. James, "Intel Ivy Bridge unveiled—the first commercial trigate, high-k, metal-gate CPU," in *Proceedings of the Custom Integrated Circuits Conference (CICC '12)*, pp. 9–12, September 2012.

[2] C. Nicopoulos, S. Srinivasan, A. Yanamandra et al., "On the effects of process variation in network-on-chip architectures," *IEEE Transactions on Dependable and Secure Computing*, vol. 7, no. 3, pp. 240–254, 2010.

[3] M. Taassori, M. Taassori, and M. Mossavi, "Adaptive data compression in NoC architectures for power optimization," *International Review on Computers and Software*, vol. 5, no. 5, pp. 540–547, 2010.

[4] D. Bertozzi and L. Benini, "Xpipes: a network-on-chip architecture for gigascale systems-on-chip," *IEEE Circuits and Systems Magazine*, vol. 4, no. 2, pp. 18–31, 2004.

[5] S. J. Lee, K. Lee, and H. J. Yoo, "Analysis and implementation of practical, cost-effective networks on chips," *IEEE Design and Test of Computers*, vol. 22, no. 5, pp. 422–433, 2005.

[6] Y. J. Yoon, N. Concer, M. Petracca, and L. Carloni, "Virtual channels versus multiple physical networks: a comparative analysis," in *Proceedings of the 47th ACM/IEEE Design Automation Conference (DAC '10)*, pp. 162–165, June 2010.

[7] L. Benini and G. de Micheli, "Networks on chips: a new SoC paradigm," *IEEE Computer*, vol. 35, no. 1, pp. 70–78, 2002.

[8] J. C. S. Palma, L. S. Indrusiak, F. G. Moraes, R. Reis, and M. Glesner, "Reducing the power consumption in networks-on-chip through data coding schemes," in *Proceedings of the 14th IEEE International Conference on Electronics, Circuits and Systems (ICECS '07)*, pp. 1007–1010, December 2007.

[9] N. Jafarzadeh, M. Palesi, A. Khademzadeh, and A. Afzali-Kusha, "Data Encoding Techniques for Reducing Energy Consumption in Network-on-Chip," *IEEE Transactions on Very Large Scale Integration (VLSI) Systems*, vol. 22, no. 3, pp. 675–685, 2014.

[10] T. Y. Lee, C. H. Huang, and X. S. Lin, "Design of buffer clock-gating architecture for network-on-chip," in *Proceedings of the 22th VLSI Design/CAD Symposium*, pp. 2–5, August 2011.

[11] R. Ezz-Eldin, M. A. El-Moursy, and A. M. Refaat, "Low leakage power NoC switch using AVC," in *Proceedings of the IEEE International Symposium on Circuits and Systems (ISCAS '12)*, pp. 2549–2552, Seoul, Republic of Korea, May 2012.

[12] T. R. da Rosa, V. Larrea, N. Calazans, and F. G. Moraes, "Power consumption reduction in MPSoCs through DFS," in *Proceedings of the 25th Symposium on Integrated Circuits and Systems Design (SBCCI '12)*, pp. 1–6, 2012.

[13] G. Huaxi, X. Jiang, and Z. Wei, "A low-power fat tree-based optical network-on-chip for multiprocessor system-on-chip," in *Proceedings of the Design, Automation and Test in Europe Conference and Exhibition (DATE '09)*, pp. 3–8, April 2009.

[14] H. Gu, K. H. Mo, J. Xu, and W. Zhang, "A low-power low-cost optical router for optical networks-on-chip in multiprocessor systems-on-chip," in *Proceedings of the IEEE Computer Society Annual Symposium on VLSI (ISVLSI '09)*, pp. 19–24, Tampa, Fla, USA, May 2009.

[15] K. Swaminathan, G. Lakshminarayanan, F. Lang, M. Fahmi, and S. B. Ko, "Design of a low power network interface for Network on chip," in *Proceedings of the 26th IEEE Canadian Conference on Electrical and Computer Engineering (CCECE '13)*, pp. 1–4, May 2013.

[16] R. Mullins, A. West, and S. Moore, "Low-latency virtual-channel routers for on-chip networks," in *Proceedings of the 31st Annual International Symposium on Computer Architecture (ISCA '04)*, pp. 188–197, 2004.

[17] H. Katabami, H. Saito, and T. Yoneda, "Design of a GALS-NoC using soft-cores on FPGAs," in *Proceeding of the Embedded Multicore Socs (MCSoC '13)*, pp. 26–28, September 2013.

[18] J. Wang, Y. Li, Q. Peng, and T. Tan, "A dynamic priority arbiter for network-on-chip," in *Proceedings of the IEEE International Symposium on Industrial Embedded Systems (SIES '09)*, pp. 253–256, July 2009.

[19] S. Saponara, L. Fanucci, and M. Coppola, "Design and coverage-driven verification of a novel network-interface IP macrocell for network-on-chip interconnects," *Journal of Microprocessors and Microsystems*, vol. 35, no. 6, pp. 579–592, 2011.

Permissions

All chapters in this book were first published in VLSI, by Hindawi Publishing Corporation; hereby published with permission under the Creative Commons Attribution License or equivalent. Every chapter published in this book has been scrutinized by our experts. Their significance has been extensively debated. The topics covered herein carry significant findings which will fuel the growth of the discipline. They may even be implemented as practical applications or may be referred to as a beginning point for another development.

The contributors of this book come from diverse backgrounds, making this book a truly international effort. This book will bring forth new frontiers with its revolutionizing research information and detailed analysis of the nascent developments around the world.

We would like to thank all the contributing authors for lending their expertise to make the book truly unique. They have played a crucial role in the development of this book. Without their invaluable contributions this book wouldn't have been possible. They have made vital efforts to compile up to date information on the varied aspects of this subject to make this book a valuable addition to the collection of many professionals and students.

This book was conceptualized with the vision of imparting up-to-date information and advanced data in this field. To ensure the same, a matchless editorial board was set up. Every individual on the board went through rigorous rounds of assessment to prove their worth. After which they invested a large part of their time researching and compiling the most relevant data for our readers.

The editorial board has been involved in producing this book since its inception. They have spent rigorous hours researching and exploring the diverse topics which have resulted in the successful publishing of this book. They have passed on their knowledge of decades through this book. To expedite this challenging task, the publisher supported the team at every step A small team of assistant editors was also appointed to further simplify the editing procedure and attain best results for the readers.

Apart from the editorial board, the designing team has also invested a significant amount of their time in understanding the subject and creating the most relevant covers. They scrutinized every image to scout for the most suitable representation of the subject and create an appropriate cover for the book.

The publishing team has been an ardent support to the editorial, designing and production team. Their endless efforts to recruit the best for this project, has resulted in the accomplishment of this book. They are a veteran in the field of academics and their pool of knowledge is as vast as their experience in printing. Their expertise and guidance has proved useful at every step. Their uncompromising quality standards have made this book an exceptional effort. Their encouragement from time to time has been an inspiration for everyone.

The publisher and the editorial board hope that this book will prove to be a valuable piece of knowledge for researchers, students, practitioners and scholars across the globe.

List of Contributors

Kai Huang
Department of Information Science and Electronic Engineering, Zhejiang University, Hangzhou 310027, China

Peng Zhu and Xiaolang Yan
Institute of Very Large Scale Integrated Circuit Design, Zhejiang University, Hangzhou 310027, China

Rongjie Yan
Laboratory of Computer Science, Institute of Software, Chinese Academy of Sciences, Beijing 100080, China

David H. K. Hoe
Department of Engineering, Loyola University Maryland, Baltimore, MD 21210, USA

Xiaoyu Jin
Department of Electrical Engineering, University of Texas at Tyler, Tyler, TX 75799, USA

Vadim Geurkov and Lev Kirischian
Department of Electrical and Computer Engineering, Ryerson University, 350 Victoria Street, Toronto, ON, Canada M5B 2K3

Deepa Yagain and A. Vijaya Krishna
Department of ECE, PESIT, Bangalore 560085, India

Xiao Wang
Shenyang Institute of Automation, Chinese Academy of Sciences, Shenyang 110016, China
University of the Chinese Academy of Sciences, Beijing 100049, China
Key Laboratory of Opto-Electronic Information Processing, Chinese Academy of Sciences, Shenyang 110016, China

Zelin Shi and Baoshu Xu
Shenyang Institute of Automation, Chinese Academy of Sciences, Shenyang 110016, China
Key Laboratory of Opto-Electronic Information Processing, Chinese Academy of Sciences, Shenyang 110016, China

Ramiro Taco, Marco Lanuzza and Domenico Albano
Department of Computer Science, Modeling, Electronics and System Engineering, University of Calabria, Via P. Bucci 42C, 87036 Rende, Italy

Christopher Bailey
Department of Computer Science, University of York, Heslington, York YO10 5DD, UK

Brendan Mullane
Circuits and Systems Research Centre, University of Limerick, Limerick, Ireland

Kaiyu Wang, Qingxin Yan, Xianwei Qi, Yudi Zhou and Zhenan Tang
Department of Electronic Engineering, Dalian University of Technology, Gaoxinyuanqu Linggong Road 2, Dalian 116024, China

Shihua Yu
School of Computer Science and Technology, Hulunbuir College, Xuefu Road, Hulunbuir 021008, China

Chi-Chia Sun
Department of Electrical Engineering, National Formosa University, Wunhua Road 64, Huwei 632, Taiwan

Jürgen Götze
Information Processing Lab, Technology University of Dortmund, Otto-Hahn-Strase 4, 44221 Dortmund, Germany

Gene Eu Jan
Institute of Electrical Engineering, National Taipei University, University Road 151, San Shia District, New Taipei City 23741, Taiwan

Kalannagari Viswanath
Pondicherry Engineering College, Puducherry 605 014, India

Ramalingam Gunasundari
Department of ECE, Pondicherry Engineering College, Puducherry 605 014, India

Jian Chen and Chien-In Henry Chen
Department of Electrical Engineering, Wright State University, Dayton, OH 45435, USA

C. John Moses and V. M. Anne Sophia
Department of ECE, St. Xavier's Catholic College of Engineering, Nagercoil 629003, India

D. Selvathi
Department of ECE, Mepco Schlenk Engineering College, Sivakasi 626005, India

Xiao Wang
Shenyang Institute of Automation, Chinese Academy of Sciences, Shenyang 110016, China
University of the Chinese Academy of Sciences, Beijing 100049, China
Key Laboratory of Opto-Electronic Information Processing, Chinese Academy of Sciences, Shenyang 110016, China

Zelin Shi
Shenyang Institute of Automation, Chinese Academy of Sciences, Shenyang 110016, China
Key Laboratory of Opto-Electronic Information Processing, Chinese Academy of Sciences, Shenyang 110016, China

Yu-Cheng Fan, Chih-Kang Lin, Shih-Ying Chou, Chun-Hung Wang, Shu-Hsien Wu and Hung-Kuan Liu
Department of Electronic Engineering, National Taipei University of Technology, Taipei 10608, Taiwan

S. Syed Ameer Abbas and S. Susithra
Department of Electronics and Communication Engineering, Mepco Schlenk Engineering College, Sivakasi 626 005, India

S. J. Thiruvengadam
Department of Electronics and Communication Engineering, Thiagarajar College of Engineering, Madurai 625 015, India

Mahshid Mojtabavi Naeini and Chia Yee Ooi
Department of Electronic Systems Engineering, Malaysia-Japan International Institute of Technology, Universiti Teknologi Malaysia, Jalan Sultan Yahya Petra, 54100 Kuala Lumpur, Malaysia

Shahzad Asif and Yinan Kong
Department of Engineering, Macquarie University, Sydney, NSW 2109, Australia

Sahar Arshad and Muhammad Ismail
Department of Electronic Engineering, University College of Engineering and Technology, The Islamia University of Bahawalpur, Bahawalpur 63100, Pakistan

Usman Ahmad
Scholar Teacher Research Alliance for Problem Solving (STRAPS), Bahawalpur 63100, Pakistan

Anees ul Husnain and Qaiser Ijaz
Department of Computer System Engineering, University College of Engineering and Technology, The Islamia University of Bahawalpur, Bahawalpur 63100, Pakistan

Ran Xiao and Chunhong Chen
Department of Electrical and Computer Engineering, University of Windsor, Windsor, ON, Canada N9B 3P4

Shikha Panwar, Mayuresh Piske and Aatreya Vivek Madgula
School of Electronics Engineering (SENSE), VIT University, Vandalur-Kelambakkam Road, Chennai 600127, India

Marwan A. Jaber and Daniel Massicotte
Laboratory of Signals and Systems Integrations, Electrical and Computer Engineering Department, Université du Québec à Trois-Rivières, QC, Canada G9A 5H7

P. K. Biswal, K. Mishra, S. Biswas and H. K. Kapoor
Department of Computer Science and Engineering, Indian Institute of Technology, Guwahati 781 039, India

Reza Faghih Mirzaee
Department of Computer Engineering, Islamic Azad University, Shahr-e-Qods Branch, Tehran 37541-374, Iran

Keivan Navi
Faculty of Electrical and Computer Engineering, Shahid Beheshti University, G.C., Tehran 1983963113, Iran

Nader Bagherzadeh
Department of Electrical and Computer Engineering, Center for Pervasive Communications and Computing (CPCC), University of California, Irvine, CA 92697, USA

Trong-Yen Lee and Chi-Han Huang
Department of Electronic Engineering, National Taipei University of Technology, Taipei 10608, Taiwan

www.ingramcontent.com/pod-product-compliance
Lightning Source LLC
Chambersburg PA
CBHW080501200326
41458CB00012B/4049